Salmonella

From Genome to Function

Edited by

Steffen Porwollik

Vaccine Research Institute of San Diego
San Diego, CA
USA

Caister Academic Press

Caister Academic Press
Norfolk, UK

www.caister.com

British Library Cataloguing-in-Publication Data
A catalogue record for this book is available from the British Library

ISBN: 978-1-904455-73-8

Cover image adapted from Figure 6.1

Printed and bound in Great Britain

Contents

Contributors

Joshua N. Adkins
Environmental Molecular Science Laboratory
Pacific Northwest National Laboratory
Richland, WA
USA

joshua.adkins@pnl.gov

Helene Andrews-Polymenis
Department of Microbial and Molecular Pathogenesis
College of Medicine
Texas A&M University
College Station, TX
USA

handrews@medicine.tamhsc.edu

Ellen T. Arena
Department of Microbiology and Immunology
Michael Smith Laboratories
University of British Columbia
Vancouver, BC
Canada

arena003@interchange.ubc.ca

Andreas J. Bäumler
Department of Medical Microbiology and Immunology
School of Medicine
University of California at Davis
Davis, CA
USA

ajbaumler@ucdavis.edu

Jennifer L. Bishop
Department of Microbiology and Immunology
Life Sciences Centre
University of British Columbia
Vancouver, BC
Canada

jlbishop@interchange.ubc.ca

Roslyn N.E. Brown
Environmental Molecular Science Laboratory
Pacific Northwest National Laboratory
Richland, WA
USA

roslyn.brown@pnl.gov

Doug Call
U.S. Department of Agriculture
Agricultural Research Service
Athens, GA
USA

drcall@vetmed.wsu.edu

Rocio Canals
Vaccine Research Institute of San Diego
San Diego, CA
Department of Pathology and Laboratory Medicine
University of California
Irvine, CA
USA

rcanals@sdibr.org

Robert W. Crawford
The Ohio State University
Center for Microbial Interface Biology
Department of Molecular Virology, Immunology, and Medical Genetics
Columbus, OH
USA

osurobbie@gmail.com

Ralf Dieckmann
National Salmonella Reference Laboratory
Federal Institute for Risk Assessment
Berlin
Germany

ralf.dieckmann@bfr.bund.de

Gordon Dougan
Wellcome Trust Sanger Institute
Wellcome Trust Genome Campus
Hinxton
Cambridge
UK

gd1@sanger.ac.uk

Ye Feng
Genomics Research Center
Harbin Medical University
Harbin
China

pandafengye@yahoo.com.cn

B. Brett Finlay
Department of Microbiology and Immunology
Michael Smith Laboratories
University of British Columbia
Vancouver, BC
Canada

bfinlay@interchange.ubc.ca

Maria C. Fookes
The Wellcome Trust Sanger Institute
Wellcome Trust Genome Campus
Hinxton
Cambridge
UK

mcf@sanger.ac.uk

Roman G. Gerlach
Robert Koch Institut
Wernigerode Branch
Wernigerode
Germany

GerlachR@rki.de

Geoffrey Gonzalez-Escobedo
The Ohio State University
Center for Microbial Interface Biology and Department of Molecular Virology, Immunology and Medical Genetics
Columbus, OH
USA

gonzalez-escobed.1@buckeyemail.osu.edu

Jean Guard
U.S. Department of Agriculture
Agricultural Research Service
Athens, GA
USA

jean.guard@ars.usda.gov

John S. Gunn
The Ohio State University
Center for Microbial Interface Biology
Department of Molecular Virology, Immunology, and Medical Genetics
Columbus, OH
USA

john.gunn@osumc.edu

Kenneth W. Harder
Department of Microbiology and Immunology
Life Sciences Centre
University of British Columbia
Vancouver, BC
Canada

kharder@interchange.ubc.ca

Rasika M. Harshey
Section of Molecular Genetics and Microbiology &
Institute of Cellular and Molecular Biology
University of Texas at Austin
Austin, TX
USA

rasika@uts.cc.utexas.edu

Elisabeth Hauser
National Salmonella Reference Laboratory
Federal Institute for Risk Assessment
Berlin
Germany

elisabeth.hauser@bfr.bund.de

Fred Heffron
Department of Molecular Microbiology and Immunology
Oregon Health and Science University
Portland, OR
USA

heffronf@ohsu.edu

Michael Hensel
Abteilung Mikrobiologie
Fachbereich Biologie/Chemie
Universität Osnabrück
Osnabrück
Germany

Michael.Hensel@biologie.uni-osnabrueck.de

Robert M. Hoffman
AntiCancer, Inc.
Department of Surgery, University of California
San Diego
San Diego, CA
USA

all@anticancer.com

Kathryn E. Holt
Wellcome Trust Sanger Institute
Wellcome Trust Genome Campus
Hinxton
Cambridge
UK

kholt@unimelb.edu.au

Sridhar Javayel
RNA Biology Group
Max Planck Institute for Infection Biology
Berlin
Germany

sridhar@mpiib-berlin.mpg.de

Afshan Kidwai
Department of Molecular Microbiology and
Immunology
Oregon Health and Science University
Portland, OR
USA

kidwai@ohsu.edu

Robert A. Kingsley
Wellcome Trust Sanger Institute
Wellcome Trust Genome Campus
Hinxton
Cambridge
UK

rak@sanger.ac.uk

Shu-Lin Liu
Genomics Research Center
Harbin Medical University
Harbin
China;
Department Microbiology and Infectious Diseases
University of Calgary
Calgary
Canada

slliu@ucalgary.ca

Wei-Qiao Liu
Department of Microbiology and Infectious Diseases
University of Calgary
Calgary
Canada

weiqiao.lui@gmail.com

Michael McClelland
Vaccine Research Institute of San Diego
San Diego, CA;
Department of Pathology and Laboratory Medicine
University of California
Irvine, CA
USA

mmcclelland@sdibr.org

Jason D. McDermott
Environmental Molecular Science Laboratory
Pacific Northwest National Laboratory
Richland, WA
USA

jason.mcdermott@pnl.gov

Burkhard Malorny
National Salmonella Reference Laboratory
Federal Institute for Risk Assessment
Berlin
Germany

burkhard.malorny@bfr.bund.de

Cesar A. Morales
U S. Department of Agriculture
Agricultural Research Service
Athens, GA
USA

Cesar.Morales@ars.usda.gov

George Niemann
Department of Molecular Microbiology and Immunology
Oregon Health and Science University
Portland, OR
USA

niemanng@ohsu.edu

Sean-Paul Nuccio
Department of Medical Microbiology and Immunology
School of Medicine
University of California at Davis
Davis, CA
USA

snuccio@ucdavis.edu

Kai Papenfort
Institute for Molecular Infection Biology
University of Würzburg
Würzburg
Germany

kai.papenfort@uni-wuerzburg.de

Tim T. Perkins
Wellcome Trust Sanger Institute
Wellcome Trust Genome Campus
Hinxton
Cambridge
UK

timothy_trevor.perkins@novartis.com

Wolfgang Rabsch
Robert Koch Institute
Wernigerode Branch
National Reference Centre for Salmonellae and other Enterics
Wernigerode
Germany

RabschW@rki.de

Roopa Rajashekar
Abteilung Mikrobiologie
Universität Osnabrück
Osnabrück
Germany

rooshree@yahoo.com

M. Megan Reynolds
Department of Microbial and Molecular Pathogenesis
College of Medicine
Texas A&M University
College Station, TX
USA

megreynolds99@neo.tamu.edu

Kenneth E. Sanderson
Department of Biological Sciences
University of Calgary
Calgary
Canada

kesander@ucalgary.ca

Devendra Shah
U.S. Department of Agriculture
Agricultural Research Service
Athens, GA
USA

dshah@vetmed.wsu.edu

Richard Smith
Environmental Molecular Science Laboratory
Pacific Northwest National Laboratory
Richland, WA
USA

richard.smith@pnl.gov

Nicholas R. Thomson
The Wellcome Trust Sanger Institute
Wellcome Trust Genome Campus
Hinxton
Cambridge
UK

nrt@sanger.ac.uk

Sandra Truepschuch
Robert Koch Institute
Wernigerode Branch
National Reference Centre for Salmonellae and other Enterics
Wernigerode
Germany

TruepschuchS@rki.de

Jörg Vogel
Institute for Molecular Infection Biology
University of Würzburg
Würzburg
Germany

joerg.vogel@uni-wuerzburg.de

Daniel Windhorst
Lohmann Animal Health GmbH & Co KG
Process Development Vaccines
Cuxhaven
Germany

Daniel.Windhorst@Lohmann.de

Hyunjin Yoon
Dept. of Molecular Microbiology and Immunology
Oregon Health and Science University
Portland, OR
USA

yoonh@ohsu.edu

Other Books of Interest

Preface

The impact of *Salmonella* research on our understanding of bacterial genetics, pathogenesis and interactions with hosts cannot be overestimated. A multitude of tools and molecular techniques had been developed using this easily tractable and cultivatable bacterium to study, manipulate and engineer bacterial traits. Aside from the related *Escherichia coli,* we know more about *Salmonella* than any other bacterium. And yet, *Salmonella* as a pathogen has not been vanquished. It remains responsible for an estimated one million deaths and 100 million human infections annually. Even in highly developed countries with excellent sanitary conditions and control measures in the food industry, *Salmonella* regularly manages to escape the rigorous screening procedures and cause multi-faceted disease in a variety of hosts including humans and livestock. In a recent case, over 500 million contaminated chicken eggs were recalled in the United States because of *Salmonella* contamination; yet again the bacterium had evaded all screenings prior to food distribution to supermarkets country-wide. Alarming trends are observed in Africa, where traditional non-systemic *Salmonella* serovars are increasingly reported to cause systemic disease. Typhoid and Paratyphoid fever is still rampant in South Asia, and pose an even greater danger during floods.

Our extensive knowledge about *Salmonella* has been documented in many excellent books, including the Holy Grail in this field, '*E. coli* and *Salmonella*', a huge and comprehensive lexicon on molecular biology of the two model bacteria. So, why another *Salmonella* book?

The reason is fairly straightforward: the advent of genomic sequencing has greatly impacted what we know about *Salmonella,* and in the past five years data accumulation has accelerated to a mind-boggling speed which makes it difficult to keep track of new developments. How many *Salmonella* genomes are decoded? The answer changes almost weekly. Soon, genomic sequencing may be a standard procedure for any strain that causes an outbreak anywhere in the world. Costs are falling rapidly, and consistently, and new technologies are emerging that enable quick and reliable sequence information at ever increasing speed and accuracy.

To date, approximately 100 *Salmonella* genomes have been sequenced or are in the process of being decoded, covering both *Salmonella* species (*S. enterica* and *S. bongori*) as well as all *S. enterica* subspecies and important human pathogenic *Salmonella* serovars. New genomes are constantly entering the databases. As an example, at this year's meeting of the American Society of Microbiology, Life Technologies reported *en masse* sequencing of 15 new *Salmonella* genomes of strains rarely associated with human disease, to facilitate comparative studies. Harvesting all this information remains one of the major challenges in the field. The different *Salmonella* genomes are mosaic puzzles – results of lateral transfer events that occur within a stable genetic background. Extraordinary diversity of host ranges and pathogenicity traits between different strains are the consequence of both specific genome insertions/deletions and minute changes in genome composition.

I put together this book to illustrate **how** and **where** this sequence information influences today's research, and to what extent genomic sequencing actually helps scientists in their quest to solve the many riddles *Salmonella* poses on a molecular level. The obtained sequence information has a dramatic impact on all aspects of *Salmonella* research, whatever its focus – be it detection, identification and typing, be it *Salmonella* pathogenic traits and processes as well as virulence determinants, be it regulatory circuitries and particular aspects of the bacterium's interaction with its hosts, or be it *Salmonella*'s surprisingly multi-faceted lifestyle characteristics.

In this book, internationally acclaimed experts review cutting-edge topics in *Salmonella* genomics and molecular biology, providing a timely snapshot of the current state of research. The book first reviews latest approaches to sub-species level classification (Chapter 1) and advances in phage typing of *Salmonella* serovars (Chapter 2), and proceeds to a discussion and illustration of comparative genomics (Chapter 3). High throughput methods aimed at identification of the genetic determinants for survival of the bacterium in different environments are discussed (Chapter 4), and the evolution of niche specialization by *Salmonella* is examined (Chapter 5). *Salmonella* Typhi pathogenesis is surveyed separately (Chapter 6) because of its extraordinary importance in human disease. The book also explores all currently identified and characterized *Salmonella* sRNAs, a field that has emerged in the past ten years and where *Salmonella* may exceed *E. coli* in terms of the depth of experimentally validated information (Chapter 7). Reviews are presented on the status of current research with respect to structures of *Salmonella* that have interactive capabilities with the environment, such as fimbriae (Chapter 8), flagella (Chapter 9) and secreted virulence factors (Chapter 10). This is followed by reviews of interactions of the host with intracellular *Salmonella* (Chapter 11), the human host's anti-*Salmonella* immune response (Chapter 12), and *Salmonella* biofilm formation (Chapter 13). The final chapter in this book illustrates one company's efforts to harness and exploit *Salmonella* for tumour therapy (Chapter 14). *Salmonella* research is currently at a very exciting juncture. In addition to their fascinating resilience in both the environment and eukaryotic hosts, *Salmonella* prefers tumours over any other location within the human host (by a factor of 1000 or more). This ability could propel *Salmonella* into future use as a therapeutic delivery agent to control and/or cure cancers.

The nomenclature of *Salmonella* is in constant flux. The latest Opinion of the Judicial Commission of the International Committee on Systematics of Prokaryotes (*International Journal of Systematic and Evolutionary Microbiology* 2005, vol. 55, pp. 519–520) suggested the classification of *Salmonella* into two species, *S. enterica* and *S. bongori,* with the former subclassified into six subspecies. While this taxonomical classification has gained wide acceptance, it is still disputed. The serological classification of *Salmonella* into >2500 serovars is historically, clinically and epidemiologically important, however, there is some controversy as to how to properly indicate serovars in written documents. J.P. Euzéby's list of prokaryotic names with standing in nomenclature (LPSN, http://www.bacterio.cict.fr/) advises to write them in the following form: '*Salmonella* London or *Salmonella* ser. London or *Salmonella* serovar London (complete name: *Salmonella enterica* subsp. *enterica* serovar London...)'. The World Health Organization's International Center for Salmonella and the Centers for Disease Control and Prevention (CDC) use this shortened serovar label, so I have decided to follow these same guidelines in this book.

Compiling a snapshot of the current state of affairs in *Salmonella* genomics and molecular biology presents a challenge because this is a fast-moving field. Every possible effort has been made to keep the information presented as recent as possible, with the last updates completed in July 2010.

In summary, the chapters of this book bring researchers working with *Salmonella* and related organisms up to date with the latest insights into the most prominent aspects of *Salmonella* biology, fuelled by the genomic information that characterized and formed molecular bacteriology in the first decade of the 21st century. The book is recommended reading for all scientists that work on bacterial genomics, molecular biology and bacterial molecular & cellular pathogenesis because the impact of genomic sequencing on *Salmonella* is generalizable to all microbial organisms.

Steffen Porwollik
San Diego, August 2010

New Approaches in Subspecies-level *Salmonella* Classification

1

Burkhard Malorny*, Elisabeth Hauser and Ralf Dieckmann

Abstract

Salmonellae form a complex group of bacteria consisting of two species, six subspecies and more than 2500 serovars (serotypes). *Salmonella* identification below species level is most often limited to phenotypic typing methods such as biochemical and serological identification, which are costly, time-consuming and do not always reflect the evolution of *Salmonella* groups. Newer methods for *Salmonella* typing and subtyping include genome-based methods such as pulsed field gel electrophoresis (PFGE), multiple loci VNTR analysis (MLVA), multilocus sequence typing (MLST) and (multiplex-) PCR-based methods. In the last years further molecular typing technologies were evaluated for this purpose. This chapter discusses some of these emerging technologies and gives an outlook on future developments with a focus on oligonucleotide microarrays, spectroscopic methods such as MALDI-TOF mass spectrometry and special developments such as bead-based suspension arrays using Luminex technology and DNA sequence-based approaches. We review these new techniques, which promise significant advantages compared to traditional culture-based methods with respect to speed, ease of use, reliability and automation and discuss their advantages and disadvantages.

Introduction

The ability to distinguish strains or clonal lineages of a bacterial pathogen is a prerequisite for addressing many questions in clinical microbiology and epidemiology. Currently, the traditional *Salmonella* serotyping scheme according to White–Kauffmann–Le Minor is accepted worldwide as a 'gold standard' for the classification of *Salmonellae* below the subspecies level. However, molecular methods are now almost universally used to characterize strains and to determine the genetic relatedness within a subset of defined strains. The choice of the method depends on the epidemiological question that needs to be answered.

To study the phylogeny or global epidemiology of *Salmonella*, a set of distantly related isolates must be selected, and approaches which reflect the variations in those strains that accumulate relatively slowly need to be applied. Sequence variation within housekeeping genes is ideal for such studies because these genes are considered to be neutral in evolution and generally their function is well understood.

Methods for outbreak studies, where it is important to differentiate certain genotypes within a bacterial species, should be rather highly discriminative. This is commonly achieved by analysis of DNA restriction sites within the test strain, e.g. using pulsed-field gel electrophoresis (PFGE) or fluorescent amplified fragment length polymorphism (fAFLP) analysis. A new high resolution approach is the multilocus variable-number tandem-repeat analysis (MLVA), which is based on the genetic analysis of various loci within a strain that consist of repeating units of approximately 1–100 base pairs in length. However, MLVA has great potential to also study global epidemiology, but methods applicable only for a few *Salmonella* serovars have so far been developed.

*Corresponding author

This chapter will focus on new approaches which are useful for global epidemiology studies and summarizes specific PCR and array-based detection assays for the identification of *Salmonella* subspecies and serovars. Methods which are applied for *Salmonella* outbreak studies were previously reviewed by Foley *et al.* (2007).

Taxonomy and traditional subspecies classification

Phenotypic classification methods

The genus *Salmonella* is composed of motile bacteria that belong to the family Enterobacteriaceae (Farmer III, 2003). Two species are currently recognized in the genus *Salmonella*, *Salmonella enterica* and *Salmonella bongori*. *S. enterica* has been subdivided into six subspecies, *S. enterica* subsp. *enterica* (designated subspecies I), *S. enterica* subsp. *salamae* (subspecies II), *S. enterica* subsp. *arizonae* (subspecies IIIa), *S. enterica* subsp. *diarizonae* (subspecies IIIb), *S. enterica* subsp. *houtenae* (subspecies IV) and *S. enterica* subsp. *indica* (subspecies VI) (Tindall *et al.*, 2005). A number of biochemical reactions are used to differentiate between the species and subspecies (Farmer III, 2003) using traditional media in tubes, commercial manual biochemical systems such as the API20E system from bioMèrieux SA (Geneve, Suisse) or Enterotube II from BD (NJ, USA), or automated systems such as VIDAS from bioMèrieux SA.

In the past, **multilocus enzyme electrophoresis (MLEE)** was frequently applied to study the evolutionary relationships of strains within subspecies and serovars of *Salmonellae* (Selander *et al.*, 1990a,b, 1996; Beltran *et al.*, 1991; Boyd *et al.*, 1993). MLEE assesses the allelic variation in multiple genes (mainly housekeeping genes) in a strain by determination of electromorphs (allozymes) of an enzyme. Distinctive allele profiles (multilocus enzyme genotypes) are then designated as electrophoretic types (ETs). Based on MLEE (24 enzyme loci) all but four of 80 ETs fall into seven clusters that correspond precisely to the six subspecies and the species *S. bongori* defined by biotyping and DNA hybridizations (Boyd *et al.*, 1996). Two exceptional ETs whose isolates have formed an additional cluster were designated subspecies VII (Selander *et al.*, 1996). Biochemically, they belonged to subspecies IV. MLEE is limited by the observation that the branching order of the subspecies depends largely on variation at just a few loci. Furthermore, silent mutations can not be assessed and multiple amino acid substitutions in individual proteins and the frequent convergence of electromorphs can produce similar effects on electrophoretic mobility.

According to the White–Kauffmann–Le Minor scheme (Grimont and Weill, 2007) subspecies are further divided into serovars by **serotyping**, which is widely used as an epidemiological standardized typing method. The primary antigens assayed are those of the polysaccharide domain of the cell surface lipopolysaccharide (O-antigens) and the flagellin proteins that form the filaments of phase 1 and phase 2 flagella (H1 and H2 antigens). Currently, 46 different O-antigens, and 114 different H antigens have been recognized. Variation in O-antigen is normally mediated by the chromosomal *rfb* cluster but might also be caused by lysogenic conversion as a result of infection with certain phages (Brüssow *et al.*, 2004). Each distinct combination of O, H1, and H2 antigens is given formal recognition as a serovar. Minor variant antigenic factors are often ignored (e.g. O5 antigen). Currently, over 2500 different serovars are differentiated by this scheme. With respect to flagella, most serovars are diphasic, alternately expressing two genes (*fliC* or *fljB*) that encode the structural unit of flagellins with various antigenic characters. Several serovars are monophasic, and one of them (Gallinarum) lacks flagella entirely.

Genome-based classification methods

Taxonomists acknowledged **DNA–DNA hybridization** as the reference method for the establishment of relationships within and between species (Vandamme, 2003). A species is defined by strains, including the type strain, that share 70% or more DNA–DNA relatedness with ΔT_m of 5°C or less (Wayne *et al.*, 1987). A limited number of genomic DNA hybridizations agreed with the biochemical reactions recognizing the six *S. enterica* subspecies, and the *S. bongori* (Crosa *et al.*, 1973; Le Minor *et al.*, 1982, 1986). However, DNA–DNA hybridizations obtained in different laboratories do not always correlate well, partly because of different conditions used. The advent of sequencing of PCR amplified DNA during the mid 1980s started a new era for taxonomists and population genetics. **Sequencing** of DNA segments from housekeeping genes permitted to estimate the genetic relationship even below the subspecies level, and gave insight into how much horizontal gene transfer and recombination contributed to evolution. The genetic relationships among strains of the six subspecies and *S. bongori* based on the combined nucleotide sequences of five housekeeping genes *putP* (proline permease), *gapA* (glyceraldehyde-3-phosphate dehydrogenase), *mdh* (malate dehydrogenase), *gnd* (phosphogluconate dehydrogenase) and *aceK* (isocitrate dehydrogenase kinase/phosphatase) were fully consistent with evidence from DNA hybridization experiments (Le Minor *et al.*, 1986). Sequencing of these loci revealed some intragenic recombination events whereas others reflected the exchange of entire genes. For more detailed information we refer to a review on evolutionary genetics and the evolution of housekeeping and invasion genes of *S. enterica* (Selander *et al.*, 1990b).

Emerging technologies

DNA microarray-based assays for *Salmonella* characterization and identification

DNA array technologies have been rapidly developed within the last ten years. Biological probes are ordered mostly in a miniaturized grid on an appropriate surface. A number of such arrays have been developed using various probes obtained from *Salmonella* DNA sequences. Whole genome DNA microarrays have been successfully applied in comparative genomic hybridizations (CGH) for *Salmonella* with the first publication of the serovar Typhimurium and Typhi genome sequences (McClelland *et al.*, 2001; Parkhill *et al.*, 2001). A phylogenetic tree based on CGH between the serovar Typhimurium strain LT2 genome and a selection of *Salmonella* strains belonging to the six subspecies was largely concordant with previous studies using sequence information from several housekeeping gene loci (Porwollik *et al.*, 2002). Further comparisons of 79 *S. enterica* subsp. *enterica* revealed that strains within the same serovar often differed by the presence and absence of hundreds of genes (Porwollik *et al.*, 2004). Gene contents sometimes differed more within a serovar than between serovars. Groups of strains that share a distinct profile of gene content can be referred to as 'genovars' to distinguish them from serovars. This new designation reflects the classification of *Salmonella* strains that differ profoundly in their gene pool and has potential with increasing availability of complete *Salmonella* genomes to substitute the classical White–Kauffmann–Le Minor serovar scheme.

A condensed selection of 104 various probes detecting virulence genes of *Salmonella* has been shown to be useful to study the epidemiology of subspecies I serovars (Huehn and Malorny, 2009). Often a serovar has a common subset of virulence genes, but different lineages are also observed within a serovar with significant differences in their virulence gene repertoire, such as in serovar Paratyphi B [D(–)tartrate positive] (Huehn *et al.*, 2009). A similar approach used unique gene presence/absence profiles targeting various core genes of *S. enterica* subspecies I specific genes, fimbrial genes, pathogenicity islands, Gifsy prophages elements and other variable genes (Scaria *et al.*, 2008).

Another DNA microarray, which exclusively targets antigen-specific sequences within the flagella (H) antigen phase 1 (*fliC*) and phase 2 (*fljB*) genes and somatic (O) antigen biosynthesis genes within the *rfb* cluster, has recently been published (Yoshida *et al.*, 2007).

PCR-based assays for *Salmonella* serovar identification

A growing number of 'molecular' serotyping approaches have been described. These include (multiplex-) PCR-based assays that directly target the O- (*rfb*) and H- (*fliC*, *fljB*) antigen genes, other serovar-specific genes, or combinations of both. Table 1.1 gives an overview of the properties of published PCR assays for the identification of *Salmonella* subspecies, serogroups and serovars. Targeting of genes responsible for serovar antigens provides a direct correlation between serovars determined by traditional and molecular

Table 1.1 Summary of properties of published PCR assays for the detection of *Salmonella* subspecies, serogroups and serovars

Specific target genes (PCR product size in bp)	Type of assay	Specificity	Validation status, strains/ serovars tested	Reference
rfbJ-B (882), *rfbJ*-C2 (820), *rfbS*-D/A (720)	PCR, size-based detection	Serogroups B, C2, D/A	15 *Salmonella* serovars, including five Typhi and only one group C2 serovar	Luk *et al.* (1993)
fliC (343)	Nested PCR, size-based detection	Serovar Typhi	10 *Salmonella* serovars, including two Typhi	Song *et al.* (1993)
Plasmid specific probe (620)	PCR, size-based detection	Serovar Enteritidis	Plasmid-free strains are not detected	Wood *et al.* (1994)
rfbJ-B (882), *rfbJ*-C2 (820), *rfbS*-D (720)	Multiplex PCR, size-based detection	Serogroups B, C2, D	485 clinical isolates	Kongmuang *et al.* (1994)
Gene for HeLa cell invasion (93)	PCR, sequenced-based detection	*S. enterica* subsp. I	228 clinical isolates including the *S. enterica* subspecies and *S.* bongori	Chevrier *et al.* (1995)
viaB (599)	Nested PCR, size-based detection	Serovar Typhi	10 *Salmonella* serovars, including six Typhi, product also present in Paratyphi C	Hashimoto *et al.* (1995)
sefA(310)	PCR, size-based detection	Enteritidis (*Salmonella* serogroup D)	No strains tested	Woodward and Kirwan (1996)
rfbS (720)	PCR, sequenced-based detection *(DIG-ELISA)*	Serogroup D	145 *Salmonellae* of different serogroups including 82 serogroup D isolates	Luk *et al.* (1997)
IS*200* element (>900)	PCR, size-based detection	Serovar Abortusovis	11 *Salmonella* serovars	Beuzon *et al.* (1997)
sipB/C (232), *cmlA/tetR* (260), *bla*PSE-1 (132), *bla*TEM (291)	Multiplex PCR, size-based detection	Serovar Typhimurium phage type DT104 (multidrug resistant)	7 different *Salmonella* serovars	Carlson *et al.* (1999)
rfbJ-B (882), *rfbJ*-C2 (820), *rfbS*-D (720) *rfb*-C1 (471)	Two Multiplex PCR, size-based detection	Serogroups B, C1, C2-C3, D	ID of rough isolates, 80 *Salmonella* strains from different serogroups	Hoorfar *et al.* (1999)

fliC (559), *sefA* (312)	Multiplex PCR, size-based detection	Serovars Typhimurium and Enteritidis	14 different *Salmonella* serovars, *sefA* serogroup D specific, 1078 swabs analysed	Soumet *et al.* (1999)
sdf I (293)	PCR, size-based detection	Serovar Enteritidis	73 non-Enteritidis serovars representing 34 different serovars, *sdf* does not detect phage types 6A,9A,11,16,20,27	Agron *et al.* (2001)
fljB (various), flagellin phase 2	Multiplex PCR, size-based detection	H:1,2, H:1,5, H:1,6, H:1,7, H:l,w, H:e,*n*,x and H:e,*n*,z15	140 *Salmonella* strains representing 49 different serovars	Echeita *et al.* (2002)
tyv (*rfbE*) (615), *prt* (*rfbS*) (258), *viaB* (439), *fliC*	Multiplex PCR, size-based detection	Serovars Typhi and Paratyphi A	45 *Salmonella* strains representing 21 different serovars	Hirose *et al.* (2002)
IE gene sequence	PCR, size-based detection	Serovar Enteritidis	122 *Salmonella* strains representing 79 different serovars	Wang and Yeh (2002)
wzx-wzy (501)	PCR, size-based detection	Serogroup O:6,14	443 *Salmonella* strains including all Top 100 serotypes and at least one isolate of the 46 O antigen serogroups	Fitzgerald *et al.* (2003)
rfbJ (663), *fliC* (183) and *fljB* (526)	Multiplex PCR size-based detection	Serovar Typhimurium	57 *Salmonella* strains representing 16 serovars	Lim *et al.* (2003)
fliC (various), flagellin phase 1	Multiplex PCR, size-based detection	H:g-complex, H:i, H:r, H:l,v, H:e,h, Hz10, H:b, and H:d	161 *Salmonella* strains expressing different H-antigens	Herrera-Leon *et al.* (2004)
sdf1 (304), STM specific (401), *rfb*-C2 (502), 4,[5],12:i:- specific sequence (705)	Multiplex PCR, size-based detection	Serovar Enteritidis, Typhimurium, 4,[5],12:i:-, Hadar, other serogroup C2	126 *Salmonella* strains representing 46 serovars from different subspecies, *sdf* did not detect phage types 6A,9A,11,16,20,27	Alvarez *et al.* (2004)

Table 1.1 continued

Specific target genes (PCR product size in bp)	Type of assay	Specificity	Validation status, strains/ serovars tested	Reference
fliC (963), *viaB*	PCR, size-based detection	Serovar Choleraesuis, Paratyphi C	97 *Salmonella* Choleraesuis strains and several non-Choleraesuis serovars	Chiu *et al.* (2005)
rfbS (147)	PCR, size-based detection	Serovar Pullorum	32 *Salmonella* strains representing 20 different serovars	Desai *et al.* (2005)
pse (156), *floR* (266), *str* (203), *sul1* (351), *spvC* (391), *tetG* (447), *int1* (various)	Multiplex PCR, size-based detection	Multidrug resistant Typhimurium DT104	104 clinical isolates of *Salmonella* serogroup B	Chiu *et al.* (2006)
sefA (330), fimbrial gene, *pefA* (497)	Multiplex PCR, size-based detection	Serovars Enteritidis, Typhimurium	288 *Salmonella* strains containing *sefA,* serogroup D specific	Cortez *et al.* (2006)
10 genomic regions derived from microarray technology	Two five-plex PCR, size-based detection, serovar specific amplification patterns	30 serovars	30 most prevalent serovars	Kim *et al.* (2006c)
38 primer pairs derived from genome sequence comparisons	Multiplex PCR, size-based detection, serovar specific amplification patterns	Subspecies and major serovars	*Salmonella* strains representing more than 90 different serovars from all subspecies	Kim *et al.* (2006a)
STM4497 (310)	PCR, size-based detection	Serovar Typhimurium	164 *Salmonella* strains tested	Kim *et al.* (2006b)
sdf I (333)	Multiplex PCR, size-based detection	Serovar Enteritidis	54 *Salmonella* strains representing 19 different serovars	Trafny *et al.* (2006)
fljB (727, 413)	PCR, size-based detection	Serovar Infantis	*Salmonella* strains representing 43 different serovars	Kardos *et al.* (2007)
tyv-D (615), *wzx*-B (230), *wzx*-C1 (483), *wzx*-C2 (154), *wzx*-E (345)	Three multiplex PCR, size-based detection	Various serovars, Enteritidis, Typhimurium, Heidelberg, Infantis, Virchow	67 *Salmonella* strains representing 32 serovars	Herrera-Leon *et al.* (2007)

prot6e (206)	Real Time PCR, TaqMan	Serovar Enteritidis	79 Enteritidis strains comprising 19 different phage types and 119 non-Enteritidis strains comprising 54 serovars.	Malorny *et al.* (2007)
rfbJ (663), *fliC* (183), *fljB* (526)	IMS, Multiplex PCR, size-based detection	Serovar Typhimurium	33 *Salmonella* isolates from 400 fecal samples representing 8 different serovars	Zahraei Salehi *et al.* (2007)
invA, *rfbJ*-B, *fliC*, *fljB*	Multiplex PCR, size-based detection	Serovar Brandenburg	72 *Salmonella* strains representing 28 serovars	Perera and Murray (2008)
Panel of 46 or 64 genes	Real time PCR, Sybr Green Multiplex PCR, size-based detection	12 serovars (16 genovars) 11 serovars (9 genovars)	24 *S. enterica* strains representing 12 different serovars	Arrach *et al.* (2008)
rfbJ (662), *tyv* (614), *viaB* (439), *prt* (256), *fliC*-a (423), *fliC*-b (551), *fliC*-d (763) d-tartrate (289)	Two multiplex PCR, size-based detection	Serovars Typhi, Paratyphi A and B	46 *Salmonella* strains representing 25 different serovars	Levy *et al.* (2008)
sefA (334), *fliC* (433), *sdf* (299)	Real Time PCR, TaqMan	Serovars Enteritidis, *Gallinarum*, Typhimurium, Kentucky and Dublin	60 *Salmonella* strains representing 26 different serovars	O'Regan *et al.* (2008)
abe1-B (561); *wbaD/ manC*-C1) (341); *abe2*-C2 (397); *prt*-A/D1 (624); *wzx/ wzy*-E1 (281); *fliC*-i (508); *fliC*-g,m (309); *fliC*-r (169); *fliC*-z10 (363); *fljB*-1,2;1,5;1,6;1,7 (294); *fljB*-e,*n*,x;e,*n*,z15 (152)	Multiplex PCR, size-based detection	Serovars Enteritidis, Hadar, Heidelberg, Typhimurium	239 *Salmonella* strains representing 43 different serovars	Hong *et al.* (2008)
stkF (159), SPA2473 (324), SPA2539 (484), *hsdM* (627)	Multiplex PCR, size-based detection	Serovar Paratyphi A	113 *Salmonella* strains representing 87 different serovars	Teh *et al.* (2008), Ou *et al.* (2007)

Table 1.1 continued

Specific target genes (PCR product size in bp)	Type of assay	Specificity	Validation status, strains/serovars tested	Reference
genomic regions derived from genome sequence comparisons	Multiplex PCR, size-based detection, real-time PCR, TaqMan	Serovars Choleraesuis, Paratyphi C	190 *Salmonella* strains representing more than 50 different serovars	Woods *et al.* (2008)
invA, prot6e, *fliC* (133–136)	Real Time PCR, molecular beacons	Enteritidis, Typhimurium	44 *Salmonella* strains, 21 *Salmonella* Enteritidis, 17 Typhimurium	Hadjinicolaou *et al.* (2009)
16 genomic targets derived from comparative genomic studies (90–264 bp)	Multiplex PCR, capillary gel electrophoresis	Serovars Typhimurium, Enteritidis, 4,[5],12:i:-, Montevideo, Saintpaul, Heidelberg, Javiana, Typhi, Paratyphi A and others	489 *Salmonella* isolates representing the most common serovars, blinded testing of further 751 isolates	Leader *et al.* (2009)
sefA, *fliC*	Multiplex real-time PCR	Serovars Enteritidis, Typhimurium	110 *Salmonella* isolates representing 13 different serovars	Lee *et al.* (2009b)
abe1-B (561); *wbaD/manC*-C1) (341); *abe2*-C2 (397); *prt*-A/D1 (624); *wzx/wzy*-E1 (281); *fliC*i (508), *fliC*-g,m, (309), *fliC*-r, (169), *fliC*-z10 (363), *fljB*-1,2;1,5;1,6;1,7 (294); *fljB*-e,*n*,x;e,*n*,z15 (152)	PCR, size-based detection	Serovars Enteritidis, Typhimurium	165 *Salmonella* strains containing 12 different serovars	Soliman *et al.* (2009)
fljB (848), *mdcA* (728), *gatD* (501), *stn* (179) and STM4057 (137)	Multiplex PCR, size-based detection	*Salmonella* subspecies	53 *Salmonella* strains representing all subspecies and *S. bongori*	Lee *et al.* (2009a)

methods. Molecular serotyping depends on the identification and detection of serospecific polymorphisms present in antigen genes, and combines the performance properties of serology (typeability, epidemiological concordance, reproducibility) with the advantages of DNA analysis (speed, can be automated and run with high throughput at reduced labour costs, DNA probes easier to produce and handle, cheaper than antisera). In addition, the time-consuming flagellar phase reversal to identify both flagellar antigens is not necessary and rough strains that are untypeable by traditional serotyping can be identified (Hoorfar *et al.*, 1999). Furthermore, DNA-based methods can be more discriminating than traditional phenotypic methods such as serotyping or MLEE. The methodologies include multiplex PCR assays targeting *rfb* genes (Luk *et al.*, 1993; Herrera-Leon *et al.*, 2007), phase 1 (*fliC*) flagellar antigen genes (Herrera-Leon *et al.*, 2004; McQuiston *et al.*, 2004) and phase 2 (*fljB*) flagellar antigen genes (Echeita *et al.*, 2002; McQuiston *et al.*, 2004).

Another approach for the identification of serovars/genovars has been developed based on the detection of multiple genes selected from comparative genomic hybridization (CHG) data. A subset of 146 genes was used for real-time PCR to identify 12 serovars (16 genovars) in 24 *S. enterica* strains (Arrach *et al.*, 2008). Another subset of 64 genes was used to identify 8 serovars (9 genovars) in 12 multiplex PCRs on 11 *S. enterica* strains. Pending optimization and automatization, this approach has great potential for subtyping *Salmonella* strains below subspecies and serovar level.

A similar study was published (Kim *et al.*, 2006c), where the authors developed two multiplex PCRs based on 12 target genes selected from the *S. enterica* serovar Typhimurium and *S. enterica* serovar Typhi genomes to differentiate 30 serovars in 22 unique amplification patterns. A blind test screening 111 clinical isolates revealed 97% correctly identified serovars.

PCR assays targeting genes other than those that are serovar antigen-encoding have to be regarded with care, because the serovar-specificity of certain genes may not have been fully validated due to the lacking of genome sequences or poor experimental validation using an insufficient number of different non-target serovars. With the advent of high throughput sequencing technologies, more and more *S. enterica* genomes are being decoded, mitigating this particular pitfall of these PCR assays in the future.

New platforms for PCR-based typing approaches

Advances in Polymerase Chain Reaction (PCR) technology resulted in the development of a number of PCR-based sub-typing approaches. Many of these are simple, rapid and some are broadly applicable independent of any knowledge of the genome sequence.

Luminex technology

An innovative commercially available Luminex xMAP® system with potential applications throughout the life science and diagnostic industries has been developed by Luminex Corporation (Texas, USA, www.luminexcorp.com). Luminex technology uses micro-beads of a defined spectral colour known as microspheres. The creation of 100 different 'coloured' beads, each with a unique spectral address allows for multiplexing of up to 100 analytes per well. Each bead set can be coupled to oligonucleotide probes specific for e.g. O-antigen or H-antigen encoding genes (Fig. 1.1). In a first step, target DNA sequences are amplified by PCR followed by hybridization of the denatured biotin-labelled PCR fragments to the pooled bead sets. In the next step, streptavidin-phycoerythrin (SAPE) binds to biotin-labelled PCR fragments, and finally spectral colour and amplicon fluorescence combine in single particles that can be detected using a fluorescence-activated cell sorter (FACS). A multiplex, bead-based suspension array for the molecular determination of *Salmonella* serogroups was developed by Fitzgerald *et al.* (Fitzgerald *et al.*, 2007). This procedure consists of the amplification of PCR fragments from genes of the *rfb* gene cluster using serogroup-specific PCR primers followed by detection and identification of the amplicons by hybridization to specific DNA probes using Luminex technology (Dunbar and Jacobson, 2007).

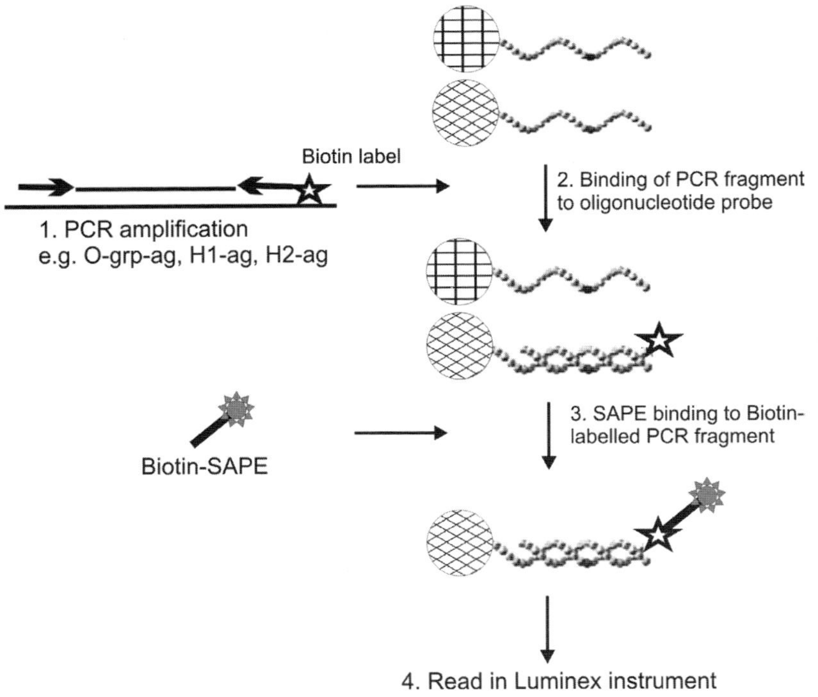

Figure 1.1 Principle of the Luminex® xMAP™ *Salmonella* typing system. 1. O-antigen (O-grp-ag), H1- (H1-ag), and H2- (H2-ag) antigen target sequences are amplified using specific primers one of which is labelled with biotin. 2. Microspheres with a unique spectral address and coupled to oligonucleotide probes are specifically hybridized to amplified PCR fragments. 3. Streptavidin–phycoerythrin (SAPE) binds to biotin-labelled PCR fragments. 4. Spectral colour of microsphere and amplicon fluorescence combine in single particles detected by a fluorescence-activated cell sorter (FACS).

Premi®Test assay

The Premi®Test *Salmonella* was recently introduced by DSM Nutritional Products (Geleen, The Netherlands, www.dsm.com), in conjunction with its partner Check-Points (Wageningen, The Netherlands; www.check-points.com). This test, a microarray format based on a low density DNA chip from Clondiag (Jena, Germany, www.clondiag.com), recognizes multiple DNA markers specific for different *Salmonella* serotypes (Wattiau *et al.*, 2008b; Wattiau *et al.*, 2008a). The test is based on a multiplexed ligation detection reaction (LDR) to generate a collection of circular DNA molecules that are subsequently amplified using PCR (Schouten *et al.*, 2002). The amplified molecules are then hybridized to the low-density DNA microarray, which contains probe-specific complementary oligonucleotides (Fig. 1.2). The array is fixed at the bottom of a 1.5 ml Eppendorf tube where the hybridization reaction takes place. Positive hybridization is detected by photometric detection using a biotin label incorporated in one of the PCR primers. Currently the system can identify more than 75 different prevalent *Salmonella* serovars. The whole procedure takes about 8 hours, enabling a result within 1 working day after enrichment.

Rep-PCR

Repetitive sequence-based PCR (rep-PCR) is a rapid method for bacterial typing that relies on amplification of sequences between various repetitive non-coding elements in the bacterial chromosome. Amplified DNA fragments can be separated by electrophoresis resulting in specific rep-PCR profiles that can be employed for subspecies discrimination. Alternatively, electrophoresis can be replaced by separation in a microfluidics chamber, and digitized gel images are analysed. Such systems for automated

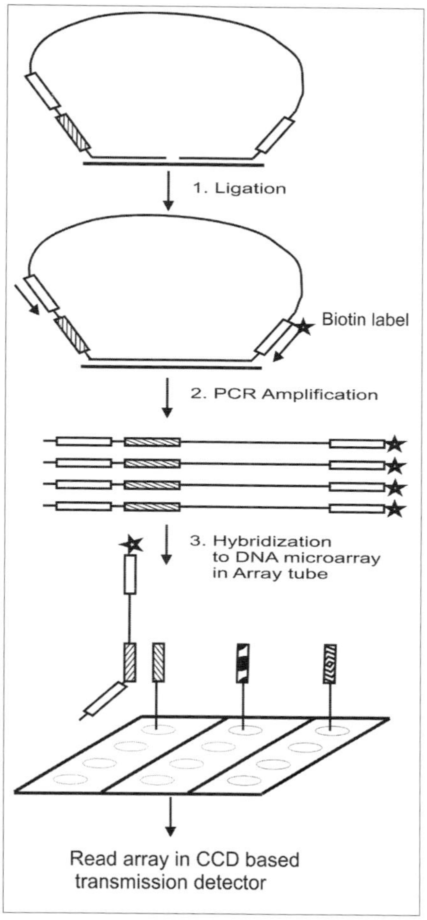

Figure 1.2 Principle of the Premi®Test *Salmonella* typing system. 1. *Salmonella* serovar specific target sequences hybridize to a ligation detection reaction (LDR) probe and the nick of the LDR arms is ligated. 2. Successful ligation products are amplified by PCR using primer sequences (white box) located on the LDR probe. One primer is labelled with biotin. A probe specific ZIP code sequence (hazed box) located between primers is also amplified with the specific *Salmonella* target sequence. 3. PCR products are hybridized to complementary ZIP sequences printed on the array tube (CLONDIAG, Jena, Germany). Detection is performed by a streptavidin–biotin enzyme-linked immunoassay.

bacterial genotyping are marketed as DiversiLab systems™ by bioMèrieux (bioMèrieux SA, Marcy L'Etoile, *France,* www.biomerieux.com*)*. The technology has been applied to various bacterial and fungal species (Shutt *et al.*, 2005; Healy *et al.*, 2005; Pounder *et al.*, 2006; Hansen *et al.*, 2008; Cleland *et al.*, 2008) and although being more expensive, it has the advantage of providing faster results than e.g. PFGE or MLST. Time to result is about 5 hours and 48 samples can be analysed per day. Wise *et al.* evaluated extragenic palindromic sequence-based PCR to predict *Salmonella enterica* serotypes using the DiversiLab *Salmonella* Kit for DNA fingerprinting (Wise *et al.*, 2009). After constructing a database from 216 *Salmonella* strains belonging to 14 serovars, the authors conducted a blind study using 44 poultry isolates, of which 36 (81.8%) were in concordance with the traditional serotyping result. As expected, six serotypes that were not included in the original library returned no result. However, a more comprehensive study including diverse strains from a larger number of different serovars has to be performed to estimate the potential of the method for *Salmonella* serotyping.

MLVA

Multilocus variable-number of tandem-repeat analysis (MLVA) is a new approach to sub-typing bacteria. The presence of repeated sequences is a fundamental feature of all genomes (Kolpakov *et al.*, 2003). These features represent the simplest form of regularity. Repeats in DNA are commonly classified as *interspersed* and *tandem repeats.* The former usually arise following replicate transposition whilst tandem repeats usually result from replication slippage. Tandem repeats produce uncharacteristic physical structures in DNA, causing DNA polymerase slippage and repeat amplification (Benson, 1999). Bacterial genomes are interspersed with short nucleotide sequences repeated many times.

Often these tandem repeats differ in copy number in different strains of the same species and provide a means of discrimination between strains with a high degree of sensitivity. Because the copy number in any specific tandem repeat is often polymorphic within a population, these elements have proven useful in linkage analyses and DNA fingerprinting. As the number of completed bacterial genomes continues to increase, MLVA is an attractive tool to subtype isolates within a species. MLVA involves the amplification and fragment size analysis of polymorphic regions of DNA containing variable numbers of tandem repeated sequences. Initial studies of

MLVA indicate improved discriminatory power compared to PFGE, but there is a need for standardization of the different equipment that is used for automated fragment detection. Currently, for the species *S. enterica* only, MLVA methods are applicable for subtyping of isolates within specific serovars including Typhimurium (Lindstedt *et al.*, 2004), Enteritidis (Malorny *et al.*, 2008; Boxrud *et al.*, 2007), Infantis (Ross and Heuzenroeder, 2008) and Typhi (Ramisse *et al.*, 2004). This technique, once refined, has great potential for subtyping below the subspecies level. For development it is important to include an appropriate subset of VNTR loci generally occurring in each *Salmonella* genome which reflect their long-term genetic relatedness.

DNA sequencing-based assays

The advent of automated DNA sequencing technology has led to the development and implementation of DNA sequence-based subtyping techniques applicable to many isolates in parallel. Principles and instruments for high-throughput sequencing have been developed during the last few years which are based on different technologies. Pyrosequencing is based on the quantitative detection of pyrophosphate released following nucleotide incorporation into a growing DNA chain ('sequencing by synthesis') and is suitable for rapid detection of single nucleotide polymorphisms (Ronaghi *et al.*, 1998). The pyrosequencing technology is the detection method used in the Whole Genome Sequencing technology from 454 Life Sciences, a company of Roche Applied Science (Branford, CT) and the second generation instrument genome Sequencer FLX (GS FLX). Mortimer *et al.* (Mortimer *et al.*, 2004) initiated a DNA sequence-based approach for serotyping *Salmonella enterica* that used a short sequence- or single nucleotide polymorphism-(SNP)-based assay targeting characteristic motifs of the *fliC* gene using pyrosequencing. Serospecific motifs were identified by comparison of amino acid sequences coding for H1-antigens. The method consists of two steps. First, a specific PCR amplifies the target DNA, the resulting amplicon is then used for the pyrosequencing assay. The whole process takes about 2 hours 45 min for 96 samples. Combination of the *fliC* assay to a *fljB* and *rfb* assay would allow for determination of the full antigenic formulae of *Salmonella* serotypes.

Principally, due to the dramatic drop of DNA sequencing costs and portable data generation typing and sequence-based detection methods will eventually replace array-based systems. The typing method called **multilocus sequence typing** (**MLST**) might be the great winner of this evolution. MLST is based on the concept of MLEE but allelic types are determined from nucleotide sequences of housekeeping genes and not electrophoretic mobilities of the enzymes they encode (Maiden *et al.*, 1998). Therefore, MLST detects changes at DNA level that are not apparent by phenotypic approaches, such as serotyping or MLEE. Data generated by MLST are non-ambigous and can be readily compared between laboratories (Maiden, 2006). A number of MLST studies have been performed for *S. enterica*. Depending on the selection of housekeeping genes the discriminative power varied leading to different subtype discrimination. The first *Salmonella* MLST was published by Kotetishvili *et al.* (2002) and was based on the genes for the 16S RNA, *pduF* (1,2-propanediol utilization factor), *glnA* (glutamate synthetase) and *manB* (phosphomannomutase). The proportions of polymorphic sites varied for the four genes, ranging from 0.6% for 16S RNA fragments to 5.6% for the *manB* locus. Sukhnanand *et al.* (Sukhnanand *et al.*, 2005) selected seven housekeeping genes and virulence genes (*panB*, *fimA*, *aceK*, *mdh*, *icdA*, *manB* and *spaN*), and found a correlation between MLST types and serovars and allowed within-serovar differentiation for 9 of the 12 serovars characterized. A study published in the same year selected the three housekeeping genes *manB*, *glnA* and *pduF* as well as the virulence gene *spaM* (Fakhr *et al.*, 2005). Sequence diversity within 85 serovar Typhimurium strains has not been detected. The largest collection of *Salmonella* MLST data is provided by the University College Cork, Ireland (http://mlst.ucc.ie/mlst/dbs/Senterica). The database has currently more than 2300 entries. Its *S. enterica* MLST scheme uses internal fragments of seven housekeeping genes, namely *thrA*, *purE*, *sucA*, *hisD*, *aroC*, *hemD*, and *dnaN*. Several studies have been published using this powerful MLST scheme (Kidgell *et al.*, 2002;

Cooke *et al.*, 2008; Wiesner *et al.*, 2009; Torpdahl *et al.*, 2005). However, most authors have shown that the MLST scheme lacked diversity and that the ability to discriminate between isolates was higher with pulsed-field gel electrophoresis (PFGE). In contrast to PFGE, MLST is therefore useful to study long-term evolution rather than short-term genetic relationships. Another MLST scheme was developed based on the four genes *atpD* (beta-subunit of ATP-synthase), *gyrB* (subunit B of DNA gyrase), *fliC* and *fljB* with the aim to generate results that are more compatible with those from the traditional serotyping scheme (Tankouo-Sandjong *et al.*, 2007).

Rapid phenotypic fingerprinting technologies

Various analytical instrumentation-based approaches often referred to as whole-organism fingerprinting are under investigation for the rapid identification of bacterial pathogens. Such methods involve the measurement of large numbers of spectral characters that together reflect the whole-cell chemical composition. These techniques include mass spectrometric approaches like pyrolysis mass spectrometry (PyMS) and MALDI-TOF mass spectrometry as well as label-free vibrational spectroscopic techniques such as Fourier transform infrared (FT-IR) spectroscopy and Raman spectroscopy. These technologies are based on cellular components and therefore considered phenotypic methods. Since phenotypic methods for bacterial identification can detect both target and unanticipated bacteria, they might be useful in early detection of an outbreak caused by bioterror agents, when little information on the agent is available. Starting from whole cells of a pure culture they can be used for detection and identification of bacterial pathogens essentially without the need for any pre-fractionations, amplifications, labelling or staining. Sample preparation is generally fast and simple and typically consists of harvesting small amounts of cells from the cultivation medium, and transferring them on special reusable sample plates either directly or as a suspension in a suitable solvent. Generally, these techniques can be automated and run at high throughput. Whole-organism fingerprinting relies on the generation of unique fingerprint spectra from whole cells. These spectra are stable across taxonomic groups and therefore allow bacterial discrimination at different taxonomic levels (serotype, strain, species, and genus) by comparing spectral patterns with databases of spectra of known bacteria using pattern recognition algorithms. The major advantages of whole-organism fingerprinting methods are their speed (typical instrument analysis times of less than one minute per sample, analysis of several hundred samples in one batch possible), and their low consumable costs. General disadvantages include spectral variations caused by parameters such as culture medium, growth time and growth temperature, the necessity to analyse pure cultures and the overlapping of signals in samples, which may lead to misinterpretation of results.

PyMS

Pyrolysis mass spectrometry (PyMS) involves the thermal degradation of a complex material such as bacterial cells in a vacuum. The smaller, volatile fragments (the pyrolysate) are then separated and analysed by mass spectrometry on the basis of their mass-to-charge (m/z) ratios to produce a pyrolysis mass spectrum, which can then be used as a chemical fingerprint of the cells. The technique has been successfully applied for the interstrain comparison of a variety of medically important organisms including *Salmonellae* (Reiner, 1972; Gutteridge, 1987; Tas *et al.*, 1987; Goodacre *et al.*, 1998). Wilkes *et al.* (2006) used pyrolysis metastable atom bombardment (MAB)-MS with an artificial neural network to characterize *Salmonella* isolates, and showed grouping of the strains according to serotypes, but only on a very limited number of strains and serovars (four). The technique is relatively labour intensive and technically challenging and problems of long-term reproducibility due to instrumental drift have not been solved satisfactorily to allow its use for routine microbial identification within clinical microbiology (Timmins and Goodacre, 2005). Therefore, PyMS has been rather restricted to the investigations of short-term outbreaks and soft ionization methods such as electrospray ionization (ESI-MS) and matrix-assisted laser desorption ionization (MALDI) were increasingly applied for taxonomic purposes.

MALDI-TOF MS

Whole-Cell MALDI-TOF MS fingerprinting for bacterial identification is based on the detection of highly abundant low molecular weight proteomic components that are directly accessible by MALDI-TOF mass spectrometry of whole cells. Sample preparation consists of a simple on-target extraction using a MALDI matrix dissolved in an appropriate organic solvent. Typically, the mass range from 2000 Da to 20,000 Da is recorded. An advantage of the method compared to the other physicochemical techniques lies in the fact that the observed mass values can be linked to the steadily growing information available in genome sequence databases. Individual marker proteins can thus be tentatively assigned to known proteins by calculating molecular weights from known protein sequences (and taking possible post-translational modifications into account) (Demirev *et al.*, 1999). Pineda *et al.* (2003) used a statistical model to calculate the significance of microorganism identification by MS-based proteome database searching in order to quantitatively assess the risk of false identification. Species identification can be readily achieved using fingerprint-based approaches by comparing whole cell mass spectra with databases of reference spectra obtained from known microbial species (Stackebrandt *et al.*, 2005; Vanlaere *et al.*, 2008; Sauer *et al.*, 2008; Hsieh *et al.*, 2008; Ilina *et al.*, 2009). Software tools including reference databases for automated MALDI-TOF MS–based species identification are commercially available (MALDI Biotyper™, Bruker Daltonik, Bremen, www.bdal.de; Saramis™, Anagnostec, Potsdam, www.anagnostec.com). Several studies showed that MALDI-TOF mass spectra of whole vegetative bacterial cells including *Salmonella* are dominated by highly abundant proteins, mainly ribosomal proteins (Demirev *et al.*, 2001; Fagerquist *et al.*, 2005; Sun *et al.*, 2006; Dieckmann *et al.*, 2008; Teramoto *et al.*, 2007). Ribosomal proteins can account for more than 20% of the total cellular protein content with a mean copy number of about 17000 copies per cell. Furthermore, ribosomal proteins are constitutively expressed and highly conserved and therefore represent ideal candidates to be used as universal protein marker ions for species identification based on rapid MALDI-TOF MS profiling (Teramoto *et al.*, 2007). Spectra comparison requires a high level of reproducibility, which is related to bacterial growth conditions (media composition and growth time), experimental setup (choice of matrix, extraction solvents, calibration standards) and instrumental setup (MALDI instrument and performance). These issues have been examined in several studies and it is now generally accepted that the method, when performed under standardized conditions, can provide reliable information for species identification despite an inherent variability, which can be accounted for by the algorithm used for the comparison (Wang *et al.*, 1998; Williams *et al.*, 2003; Valentine *et al.*, 2005; Wunschel *et al.*, 2005). The bioinformatic approach should enhance the reliability of the method compared to pattern matching approaches, especially when discrimination of closely related microorganisms, e.g. below the species level, is desired, that often relies on differences in only a single or a few protein marker ions. The method is based on the mass spectrometric detection of non-synonymous polymorphisms present in highly expressed housekeeping genes that result in amino acid exchanges at the protein level. Different studies indicated that such variations of amino acid sequences of e.g. ribosomal subunit proteins can also be observed below the species level, allowing for the identification of long-term clonal lineages within bacterial species, such as *Salmonella* subspecies or serovars. Rather than comparing whole cell fingerprints, this approach relies on the identification of characteristic biomarker ions that are diagnostic (specific) for the bacterial genus, species, subspecies or strain. Leuschner *et al.* (Leuschner *et al.*, 2004) analysed several strains of six different *S. enterica* serotypes to identify serovar-specific consensus-patterns. However, only a small number of consensus peaks could be identified for some prevalent *Salmonella* serovars, overlaps were observed between serovars and apparently no serovar-identifying biomarker ions could be extracted. Differentiation of closely related bacteria can be enhanced by using a weighted pattern matching approach: Replicate measurements of multiple strains from different taxonomic groups are performed and the most discriminating and most reproducible biomarker ions are identified and upweighted in the subsequent computer-aided identification routines. The discriminative value of a biomarker ion is mainly indicated by its frequency of appearance and its specificity. Using this approach, the two *Salmonella* species, *S. enterica* and *S. bongori* were shown to be clearly discriminated by Whole-Cell MALDI-TOF MS. Moreover, identifying biomarker ions were detected to allow classification and identification of the six *S. enterica* subspecies (Fig. 1.3A) (Dieckmann *et al.*, 2008). Fig. 1.3B shows an overlay of whole-cell MALDI-TOF mass spectra in a narrow mass range obtained from several strains representing the six *Salmonella enterica* subspecies. The overall patterns of the spectra are very similar, but small group-specific differences can be observed. The whole-cell MALDI-TOF mass spectra of two different *S. enterica* subsp. I serovars such as Typhimurium (STM) and Enteritidis (SE) typically contain more than 160 distinct protein signals, of which more than 90% are common to both serovars. Due to the limited number of discriminating peaks and due to the fact that different serovars often display the same sequence variations, it can not be expected that this method could be applied as a universal technique for *Salmonella* serovar identification. However, mass variations between the two most prevalent serovars were detected in about 15 protein signals indicating that this approach could be useful for *Salmonella* serovar discrimination and pre-screening of prevalent serovars such as Typhimurium and Enteritidis (Fig. 1.3B).

FT-IR

Fourier transform infrared spectroscopy (FT-IR) permits differentiation between bacterial species and strains by generating distinctive FT-IR spectra for each microbe due to differences in the structure and quantity of cellular constituents like cell wall polysaccharides, lipids and proteins (Naumann *et al.*, 1991). The FT-IR spectra of intact cells reflect the vibrational or rotational motions of specific functional groups or bonds in cell wall components. Due to the complex nature of cell membranes, FT-IR spectra of intact bacteria

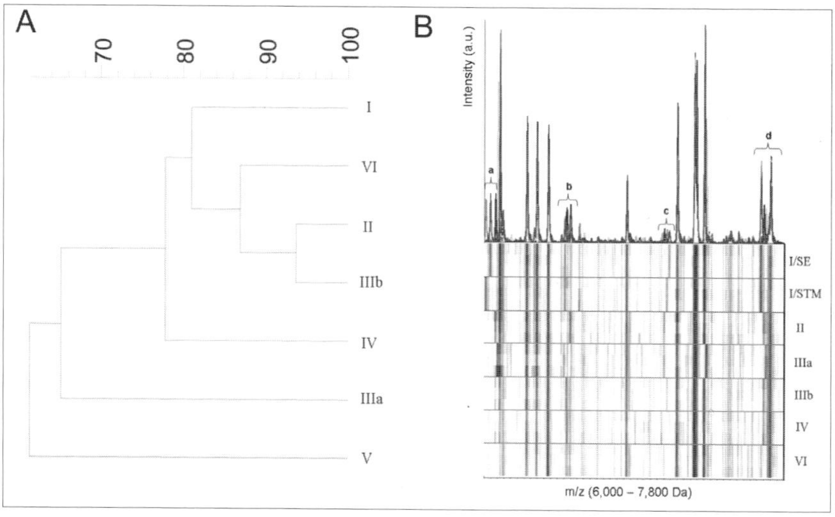

Figure 1.3 (A) Phylogenetic classification of *S. enterica* subsp. and *S. bongori* (V) based on variations in whole-cell MALDI-TOF mass data of multiple protein markers. (B) Whole-Cell MALDI-TOF mass spectra of *Salmonella* strains representing the six *Salmonella enterica* subspecies. Overlay of mass spectra in the mass range from 6000 to 7800 Da (top) and gel view representation of the individual spectra (bottom). *S. enterica* subsp. *enterica* serovar Enteritidis, I/SE; *S. enterica* subsp. *enterica* serovar Typhimurium, I/STM; *S. enterica* subsp. *salamae* (II); *S. enterica* subsp. *arizonae* (IIIa); *S. enterica* subsp. *diarizonae* (IIIb); *S. enterica* subsp. *houtenae* (IV) and *S. enterica* subsp. *indica* (VI); (A–D), potentially discriminative peak areas used for *Salmonella* differentiation below the species level.

typically show broad bands rather than distinct peaks (Helm *et al.*, 1991). Samples can be scanned and analysed in less than five minutes requiring only very little samples (nanogram-microgram). Following the pioneering studies of Naumann *et al.*, numerous studies demonstrated the application of FT-IR spectroscopy for identification and classification of bacterial species and subspecies as well as antibiotic resistance phenotypes (for a recent review see Burgula *et al.*, 2007). However, only a few studies were published that tried to evaluate the potential of the method for *Salmonella* identification below species level. The key to success in identifying microorganisms below species level with FT-IR spectroscopy is achieving excellent reproducibility in sample growth conditions, sample preparation, and collection of the infrared spectrum. Kim *et al.* (Kim *et al.*, 2005) analysed six different *Salmonella* serovars by FT-IR of intact cells cultured in broth and found that this approach failed to differentiate the *Salmonella* serovars with the exception of serovar Brandenburg, which showed distinct bands in the amide I/II spectral region of 1800–1500 cm^{-1}. A better sub-grouping of *Salmonella* serotypes using FT-IR was achieved only by using more laborious sample preparation procedures that relied on either the analysis of LPS extracts in the 1200–900 cm^{-1} region (Kim *et al.*, 2005) or outer membrane protein preparations in the 1800–1500 cm^{-1} region (Kim *et al.*, 2006d). Baldauf *et al.* used intact cells grown on selective agar plates and showed that a SIMCA model (Soft Independent Modeling of Class Analogy) could distinguish 10 *Salmonella* strains belonging to five different serovars (Enteritidis, Typhimurium, Heidelberg, Kentucky, Anatum) and group the *S. enterica* strains according to serovar (Baldauf *et al.*, 2006; Baldauf *et al.*, 2007). Major discrimination was achieved between 1000 and 970 cm^{-1} indicating that the signal responsible for the differentiation among selected serovars was associated with IR frequencies of the O-specific (lipopolysaccharides) antigens from the bacterial cell envelope.

Raman spectroscopy

While IR spectroscopy measures the absorption of infrared light by the sample, Raman spectroscopy relies on inelastic scattering, or Raman scattering, of monochromatic light, following excitation of the sample by a laser, in the visible, near infrared, or near ultraviolet range. Whole-cell Raman fingerprints are integrated spectra of the total chemical composition of the cell. Like IR fingerprints, Raman whole-organism fingerprints can be used for bacterial identification at the species and strain levels (Goodacre *et al.*, 1998; Maquelin *et al.*, 2006). Normal Raman spectroscopy and the more sensitive Surface-enhanced *Raman* scattering (*SERS*) have been applied for the rapid screening of various species including *Salmonella* (Montoya *et al.*, 2003; Liu *et al.*, 2008), however, a successful differentiation of *Salmonella* subspecies or serovars has not been demonstrated yet.

Phenotypic arrays

Another new approach is the phenotypic characterization of bacteria which attempts to give a global view of cellular phenotypes (Bochner, 2009). Innovative high-throughput 'Phenotype MicroArrays' have been developed by BIOLOG, Inc (Hayward, CA, www.biolog.com) consisting of hundreds of assays of C-source, N-source, P-source and S-source metabolisms, 100 assays of biosynthetic pathways, 10 assays of ion effects and osmolarity, 100 assays of pH effects and pH control with deaminases and decarboxylases, and 1000 assays of chemical sensitivity. Assays can provide not only qualitative but also quantitative results. The technique is powerful to investigate the phenotypic changes in gene knockout strains using a wide range of microbial species and has high potential to detect small differences between closely related strains. For *Salmonella,* a comprehensive profiling study has been not yet performed. Some specific studies indicate the discriminative power for *S. enterica* within the serovar Enteritidis (Morales *et al.*, 2005; Guard-Bouldin *et al.*, 2007). However, the usefulness for robust typing of many strains remains to be shown, and costs are currently prohibitively high for routine use.

Outlook

Long-term or global epidemiology requires methods with high discriminatory power using slowly accumulating genetic variation. In the past two decades, DNA-based techniques have been generally better appreciated and more frequently applied in microbial epidemiology and evolutionary studies than phenotypic approaches. A given phenotype does not necessarily reflect evolutionary history. Small DNA changes in single phenotypic markers can result in dramatically different phenotypes when the evolutionarily more relevant genotypes may be very similar, and vice versa. In addition, traditional phenotypic methods such as biochemical typing, phage typing or serotyping are often quite laborious and cannot easily be automated. MLEE had been an exception for several years, a phenotypic method that proved to be very useful for analysis of bacterial population genetics. Meanwhile, MLST, the DNA-based successor of MLEE, proved to be much more practical and is now more widely used. Some of the physicochemical whole-cell fingerprinting methods might lead to a revival of phenotypic methods since they address a large number of spectral characters simultaneously and have the advantage of being cost-effective, easily performed, very rapid even compared to DNA-based methods, and can be easily automatized. Additionally, these methods have the advantage that they can provide target identification without any *a priori* knowledge of the sample, provided an appropriate reference database is available. The MALDI-TOF MS technique in particular seems to be on the way of becoming a routine diagnostic tool for microbial species identification. Since the method targets sequence variation mainly within multiple housekeeping proteins, it could also provide valuable information for characterization of microorganisms below the (sub-) species level. However, the potential of whole-cell fingerprinting methods for subspecies typing has not yet been fully evaluated. A disadvantage of whole-cell fingerprinting methods compared to PCR methods lies in the fact that these methods require pure cultures. PCR-based methods on the other hand can only detect a limited spectrum of targets defined by the oligonucleotide probes used, but in principle allow the detection of small amounts of DNA also in complex samples. Recent advances that combine PCR amplification with innovative technologies such as micro-bead detection (Luminex), low-density microarrays (PremiTest) or microfluidics chip devices (DiversiLab™), that have been recently brought to the market, still have to prove their performance and robustness in the environment of a real-world microbiology laboratory. The same holds true for diagnostic approaches based on DNA microarrays, which are and will be of great value as research tools, but may not always prove to be applicable to the diagnostics market due to cost considerations. The ultimate form of typing, providing the most detailed information on DNA diversity, is full-genome sequencing. DNA sequencing costs have dropped several orders of magnitude during the last years; and DNA sequencing-based methods can be expected to play a major role in bacterial subspecies typing and future clinical microbiology in general.

Acknowledgements

RD was financially supported by the Bundesministerium für Wirtschaft und Technologie (BMWI), grant PROINNO KF0350103MD8. EH was financially supported by the Bundesministerium für Bildung und Forschung (BMBF), grant FBI-Zoo 01KI 07123.

References

Agron, P.G., Walker, R.L., Kinde, H., Sawyer, S.J., Hayes, D.C., Wollard, J., and Andersen, G.L. (2001). Identification by subtractive hybridization of sequences specific for *Salmonella enterica* serovar Enteritidis. Appl. Environ. Microbiol. *67*, 4984–4991.

Alvarez, J., Sota, M., Vivanco, A.B., Perales, I., Cisterna, R., Rementeria, A., and Garaizar, J. (2004). Development of a multiplex PCR technique for detection and epidemiological typing of *Salmonella* in human clinical samples. J. Clin. Microbiol. *42*, 1734–1738.

Arrach, N., Porwollik, S., Cheng, P., Cho, A., Long, F., Choi, S.H., and McClelland, M. (2008). *Salmonella* serovar identification using PCR-based detection of gene presence and absence. J. Clin. Microbiol. *46*, 2581–2589.

Baldauf, N.A., Rodriguez-Romo, L.A., Männig, A., Yousef, A.E., and Rodriguez-Saona, L.E. (2007). Effect of selective growth media on the differentiation of *Salmonella enterica* serovars by Fourier-Transform Mid-Infrared Spectroscopy. J. Microbiol. Methods *68*, 106–114.

Baldauf, N.A., Rodriguez-Romo, L.A., Yousef, A.E., and Rodriguez-Saona, L.E. (2006). Differentiation of selected *Salmonella enterica* serovars by Fourier transform mid-infrared spectroscopy. Appl. Spectrosc. *60*, 592–598.

Beltran, P., Plock, S.A., Smith, N.H., Whittam, T.S., Old, D.C., and Selander, R.K. (1991). Reference collection of strains of the *Salmonella* Typhimurium complex from natural populations. J. Gen. Microbiol. *137*, 601–606.

Benson, G. (1999). Tandem repeats finder: a program to analyze DNA sequences. Nucleic Acids Res. *27*, 573–580.

Beuzon, C.R., Schiaffino, A., Leori, G., Cappuccinelli, P., Rubino, S., and Casadesús, J. (1997). Identification of *Salmonella abortusovis* by PCR amplification of a serovar-specific IS*200* element. Appl. Environ. Microbiol. *63*, 2082–2085.

Bochner, B.R. (2009). Global phenotypic characterization of bacteria. FEMS Microbiol. Rev. *33*, 191–205.

Boxrud, D., Pederson-Gulrud, K., Wotton, J., Medus, C., Lyszkowicz, E., Besser, J., and Bartkus, J.M. (2007). Comparison of multiple-locus variable-number tandem repeat analysis, pulsed-field gel electrophoresis, and phage typing for subtype analysis of *Salmonella enterica* serotype Enteritidis. J. Clin. Microbiol. *45*, 536–543.

Boyd, E.F., Wang, F.S., Beltran, P., Plock, S.A., Nelson, K., and Selander, R.K. (1993). *Salmonella* reference collection B (SARB): strains of 37 serovars of subspecies I. J. Gen. Microbiol. *139 Pt 6*, 1125–1132.

Boyd, E.F., Wang, F.S., Whittam, T.S., and Selander, R.K. (1996). Molecular genetic relationships of the *Salmonellae*. Appl. Environ. Microbiol. *62*, 804–808.

Brüssow, H., Canchaya, C., and Hardt, W.D. (2004). Phages and the evolution of bacterial pathogens: from genomic rearrangements to lysogenic conversion. Microbiol. Mol. Biol. Rev. *68*, 560–602.

Burgula, Y., Khali, D., Kim, S., Krishnan, S.S., Cousin, M.A., Gore, J.P., Reuhs, B.L., and Mauer, L.J. (2007). Review of mid-infrared fourier transform-infrared spectroscopy applications for bacterial detection. J. Rapid Methods Autom. Microbiol. *15*, 146–175.

Carlson, S.A., Bolton, L.F., Briggs, C.E., Hurd, H.S., Sharma, V.K., Fedorka-Cray, P.J., and Jones, B.D. (1999). Detection of multiresistant *Salmonella* Typhimurium DT104 using multiplex and fluorogenic PCR. Mol. Cell. Probes *13*, 213–222.

Chevrier, D., Popoff, M.Y., Dion, M.P., Hermant, D., and Guesdon, J.L. (1995). Rapid detection of *Salmonella* subspecies I by PCR combined with non-radioactive hybridisation using covalently immobilised oligonucleotide on a microplate. FEMS Immunol. Med. Microbiol. *10*, 245–501.

Chiu, C.H., Su, L.H., Chu, C.H., Wang, M.H., Yeh, C.M., Weill, F.X., and Chu, C. (2006). Detection of multidrug-resistant *Salmonella enterica* serovar Typhimurium phage types DT102, DT104, and U302 by multiplex PCR. J. Clin. Microbiol. *44*, 2354–2358.

Chiu, T.H., Pang, J.C., Hwang, W.Z., and Tsen, H.Y. (2005). Development of PCR primers for the detection of *Salmonella enterica* serovar Choleraesuis based on the *fliC* gene. J. Food Prot. *68*, 1575–1580.

Cleland, D., Krader, P., and Emerson, D. (2008). Use of the DiversiLab repetitive sequence-based PCR system for genotyping and identification of Archaea. J. Microbiol. Methods 73, 172–178.

Cooke, F.J., Brown, D.J., Fookes, M., Pickard, D., Ivens, A., Wain, J., Roberts, M., Kingsley, R.A., Thomson, N.R., and Dougan, G. (2008). Characterization of the genomes of a diverse collection of *Salmonella enterica* serovar Typhimurium definitive phage type 104. J. Bacteriol. *190*, 8155–8162.

Cortez, A.L.L., Carvalho, A.C.F.B., Ikuno, A.A., Bürger, K.P., and Vidal-Martins, A.M.C. (2006). Identification of *Salmonella* spp. isolates from chicken abattoirs by multiplex-PCR. Res. Vet. Sci. *81*, 340–344.

Crosa, J.H., Brenner, D.J., Ewing, W.H., and Falkow, S. (1973). Molecular relationships among the Salmonelleae. J. Bacteriol. *115*, 307–315.

Demirev, P.A., Ho, Y.P., Ryzhov, V., and Fenselau, C. (1999). Microorganism identification by mass spectrometry and protein database searches. Anal. Chem. *71*, 2732–2738.

Demirev, P.A., Lin, J.S., Pineda, F.J., and Fenselau, C. (2001). Bioinformatics and mass spectrometry for microorganism identification: proteome-wide post-translational modifications and database search algorithms for characterization of intact *H. pylori*. Anal. Chem. *73*, 4566–4573.

Desai, A.R., Shah, D.H., Shringi, S., Lee, M.J., Li, Y.H., Cho, M.R., Park, J.H., Eo, S.K., Lee, J.H., and Chae, J.S. (2005). An allele-specific PCR assay for the rapid and serotype-specific detection of *Salmonella* Pullorum. Avian Dis. *49*, 558–561.

Dieckmann, R., Helmuth, R., Erhard, M., and Malorny, B. (2008). Rapid classification and identification of *Salmonellae* at the species and subspecies levels by whole-cell matrix-assisted laser desorption ionization-time of flight mass spectrometry. Appl. Environ. Microbiol. *74*, 7767–7778.

Dunbar, S.A., and Jacobson, J.W. (2007). Quantitative, multiplexed detection of *Salmonella* and other pathogens by Luminex xMAP suspension array. Methods Mol. Biol. *394*, 1–19.

Echeita, M.A., Herrera, S., Garaizar, J., and Usera, M.A. (2002). Multiplex PCR-based detection and identification of the most common *Salmonella* second-phase flagellar antigens. Res. Microbiol. *153*, 107–113.

Fagerquist, C.K., Miller, W.G., Harden, L.A., Bates, A.H., Vensel, W.H., Wang, G., and Mandrell, R.E. (2005). Genomic and proteomic identification of a DNA-binding protein used in the 'fingerprinting' of *Campylobacter* species and strains by MALDI-TOF-MS protein biomarker analysis. Anal. Chem. *77*, 4897–4907.

Fakhr, M.K., Nolan, L.K., and Logue, C.M. (2005). Multilocus sequence typing lacks the discriminatory ability of pulsed-field gel electrophoresis for typing *Salmonella enterica* serovar Typhimurium. J. Clin. Microbiol. *43*, 2215–2219.

Farmer I.I.I., J.J. (2003). *Enterobacteriaceae*: Introduction and identification. In Manual of Clinical Microbiology, Murray, P.R., Baron, E.J., Jorgensen, J.H., Pfaller,

M.A., and Yolken, R.H., eds (Washington D.C., ASM Press), pp. 636–653.

Fitzgerald, C., Collins, M., van Duyne, S., Mikoleit, M., Brown, T., and Fields, P. (2007). Multiplex, bead-based suspension array for molecular determination of common *Salmonella* serogroups. J. Clin. Microbiol. *45*, 3323–3334.

Fitzgerald, C., Sherwood, R., Gheesling, L.L., Brenner, F.W., and Fields, P.I. (2003). Molecular analysis of the *rfb* O antigen gene cluster of *Salmonella enterica* serogroup O:6,14 and development of a serogroup-specific PCR assay. Appl. Environ. Microbiol. *69*, 6099–6105.

Foley, S.L., Zhao, S., and Walker, R.D. (2007). Comparison of molecular typing methods for the differentiation of *Salmonella* foodborne pathogens. Foodborne Pathog. Dis. *4*, 253–276.

Goodacre, R., Timmins, É.M., Burton, R., Kaderbhai, N., Woodward, A.M., Kell, D.B., and Rooney, P.J. (1998). Rapid identification of urinary tract infection bacteria using hyperspectral whole-organism fingerprinting and artificial neural networks. Microbiology *144*, 1157–1170.

Grimont, P.A.D., and Weill, F.-X. (2007). Antigenic formulae of the *Salmonella* serovars. (Paris, France: WHO Collaborating Centre for Reference and Research on *Salmonella*, Institut Pasteur).

Guard-Bouldin, J., Morales, C.A., Frye, J.G., Gast, R.K., and Musgrove, M. (2007). Detection of *Salmonella enterica* subpopulations by phenotype microarray antibiotic resistance patterns. Appl. Environ. Microbiol. *73*, 7753–7756.

Gutteridge, C.S. (1987). Characterization of microorganisms by pyrolysis mass spectrometry. Methods Microbiol. *19*, 227–272.

Hadjinicolaou, A.V., Demetriou, V.L., Emmanuel, M.A., Kakoyiannis, C.K., and Kostrikis, L.G. (2009). Molecular beacon-based real-time PCR detection of primary isolates of *Salmonella* Typhimurium and *Salmonella* Enteritidis in environmental and clinical samples. BMC Microbiol. *9*., art. no. 97.

Hansen, D., Healy, M., Reece, K., Smith, C., and Woods, G.L. (2008). Repetitive-sequence-based PCR using the DiversiLab system for identification of *Aspergillus* species. J. Clin. Microbiol. *46*, 1835–1839.

Hashimoto, Y., Itho, Y., Fujinaga, Y., Khan, A.Q., Sultana, F., Miyake, M., Hirose, K., Yamamoto, H., and Ezaki, T. (1995). Development of nested PCR based on the ViaB sequence to detect *Salmonella* Typhi. J. Clin. Microbiol. *33*, 775–777.

Healy, M., Reece, K., Walton, D., Huong, J., Frye, S., Raad, I.I., and Kontoyiannis, D.P. (2005). Use of the DiversiLab system for species and strain differentiation of *Fusarium* species isolates. J. Clin. Microbiol. *43*, 5278–5280.

Helm, D., Labischinski, H., Schallehn, G., and Naumann, D. (1991). Classification and identification of bacteria by Fourier-transform infrared spectroscopy. J. Gen. Microbiol. *137*, 69–79.

Herrera-Leon, S., McQuiston, J.R., Usera, M.A., Fields, P.I., Garaizar, J., and Echeita, M.A. (2004). Multiplex PCR for distinguishing the most common phase-1 flagellar antigens of *Salmonella* spp. J. Clin. Microbiol. *42*, 2581–2586.

Herrera-Leon, S., Ramiro, R., Arroyo, M., Diez, R., Usera, M.A., and Echeita, M.A. (2007). Blind comparison of traditional serotyping with three multiplex PCRs for the identification of *Salmonella* serotypes. Res. Microbiol. *158*, 122–127.

Hirose, K., Itoh, K., Nakajima, H., Kurazono, T., Yamaguchi, M., Moriya, K., Ezaki, T., Kawamura, Y., Tamura, K., and Watanabe, H. (2002). Selective amplification of *tyv* (*rfbE*), *prt* (*rfbS*), *viaB*, and *fliC* genes by multiplex PCR for identification of *Salmonella enterica* serovars Typhi and Paratyphi A. J. Clin. Microbiol. *40*, 633–636.

Hong, Y., Liu, T., Lee, M.D., Hofacre, C.L., Maier, M., White, D.G., Ayers, S., Wang, L., Berghaus, R., and Maurer, J.J. (2008). Rapid screening of *Salmonella enterica* serovars Enteritidis, Hadar, Heidelberg and Typhimurium using a serologically-correlative allelotyping PCR targeting the O and H antigen alleles. BMC Microbiol. *8*, 178.

Hoorfar, J., Baggesen, D.L., and Porting, P.H. (1999). A PCR-based strategy for simple and rapid identification of rough presumptive *Salmonella* isolates. J. Microbiol. Methods *35*, 77–84.

Hsieh, S.Y., Tseng, C.L., Lee, Y.S., Kuo, A.J., Sun, C.F., Lin, Y.H., and Chen, J.K. (2008). Highly efficient classification and identification of human pathogenic bacteria by MALDI-TOF MS. Molec. Cell. Prot. *7*, 448–456.

Huehn, S., Helmuth, R., Bunge, C., van Pelt, W., and Malorny, B. (2009). Characterization of pathogenic and resistant genome repertoire reveals two clonal lines in *Salmonella enterica* subsp. *enterica* serovar Paratyphi B (+)-tartrate positive. Foodborne Path. Dis. *6*, 431–443.

Huehn, S., and Malorny, B. (2009). DNA microarray for molecular epidemiology of *Salmonella*. Methods Mol. Biol. *551*, 249–285.

Ilina, E.N., Borovskaya, A.D., Malakhova, M.M., Vereshchagin, V.A., Kubanova, A.A., Kruglov, A.N., Svistunova, T.S., Gazarian, A.O., Maier, T., Kostrzewa, M., and Govorun, V.M. (2009). Direct bacterial profiling by matrix-assisted laser desorption-ionization time-of-flight mass spectrometry for identification of pathogenic *Neisseria*. J. Mol. Diagn. *11*, 75–86.

Kardos, G., Farkas, T., Antal, M., Nogrady, N., and Kiss, I. (2007). Novel PCR assay for identification of *Salmonella enterica* serovar Infantis. Lett. Appl. Microbiol. *45*, 421–425.

Kidgell, C., Reichard, U., Wain, J., Linz, B., Torpdahl, M., Dougan, G., and Achtman, M. (2002). *Salmonella* typhi, the causative agent of typhoid fever, is approximately 50,000 years old. Infect. Genet. Evol. *2*, 39–45.

Kim, H.J., Park, S.H., and Kim, H.Y. (2006a). Comparison of *Salmonella enterica* serovar Typhimurium LT2 and non-LT2 *Salmonella* genomic sequences, and genotyping of *Salmonellae* by using PCR. Appl. Environ. Microbiol. *72*, 6142–6151.

Kim, H.J., Park, S.H., Lee, T.H., Nahm, B.H., Chung, Y.H., Seo, K.H., and Kim, H.Y. (2006b). Identification of

Salmonella enterica serovar Typhimurium using specific PCR primers obtained by comparative genomics in *Salmonella* serovars. J. Food Prot. *69*, 1653–1661.

Kim, S., Frye, J.G., Hu, J., Fedorka-Cray, P.J., Gautom, R., and Boyle, D.S. (2006c). Multiplex PCR-based method for identification of common clinical serotypes of *Salmonella enterica* subsp. *enterica*. J. Clin. Microbiol. *44*, 3608–3615.

Kim, S., Kim, H., Reuhs, B.L., and Mauer, L.J. (2006d). Differentiation of outer membrane proteins from *Salmonella enterica* serotypes using Fourier transform infrared spectroscopy and chemometrics. Lett. Appl. Microbiol. *42*, 229–234.

Kim, S., Reuhs, B.L., and Mauer, L.J. (2005). Use of Fourier transform infrared spectra of crude bacterial lipopolysaccharides and chemometrics for differentiation of *Salmonella enterica* serotypes. J. Appl. Microbiol. *99*, 411–417.

Kolpakov, R., Bana, G., and Kucherov, G. (2003). mreps: Efficient and flexible detection of tandem repeats in DNA. Nucleic Acids Res. *31*, 3672–3678.

Kongmuang, U., Luk, J.M., and Lindberg, A.A. (1994). Comparison of three stool-processing methods for detection of *Salmonella* serogroups B, C2, and D by PCR. J. Clin. Microbiol. 32, 3072–3074.

Kotetishvili, M., Stine, O.C., Kreger, A., Morris, J.J., Jr., and Sulakvelidze, A. (2002). Multilocus sequence typing for characterization of clinical and environmental *Salmonella* strains. J. Clin. Microbiol. *40*, 1626–1635.

Le Minor, L., Popoff, M.Y., Laurent, B., and Hermant, D. (1986). Characterization of a 7th subspecies of *Salmonella*: *Salmonella* Choleraesuis subsp. *indica* subsp. Nov. Ann. Inst. Pasteur Microbiol. *137B*, 211–217.

Le Minor, L., Veron, M., and Popoff, M. (1982). The taxonomy of *Salmonella*. Ann. Microbiol. (Paris) 133, 223–243.

Leader, B.T., Frye, J.G., Hu, J., Fedorka-Cray, P.J., and Boyle, D.S. (2009). High-throughput molecular determination of *Salmonella enterica* serovars by use of multiplex PCR and capillary electrophoresis analysis. J. Appl. Microbiol. *47*, 1290–1299.

Lee, K., Iwata, T., Shimizu, M., Taniguchi, T., Nakadai, A., Hirota, Y., and Hayashidani, H. (2009a). A novel multiplex PCR assay for *Salmonella* subspecies identification. J. Appl. Microbiol. *107*, 805–811.

Lee, S.H., Jung, B.Y., Rayamahji, N., Lee, H.S., Jeon, W.J., Choi, K.S., Kweon, C.H., and Yoo, H.S. (2009b). A multiplex real-time PCR for differential detection and quantification of *Salmonella* spp., *Salmonella enterica* serovar Typhimurium and Enteritidis in meats. J. Vet. Sci. *10*, 43–51.

Leuschner, R.G.K., Beresford-Jones, N., and Robinson, C. (2004). Difference and consensus of whole cell *Salmonella enterica* subsp. *enterica* serovars matrix-assisted laser desorption/ionization time-of-flight mass spectrometry spectra. Lett. Appl. Microbiol. *38*, 24–31.

Levy, H., Diallo, S., Tennant, S.M., Livio, S., Sow, S.O., Tapia, M., Fields, P.I., Mikoleit, M., Tamboura, B., Kotloff, K.L., Lagos, R., Nataro, J.P., Galen, J.E., and Levine, M.M. (2008). PCR method to identify *Salmonella enterica* serovars Typhi, Paratyphi A, and Paratyphi B among *Salmonella* isolates from the blood of patients with clinical enteric fever. J. Clin. Microbiol. *46*, 1861–1866.

Lim, Y.H., Hirose, K., Izumiya, H., Arakawa, E., Takahashi, H., Terajima, J., Itoh, K., Tamura, K., Kim, S.I., and Watanabe, H. (2003). Multiplex polymerase chain reaction assay for selective detection of *Salmonella enterica* serovar Typhimurium. Jpn. J. Infect. Dis. *56*, 151–155.

Lindstedt, B.A., Vardund, T., Aas, L., and Kapperud, G. (2004). Multiple-locus variable-number tandem-repeats analysis of *Salmonella enterica* subsp. *enterica* serovar Typhimurium using PCR multiplexing and multicolor capillary electrophoresis. J. Microbiol. Methods 59, 163–172.

Liu, Y., Chen, Y.R., Nou, X., Kim, M.S., and Chao, K. (2008). Label-free SERS for rapid species identification of *Escherichia coli, Listeria monocytogenes,* and *Salmonella* Typhimurium bacteria. Spectroscopy 23, 48–54.

Luk, J.M., Kongmuang, U., Reeves, P.R., and Lindberg, A.A. (1993). Selective Amplification of Abequose and Paratose Synthase Genes (*rfb*) by Polymerase Chain Reaction for Identification of *Salmonella* Major Serogroups (A, B, C2, and D). J. Clin. Microbiol. *31*, 2118–2123.

Luk, J.M., Kongmuang, U., Tsang, R.S.W., and Lindberg, A.A. (1997). An enzyme-linked immunosorbent assay to detect PCR products of the *rfbS* gene from serogroup D *Salmonellae*: A rapid screening prototype. J. Clin. Microbiol. *35*, 714–718.

Maiden, M.C. (2006). Multilocus sequence typing of bacteria. Annu. Rev. Microbiol. *60*, 561–588.

Maiden, M.C., Bygraves, J.A., Feil, E., Morelli, G., Russell, J.E., Urwin, R., Zhang, Q., Zhou, J., Zurth, K., Caugant, D.A., *et al.* (1998). Multilocus sequence typing: a portable approach to the identification of clones within populations of pathogenic microorganisms. Proc. Natl. Acad. Sci. U.S.A. *95*, 3140–3145.

Malorny, B., Bunge, C., and Helmuth, R. (2007). A real-time PCR for the detection of *Salmonella* Enteritidis in poultry meat and consumption eggs. J. Microbiol. Methods *70*, 245–251.

Malorny, B., Junker, E., and Helmuth, R. (2008). Multi-locus variable-number tandem repeat analysis for outbreak studies of *Salmonella enterica* serotype Enteritidis. BMC Microbiol. *8*, 84.

Maquelin, K., Dijkshoorn, L., Van Der Reijden, T.J.K., and Puppels, G.J. (2006). Rapid epidemiological analysis of *Acinetobacter* strains by Raman spectroscopy. J. Microbiol. Methods *64*, 126–131.

McClelland, M., Sanderson, K.E., Spieth, J., Clifton, S.W., Latreille, P., Courtney, L., Porwollik, S., Ali, J., Dante, M., Du, F., Hou, S., Layman, D., Leonard, S., Nguyen, C., Scott, K., Holmes, A., Grewal, N., Mulvaney, E., Ryan, E., Sun, H., Florea, L., Miller, W., Stoneking, T., Nhan, M., Waterston, R., and Wilson, R.K. (2001). Complete genome sequence of *Salmonella enterica* serovar Typhimurium LT2. Nature *413*, 852–856.

McQuiston, J.R., Parrenas, R., Ortiz-Rivera, M., Gheesling, L., Brenner, F., and Fields, P.I. (2004).

Sequencing and comparative analysis of flagellin genes *fliC*, *fljB*, and *flpA* from *Salmonella*. J. Clin. Microbiol. *42*, 1923–1932.

Montoya, J.R., Armstrong, R.L., and Smith, G.B. (2003). *Salmonella* using surfaced enhanced Raman scattering. Proceedings of SPIE – The International Society for Optical Engineering *5085*, 144–152.

Morales, C.A., Porwollik, S., Frye, J.G., Kinde, H., McClelland, M., and Guard-Bouldin, J. (2005). Correlation of phenotype with the genotype of egg-contaminating *Salmonella enterica* serovar Enteritidis. Appl. Environ. Microbiol. *71*, 4388–4399.

Mortimer, C.K., Peters, T.M., Gharbia, S.E., Logan, J.M., and Arnold, C. (2004). Towards the development of a DNA-sequence-based approach to serotyping of *Salmonella enterica*. BMC Microbiol. *4*, 31.

Naumann, D., Helm, D., and Labischinski, H. (1991). Microbiological characterizations by FT-IR spectroscopy. Nature *351*, 81–82.

O'Regan, E., McCabe, E., Burgess, C., McGuinness, S., Barry, T., Duffy, G., Whyte, P., and Fanning, S. (2008). Development of a real-time multiplex PCR assay for the detection of multiple *Salmonella* serotypes in chicken samples. BMC Microbiol. *8*.

Ou, H.-Y., Ju, C.T.S., Thong, K.-L., Ahmad, N., Deng, Z., and Barer, M. (2007). Translational genomics to develop a *Salmonella enterica* Serovar Paratyphi A multiplex polymerase chain reaction assay. J. Mol. Diagn. *9*, 624–630.

Parkhill, J., Dougan, G., James, K.D., Thomson, N.R., Pickard, D., Wain, J., Churcher, C., Mungall, K.L., Bentley, S.D., Holden, M.T., *et al.* (2001). Complete genome sequence of a multiple drug resistant *Salmonella enterica* serovar Typhi CT18. Nature *413*, 848–852.

Perera, K., and Murray, A. (2008). Development of a PCR assay for the identification of *Salmonella enterica* serovar Brandenburg. J. Med. Microbiol. *57*, 1223–1227.

Pineda, F.J., Antoine, M.D., Demirev, P.A., Feldman, A.B., Jackman, J., Longenecker, M., and Lin, J.S. (2003). Microorganism identification by matrix-assisted laser/desorption ionization mass spectrometry and model-derived ribosomal protein biomarkers. Anal. Chem. *75*, 3817–3822.

Porwollik, S., Boyd, E.F., Choy, C., Cheng, P., Florea, L., Proctor, E., and McClelland, M. (2004). Characterization of *Salmonella enterica* subspecies I genovars by use of microarrays. J. Bacteriol. *186*, 5883–5898.

Porwollik, S., Wong, R.M., and McClelland, M. (2002). Evolutionary genomics of *Salmonella*: gene acquisitions revealed by microarray analysis. Proc. Natl. Acad. Sci. U.S.A. *99*, 8956–8961.

Pounder, J.I., Shutt, C.K., Schaecher, B.J., and Woods, G.L. (2006). Clinical evaluation of repetitive sequence-based polymerase chain reaction using the DiversiLab System for strain typing of vancomycin-resistant enterococci. Diagn. Microbio. Infect. Dis. *54*, 183–187.

Ramisse, V., Houssu, P., Hernandez, E., Denoeud, F., Hilaire, V., Lisanti, O., Ramisse, F., Cavallo, J.D., and Vergnaud, G. (2004). Variable number of tandem repeats in *Salmonella enterica* subsp. *enterica* for typing purposes. J. Clin. Microbiol. *42*, 5722–5730.

Reiner, E. (1972). Rapid characterization of *Salmonella* organisms by means of pyrolysis-gas-liquid chromatography. Anal. Chem. *44*, 1058–1061.

Ronaghi, M., Uhlen, M., and Nyren, P. (1998). A sequencing method based on real-time pyrophosphate. Science 281, 363, 365.

Ross, I.L., and Heuzenroeder, M.W. (2008). A comparison of three molecular typing methods for the discrimination of *Salmonella enterica* serovar Infantis. FEMS Immunol. Med. Microbiol. *53*, 375–384.

Sauer, S., Freiwald, A., Maier, T., Kube, M., Reinhardt, R., Kostrzewa, M., and Geider, K. (2008). Classification and identification of bacteria by mass spectrometry and computational analysis. PLoS ONE *3*.

Scaria, J., Palaniappan, R.U., Chiu, D., Phan, J.A., Ponnala, L., McDonough, P., Grohn, Y.T., Porwollik, S., McClelland, M., Chiou, C.S., Chu, C., and Chang, Y.F. (2008). Microarray for molecular typing of *Salmonella enterica* serovars. Mol. Cell. Probes *22*, 238–243.

Schouten, J.P., McElgunn, C.J., Waaijer, R., Zwijnenburg, D., Diepvens, F., and Pals, G. (2002). Relative quantification of 40 nucleic acid sequences by multiplex ligation-dependent probe amplification. Nucl. Acids Res. *30*.

Selander, R.K., Beltran, P., Smith, N.H., Barker, R.M., Crichton, P.B., Old, D.C., Musser, J.M., and Whittam, T.S. (1990a). Genetic population structure, clonal phylogeny, and pathogenicity of *Salmonella* paratyphi B. Infect. Immun. *58*, 1891–1901.

Selander, R.K., Beltran, P., Smith, N.H., Helmuth, R., Rubin, F.A., Kopecko, D.J., Ferris, K., Tall, B.D., Cravioto, A., and Musser, J.M. (1990b). Evolutionary genetic relationships of clones of *Salmonella* serovars that cause human typhoid and other enteric fevers. Infect. Immun. *58*, 2262–2275.

Selander, R.K., Li, J., and Nelson, K. (1996). Evolutionary genetics of *Salmonella enterica*. In *Escherichia coli* and *Salmonella*, Neidhardt, F.C. *et al.*, ed. (Washington, D.C., ASM Press), pp. 2691–2707.

Shutt, C.K., Pounder, J.I., Page, S.R., Schaecher, B.J., and Woods, G.L. (2005). Clinical evaluation of the DiversiLab microbial typing system using repetitive-sequence-based PCR for characterization of *Staphylococcus aureus* strains. J. Clin. Microbiol. *43*, 1187–1192.

Soliman, S.E., Kilpatrick, C., Ahmed, S.M., Abouelhassan, M.E., Nimmanapelli, R., and Reddy, P.G. (2009). Allelotyping PCR for detection and screening of *Salmonella enterica* serovar Enteritidis and Typhimurium. Int. J. Poultry Sci. *8*, 156–160.

Song, J.H., Cho, H., Park, M.Y., Na, D.S., Moon, H.B., and Pai, C.H. (1993). Detection of *Salmonella* Typhi in the blood of patients with typhoid fever by polymerase chain reaction. J. Clin. Microbiol. *31*, 1439–1443.

Soumet, C., Ermel, G., Rose, V., Drouin, P., Salvat, G., and Colin, P. (1999). Identification by a multiplex PCR-based assay of *Salmonella* Typhimurium and *Salmonella* Enteritidis strains from environmental swabs of poultry houses. Lett. Appl. Microbiol. *29*, 1–6.

Stackebrandt, E., Päuker, O., and Erhard, M. (2005). Grouping myxococci (Corallococcus) strains by matrix-assisted laser desorption ionization time-of-flight (MALDI TOF) mass spectrometry: comparison with gene sequence phylogenies. Current Microbiol. *50*, 71–77.

Sukhnanand, S., Alcaine, S., Warnick, L.D., Su, W.L., Hof, J., Craver, M.P., McDonough, P., Boor, K.J., and Wiedmann, M. (2005). DNA sequence-based subtyping and evolutionary analysis of selected *Salmonella enterica* serotypes. J. Clin. Microbiol. *43*, 3688–3698.

Sun, L., Teramoto, K., Sato, H., Torimura, M., Tao, H., and Shintani, T. (2006). Characterization of ribosomal proteins as biomarkers for matrix-assisted laser desorption/ionization mass spectral identification of *Lactobacillus plantarum*. Rapid Commun. Mass Spectrom. *20*, 3789–3798.

Tankouo-Sandjong, B., Sessitsch, A., Liebana, E., Kornschober, C., Allerberger, F., Hachler, H., and Bodrossy, L. (2007). MLST-v, multilocus sequence typing based on virulence genes, for molecular typing of *Salmonella enterica* subsp. *enterica* serovars. J. Microbiol. Methods *69*, 23–36.

Tas, A.C., De Waart, J., Bouwman, J., De Brauw, M.C.T.N., and Van Der Greef, J. (1987). Rapid characterization of *Salmonella* strains with direct chemical ionization pyrolysis. J. Analyt. Appl. Pyrolysis *11*, 329–340.

Teh, C.S.J., Chua, K.H., Puthucheary, S.D., and Thong, K.L. (2008). Further evaluation of a multiplex PCR for differentiation of *Salmonella* paratyphi A from other *Salmonellae*. Jap. J. Inf. Dis. *61*, 313–314.

Teramoto, K., Sato, H., Sun, L., Torimura, M., Tao, H., Yoshikawa, H., Hotta, Y., Hosoda, A., and Tamura, H. (2007). Phylogenetic classification of *Pseudomonas putida* strains by MALDI-MS using ribosomal subunit proteins as biomarkers. Anal. Chem. *79*, 8712–8719.

Timmins, É., and Goodacre, R. (2005). Discrimination and identification of microorganisms by pyrolysis mass spectrometry: From burning ambitions to cooling embers – A historical perspective. Chemical Analysis *169*, 319–343. 2005.

Tindall, B.J., Grimont, P.A.D., Garrity, G.M., and Euzeby, J.P. (2005). Nomenclature and taxonomy of the genus *Salmonella*. Int. J. Syst. Evol. Microbiol. *55*, 521–524.

Torpdahl, M., Skov, M.N., Sandvang, D., and Baggesen, D.L. (2005). Genotypic characterization of *Salmonella* by multilocus sequence typing, pulsed-field gel electrophoresis and amplified fragment length polymorphism. J. Microbiol. Methods *63*, 173–184.

Trafny, E.A., Kozlowska, K., and Szpakowska, M. (2006). A novel multiplex PCR assay for the detection of *Salmonella enterica* serovar Enteritidis in human faeces. Lett. Appl. Microbiol. *43*, 673–679.

Valentine, N., Wunschel, S., Wunschel, D., Petersen, C., and Wahl, K. (2005). Effect of culture conditions on microorganism identification by matrix-assisted laser desorption ionization mass spectrometry. Appl. Environ. Microbiol. *71*, 58–64.

Vandamme, P.A.R. (2003). Taxonomy and classification of bacteria. In Manual of Clinical Microbiology, Murray, P.R., Baron, E.J., Jorgensen, J.H., Pfaller, M.A., and Yolken, R.H., eds (Washinghton D.C., ASM Press), pp. 271–285.

Vanlaere, E., Sergeant, K., Dawyndt, P., Kallow, W., Erhard, M., Sutton, H., Dare, D., Devreese, B., Samyn, B., and Vandamme, P. (2008). Matrix-assisted laser desorption ionisation-time-of of-flight mass spectrometry of intact cells allows rapid identification of *Burkholderia cepacia* complex. J. Microbiol. Methods *75*, 279–286.

Wang, S.J., and Yeh, D.B. (2002). Designing of polymerase chain reaction primers for the detection of *Salmonella* Enteritidis in foods and faecal samples. Lett. Appl. Microbiol. *34*, 422–427.

Wang, Z., Russon, L., Li, L., Roser, D.C., and Long, S.R. (1998). Investigation of spectral reproducibility in direct analysis of bacteria proteins by matrix-assisted laser desorption/ionization time-of-flight mass spectrometry. Rapid Commun. Mass Spectrom. *12*, 456–464.

Wattiau, P., Van Hessche, M., Schlicker, C., Vander, V.H., and Imberechts, H. (2008a). Comparison of classical serotyping and PremiTest assay for routine identification of common *Salmonella enterica* serovars. J. Clin. Microbiol. *46*, 4037–4040.

Wattiau, P., Weijers, T., Andreoli, P., Schliker, C., Veken, H.V., Maas, H.M., Verbruggen, A.J., Heck, M.E., Wannet, W.J., Imberechts, H., and Vos, P. (2008b). Evaluation of the Premi®Test *Salmonella*, a commercial low-density DNA microarray system intended for routine identification and typing of *Salmonella enterica*. Int. J. Food Microbiol. *123*, 293–298.

Wayne, L.G., Brenner, J., Colwell, R.R., Grimont, A.D., Kandler, P., Krichevsky, I., Moore, L.H., Moore, W.E.C., Murray, G.E., Stackebrandt, E., *et al.* (1987). Report of the Ad Hoc Committee on reconciliation of approaches to bacterial systematics. Int. J. Syst. Bacteriol. *37*, 463–464.

Wiesner, M., Zaidi, M.B., Calva, E., Fernandez-Mora, M., Calva, J.J., and Silva, C. (2009). Association of virulence plasmid and antibiotic resistance determinants with chromosomal multilocus genotypes in Mexican *Salmonella enterica* serovar Typhimurium strains. BMC Microbiol. *9*, 131.

Wilkes, J.G., Rafii, F., Sutherland, J.B., Rushing, L.G., and Buzatu, D.A. (2006). Pyrolysis mass spectrometry for distinguishing potential hoax materials from bioterror agents. Rapid Commun. Mass Spectrom. *20*, 2383–2386.

Williams, T.L., Andrzejewski, D., Lay, J., and Musser, S.M. (2003). Experimental factors affecting the quality and reproducibility of MALDI TOF mass spectra obtained from whole bacteria cells. J. Am. Soc. Mass Spectrom. *14*, 342–351.

Wise, M.G., Siragusa, G.R., Plumblee, J., Healy, M., Cray, P.J., and Seal, B.S. (2009). Predicting *Salmonella enterica* serotypes by repetitive sequence-based PCR. J. Microbiol. Methods *76*, 18–24.

Wood, M.W., Mahon, J., and Lax, A.J. (1994). Development of a probe and PCR primers specific to the virulence plasmid of *Salmonella* Enteritidis. Mol. Cell. Probes *8*, 473–479.

Woods, D.F., Reen, F.J., Gilroy, D., Buckley, J., Frye, J.G., and Boyd, E.F. (2008). Rapid multiplex PCR and real-time TaqMan PCR assays for detection of *Salmonella enterica* and the highly virulent serovars Choleraesuis and Paratyphi C. J. Clin. Microbiol. *46*, 4018–4022.

Woodward, M.J., and Kirwan, S.E. (1996). Detection of *Salmonella* Enteritidis in eggs by the polymerase chain reaction. Vet. Rec. *138*, 411–413.

Wunschel, S.C., Jarman, K.H., Petersen, C.E., Valentine, N.B., Wahl, K.L., Schauki, D., Jackman, J., Nelson, C.P., and White, V. (2005). Bacterial analysis by MALDI-TOF mass spectrometry: An inter-laboratory comparison. J. Am. Soc. Mass Spectrom. *16*, 456–462.

Yoshida, C., Franklin, K., Konczy, P., McQuiston, J.R., Fields, P.I., Nash, J.H., Taboada, E.N., and Rahn, K. (2007). Methodologies towards the development of an oligonucleotide microarray for determination of *Salmonella* serotypes. J. Microbiol. Methods *70*, 261–271.

Zahraei Salehi, T., Tadjbakhsh, H., Atashparvar, N., Nadalian, M.G., and Mahzounieh, M.R. (2007). Detection and identification of *Salmonella* Typhimurium in bovine diarrhoeic fecal samples by immunomagnetic separation and multiplex PCR assay. Zoonoses Public Health *54*, 231–236.

Typing Phages and Prophages of *Salmonella*

2

Wolfgang Rabsch*, Sandra Truepschuch, Daniel Windhorst and Roman G. Gerlach

Abstract

Most *Salmonella* strains contain prophages or remnant phages and release them spontaneously. Special bacteriophages were developed and used in phage typing systems for epidemiological work all over the world since 1947 to control salmonellosis. This method provides fast and inexpensive characterization of frequent serovars such as *Salmonella* Typhimurium or Typhi on the sub-serovar level and is especially useful for primary analysis before investigation by other, more expensive molecular techniques such as sequencing. Prophages are themselves not only variable elements in a chromosome but also variable by module exchange within the prophage genome, thus providing a high discriminating power.

The availability of several genome sequences of different *Salmonella* serovars has recently led to the identification of new prophage-like elements. The prophages present in serovars Typhimurium, Enteritidis and Typhi are discussed. *Salmonella* phages frequently carry foreign DNA, so called morons. These morons are not necessary for phage functions but provide a benefit for the host. A list of some new morons found in different *Salmonella* serovars is presented.

Recently, a monophasic variant of *Salmonella* Typhimurium mainly belonging to Anderson phage type DT193 has become one of the dominant causes of salmonellosis in Germany and other European countries. These strains with the antigenic formula 4,[5],12:i:- do not express the H2 or 2nd phase flagellum. Investigation of their prophage attachment sites showed that the sites for Gifsy-1, Gifsy-2 and ST64B were occupied by the respective prophages. In about 90% of the monophasic DT193 strains the P22/ST64T attachment site was occupied by a novel 18.4 kb fragment, containing several open reading frames with significant similarity to phage-related genes.

Salmonella phage typing schemes

Communicable bacterial diseases present epidemiologists with the task of determining the avenues by which particular epidemic strains of bacteria have infected patients. Moreover, in a given outbreak of an infectious disease such as typhoid fever, it is important to know whether the organisms isolated from patients have a common origin, that is, whether the epidemic originates from one or multiple sources. In an infection caused by a single bacterial species, a method of classification at the sub-serovar level is needed to facilitate epidemiological studies. In other words, the species should be divided into 'types'. In an efficient typing scheme, the organisms of the same outbreak, whether isolated from patients, from vehicles of infection such as food or water or from the initial carrier, should all react identically, whereas similar isolates from a different source should display a distinct pattern. If an organism is widely disseminated and commonly carried by otherwise healthy persons, epidemiological studies and efficient treatment are impossible without such a typing method.

The practicability of phage typing systems primarily depends on the host specificity of phages, which is affected by host controlled modification and restriction systems to protect from phages,

*Corresponding author

adsorption to lipopolysaccharides and outer membranes, superinfection exclusion systems and the ability to influence the host's immune system. All these properties are involved in the biological basis of phage typing (Rabsch, 2007). *Salmonella* strains within a particular serovar are essentially differentiated into a number of phage types by their pattern of susceptibility to lysis by a set of phages with different specificity. The phages must have well-defined propagation strains that allow reproducible discrimination between different *Salmonella* strains. There is a high degree of correlation between the phage type and the epidemic source. However, phage typing is a phenotypical method that depends very much on the experience of the individual laboratory and on support from the reference centre that coordinates the maintenance of phages and the updating of the system. Only when the phage typing method is harmonized and the performance in different laboratories is controlled, the results can be regarded as definitive and are comparable between laboratories (Baggesen *et al.*, 2010).

Sonnenschein (1928) isolated specific phages for *Salmonella* Paratyphi B and *Salmonella* Typhi, and suggested their use for the rapid identification of these organisms. In 1938 Craigie and Yen published the first Vi-phage typing scheme for *Salmonella* Typhi. This scheme had four Vi-phages designated by the Roman numerals I–IV. However, they found that Vi-phage II possessed the unusual property of acquiring a high specificity for the last strain (Vi-positive *Salmonella* Typhi) on which it had grown. Progressive adaption of this phage by Craigie and Felix (1947) led to the development of the Vi-phage typing scheme. Over the years, this scheme has been improved and extended and now recognises 110 Vi-phage types with 110 Vi-phages, which is currently in use worldwide. In 1947 the International Committee for Enteric Phage Typing was founded. In 1974 it was renamed the International Federation for Enteric Phage Typing (IFEPT). IFEPT is responsible for the development of the Vi-phage typing scheme and the distribution of the Vi-phages to member laboratories worldwide.

Different phage typing schemes have been developed for other frequently isolated serovars, such as *Salmonella* Typhimurium (Anderson *et al.*, 1977), *Salmonella* Enteritidis (Ward *et al.*, 1987), *Salmonella* Derby (Michael *et al.*, 2006), *Salmonella* Bovismorbificans (Liesegang *et al.*, 2002), *Salmonella* Agona (Rabsch *et al.*, 2005) and, most recently, *Salmonella* Infantis (Miller, 2009). All phage typing schemes and combinations thereof currently used at our institute are listed in Table 2.1.

Some typing phages have been investigated by electron microscopy to characterize their structure and family. Eisenstark and Ackermann investigated the phages of the Typhimurium phage typing scheme of Lilleengen, and found them to consist of podovirus and siphovirus families (Eisenstark *et al.*, 2009).

Long-term surveillance of *Salmonella* by phage typing

Salmonella *Typhimurium*

S. enterica serovar Typhimurium is the dominant serovar of the Group B-*Salmonella* that carry the somatic antigens [O 4,(5),12] and diphasic flagellar antigens (H i:1,2).

Three phage typing schemes have been widely used for this serovar – the Lilleengen scheme (Lilleengen, 1948), the Felix–Callow scheme (Felix and Callow, 1943) and the Anderson scheme (Anderson *et al.*, 1977). From the 1960s until now, the national epidemiological surveillance programme at the Robert Koch Institute was performed using a combination of the Felix–Callow and Lilleengen typing systems (Kühn *et al.*, 1973). Since 1966, the extended phage typing system of Anderson has also been used. This phage typing system is now used in Europe, Canada and Australia. The complete genome sequence of *Salmonella* Typhimurium LT2 (Lilleengen phage type) or DT4 (Anderson phage type) contains the four functional prophages Gifsy-1, Gifsy-2, Fels-1 and Fels-2, and previously unknown phage or phage remnants (McClelland *et al.*, 2001). In a screen for phage-like genes in strain LT2, four new remnant phages were identified designated Def–1 to 4 (Table 2.2; Casjens, 2010). Fine discrimination of *Salmonella* Typhimurium by phage typing allows surveillance over long time periods and uncovers trends regarding, for example, differences in plasmid-borne and chromosomal

Table 2.1 Summary of phage typing schemes* used at the *Salmonella* Reference Centre for *Salmonella* and other Enteric pathogens at the Robert Koch Institute, Wernigerode, Germany

Salmonella enterica serovar	No. of phages	No. of phage types	Scheme developed by	Reference
Typhi	110	110	National Salmonella Reference Laboratory (NSRL), London, (phage E1a and E1b at Robert Koch Institute, Wernigerode)	Anderson and Felix (1953); Ward added more phages (unpublished)
Paratyphi A	10	19	NSRL, London	Banker (1955); Ward added more phages (unpublished)
Paratyphi B and Java	12	53	NSRL, London	Felix and Callow (1951); Ward added more phages (unpublished)
	10	47	Scholtens, National Institute for Public Health and the Environment Bilthoven (RIVM)	Scholtens (1950)
Enteritidis	17	96	NSRL, London	Ward *et al.* (1987)
	10	42	Veterinary Medical Research Institute of the Hungarian Academy of Sciences, Budapest, Hungary	Laszlo *et al.* (1985)
Typhimurium	39	280	NSRL, London	Anderson *et al.* (1977)
	25	137	Robert Koch Institute, Wernigerode	Felix and Callow (1951); Lilleengen (1948); Rabsch (2007)
Heidelberg	14	66	National Microbiology Laboratory, Public Health Agency, Winnipeg, Canada	R. Ahmed (unpublished)
Oranienburg	12	17	National Microbiology Laboratory, Public Health Agency, Winnipeg, Canada	R. Ahmed (unpublished)
Hadar	9	62	NSRL, London	De Sa *et al.* (1980)
Virchow	15	67	NSRL, London	Chambers *et al.* (1987)
Agona	14	55	Robert Koch Institute, Wernigerode	Rabsch *et al.* (2005)
Bovismorbificans	10	36	Robert Koch Institute, Wernigerode; International Union of Microbiological Societies, Adelaide	Liesegang *et al.* (2005); Davos (unpublished)
Derby	16	59	Robert Koch Institute, Wernigerode	Michael *et al.* (2002)
Infantis	17	61	Robert Koch Institute, Wernigerode	Miller (2009)

*Two rough phages and the O1 phage are used for further characterization in all typing schemes.

Table 2.2 *Salmonella* prophage genes that modify phenotypes in host infection

Region, phage (strain)	Gene	Link to virulence or modification	References
STM2584-STM2637, Gifsy-1 (Typhimurium LT2)	*gogA* (STM2614)	Similar to *pipA* of SPI-5	Bossi and Figueroa-Bossi (2005)
	gipA (STM2599)	Needed for growth in Peyer's patches	
	gogD (STM2585A)	Similar to *pagK* (STM1867,remnant Def2) and *pagJ* (STM14_1480)	
	gogB (STM 2584)	Similar to type III secreted proteins of LRR family	
STM1005-STM1056, Gifsy-2 (Typhimurium LT2)	*gtgA* (STM1026)	Similar to *pipA*	
	grvA (between STM1033-1036)	Antivirulence gene	
	ailT (STM1043)	Similar to attachment-invasion locus (*ail*) and to serum resistance proteins	
	sodC1 (STM1044)	Periplasmic [Cu, Zn] superoxidase dismutase; protects against macrophage oxidative burst	
	sseI (*srfH*, *gtgB*) (STM1051)	Type III translocated protein under SPI-2 control	
	gtgE (STM1055)	Needed for mouse virulence	
	gtgF (STM1056)	Similar to macrophage survival gene (*msgA*)	
Gifsy-3 (Typhimurium 14028)	*pagJ* (STM14_1480)	PhoP/PhoQ-activated locus	
	sspH1 (STM14_1483)	Type III translocated protein; downregulates interleukin-8	
STM0893-STM0929, Fels-1 (Typhimurium LT2)	*sodC* (STM0924)	Periplasmic [Cu, Zn] superoxide dismutase; similar to *sodC1*	
	nanH (STM0928)	Neuraminidase; involved in nutrient scavenging, host cell adhesion, and toxin action	
SopEϕ (Typhimurium SL1344)	*sopE* (SL2674)	Type III translocated G nucleotide exchange factor, promotes epithelial cell invasions	
STM0554-STM0559, Def1 (Typhimurium LT2)	*gtrABC*	Adjacent *argU*	Casjens *et al.* (2010); Lawley *et al.* (2006)

STM1855-STM1871, Def2 (Typhimurium LT2)	*sopE2* (STM1855), *pagK* (STM1867)	Type III translocated G nucleotide exchange factor, promotes epithelial cell invasions	
STM2230-STM2245, Def3 (Typhimurium LT2)	*oafA* (STM2232), *sspH2* (STM2241)	Acetylation of abequose, O5 antigen	
STM4196-STM4219, Def4 (Typhimurium LT2)	*gtrABC* cluster (STM4204-4206)	Glycosyltransferase, glucose translocase, necessary for long term murine systemic infection in Typhimurium SL1344	
ES18	Genes 17-20	Unknown, gene19 homolog adjacent to Pasteurella multicida phage carried toxin gene	Casjens *et al.* (2005)
ST64B (Typhimurium DT104)	*artA* (SDT2667), *artB* (SDT2666)	Pertussis-like toxin	Saitho *et al.* (2005); Pelludat *et al.* (2005)
ε 34 , P22, ST64T, ST104	*gtrABC*	Linked glucose translocase (flippase), antigen conversion 3,15 (serogroup E2) to 3,15,34 (serogroup E3), phage mediated alteration of LPS	Villafane *et al.* (2008); Brüssow *et al.* (2004)
P14	unknown	LPS elongation in serogroup C	Nnalue *et al.* (1990)
Gifsy-2 (Abortusovis)	*astA*	Heat stable toxin, involved in acquisition of IS1414	Bacciu *et al.* (2004)

Adapted from Bossi and Figueroa-Bossi (2005).

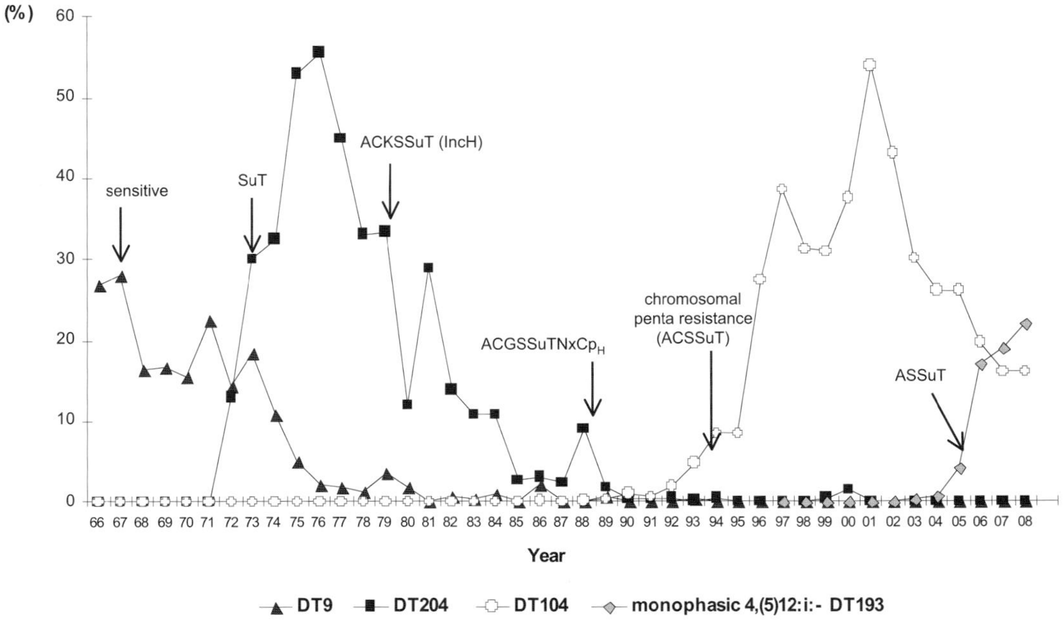

Figure 2.1 Temporal occurrence patterns of *Salmonella* Typhimurium DT9, DT204, DT104 and monophasic DT193 strains in humans in Germany. Resistances are abbreviated as follows: A, ampicillin; C, chloramphenicol; G, gentamicin; K, kanamycin; S, streptomycin; T, oxytetracylines; Su, sulphonamides; Nx, nalidixic acid; Cp_H, ciprofloxacin high.

resistance genes. The dominant phage types in Germany since the 1960s are presented in Fig. 2.1 (Rabsch, 2007), and include four types: DT9, DT204, DT104 and DT193.

Salmonella Typhimurium DT9 is a highly virulent phage type, which prompted the development of a live vaccine for veterinary application (Linde *et al.*, 1990; Rabsch *et al.*, 2001).

DT204 is characterized by the presence of *sopE*, a gene encoding an effector protein involved in bacterial invasion of intestinal epithelial cells (Hardt *et al.*, 1998). The *sopE* gene was present in phage type DT204 in addition to the remnant SopE2 phage detected predominantly in the 1970s in Germany. Multiple antibiotic resistant strains of DT204 (most commonly resistance type CSSuT, i.e. chloramphenicol, streptomycin, sulfonamides, tetracycline) spread rapidly among cattle populations in England and Wales in 1977 (Threlfall *et al.*, 1978).

DT104 contains a region that displays 61% identity with a pertussis-like toxin (Pelludat *et al.*, 2005). Isolates of this phage type often possess resistance to ampicillin, chloramphenicol, streptomycin, sulfonamides, and tetracycline (ACSSuT; Briggs and Fratamico, 1999).

Since 1990, monophasic *Salmonella* isolates with the antigenic formula 4,[5],12:i:- have been reported worldwide: in Thailand in chicken meat and pork (Boonmar *et al.*, 1998), in Spain in humans and pigs (Echeita *et al.*, 1999) and in the USA in humans and raw chicken meat (Agasan *et al.*, 2002). Most of the monophasic strains in Germany belong to phage type DT193 (untypeable by Felix/Callow and Lilleengen typing schemes). Monophasic *Salmonella* DT193 is most likely a *Salmonella* Typhimurium variant, although they formally could represent a novel serotype or variants of either *Salmonella* Typhimurium [4,(5),12:i:1,2] or *Salmonella* Lagos [4,(5),12:i:1,5] which have lost, either by mutation or deletion, the *fljB* gene that encodes the phase 2 flagella antigen (Garaizar *et al.*, 2002). The origin of these strains is still enigmatic. It is assumed that serovar 4,[5],12:i:- is related to, but distinct from, serovar Typhimurium; however, without whole genome sequencing the existing tests are suggestive, but not conclusive. Some strains of serotype 4,[5],12:i:- exhibit enhanced pathogenicity among otherwise healthy people (Agasan *et al.*, 2002), and are therefore representing pathogens of clinical significance.

The factors responsible for the temporal prevalence of individual multidrug-resistant phage types are not fully understood. The rapid spread of a new strain might be in some cases linked to animal trading. One possible factor contributing to the temporal dominance of certain phage types could be the horizontal transfer of virulence genes (Bäumler, 1997). In 2006, monophasic *Salmonella enterica* serovar 4,[5],12:i:- DT193 strains caused some large diffuse outbreaks with increased need for hospitalization. The majority of the strains exhibited an ASSuT-resistance pattern. These strains were shown to have special epidemiological markers including PFGE-pattern STYMXBA.0131 and the presence of an 18 kb genomic island (phylogenetically probably a remnant prophage, see below).

Other monophasic strains with the antigenic formula 4,12:i:- have been reported with increasing frequency over the last 2–3 years. However, most of the monophasic ASSuT resistant strains still belong to the DT193 phage type. DT193 is a diverse phage type in terms of genetic relatedness to other DTs. Most of these reported strains had a variable number of plasmids with molecular masses from 2.6–80 MDa (Baquar *et al.*, 1994). Outbreaks of monophasic DT193 strains with PFGE type STYMXB.0131, chromosomal ASSuT drug resistance pattern (Lucarelli *et al.*, 2010) and an 18.4 kb island have been reported in Denmark, Luxemburg (Mossong *et al.*, 2007), Ireland and Austria. Single cases have been reported in France, Italy, the Netherlands and Finland. Since the pathogenicity repertoire is highly similar to that of *Salmonella* Typhimurium, it is essential that interventions are introduced in order to limit human infections (Hauser *et al.*, 2010).

Salmonella *Typhi and Paratyphi A*

Salmonella enterica serovar Typhi is the causative agent of typhoid fever, an acute systemic illness with a high proportion of deaths, especially in south-central and south-east Asia, particularly in India, South and Central America, Africa (50–80% multidrug resistant isolates), Saudi Arabia, Oman, and Yemen (5–30% MDR isolates). In Europe, *S. enterica* serovar Typhi infections occur among travellers returning from areas where the disease is endemic. For standardization, the Typhi typing phages were distributed to the national reference centres all over the world. The frequency of distribution of the (Vi)-phage types of Typhi and Paratyphi A, examined by the National Centres represented in IFEPT, are analysed in reports every three years. The last three IFEPT reports (1999–2007) and our own data show that the most common type detected in these international studies is *Salmonella* Typhi phage type E1 (= E1a, see Table 2.1). There was, however, a wide range of other types, many of which are characteristic of particular geographical zones, exemplified by Malaysia (phage type B1), Iraq (phage type E9), Ghana (phage type D1), Senegal (phage types A and D6), Tunisia (phage type N), and others. The international studies show that the dominant *Salmonella* Paratyphi A phage type in India is PT1, but in Turkey and China strains of the PT6 type prevail. The occurrence of such types out of their habitual context indicates their 'exotic' origin, providing a useful tool to trace back the origin. One important trend observed is that fluoroquinolone resistance in *Salmonella* Paratyphi A strains is increasing more rapidly compared with *Salmonella* Typhi. Currently available typhoid vaccines do not protect against *Salmonella* Paratyphi A, and it appears that antibiotic resistant Paratyphi A is an emerging problem (Ochiai *et al.*, 2005).

The first description of a strain of *Salmonella* Typhi phage type E9 with resistance to extended beta-lactam antibiotics (ESBL) and quinolone resistance was investigated at our institute (Pfeifer *et al.*, 2009). Further characterization of the genetic determinants of resistance showed that they belonged to TEM-1, CTX-M-15 and qnrB2 classes. In addition, the nucleotide sequences of the quinolone resistance-determining regions of the *gyrA*, *gyrB*, *parC*, and *parE* genes were determined as well as plasmid-borne quinolone resistance genes (*qnrB*, *qnrS* and *qnrA*). This isolate of *Salmonella* Typhi contained blaCTX-15 and qnrB2, mediating resistance to cephalosporins and reduced quinolone susceptibility which could be easily transferred by conjugation into *E. coli*. This occurrence is alarming because the dissemination of such strains with acquired resistances will further limit the therapeutic options for treatment of typhoid fever (Pfeifer *et al.*, 2009).

Salmonella *Infantis*

For the new *Salmonella* Infantis typing scheme, 154 typing phages were tested on 1008 *Salmonella* Infantis strains isolated from outbreaks and sporadic cases of gastroenteritis in humans, from animals, food, feed and the environment during 1973–2009, and evaluated according to their discriminating capacity. The final typing scheme was established using 17 typing phages. This scheme discriminated 61 phage types (Table 2.1) in 50 outbreaks (187 strains in 40 food poisoning and 10 hospital outbreaks) and single occurrences. Phage typing in combination with *Xba*I macrorestriction analysis was subsequently performed for better sub-differentiation of the outbreaks. The most frequently observed clone PT 29/*Xba*I 27 was detected in broiler and human outbreaks, which indicates a clonal distribution. Other phage types of *Salmonella* Infantis were found in different geographical locations such as Central Europe, Iceland and Australia. *Salmonella* Infantis recently emerged as a nosocomial pathogen in several German hospitals. Studies of antibiotic resistance (against 17 antibiotics) in *Salmonella* Infantis strains showed 21% multi-resistant strains, including four isolates with ESBL, CTX-M-1 and TEM-52 determinants (Miller, 2009).

Outer membranes as receptors for typing phages

The outer membrane (OM) is much more permeable than the inner membrane, mainly due to its variety of porins (OMPs). Phages use these OMPs and IROMPs (Iron Regulated Outer Membrane Proteins) as recognition sites. It is known that OmpC is the receptor for Gifsy-1 and Gifsy-2 phages of *Salmonella* Typhimurium (Ho and Slauch, 2001). Ligand-gated porins (LGP), which often function in the uptake of siderophores, show broad multifunctionality by also acting as receptors for bacteriophages, toxins, and antibiotics. FhuA, for example, recognizes the hydroxamate siderophore ferrichrome; but also phages T1, T5, ES-18 as well as colicin M and the antibiotics albomycin and microcin 25 (Braun, 2009).

The H8 phage is part of an existing collection of *Salmonella* Enteritidis bacteriophages (Lalko and Laszlo scheme, see Table 2.1). This phage distinguishes 7 different *Salmonella* Enteritidis phage types and was identified by close similarity in morphology and genomic structure to the phage T5 of the Siphoviridae family (Rabsch *et al.*, 2007). The phage's recognition site is a unique IROMP (Iron Regulated Outer Membrane Protein), FepA. This protein is the receptor for enterobactin and other catecholate type siderophores. The phage infects *S. enterica* and *E. coli* by initial adsorption to FepA. Ferric enterobactin inhibited the infection of *E. coli* by H8. Other ferric catecholate receptors (Fiu, Cir, IroN) do not participate in phage adsorption. Penetration of the cell by H8 is TonB-dependent. Mutations in *exbB*, encoding a prominent TonB-like outer membrane protein, do not confer resistance to the bacteriophage, but *exbB tolQ/R*-double mutants were resistant to H8. Experiments with deletion and substitution mutants of *E. coli* FepA showed that bacteriophage adsorption and penetration is facilitated by many of the same charged residues in the receptor protein that promote the binding and transport of ferric enterobactin and colicins B and D (Rabsch *et al.*, 2007). The greatest reductions in H8 susceptibility originated from site-directed single substitutions deep within the vestibule of FepA that eliminated either basic (R316A) or aromatic residues (Y260A). DNA sequence determinations of the complete H8 genome revealed that it is approximately 100 kb in size, and extensively homologous to bacteriophage T5 (Rabsch *et al.*, 2007).

Lysogenic conversion and moron genes of prophages

Most *Salmonella* infecting warm blooded animals belong to a single species, *Salmonella enterica*, and are classified into more than 2500 serovars according to their antigenic formula (White–Kaufmann–LeMinor Scheme). An even greater diversity exists at strain level as a result of countless combinations of genomic differences in individual isolates. Strain-specific variations are likely to influence aspects of the organisms biology, such as adaptation to specific hosts or environments, tropism for certain organs or tissues, and the degree of virulence. Analysis of prophage genomes indicated the presence of both selfish and mutualistic properties with respect to their host (Casjens, 2003). It was found that prophage sequences may significantly contribute to the pathogenicity of their bacterial host. Prophages are the major source of horizontally transferred DNA in bacteria and contribute significantly to bacterial inter-strain variability.

Some phages carry extra DNA that is necessary for phage functions. This DNA might originate from other hosts where it entered the phage genome during its replication cycle. If these genes provide a benefit for the host, but not for the phage they are termed 'morons' (because one phage may contain 'more DNA' than another). Morons are thought to be the basis by which phages influence the evolution of pathogenic bacteria. Some moron-encoded functions increase the fitness of the host thereby enhancing the chance of (pro-) phage propagation. The phage-encoded extra functions might allow the host bacterium to increase its fitness within its current ecological niche or even to conquer a new niche. These changes ultimately make the moron and probably the surrounding phage indispensable for the bacterium thus stabilizing the prophage (Brüssow *et al.*, 2004). To benefit the host, morons have to clear some hurdles. First, they have to be integrated into existing regulatory circuits. The moron can only be functional if its expression is precisely coordinated with other functions of the host bacterium. Second, in most cases the bacterium must possess certain (virulence) functions required for the moron to function. For example, a moron-encoded type III effector needs a functional type III secretion system (TTSS) as well as a proper chaperone. Although prophage genes are predominantly transcriptionally silent, a moron has to be actively transcribed and this feature can be used to identify these horizontally acquired genes. High-density strand-specific cDNA sequencing (ssRNA-seq) allows the identification and quantification of novel transcripts. Application of this technique to *Salmonella* Typhi Ty2 confirmed known but also suspected morons in different prophages (Perkins *et al.*, 2009).

On the other hand, bacteria can alter their properties by lysogenic conversion. Lysogenic conversion has a great impact on the evolution of pathogenic bacteria and results in a very interesting relationship of bacterium-phage coevolution. Per definition, lysogenic conversion is a change in the phenotype of the bacteria after infection with a temperate phage. After integration of the phage DNA into the bacterial chromosome, changes often involve modifications of the outer membrane of the cell, making it impervious to other phages. It is important to note that temperate phages can affect ecological fitness of bacteria in multiple ways: (i), as anchor points for genome rearrangements; (ii), via gene disruption; (iii), by protection from lytic infection; and (iv), by lysis of competing strains through prophage induction (Brüssow *et al.*, 2004). Sequencing projects have brought more and more insight into the genetic bases of these phenotypic changes (Cooke *et al.*, 2007).

Prominent morons in *Salmonella* Typhimurium

The ***gtrABC*** genes are often found in prophages (Villafane *et al.*, 2008). STM0557 and two neighbouring genes, located on a pathogenicity island termed SPI-16, resemble genes of the *gtr* cluster in seroconverting bacteriophages (Bogomolnaya *et al.*, 2008). It is known that STM0557, an *S. enterica* subspecies I-specific gene encoding an inner membrane protein, is critical for faecal shedding and intestinal persistence of Typhimurium ATCC14028 in *Salmonella*-resistant mice, but mutations in this gene do not diminish short-term intestinal colonization or invasion of cultured epithelial cells. The *gtr* genes can be defined as morons because they exert a regulatory effect on host virulence. But *gtr* genes are also responsible for phenotypic changes in LPS epitopes. The fact that *gtrABC* operons are associated with different temperate phage types suggests that these operons have been evolutionary important for the *Salmonella* temperate phages and the hosts that harbour them. The vast majority of these prophage-borne *gtrABC* operons appear to be intact, but in most cases functionality has not been proven experimentally. LPS-modification via *gtr*-genes of prophages might be involved in *Salmonella* serotype conversion. As with *Shigella* serotypes, O-antigen formation is frequently modified by the prophages carried by the bacterium. Presumably, phages carry these genes as a mechanism of superinfection exclusion or because they give a selective advantage to their host.

Another well-characterized moron gene is ***sopE***. SopE is translocated via the SPI-1-encoded type three secretion system (TTSS) into the host cell cytoplasm where it acts as a G-nucleotide exchange factor (GEF), able to activate small Rho GTPases like Rac1 and Cdc42 (Hardt *et al.*, 1998). Activation of these central switches stimulates actin polymerization, promoting uptake of the bacterium. The vector for SopE is the SopEΦ of *Salmonella* Typhimurium, which is P2 and Fels-2-like. In other serovars, a lambdoid-like phage was identified as a vector for SopE, demonstrating transfer of virulence genes between different phage families (Mirold *et al.*, 2001). Furthermore, it is possible to lysogenize wild strains with a genetically engineered *sopE::aphT*-bacteriophage. Upon lysogenic conversion with SopE$\Phi^{sopE::aphT}$, Typhimurium DT36, which shows lysis with all 30 typing phages of three schemes (Felix–Callow, Lilleengen and Anderson), completely lost susceptibility to 13 *Salmonella* Typhimurium phages, and became partially resistant to 10 (Rabsch *et al.*, 2002).

Another phage remnant encodes **SopE2**, a homologue of SopE. Both of the *sopE* morons in *Salmonella* Typhimurium are embedded in the SPI-1-mediated invasion mechanism in two ways: (i) their expression is controlled by the InvF/SicA transcription factors ensuring co-regulation with the other components and effectors of the TTSS, and (ii) the type III chaperone InvB allows their proper recognition and transport via the SPI-1 TTSS (Brüssow *et al.*, 2004). None of the components required for SopE function are encoded by SopEΦ, hence SopE travels without this 'mortal freight' (Brüssow *et al.*, 2004).

A list of important morons found in *Salmonella* Typhimurium is presented in Table 2.2.

CRISPR and phage content

CRISPRs, clustered regularly interspaced short palindromic repeats, are genetic elements that have recently been identified in approximately 40% and 90% of bacterial and archaeal genomes, respectively (Grissa *et al.*, 2007). A CRISPR consists of several identical repeats, separated by non-identical spacer sequences (Sorek *et al.*, 2008). These repeat and spacer sequences are typically 25–40 base pairs in size, while the number of repeats in a single CRISPR varies widely from 2 to 250 (Grissa *et al.*, 2007). Recent research has shown that these elements, along with CRISPR-associated genes, are involved in a silencing mechanism that can provide protection against phages and possibly other mobile elements (Barrangou *et al.*, 2007, see also Chapter 7 of this book). It is remarkable that many of the virulence factors in genomic islands are involved in active invasion that harms the host without an obvious functionality for these virulence factors outside of the host environment. Therefore, maintaining these virulence factors on genomic islands, and probably other mobile elements like phages, appears to provide important evolutionary flexibility for these pathogens. Some data in the literature indicate that phages might influence superinfection resistance by hosting these CRISPR repeat elements.

The presence of prophages within one serovar can be variable. One example is given by Figueroa-Bossi *et al.* where deletion of *sbeC*, a gene involved in DNA repair, not only derepresses *recE*, a gene implicated in the SOS response in *Salmonella*, but also activates the transcription of the Gifsy-1 *xis* (excisionase) gene, thus causing the prophage to become unstable and to excise at a high frequency. This mutation destabilizes Gifsy-2 as well, but to a lesser extent, making it possible to isolate strains that are cured of either or both prophages (Figueroa-Bossi *et al.*, 1997). The differences in prophage content can explain the observation that different phage types exist for LT2 siblings after mutagenesis. Among 53 LT2 archived strains, 16 had the same phage type as the non-archival sequenced LT2 strain. The other

37 archived cultures differed in phage typing patterns (Rabsch *et al.*, 2004).

Prophages found at identical locations in closely related strains often contain different modules and encode different types of immunity. Conversely, prophages located at different chromosomal loci can share extensive sequence identity throughout most of their genomes. Formation of hybrid phages, as shown by a hybrid phage of Fels-2 and P22, is also possible (Yamamoto, 1969).

Salmonella Enteritidis prophages

The prophage content of the phage type 4 *Salmonella* Enteritidis strain No.125109 was analysed in an *in silico* genome analysis by comparison with the genomes of *Salmonella* Typhi and *Salmonella* Typhimurium (Thomson *et al.*, 2008). Five prophage-like elements were identified which were named ΦSE10, ΦSE12, ΦSE12A, ΦSE14 and ΦSE20 after their position in the genome (Thomson *et al.*, 2008; Windhorst, 2010). Detailed information on size, location, number of genes/pseudogenes and the G+C content are given in Table 2.3. The putative virulence genes carried by these prophages are also listed in Table 2.3. The similarities to phages found in *Salmonella* Typhimurium is remarkable.

ΦSE10 is similar in parts to regions of the lambda-like phage Gifsy-2 which has been shown to significantly contribute to pathogenesis in *Salmonella* Typhimurium (Figueroa-Bossi and Bossi, 1999; Ho *et al.*, 2002). Like Gifsy-2, ΦSE10 was found to encode the highly conserved virulence determinants *sseI*, *gtgE* and *gtgF* (Ho *et al.*, 2002).

Like ΦSE10, ΦSE12 is similar in parts to regions of Gifsy-2, while its 5′-end shows more similarity to ST18. ΦSE12 carries the virulence determinants *hokW*, *pipA*, *ompX*, *sodCI* and *sopE*. Interestingly, the virulence determinants *sodCI* and *gtgA*, carried by Gifsy-2 and not conserved in ΦSE10, are present in ΦSE12, completing the set of virulence genes found in Gifsy-2 for *Salmonella* Enteritidis PT4. ΦSE12A displays a mosaic of genes from related bacteriophages (ST10, ST18, Gifsy-1, Gifsy-2 and Fels-1). Homologues of the *mig-3*, *pagK* and *pagM* genes are present in ΦSE12A as putative genes or pseudogenes, and *pagO* and *sopE2* were identified in the direct vicinity of its 3′-end.

Most of the genes in ΦSE14 encode putative phage proteins with similarities to genes of the lambda-like phage ST18 from *Salmonella* Typhi. The virulence determinants *hok/mok* and *kil* are present in ΦSE14. The SdfI region previously identified in *Salmonella* Enteritidis constitutes the 5′-end of ΦSE14 (Agron *et al.*, 2001).

ΦSE20 is highly similar in sequence and gene order to the *Salmonella* Typhimurium DT64 lambda-like phage ST64B (Mmolawa *et al.*, 2002). Like in ST64B, fragments of the virulence-associated genes *sopE*, *sspH2* and *orgA* are present in putative tail and tail fibre genes in ΦSE20. A homologue of *sseK3* is also present in ΦSE20. *Salmonella* Enteritidis isolates comprising different phage types were used in a microarray analysis for the presence of the individual genes of ΦSE20 (Windhorst, 2010). ΦSE20 is present in the isolates belonging to the phage types 1, 4 and 21 and absent in those belonging to the phage types 8, 9b and 21c according to the microarray data. For the isolates belonging to the phage types 11, 13a and 20, a variable distribution of genes was seen (Windhorst, 2010).

In another microarray-based study, two groups of *Salmonella* Enteritidis isolates were determined based on the presence of a subset of phage regions (Porwollik *et al.*, 2005). They were grouped based on absence/presence patterns for regions similar to the phages ST64B, Fels-2, ST27 and ST35. These two groups were seen to almost perfectly overlap with the separation into phage type 4-like isolates (including phage types 1, 4, 4b, 6, 7 and 24) and phage type 8-like strains (including phage types 2, 8, 13a and 23) based on LPS core structure (Guard-Petter, 1999; Liebana *et al.*, 2002; Porwollik *et al.*, 2005).

The results for the phage type 4 and 8 isolates match very well between the two studies. For phage type 1 and 13a, however, differing results for the presence of ΦSE20 were obtained. As different isolates were used in these two studies it remains unclear whether this was caused by differences in the prophage content of the isolates or by the different microarrays used.

The prophage regions identified in strain SE125109 display a mosaic of genes from related

Table 2.3 The *Salmonella* Enteritidis prophages carried by the phage type 4 strain SE125109

					Putative morons			
Phage	Size (location)	CDS/pseudogenes	G+C content%	Orthologous genes present in	Gene No.	Gene	Orthologous gene	Carried by
ΦSE10 (prophage remnant)	8,186 kb (1013381–1021566)	8/5	45,81	Gifsy-1, Gifsy-2, Fels-1, Fels-2, ST10, ST18	SEN0916	*sseI*	STM1051	Gifsy-2
					SEN0920	*gtgE*	STM1055	Gifsy-2
					SEN0921	*gtgF*	STM1056	Gifsy-2
ΦSE12 (prophage remnant)	17,753 kb (1226471–1244223)	17/8	48,86	Gifsy-2, ST18, ST10, Gifsy-1	SEN1135	*hokW*	STY2054	ST18
					SEN1143	*pipA*	STM1026 (*gtgA*) STM2614 (*gogA*)	Gifsy-2 Gifsy-1
					SEN1148 (pseudogene)	*ompX*	STM1043 STM0920	Gifsy-2 Fels-1
					SEN1149	*sodCI*	STM1044	Gifsy-2
					SEN1155	*sopE*	STY4609	SopE
ΦSE12A (prophage remnant)	8040 kb (1246083–1254122)	4/8	48,74	Gifsy-1, Gifsy-2, Fels-1 ST10, ST18	SEN1170	*pagk*	STM25851	Gifsy-1
ΦSE14 (prophage remnant)	12,642 kb (1469390–1482031)	18/3	45,15	ST10, ST18	SEN1381	*kil*	STY2070	ST18
					SEN1387	*hok/mok*	STY2054 STY2054A	ST18
ΦSE20 (prophage)	40,664 kb (2018460–2059123)	51/0	50,59	Gifsy-1, ST64B, ST10	SEN1920	*sseK3*	*sb26*	ST64B

bacteriophages. Virulence-associated genes that are carried by these prophages are also distributed across a wide range of other bacteriophages isolated from other *Salmonella* serovars. The virulence genes found within the prophage regions support the important role of bacteriophages in the movement of virulence factors among bacteria. While in many cases the virulence-associated effector proteins were highly conserved, the genes required for phage proliferation were degenerated.

Salmonella Typhi prophage content

Salmonella Typhi is a human-restricted pathogen able to cause a systemic disease known as typhoid fever. This serovar shares its adaptation to human hosts with some other serovars such as Paratyphi A, Paratyphi B, the systemic pathovar of Paratyphi C and Sendai. The genetic basis of the differences in host tropism between these serovars and broad host range *Salmonella* such as serovar Typhimurium are still not fully understood. The genetic changes accompanied with such an evolutionary change include an accumulation of mobile elements and pseudogenes as well as chromosomal rearrangements and deletions leading to genome reduction (Moran and Plague, 2004, see also Chapter 3 of this book). These tendencies can be observed in the genome of *Salmonella* Typhi CT18 which revealed the presence of ~220 pseudogenes (Parkhill *et al.*, 2001). Comparative genomics of the host-restricted serovars Typhi and Paratyphi A underscored pseudogene formation as a hallmark for successful specialization to a new ecological niche. Whereas 27 pseudogenes were shared between the serovars through recent recombination events, 22 resulted from independent converging mutations (Holt *et al.*, 2009).

Although pseudogene formation is a clear expression of host adaptation, the acquisition of additional genes through horizontal gene transfer is another mechanism to increase fitness in a limited environment. The genome of *Salmonella* Typhi CT18 harbours more than 600 genes which are absent in serovar Typhimurium, most of them associated with genomic islets, genomic islands or putative prophages (Parkhill *et al.*, 2001). Within the CT18 genome there are six distinct prophage-like regions which might encode functional phages: ST10, ST15, ST18, ST35, ST44/SopEST and ST46 (Thomson *et al.*, 2004). There are additional gene clusters present clearly resembling phage remnants: Sti2, Sti5, Sti6 and ST27 (Casjens, 2003; Thomson *et al.*, 2004). Furthermore, using a bioinformatics approach, many more genes of putative phage origin have been identified (Canchaya *et al.*, 2004). In a recent study by Holt and colleagues the genome sequences of 19 *Salmonella* Typhi strains representing a broad spectrum of the Typhi phylogenetic tree were compared (Holt *et al.*, 2008). These data revealed the presence of further prophage-like regions: ST16, ST20a/b, ST2–27, ST36 and ST46a. An overview about the identified prophage regions is given in Table 2.4.

The prophages identified so far share significant homologies to phages found in the genomes of other Enterobacteriaceae. According to these similarities in genetic content and/or genetic organization Typhi prophages can be assigned to known phage families. The putative prophages ST27, ST35 and SopEST are P2-family members. Their closest relative is *E. coli* phage 186 (Thomson *et al.*, 2004). ST36 is a 28 kbp prophage with high similarities to the P2-related phage WPhi (GenBank: AY135739). The independent acquisition of ST36 by isolates 404ty (haplotype H59) and E01–6750 (haplotype H52) underscores the limited use of phages as markers of genetic relatedness among Typhi isolates (Holt *et al.*, 2008). As a hallmark of the P2-family, these phages can carry morons within the early region, adjacent to their *cos* site or the tail fibre region. Both ST35 and SopEST encode for proteins of unknown function in these regions (Table 2.4). The gene content of the tail fibre regions can be determined using primers which recognize adjacent conserved DNA sequences. Applying this methodology to a set of *Salmonella* Typhi strains, the isolates could be categorized into two groups: Strains KT516, PR39, 9541 and 422mar92 yielded one PCR product corresponding to the tail fibre region of ST35. Isolates SARB63, SARB64 and CT18 gave one additional PCR product which was identical with the region of the SopEST phage (Thomson *et al.*, 2004). These results indicate that the additional gene encoding the type III effector protein SopE in the SopEST phage might not be

Table 2.4 Prophage regions and putative morons associated with *Salmonella* Typhi phages

Name[1] (synonym(s))	Insertion site	From – To (CT18, NC_003198)	Homology	Morons (putative and proven)	Comment
ST10 (CT18–01[2], Sti1[3])		STY1011 – STY1075	Lambdoid phages	–	Similarities to Fels-1, Gifsy-1 and Gifsy-2 phages (LT2)
Sti2[3]		STY1358 – STY1365	Lambdoid phages	STY1364; similar to pertussis-like toxin subunit ArtB (Saitoh *et al.*, 2005)	Remnant
ST15 (CT18–02[2], Sti3[3])		STY1591 – STY1643	Chimera: P2-family, Mu and lambda	STY1597; putative enterohaemolysin-2 STY1603; putative ExeA protein (type II secretion component)	Weak similarities to prophage-like locus in LT2 (4.421.647–4.441,087)
ST16		–	Mu-like element	–	Only found in Typhi strain E98–2068
ST18 (CT18–03[2], Sti4/b[3])		STY2008 – STY2077	Lambdoid phages	STY2074; putative enterohaemolysin-1	Only found in Typhi CT18; Similarities to Fels-1, Gifsy-1 and Gifsy-2 phages (*Salmonella* Typhimurium LT2)
Sti5[3]	tRNA-*pro*	STY2467 – STY2470	Lambdoid phages	*sspH2*; Type III effector (Miao *et al.*, 1999) *msgA* (STY2470); Intramacrophage survival (Gunn *et al.*, 1995)	Remnant
Sti6[3]	tRNA-*arg*	STY2627 – STY2631	P22-like	–	Remnant
ST20a/b	tRNA-*asn* (a) or adjacent copy (b)	t0869 – t0872(Ty2)	P4-family members	ORFs '7' and '9' (Fig. 2.4); putative PilV-like type IV prepilins	Typhi Ty2, E01–6750, H52-originating haplotypes (a) M223, haplotype H8 (b), similarities to *E. coli* strain ED1a phage
ST27 (CT18–04[2], Sti7[3])	10Sa RNA (*ssrA*)	STY2879–STY2899	P2-family members	*iroBCDEN;* iron uptake via salmochelin (Bäumler *et al.*,1998)	Remnant, similarities to Fels-2 phage (LT2)
ST2–27		t2647–t2656 (Ty2)	P4-family members	t2647a, t2647b; function unknown	Found in Typhi Ty2 and E01–6750
ST35 (CT18–05[2], Sti8[3])	tRNA-*trp*, tRNA-*asp* rRNA operon	STY3657–STY3703	P2-family members	STY3659, STY3664, STY3672–73; function unknown STY3670–71; putative lipoproteins	Similarities to Fels-2 phage (LT2)

Table 2.4 continued

Name[1] (synonym(s))	Insertion site	From – To (CT18, NC_003198)	Homology	Morons (putative and proven)	Comment
ST36		–	Similar to P2-family member WPhi	–	Found in Typhi isolates 404ty and E01–6750
ST44 (SopEST[1], CT18–06[2], Sti9[3])	SPI-7; tRNA-*pheU*	STY4599 – STY4645	P2-family members	*sopE*; type III effector (Hardt *et al.*, 1998) STY4629–31, STY4640; function unknown	Similarities to Fels-2 phage (LT2)
ST46 (Sti10[3])	SPI-10; tRNA-*leuX*	STY4821 – STY4832	Satellite phage of P4-family	STY4822/23; putative Ser/Thr kinases STY4824; putative 2c-type protein phosphatase	H55- and H42-originating haplotype lineages
ST46a	SPI-10; tRNA-*leuX*	–	P4-family members	ZP_03348532*; secondary glycine betaine transporter ZP_03361198*; GTP-binding protein, HSR1-related ZP_03361205*; sigma-54 factor interaction domain-containing protein; *NCBI protein accession numbers	Typhi isolates: E02–1180, E00–7866 and M223

[1]Thomson *et al.* (2004) and Holt *et al.* (2008).

[2]Canchaya *et al.* (2003).

[3]Casjens (2003).

present in all *Salmonella* Typhi strains. SopEST is associated with the SPI-7 Vi-pathogenicity island which is missing in the SARB64 isolate (Boyd *et al.*, 2003). The SopE moron occupies the tail fibre region, inactivating the ability of the phage to express different versions of tail fibre proteins which might restrict its host range. Another moron is the ST27-associated *iroBCDEN* (*iroA*) gene cluster. It was shown that these genes are under control of the iron-responsive regulator Fur. The *iroA* locus encodes for a high-affinity iron uptake system based on a glycosylated enterochelin, called salmochelin. Salmochelin is synthesized and secreted depending on *iroBCDE*. IroN acts as an outer membrane siderophore receptor able to transport salmochelin and several catecholate siderophores (Hantke *et al.*, 2003).

The prophages ST10 and ST18 are homologous, with respect to sequence as well as genetic organization, to lambda-like *Salmonella* Typhimurium phage Gifsy-2. Whereas phage ST10 is well conserved in *Salmonella* Typhi isolates, ST18 seems only to be present in strain CT18. It is thought that in the CT18 isolate ST10 and ST18 phages recombined, which modified the ST10 phage in this strain. This recombination event affected the central part of ST18, which exhibits great homology to ST10 (Holt *et al.*, 2008). The sequence of ST18 has been disrupted by STY2056, an IS-element identical to the

IS10-left transposase. It is unclear, whether this insertion influences functionality of the phage (Thomson *et al.*, 2004). The early region of ST18 harbours a putative moron, an orthologue of the enterohemolysin-associated protein Ehly-1, which is shared with phage Gifsy-2. Two morons shown to have a virulence function in serovar Typhimurium are carried by the lambdoid phage remnant Sti5: SspH2 is a leucine-rich repeat protein homologous to the *Shigella flexneri* IpaH-family, and is translocated into host cells via the SPI-2-encoded TTSS (Miao *et al.*, 1999). The SspH2 effector interferes with the eukaryotic ubiquitination pathway functioning as a novel type of E3 ubiquitin ligase (Quezada *et al.*, 2009). The *msgA* gene is the second moron which was shown to be a member of the PhoP regulon. A strain deficient for *msgA* is attenuated for survival within macrophages (Gunn *et al.*, 1995). The Sti2 phage remnant exhibits also some homologies to lambda-family members and carries one pseudogene (STY1362) and a true gene (STY1364) with some similarities to toxin-encoding genes (Canchaya *et al.*, 2004).

In the genome of *Salmonella* Typhi strain E98–2068, the Mu-family member ST16 was identified (Holt *et al.*, 2008). ST16 is about 38 kb and shares significant homologies to a CP-933T-like prophage within the *E. coli* UTI89 genome (NC_007946: 911,530 to 948,388). Identical to genes encoded by bacteriophage Mu are some of the early and late region genes of phage ST15. The early region of ST15 carries a set of uncharacterized genes which might resemble morons: STY1603 is similar to the *Aeromonas hydrophila* type II secretion protein ExeA, although it carries a frameshift mutation in strain CT18 and might not be functional. A putative enterohemolysin-2 with similarities to that from phage phi-3208 is encoded by STY1597. The ST15 phage is of mosaic nature, combining Mu-homologous genes with elements of P2 as well as lambdoid phages (Thomson *et al.*, 2004). The genetic organization more closely resembles that of phage Mu, the tail region is P2-like, whereas its tail fibre invertible region exhibits sequence similarities to lambda phages.

A group of Typhi phages shares significant sequence homologies with P4-like phages. P4-family member ST20 was first described in Ty2 (Deng *et al.*, 2003) and can be found in further isolates originating from haplotype H52. ST20 is designated ST20a if inserted at tRNA-*Asn* or ST20b if a neighbouring copy of this tRNA is used by the phage (Holt *et al.*, 2008). Sequence comparison using ACT (*www.sanger.ac.uk/Software/ACT/*) revealed significant sequence similarities of ST20 to a phage encoded by *E. coli* ED1a (NC_011745: 2,237,640 to 2,266,944; Fig. 2.2). Two previously non-annotated genes between t0870 and t0871 show sequence homologies to type IVb prepilin genes of the PilV-type. These ORFs, labelled '7' and '9' in Fig. 2.2, are not present in the corresponding *E. coli* ED1a phage and might resemble morons. Intriguingly, *Salmonella* Typhi expresses functional PilS type IVb pili which mediate contact to intestinal epithelial cells (Zhang *et al.*, 2000). Transcriptional activity of these non-annotated genes was recently demonstrated using strand-specific cDNA sequencing (Perkins *et al.*, 2009). Whether ST20-encoded subunits can be utilized by the PilS pilus apparatus is unknown.

P4-family members have a hotspot for carrying foreign DNA between their *cos* site and the integrase gene because integration can occur without affecting phage function (Thomson *et al.*, 2004). In the TY2 phage ST2–27, the region between t2647 and t2648 is occupied by two non-annotated ORFs (t2647a and t2647b) of unknown function. In mid-log cultures grown in LB both genes are actively transcribed (Perkins *et al.*, 2009). The P4-like phages ST46 and ST46a are both associated with SPI-10 and integrated near tRNA-*leuX*, a common site for P4 integration (Bishop *et al.*, 2005). ST46a was acquired within SPI-10 by a common ancestor and has been replaced by ST46 in the haplotypes originating from H55 and H42 lineages (Holt *et al.*, 2008). In ST46, between integrase and *cos*, a ~3.5 kbp fragment of significantly lower G+C content compared to the CT18 core genome can be found. Within this fragment, three ORFs encode for putative Ser/Thr kinases (STY4822–23) or a 2c-type protein phosphatase (STY4824) which was also shown to be transcriptionally active (Perkins *et al.*, 2009). ST46a is a prophage of ~32 kbp and shows in its central part significant homologies

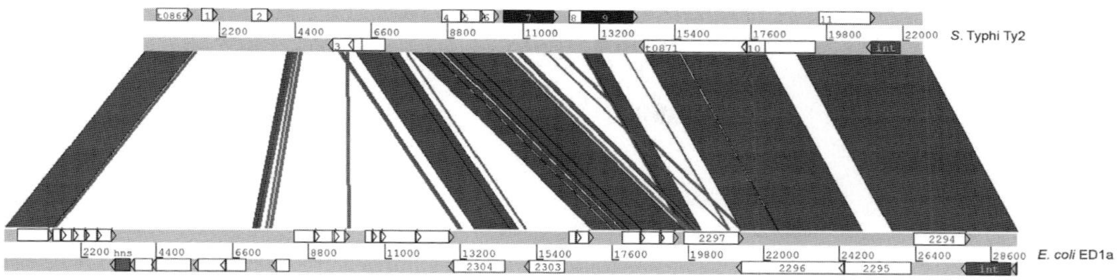

Figure 2.2 Sequence comparison of *Salmonella* Typhi phage ST20 with a homologous region in *E. coli* strain ED1a.

to the integrative element IE07 present in *E. coli* O111:H⁻ strain 11128 (NC_012264). ST46a seems to lack essential phage genes required for replication and several ORFs show similarities to genes of the defective *E. coli* prophage CP4–57. Interestingly, ST46a harbours several ORFs of putative moron nature (Table 2.4). One ORF encodes a betaine-carnitine-choline-transporter of the BetU family which is widely distributed in *E. coli* and involved in osmoregulation (Ly *et al.*, 2004). Two other ORFs might encode proteins of regulatory function: one is a GTP-binding protein of the HRS1-subfamily and the other contains a domain for interaction with sigma-54 (*rpoN*). The last example highlights an aspect of phage biology which is largely unappreciated: the influence of the prophage exerted on regulatory processes of the lysogen.

Investigation of prophage occupation sites in *Salmonella* Typhimurium

Some studies have used microarray technology to assess the presence of known phage sequences in *Salmonella* isolates from a variety of sources (Cooke *et al.*, 2007; Garaizar *et al.*, 2002). Although very powerful, this method depends on sequenced reference genomes and does not consider novel genes like prophages that may occur specifically in the sampled strains. As an alternative approach, Bossi and Figueroa-Bossi analysed the prophage content of strains by scoring the occupancy of a set of predictable phage attachment sites (*att*) by polymerase chain reaction (PCR). They assumed that prophage sequences near the ends of the element would be conserved even if the prophage had suffered deletions or rearrangements. This proved to be the case with all eight prophages analysed (Bossi and Figueroa-Bossi, 2005). The general method for assessing the presence or absence of a given prophage is depicted in Fig. 2.3. The method is based on a three primer PCR approach: two primers anneal to chromosomal sequences on either side of the *att* site, and the third primer anneals within the prophage in proximity to one boundary with the core genome. Depending on whether the *att* site is occupied or empty, the reaction yields different products of variable lengths. The distribution of some or all of these elements was examined in a collection of Typhimurium strains. Analysis of more than 60 isolates allowed us to differentiate profiles. We found that prophages Gifsy-1, Gifsy-2 or ST64B are present in around 90% of all Typhimurium strains (Table 2.5). 77.1% of the strains carry all three of them. In contrast, isolates carrying Gifsy-3 and Fels-1 seem to be rare (1.4% and 2.9% in our study). The Fels-2/SopEΦ prophage was detected in 22.9% of the *Salmonella* Typhimurium strains, and ΦW104 and P22/ST64T in 12.9% and 24.3%, respectively. Fels-2/SopEΦ was frequently detected in DT8, DT9 and DT10 strains, while ΦW104 was observed mainly in DT104. We also found one unusual DT29 strain carrying Gifsy-3, Fels-1 and Fels-2/SopEΦ, but none of the common prophages. Altogether we obtained 20 phage patterns from the 69 isolates tested. For the P22/ST64T *att* site, 28 strains gave none of the expected PCR products. Further analyses revealed that 15 of them carried a remnant prophage (see below) at this attachment site that is not related to P22/ST64T or ST104. The three-primer PCR approach also failed to

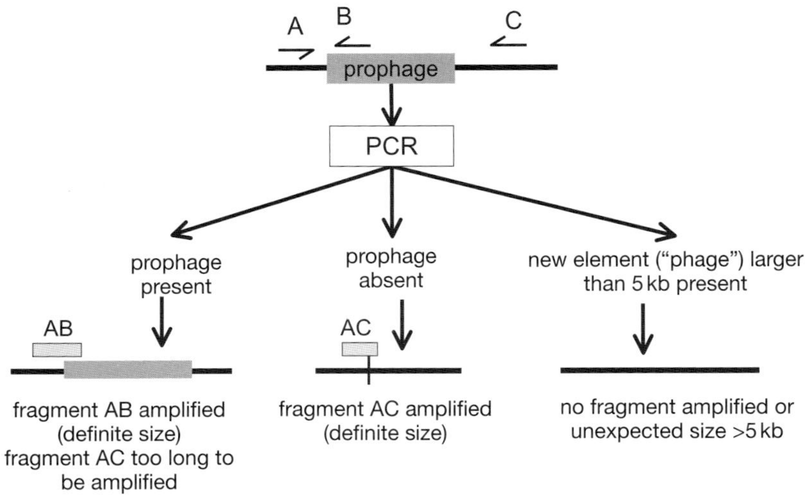

Figure 2.3 Scheme of three-primer PCR analysis of *att*-site occupancy.

give any product for Gifsy-3 in one strain, and Fels-2/SopEΦ in four strains, indicating either variations within the primer binding sites or the presence of an insertion too large to be amplified by generic PCR. From our study we conclude that the prophage occupancy pattern within phage types is heterogeneous. On the other hand, we found identical prophage profiles in different phage types.

A new remnant prophage in monophasic *Salmonella* Typhimurium

Recently, a monophasic variant of *Salmonella* Typhimurium [4,(5),12:i:-], not expressing the 2nd phase flagellum, has become one of the dominant causes of salmonellosis in Germany and other European countries (see above).

We investigated the prophage attachment sites and found that the sites for Gifsy-1, Gifsy-2 and ST64B were occupied by the respective prophages (Table 2.5). However, the sites for Fels-1, Fels-2/SopEΦ, Gifsy-3 and ΦW104 were free. In about 90% of the monophasic DT193 strains the P22/ST64T attachment site was occupied by a novel 18.4 kb fragment. This phenomenon was also observed in 4,[5],12:i:- DT193 strains from several other European countries. Sequence analysis revealed 17 open reading frames (Fig. 2.4), including phage-related genes such as a putative integrase similar to the prophage CPS-53 in *E. coli*, a putative phage injection protein from a bacteriophage of APEC O1 and the gene 13 protein of the *Shigella* phage Sf6 which belongs to the podoviridae. Using the BLASTn method, homologous sequences of 16 of the 17 ORFs were found in *Shigella* or *E. coli* genomes. The significantly lower G+C content of the novel fragment with 47.4% compared with. 52.2% in the *Salmonella* Typhimurium LT2 core genome indicates its acquisition via horizontal gene transfer (Truepschuch *et al.*, 2010).

Conclusion and outlook

Salmonella typing phages are mostly temperate phages which are adapted to different propagating strains. The application of phage typing allows easy surveillance of changes in dominant phage types in different continents. Prophages and remnant phages have been shown to carry many different additional virulence genes for *Salmonella*. At least some of these phages may become important for biocontrol in food and live stocks. Obviously, phage treatment should only be applied after sequencing and careful tests for absence of virulence genes or morons. Therefore, genetically engineered phages may be of even better use for *Salmonella* control, and this avenue of research should become more prominent in the following years.

Table 2.5 Prophage site occupancy PCR test patterns observed in 72 *Salmonella* Typhimurium strains

Number of isolates	Anderson phage type	Gifsy 1	Gifsy 2	Gifsy 3	Fels 1	Fels2/ SopE	P22/ST64T	ST64B	ϕW104
LT2*	DT4	+	+	–	+	+	–	–	–
SL1344*	DT44	+	+	–	–	+	–	+	–
ATCC14028	DT126	+	+	+	–	–	–	+	–
CVM 23701* monophasic	PTU291	+	+	–	–	No product	–	+	–
1	DT12	+	+	–	–	–	–	+	–
2	DT120	+	+	–	–	–	Remnant prophage**	+	–
1	DT120	+	+	–	–	–	+	+	–
1	DT126	+	+	–	–	–	–	+	–
1	DT146	+	+	–	–	–	+	+	–
11	DT193	+	+	–	–	–	Remnant prophage	+	–
1	DT193	+	+	–	–	–	+	+	–
2	DT2	+	+	–	–	–	–	+	–
2	DT208	+	+	–	–	–	+	+	–
1	DT20a	+	+	–	–	–	–	+	–
1	DT36	+	+	–	–	–	–	–	–
1	DT40	+	+	–	–	–	–	+	–
1	DT63	+	+	–	–	–	–	–	–
1	DT7	+	+	–	–	–	Remnant prophage	+	–
1	DT7	+	+	–	–	–	–	+	–
1	DT92	+	+	–	–	–	+	+	–
1	PTU311	+	+	–	–	–	–	+	–
1	RDNC***	+	+	–	–	–	Remnant prophage	+	–
1	ut	+	+	–	–	–	–	+	–
1	ut	+	+	–	–	–	+	+	–
2	DT10	+	+	–	–	+	–	+	–
2	DT104	+	+	–	–	–	+	+	+
1	DT104	–	+	–	–	–	+	+	+
1	DT104	–	–	–	–	–	+	+	+
1	DT104H	+	+	–	–	–	+	+	+
1	DT104H	+	+	–	–	No product	+	–	–

Table 2.5 continued

Number of isolates	Anderson phage type	Gifsy 1	Gifsy 2	Gifsy 3	Fels 1	Fels2/ SopE	P22/ST64T	ST64B	ϕW104
1	DT104L	+	+	–	–	–	+	+	+
1	DT135	+	+	–	–	+	–	+	–
1	DT193	+	+	–	–	No product	–	+	–
1	DT193	+	+	–	–	No product	+	–	–
1	DT204	+	+	–	–	+	–	+	–
1	DT36	+	–	–	–	+	No product	+	–
1	DT36	–	+	No product	–	–	No product	–	–
1	DT44	+	+	–	–	+	–	+	–
1	DT46a	+	+	–	–	–	No product	+	–
1	DT52	+	+	–	–	+	–	–	–
2	DT57	+	+	–	–	–	–	+	+
1	DT64	+	–	–	–	+	+	+	–
1	DT7	+	+	–	–	–	No product	+	–
1	DT8	+	+	–	–	+	No product	+	–
1	DT9	+	+	–	–	–	No product	+	–
3	DT9	+	+	–	–	+	No product	+	–
1	DT9	+	–	–	–	+	No product	+	–
1	DT9	–	+	–	–	+	No product	+	–
2	DT99	+	+	–	–	–	No product	+	–
1	PTU302	+	+	–	–	–	+	+	+
1	DT29	–	–	+	+	+	–	–	–
2	DT36	–	–	–	–	–	–	+	–

*Sequenced strains.

**Remnant prophage is identical with the 18.4 kb island.

***RDNC = react but do not conform with any phage type.

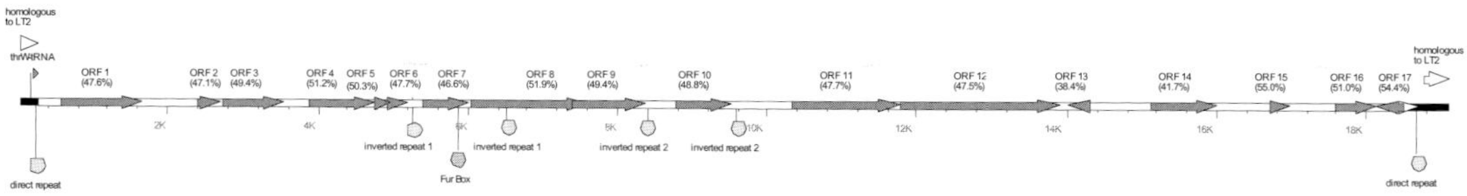

Figure 2.4 Remnant phage of *Salmonella* Typhimurium DT193.

Acknowledgements

We thank W. Jacobi and Dr. A. Fruth, Robert Koch Institute, for providing the OMP gels. Lionello Bossi, we thank for providing the primer sequences for prophage occupation sites. The authors are grateful to Peter Williams for critical reading.

References

Agasan, A., Kornblum, J., Williams, G., Pratt, C.C., Fleckenstein, P., Wong, M., and Ramon, A. (2002). Profile of *Salmonella enterica* subsp. *enterica* (subspecies I) serotype 4,5,12:i:- strains causing food-borne infections in New York City. J. Clin. Microbiol. *40*, 1924–1929.

Agron, P.G., Walker, R.L., Kinde, H., Sawyer, S.J., Hayes, D.C., Wollard, J., and Andersen, G.L. (2001). Identification by subtractive hybridization of sequences specific for *Salmonella enterica* serovar Enteritidis. Appl. Environ. Microbiol. *67*, 4984–4991.

Anderson, E.S., and Felix, A. (1953). The Vi type-determining phages carried by *Salmonella typhi*. J. Gen. Microbiol. *9*, 65–88.

Anderson, E.S., Ward, L.R., Saxe, M.J., de Sa, J.D. (1977). Bacteriophage-typing designations of *Salmonella* Typhimurium. *Hyg, J. (Lond)*. 78, 297–300.

Bacciu, D., Falchi, G., Spazziani, A., Bossi, L., Marogna, G., Leori, G.S., Rubino, S., and Uzzau, S. (2004). Transposition of the heat-stable toxin *astA* gene into a gifsy-2-related prophage of *Salmonella enterica* serovar Abortusovis. J. Bacteriol. *186*, 4568–4574.

Baggesen, D.L., Sørensen, G., Nielsen, E M., Wegener, H.C., (2010). Phage typing of *Salmonella* Typhimurium – is it still a useful tool for surveillance and outbreak investigation? Eurosurveillance, 15, issue 4, 19471.

Banker, D.D. (1955). Paratyphoid A phage typing. Nature *175*, 309–310.

Baquar, N., Threlfall, E.J., Rowe, B., and Stanley, J. (1994). Phage type 193 of *Salmonella* Typhimurium contains different chromosomal genotypes and multiple IS200 profiles. FEMS Microbiol. Lett. *115*, 291–295.

Barrangou, R., Fremaux, C., Deveau, H., Richards, M., Boyaval, P., Moineau, S., Romero, D.A., and Horvath, P. (2007). CRISPR provides acquired resistance against viruses in prokaryotes. Science *315*, 1709–1712.

Bäumler, A.J. (1997). The record of horizontal gene transfer in *Salmonella*. Trends Microbiol. *5*, 318–322.

Bishop, A.L., Baker, S., Jenks, S., Fookes, M., Gaora, P.O., Pickard, D., Anjum, M., Farrar, J., Hien, T.T., Ivens, A., *et al.* (2005). Analysis of the hypervariable region of the *Salmonella enterica* genome associated with tRNA(*leuX*). J. Bacteriol. *187*, 2469–2482.

Bogomolnaya, L.M., Santiviago, C.A., Yang, H.J., Bäumler, A.J., and Andrews-Polymenis, H.L. (2008). 'Form variation' of the O12 antigen is critical for persistence of *Salmonella* Typhimurium in the murine intestine. Mol. Microbiol. *70*, 1105–1119.

Boonmar, S., Bangtrakulnonth, A., Pornrunangwong, S., Terajima, J., Watanabe, H., Kaneko, K., and Ogawa, M. (1998). Epidemiological analysis of *Salmonella* Enteritidis isolates from humans and broiler chickens in Thailand by phage typing and pulsed-field gel electrophoresis. J. Clin. Microbiol. *36*, 971–974.

Bossi, L., and Figueroa-Bossi, N. (2005). Prophage of *Salmonella* Typhimurium. In Phages: Their Role in Bacterial Pathogenesis and Biotechnology, Waldor, M.K., Friedman, D.I., and Adhya, S.L., eds (Washington D.C., ASM Press), p. 450.

Boyd, E.F., Porwollik, S., Blackmer, F., and McClelland, M. (2003). Differences in gene content among *Salmonella enterica* serovar Typhi isolates. J. Clin. Microbiol. *41*, 3823–3828.

Braun, V. (2009). FhuA (TonA), the career of a protein. J. Bacteriol. *191*, 3431–3436.

Briggs C., E., Fratamico, P.M.(1999). Molecular characterization of an antibiotic resistance gene cluster of *Salmonella* Typhimurium DT104. Antimicrob. Agents Chemother. 43,846–849.

Brüssow, H., Canchaya, C., and Hardt, W.D. (2004). Phages and the evolution of bacterial pathogens: from genomic rearrangements to lysogenic conversion. Microbiol. Mol. Biol. Rev. *68*, 560–602.

Canchaya, C., Fournous, G., and Brüssow, H. (2004). The impact of prophages on bacterial chromosomes. Mol. Microbiol. *53*, 9–18.

Casjens, S. (2003). Prophages and bacterial genomics: what have we learned so far? Mol. Microbiol. *49*, 277–300.

Casjens, S. (2010). A plethora of putative phages and prophages. In Gone Fishing: John Roth's Contributions to Bacterial Genetics, Casadesús, J., Hughes, K., and Maloy, S., eds (Washington D.C., ASM Press), in press.

Chambers, R.M., McAddam, P., de Sa, J.D.H., Ward, L.R., and Rowe, B. (1987). A phage typing scheme for *Salmonella virchow*. FEMS Microbiol. Lett. *40*, 155–157.

Cooke, F.J., Wain, J., Fookes, M., Ivens, A., Thomson, N., Brown, D.J., Threlfall, E.J., Gunn, G., Foster, G., and Dougan, G. (2007). Prophage sequences defining hot spots of genome variation in *Salmonella enterica* serovar Typhimurium can be used to discriminate between field isolates. J. Clin. Microbiol. *45*, 2590–2598.

Craigie, J., and Felix, A.(1947). Typing of typhoid bacilli with Vi bacteriophage. *Lancet*, i, 823–827.

De Sa, J.D.H., Ward, L.R., and Rowe, B. (1980). A scheme for the phage typing of *Salmonella hadar*. FEMS Microbiol. Lett. *9*, 175–177.

Deng, W., Liou, S.R., Plunkett, G., 3rd, Mayhew, G.F., Rose, D.J., Burland, V., Kodoyianni, V., Schwartz, D.C., and Blattner, F.R. (2003). Comparative genomics of *Salmonella enterica* serovar Typhi strains Ty2 and CT18. J. Bacteriol. *185*, 2330–2337.

Echeita, M.A., Aladuena, A., Cruchaga, S., and Usera, M.A. (1999). Emergence and spread of an atypical *Salmonella enterica* subsp. *enterica* serotype 4,5,12:i:- strain in Spain. J. Clin. Microbiol. *37*, 3425.

Eisenstark, A., Rabsch, W., and Ackermann, H.W. (2009). Morphology of *Salmonella* Typhimurium typing phages of the Lilleengen set. Can. J. Microbiol. *55*.

Felix, A., and Callow, B.R. (1943). Typing of paratyphoid-B bacilli by means of Vi bacteriophage. Brit. Med. J. ii, 127.

Felix, A., and Callow, B.R. (1951). Paratyphoid-B Vi-phage typing. Lancet *2*, 10–14.

Figueroa-Bossi N., Coissac, E., Netter, P., and Bossi, L. (1997). Unsuspected prophage-like elements in *Salmonella* Typhimurium. Mol. Microbiol. *25*, 161–173.

Figueroa-Bossi, N., and Bossi, L. . (1999). Inducible prophages contribute to *Salmonella* virulence in mice. Mol. Microbiol. *33*, 167–176.

Garaizar, J., Porwollik, S., Echeita, A., Rementeria, A., Herrera, S., Wong, R.M., Frye, J., Usera, M.A., and McClelland, M. (2002). DNA microarray-based typing of an atypical monophasic *Salmonella enterica* serovar. J. Clin. Microbiol. *40*, 2074–2078.

Grissa, I., Vergnaud, G., and Pourcel, C. (2007). The CRISPRdb database and tools to display CRISPRs and to generate dictionaries of spacers and repeats. BMC Bioinformatics *8*, 172.

Guard-Petter, J. (1999). Phage type and other outer-membrane characteristics of *Salmonella enterica* serovar Enteritidis associated with virulence. In *Salmonella Enterica* Serovar Enteritidis in Humans and Animals, Saeed, A.M., Gast, R.K., Potter, M.E., and Wall, P.G., eds (Iowa State University Press, Ames, USA), vol. 1, pp. 221–232.

Gunn, J.S., Alpuche-Aranda, C.M., Loomis, W.P., Belden, W.J., and Miller, S.I. (1995). Characterization of the *Salmonella* Typhimurium *pagC/pagD* chromosomal region. J. Bacteriol. *177*, 5040–5047.

Hantke, K., Nicholson, G., Rabsch, W., and Winkelmann, G. (2003). Salmochelins, siderophores of *Salmonella enterica* and uropathogenic *Escherichia coli* strains, are recognized by the outer membrane receptor IroN. Proc. Natl. Acad. Sci. U.S.A. *100*, 3677–3682.

Hardt, W.D., Chen, L.M., Schuebel, K.E., Bustelo, X.R., and Galan, J.E. (1998). *S.* Typhimurium encodes an activator of Rho GTPases that induces membrane ruffling and nuclear responses in host cells. Cell. *93*, 815–826.

Hauser, E., Tietze, E., Helmuth, R., Junker, E., Blank, K., Prager, R., Rabsch, W., Appel, B., Fruth, A., and Malorny, B. (2010). Pork contaminated with *Salmonella enterica* serovar 4,[5], 12:i:–, an emerging health risk for humans. Appl. Environ. Microbiol. *76*, 4601–4610.

Ho, T.D., and Slauch, J.M. (2001). OmpC is the receptor for Gifsy-1 and Gifsy-2 bacteriophages of *Salmonella*. J. Bacteriol. *183*, 1495–1498.

Ho, T.D., Figueroa-Bossi, N., Wang, M., Uzzau, S., Bossi, L., and Slauch, J.M. (2002). Identification of GtgE, a novel virulence factor encoded on the Gifsy-2 bacteriophage of *Salmonella enterica* serovar Typhimurium. J. Bacteriol. *184*, 5234–5239.

Holt, K.E., Parkhill, J., Mazzoni, C.J., Roumagnac, P., Weill, F.X., Goodhead, I., Rance, R., Baker, S., Maskell, D.J., Wain, J., *et al.* (2008). High-throughput sequencing provides insights into genome variation and evolution in *Salmonella* Typhi. Nat. Genet. *40*, 987–993.

Holt, K.E., Thomson, N.R., Wain, J., Langridge, G.C., Hasan, R., Bhutta, Z.A., Quail, M.A., Norbertczak, H., Walker, D., Simmonds, M., *et al.* (2009). Pseudogene accumulation in the evolutionary histories of *Salmonella enterica* Serovars Paratyphi A and Typhi. BMC Genomics *10*, 36.

Kühn, H., Falta, R., and Rische, H. (1973). *Salmonella Typhimurium*. In Lysotypie und andere spezielle epidemiologische Laboratoriumsmethoden, Rische, H., ed. (Jena, VEB Gustav Fischer Verlag), pp. 101–139.

László, V.G., Csórián, E.S., and Pászti, J. (1985). Phage types and epidemiological significance of *Salmonella enteritidis* strains in Hungary between 1976 and 1983. Acta. Microbiol. Hung. *32*, 321–340.

Liebana, E., Garcia-Migura, L., Guard-Petter, J., McDowell, S.W., Rankin, S., Opitz, H.M., Clifton-Hadley, F.A., and Davies, R.H. 2002. *Salmonella enterica* serotype Enteritidis phage types 4, 7, 6, 8, 13a, 29 and 34: a comparative analysis of genomic fingerprints from geographically distant isolates. J. Appl. Microbiol. *92*, 196–209.

Liesegang, A., Davos, D., Balzer, J.C., Rabsch, W., Prager, R., Lightfoot, D., Siitonen, A., Claus, H., and Tschäpe, H. (2002). Phage typing and PFGE pattern analysis as tools for epidemiological surveillance of *Salmonella enterica* serovar Bovismorbificans infections. Epidemiol. Infect. *128*, 119–130.

Lilleengen, K.(1948). Typing of *Salmonella* Typhimurium by means of bacteriphage. Acta. Path. Microbiol. Scand. Supplement *77*, 1–39.

Linde K., Beer, J., Bondarenko, V., (1990). Stable *Salmonella* live vaccine strains with two or more attenuating mutations and any desired level of attenuation. Vaccine *8*, 278–282.

Lucarelli, C., Dionisi, A.M., Torpdahl, M., Villa, L., Graziani, C., Hopkins, K., Threlfall, J., Caprioli, A., and Luzzi, I. (2010). Evidence for a second genomic island conferring multidrug resistance in a clonal group of strains of *Salmonella enterica* serovar Typhimurium and its monophasic variant circulating in Italy, Denmark, and the United Kingdom. J. Clin. Microbiol. *48*, 2103–2109.

Malorny B., Schroeter, A., Bunge, C., Helmuth, H., (2002). Prevalence of Escherichia coli O157:H7 prophage-like sequences among German *Salmonella enterica* serotype Typhimurium phage types and their use in detection of phage type DT104 by the polymerase chain reaction. Vet. Microbiol. *87*, 253–265.

Miao, E.A., Scherer, C.A., Tsolis, R.M., Kingsley, R.A., Adams, L.G., Baumler, A.J., and Miller, S.I. (1999). *Salmonella* Typhimurium leucine-rich repeat proteins are targeted to the SPI1 and SPI2 type III secretion systems. Mol. Microbiol. *34*, 850–864.

Michael, G.B., Cardoso, M., Rabsch, W., and Schwarz, S. (2006). Phenotypic and genotypic differentiation of porcine *Salmonella enterica* subsp. *enterica* serovar Derby isolates. Vet. Microbiol. *118*, 312–318.

McClelland M., Sanderson, K.E., Spieth, J., Clifton, S.W., Latreille, P., Courtney, L., Porwollik, S., Ali, J., Dante, M., Du, F., *et al.* (2001). Complete genome sequence of *Salmonella enterica* serovar Typhimurium LT2. Nature. *413*, 852–856.

Miller, T.(2009). Development of a phage typing system for *Salmonella* Infantis for epidemiological purposes (German). Dissertation, Faculty of Veterinary Medicine University of Leipzig.

Mirold, S., Rabsch, W., Tschäpe, H., and Hardt, W.D. (2001). Transfer of the *Salmonella* type III effector SopE between unrelated phage families. J. Mol. Biol. *312*, 7–16.

Mmolawa, P.T., Wilmore, R., Thomas, C.J., and Heuzenroeder, M.W. (2002). Temperate phages in *Salmonella enterica* serovar Typhimurium: Implications for epidemiology. Int. J. Med. Microbiol. *291*, 633–644.

Moran, N.A., and Plague, G.R. (2004). Genomic changes following host restriction in bacteria. Curr. Opin. Genet. Dev. *14*, 627–633.

Mossong, J., Marques, P., Ragimbeau, C., Huberty-Krau, P., Losch, S., Meyer, G., Moris, G., Strottner, C., Rabsch, W., and Schneider, F. (2007). Outbreaks of monophasic *Salmonella enterica* serovar 4,[5],12:i:- in Luxembourg, 2006. Euro. Surveill. *12*, E11–12.

Nnalue, N.A., Newton, S., and Stocker, B.A. (1990). Lysogenization of *Salmonella* Choleraesuis by phage 14 increases average length of O-antigen chains, serum resistance and intraperitoneal mouse virulence. Microb Pathog *8*, 393–402.

Ochiai R.,L., Wang, X., von Seidlein L., Yang, J., Bhutta, Z.A., Bhattacharya, S.K., Agtini, M., Deen, J.L., Wain, J., Kim, D.R., *et al.* (2005). *Salmonella* Paratyphi A rates, Asia. Emerg. Infect. Dis. *11*, 1764–1766.

Parkhill, J., Dougan, G., James, K.D., Thomson, N.R., Pickard, D., Wain, J., Churcher, C., Mungall, K.L., Bentley, S.D., Holden, M.T., *et al.* (2001). Complete genome sequence of a multiple drug resistant *Salmonella enterica* Serovar Typhi CT18. Nature *413*, 848–852.

Pelludat, C., Prager, R., Tschäpe, H., Rabsch, W., Schuchhardt, J., and Hardt, W.D. (2005). Pilot study to evaluate microarray hybridization as a tool for *Salmonella enterica* serovar Typhimurium strain differentiation. J. Clin. Microbiol. *43*, 4092–4106.

Perkins, T.T., Kingsley, R.A., Fookes, M.C., Gardner, P.P., James, K.D., Yu, L., Assefa, S.A., He, M., Croucher, N.J., Pickard, D.J., *et al.* (2009). A strand-specific RNA-Seq analysis of the transcriptome of the typhoid bacillus *Salmonella* Typhi. PLoS Genet. *5*, e1000569.

Pfeifer, Y., Matten, J., and Rabsch, W. (2009). *Salmonella enterica* serovar Typhi with CTX-M beta-lactamase, Germany. Emerg. Infect. Dis. *15*, 1533–1535.

Porwollik, S., Santiviago, C.A., Cheng, P., Florea, L., Jackson, S., and McClelland, M. (2005). Differences in gene content between *Salmonella enterica* serovar Enteritidis isolates and comparison to closely related serovars *Gallinarum* and Dublin. J. Bacteriol. *187*, 6545–6555.

Quezada, C.M., Hicks, S.W., Galan, J.E., and Stebbins, C.E. (2009). A family of *Salmonella* virulence factors functions as a distinct class of autoregulated E3 ubiquitin ligases. Proc. Natl. Acad. Sci. U.S.A. *106*, 4864–4869.

Rabsch, W. (2007). *Salmonella* Typhimurium phage typing for pathogens. Methods Mol. Biol. *394*, 177–211.

Rabsch, W., Andrews, H.L., Kingsley, R.A., Prager, R., Tschäpe, H., Adams, L.G., and Bäumler, A.J. (2002). *Salmonella enterica* serotype Typhimurium and its host-adapted variants. Infect. Immun. *70*, 2249–2255.

Rabsch, W., Liesegang, A., and Tschäpe, H. (2001). [Laboratory-based surveillance of salmonellosis of humans in Germany–safety of *Salmonella Typhimurium* and *Salmonella* Enteritidis live vaccines] (article in German, English available). Berl. Münch. Tierärztl. Wochenschr *114*, 433–437.

Rabsch W., Helm, R.A., and Eisenstark, A. (2004). Diversity of phage types among archived cultures of the Demerec collection of *Salmonella enterica* serovar Typhimurium strains. Appl. Environ. Microbiol. *70*, 664–669.

Rabsch, W., Ma, L., Wiley, G., Najar, F.Z., Kaserer, W., Schuerch, D.W., Klebba, J.E., Roe, B.A., Laverde Gomez, J.A., Schallmey, M., *et al.* (2007). FepA- and TonB-dependent bacteriophage H8: receptor binding and genomic sequence. J. Bacteriol. *189*, 5658–5674.

Rabsch, W., Prager, R., Koch, J., Stark, K., Roggentin, P., Bockemuhl, J., Beckmann, G., Stark, R., Siegl, W., Ammon, A., *et al.* (2005). Molecular epidemiology of *Salmonella enterica* serovar Agona: characterization of a diffuse outbreak caused by aniseed-fennel-caraway infusion. Epidemiol. Infect. *133*, 837–844.

Saitoh, M., Tanaka, K., Nishimori, K., Makino, S., Kanno, T., Ishihara, R., Hatama, S., Kitano, R., Kishima, M., Sameshima, T., *et al.* (2005). The *artAB* genes encode a putative ADP-ribosyltransferase toxin homologue associated with *Salmonella enterica* serovar Typhimurium DT104. Microbiology *151*, 3089–3096.

Scholtens, R.T. (1950). Phage typing of *Salmonella paratyphi* B in the Netherlands. Antonie van Leeuwenhoek. *16*, 256–268.

Sonnenschein, C. (1928). Bakteriendiagnose mit Bakteriophagen. Beschleunigung der bakteriologischen

Typhusdiagnose. Deut. Med. Wochschr. 25 (article in German).

Sorek, R., Kunin, V., and Hugenholtz, P. (2008). CRISPR – a widespread system that provides acquired resistance against phages in bacteria and archaea. Nat. Rev. Microbiol. *6*, 181–186.

Thomson, N., Baker, S., Pickard, D., Fookes, M., Anjum, M., Hamlin, N., Wain, J., House, D., Bhutta, Z., Chan, K., *et al.* (2004). The Role of Prophage-like Elements in the Diversity of *Salmonella enterica* Serovars. J. Mol. Biol. *339*, 279–300.

Thomson, N.R., Clayton, D.J., Windhorst, D., Vernikos, G., Davidson, S., Churcher, C., Quail, M.A., Stevens, M., Jones, M.A., Watson, M., *et al.* (2008). Comparative genome analysis of *Salmonella* Enteritidis PT4 and *Salmonella Gallinarum* 287/91 provides insights into evolutionary and host adaptation pathways. Genome Res. *18*, 1624–1637.

Threlfall, E.J., Ward, L.R., and Rowe, B. (1978). Spread of multiresistant strains of *Salmonella* Typhimurium phage types 204 and 193 in Britain. Br. Med. J. *2*, 997.

Truepschuch, S., Laverde Gomez, J.A., Ediberidze, I., Flieger, A., and Rabsch, W. (2010). Characterisation of multidrug resistant *Salmonella* Typhimurium 4,[5],12:i:- DT193 strains carrying a novel genomic island adjacent to the *thrW* tRNA locus. Int. J. Med. Microbiol. *300*, 279–288.

Villafane, R., Zayas, M., Gilcrease, E.B., Kropinski, A.M., and Casjens, S.R. (2008). Genomic analysis of bacteriophage epsilon 34 of *Salmonella enterica* serovar Anatum (15+). BMC Microbiol. *8*, 227.

Ward, L.R., de Sa, J.D., and Rowe, B. (1987). A phage-typing scheme for *Salmonella* Enteritidis. Epidemiol. Infect. *99*, 291–294.

Windhorst, D. 2010. Analysis and characterization of the prophage content in *Salmonella* Enteritidis. Dissertation, Leibnitz University of Hannover, Germany.

Yamamoto, N. (1969).Genetic evolution of bacteriophage. I. Hybrids between unrelated bacteriophages P22 and Fels-2. Proc. Natl. Acad. Sci. U.S.A. 62,63–69.

Zhang, X.L., Tsui, I.S., Yip, C.M., Fung, A.W., Wong, D.K., Dai, X., Yang, Y., Hackett, J., and Morris, C. (2000). *Salmonella enterica* Serovar Typhi uses type IVB pili to enter human intestinal epithelial cells. Infect. Immun. *68*, 3067–3073.

Comparison of *Salmonella* Genomes

3

Ye Feng, Wei-Qiao Liu, Kenneth E. Sanderson and Shu-Lin Liu*

Abstract

Salmonella contains over 2600 known lineages, each with distinct biologic characteristics, including differences in the niche in which they dwell and the nature of diseases they may cause in their hosts. Genomic sequence analysis is beginning to reveal the genetic basis that determines the phenotypic differences among them. Comparison of eight sequenced genomes of *Salmonella* subspecies I lineages, which infect warm-blooded animals including humans, demonstrates that these pathogens share about 90% of their genes (the 'core' genome), with the remaining ca. 10% genes being unique to each of the lineages (the 'accessory' genome). Prophages and Salmonella Pathogenicity Islands (SPIs) are the main components of the accessory genome. Insertion of large DNA segments, such as SPI7 in *Salmonella* Typhi, may disrupt physical balance of the genome between replication origin and terminus and rearrangements of the genome, such as inversions or translocations mediated by homologous sites (*rrn* operons, prophages, IS200, etc.), may accelerate rebalancing of the genome. Laterally transferred genes are the main driving force in *Salmonella* evolution and speciation; evidence already exists indicating that mismatch repair genes may spontaneously regulate bacterial mutability through allele conversion to facilitate or inhibit incorporation of foreign DNA. Further studies may help elucidate genetic bases of distinct pathogeneses and host ranges among the *Salmonella* pathogens.

The first determination of gene order on the genome of *Salmonella enterica* serovar Typhimurium LT2 by genetic linkage methods started the era of comparative studies of genomes between *Salmonella* and *Escherichia coli* and among *Salmonella* lineages, leading, through technical advances over the past half century, to the discoveries of the main genetic differences associated with the divergence between *Salmonella* and *Escherichia coli* and among the most important *Salmonella* pathogens infecting man and livestock. This chapter begins with a brief historical review of genetic mapping of the *Salmonella* Typhimurium LT2 genome, goes into detailed discussions about *Salmonella* genomes with a focus on genomic differences among representative *Salmonella* lineages, continues with divergence and evolution of *Salmonella* genomes, and concludes with remaining questions and perspectives about *Salmonella* genomic studies.

A historical review of *Salmonella* genomic studies

Genetic linkage maps of *Salmonella* Typhimurium LT2

The first analysis of a *Salmonella* genome was through the construction of linkage maps, which were based entirely on genetic analysis, with the genes being located along the map like beads on a string. On the linkage map, individual genes were points representing the mutation that defined them, with no revelation of the size or nucleotide sequence of the gene. The first detailed linkage map of an enteric bacterium was based on F-factor-mediated interrupted conjugation in *Escherichia coli* K12, with 100 genes. To the surprise of the researchers, the map was circular (Wollman

*Corresponding author

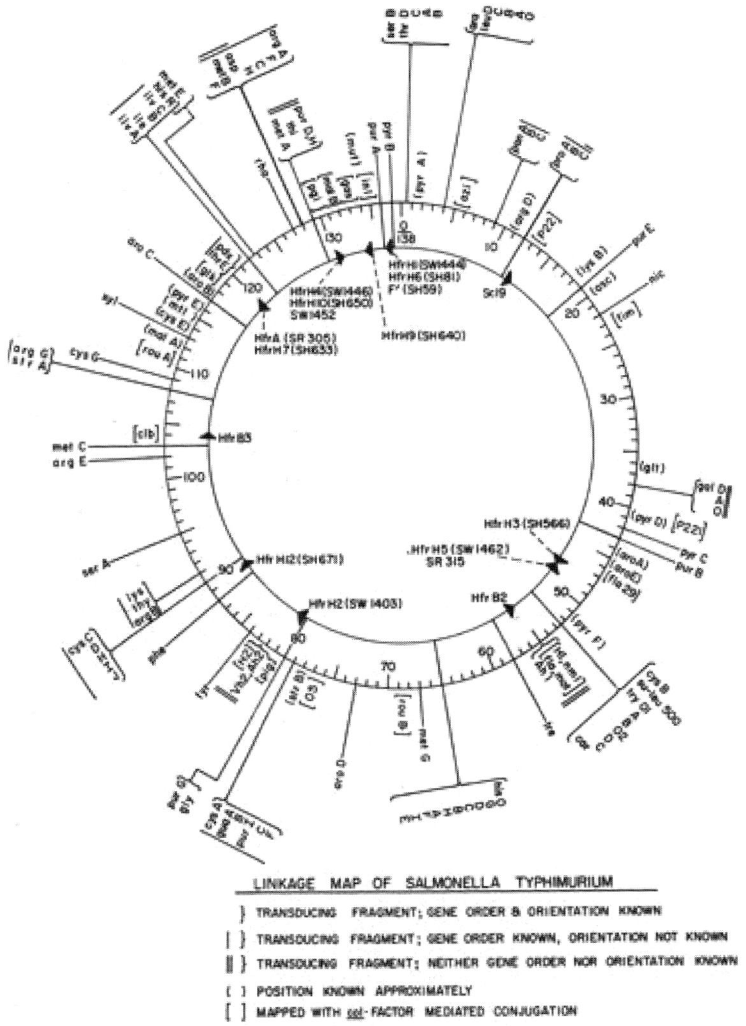

Figure 3.1. A linkage map (version 1) of *Salmonella* Typhimurium LT2 (Sanderson and Demerec, 1965). The map is marked in one-minute intervals representing the time of transfer in interrupted conjugation studies. Hfr strains are shown on the inner circle, with the arrow indicating the point-of-origin and direction of transfer. Genes are shown on the outer circle; those marked by a line were mapped by interrupted conjugation, while the groups of genes usually represent P22-mediated transduction linkage groups.

et al., 1956). P22-mediated transduction analysis of *Salmonella* Typhimurium strain LT2 in the 1950s, in the lab of Miloslav Demerec, showed several transduction linkage groups; these linkage groups are now realized to be smaller than the 46 kb size of P22, thus less than 1% of the size of the chromosome. The F-factor of *E. coli* K12 was transferred into *Salmonella* Typhimurium LT2, Hfr stains were isolated, and a circular linkage map locating 133 genes was constructed based on interrupted conjugation and the already-known transduction linkage groups (Sanderson and Demerec, 1965); the order of genes in *Salmonella* Typhimurium LT2 was almost completely identical to that in *E. coli* K12 (Fig. 3.1). The physical length of the chromosome was not known; in *E. coli* K12, the chromosome was divided into 100 centisomes, since the transfer of the whole set of genes from Hfr strains to F- minus recipient strains by conjugation took roughly 100 minutes. The transfer in *Salmonella* Typhimurium LT2 took a longer time, but for convenience of comparison with *E. coli* K12, the chromosome of LT2 was in later editions of the map similarly divided into roughly equivalent 100 units, called minutes (min) or centisomes (cs). As a result, gene locations in both *E. coli* K12 and *Salmonella* Typhimurium LT2 were referred to as 'min' on both maps, starting from *thrA* (0.0 min) clockwise through *pyrA* (1.7 min), *leu* (2.9 min), *bio* (18.1 min) to *serB* (99.5 min) and back to *thrA* (0.0 min). The maps were refined by transduction with bacteriophage P1 in *E. coli* K12 and P22 in *Salmonella* Typhimurium LT2. The length of the chromosome in kbs of DNA was predicted from analysis of joint transduction of genes by the bacteriophage P22, based on the fact that P22 packages about 45 kb of DNA in its capsid (Sanderson and Roth, 1988).

During the next three decades after the publication of the first *Salmonella* Typhimurium LT2 map, genetic and biochemical studies on *Salmonella*, mostly on the genes of *Salmonella* Typhimurium, accumulated from many different labs. This information was summarized in successive editions of the linkage map. Edition VIII was the last, published in 1995 (Sanderson *et al.*, 1995). The skeleton of Edition VIII was still similar to the interrupted conjugation linkage map developed for Edition I, but the intervals between genes was corrected by data from the physical map of LT2, the first version of which was published in 1992 (Liu and Sanderson, 1992), followed by a significantly refined version one year later (Liu *et al.*, 1993c). In addition, Edition VIII incorporated the then available nucleotide sequencing data; the non-redundant genomic DNA sequences comprised 197 contigs, with a total of 548 kb altogether, representing ~11% of the *Salmonella* chromosome. Edition VIII also presented a map of pSLT, the 93 kb plasmid, which has been called the virulence plasmid or the cryptic plasmid; 29 genes were mapped on the plasmid in this edition. Plasmid pSLT was considered part of the normal genome of *Salmonella* Typhimurium LT2, because it was almost invariably found in LT2 lines and most analysed *Salmonella* Typhimurium wild type strains. This plasmid has important effects on the phenotype of the bacteria.

The physical map of the *Salmonella* Typhimurium LT2 genome

When the pulsed field gel electrophoresis technique became available, physical mapping of bacterial genomes by the use of rare-cutting endonucleases became possible. The first rare-cutter physical map of *E. coli* was published by Smith *et al.* in 1987 (Smith *et al.*, 1987). Later, Liu, Sanderson, Wong and McClelland made physical maps of *Salmonella* Typhimurium LT2, using *Xba*I and *Bln*I, respectively (Liu and Sanderson, 1992; Wong and McClelland, 1992). The *Xba*I map was greatly refined the following year (Liu *et al.*, 1993c), by data from additional endonucleases and more Tn*10* insertions. This map shows a genome of 4808 kb, which is very close to the real size of 4857 kb later determined by sequencing, and reveals that a centisome on the linkage map is about 48 kb. More importantly, all seven *rrn* operons were precisely located on this version of the map. This work also led to the discovery that the seven *rrn* operons were similarly located on the genomes of all analysed *Salmonella* lineages and *E. coli* (Liu *et al.*, 1993a; Liu and Sanderson, 1998; Liu *et al.*, 1999).

Discovery of genomic rearrangements: a large inversion in *Salmonella* Enteritidis

When the physical mapping techniques were used to analyse the genome of *Salmonella* Enteritidis, a large inversion was located between two sites, one upstream of *flgL* and the other downstream of *fliC* (Liu *et al.*, 1993b), relative to the genome of *Salmonella* Typhimurium LT2. Other than that, *Salmonella* Enteritidis and *Salmonella* Typhimurium LT2 showed very high genomic similarity, in terms of size and gene order. This was the first ever genomic comparison between distinct *Salmonella* lineages on a physical mapping basis.

The highly plastic genome of *Salmonella* Typhi and other host-adapted *Salmonella* pathogens

Construction and comparison of the *Salmonella* Typhi physical map with that of *Salmonella*

Typhimurium LT2 revealed two striking features of the *Salmonella* Typhi genome: an over 100 kb insertion (Liu and Sanderson, 1995b), which was later determined to be a *Salmonella* pathogenicity island, SPI7 (Nair *et al.*, 2004; Parkhill *et al.*, 2001), and high plasticity in physical structure, i.e. large genomic segments are organized differently in individual isolates (Liu and Sanderson, 1995d, 1996). The genomic plasticity phenomenon was interpreted to be a mechanism to compensate for the disruption of the original genomic balance between *oriC* and *terC* on the genome by the insertion of SPI7. Indeed, all analysed *Salmonella* Typhi genomes are rearranged relative to that of *Salmonella* Typhimurium LT2 and all are better balanced than otherwise, as deduced by calculation of the lengths of the two halves of the genome between *oriC* and *terC* (Kothapalli *et al.*, 2005; Liu *et al.*, 2006). Screening of a large number of *Salmonella* genomes reveals that host-adapted *Salmonella* lineages tend to have large genomic insertions (Liu and Sanderson, 1998), such as *Salmonella* Paratyphi A having 100 kb insertions (Liu and Sanderson, 1995a; McClelland *et al.*, 2004) and *Salmonella* Paratyphi C having the 90 kb SPI7 (Liu *et al.*, 2009; Liu *et al.*, 2007), on one half of the genome between *oriC* and *terC* in both cases, and these genomes are rearranged or even plastic (frequently rearranging) if simple rearrangements of large genomic segments do not restore a balance.

Genomic sequencing of *Salmonella* strains

The publication of the completely sequenced genome of *Haemophilus influenzae* in 1995 (Fleischmann *et al.*, 1995) set a new milestone of bacterial genomics research, signalling the beginning of an era of comparative studies of bacteria based on whole genome sequencing. In 2009 there were over a thousand complete genomes of bacteria in Genbank. The first sequenced *Salmonella* genomes were published in 2001, i.e. those of *Salmonella* Typhimurium LT2 (STM) (McClelland *et al.*, 2001) and *Salmonella* Typhi CT18 (STY) (Parkhill *et al.*, 2001). Since then, genomes representing lineages *Salmonella* Paratyphi A (SPA) (McClelland *et al.*, 2004), *Salmonella* Choleraesuis (SCS) (Chiu *et al.*, 2005), *Salmonella* Paratyphi C (SPC) (Liu *et al.*, 2009), *Salmonella* Enteritidis (SEN) and *Salmonella* Gallinarum (SGA) (Thomson *et al.*, 2008) have been completed and published, and more are either finished or in the process of finishing and annotation (Andrews-Polymenis *et al.*, 2009) (http://www.sanger.ac.uk/Projects/Salmonella/; http://www.ncbi.nlm.nih.gov/genomes/lproks.cgi?view=1). Almost all of these strains are representatives of *Salmonella* subspecies I; in addition, representative strains of *Salmonella* subspecies IIIa (*Salmonella enterica* subsp. *arizonae*) and what is now recognised as the only other *Salmonella* species, *Salmonella bongori*, have been completed but not yet published, and representatives of the other four subspecies are in the process of being sequenced.

The first eight published *Salmonella* genomes were sequenced entirely by the traditional Sanger technology (Table 3.1). Recently, new sequencing technologies, called Next Generation Sequencing (NGS) techniques, e.g. 454 (Roche), Solexa (Illumina), SOLiD (AB), etc., have increasingly been used, such as in sequencing of *Salmonella* Paratyphi A AKU_12601 and multiple strains of *Salmonella* Typhi (Holt *et al.*, 2008). Some of the NGS techniques generate short read lengths, for example 70 bp by Solexa or 50 bp by SOLiD, therefore requiring fairly high sequence coverage (100 X or higher) to complete a genome project, even if reference genomes are available. However, a combination of physical mapping and NGS may require as low as about 20 X coverage, reducing cost five times, because of the relatively low cost and yet sufficient accuracy of physical mapping; comparison of genomic information revealed by physical mapping and sequencing is exemplified by three strains in Table 3.2.

Salmonella lineages: all sharing a core genome and each possessing unique genes

To date, more than 2500 *Salmonella* lineages have been recognized. They are classified essentially as serovars by the Kauffman-White Scheme according to their O and H antigenic formulae (Kauffmann and Edwards, 1957; Popoff *et al.*, 2004). Here we choose to use the term 'lineages'

Table 3.1 Summary of published genomic information of *Salmonella* and of *E. coli* strains

	Strain	Abbreviation	Size (kb)	No. of CDSs	No. of Pseudogene (%)	Plasmid	Host range/ pathogenicity	References
Salmonella I	Choleraesuis SC-B67	SCS	4755	4413	151 (3.4)	pSCV50, 50 kb; pSC138, 138kb	Swine-restricted, swine-typhoid; sometimes invasive in humans	Chiu *et al.*, 2005
Salmonella I	Enteritidis PT4	SEN	4685	4206	113 (2.7)		Host-generalists, cause gastroenteritis	Thomson *et al.*, 2008
Salmonella I	Gallinarum 287/91	SGA	4658	3965	309 (7.8)		Avian-restricted, cause fowl-typhoid	Thomson *et al.*, 2008
Salmonella I	Paratyphi A ATCC9150	SPAS	4585	4093	206 (5.0)		Human-restricted, cause enteric fevers	McClelland *et al.*, 2004
Salmonella I	Paratyphi A AKU_12601	SPAA	4581	4078	204 (5.0)		Human-restricted, cause enteric fevers	Holt *et al.*, 2009
Salmonella I	Paratyphi C RKS4594	SPC	4833	4577	149 (3.3)	pSCPV, 55 kb	Human-restricted, cause enteric fevers	Liu *et al.*, 2009
Salmonella I	Typhi CT18	STYC	4809	4395	205 (4.7)	pHCM1, 218 kb; pHCM2, 106 kb	Human-restricted, cause enteric fevers	Parkhill *et al.*, 2001
Salmonella I	Typhi Ty2	STYT	4791	4308	218 (5.1)		Human-restricted, cause enteric fevers	Deng *et al.*, 2003
Salmonella I	Typhimiurium LT2	STM	4857	4423	40 (0.89)	pSLT, 93Kb	Host-generalists, cause gastroenteritis	McClelland *et al.*, 2001
Salmonella IIIa	subspecies *arizonae* RSK 2980	SAR	4600	4498				WashU-GSC website
Salmonella	*bongori* 12419	SBO	4460	4005				Sanger Institute website
E. coli	E. coli K-12	ECOK	4639	4149	177 (4.3)		Avirulent	Blattner *et al.*, 1997
E. coli	E. coli O157:H7 EDL933	ECOO	5528	5312	None described		Host-generalist, enterohemorrhagic in man	Perna *et al.*, 2001

Table 3.2 Comparison of genomic features of representative *Salmonella* strains revealed by physical mapping using PFGE, and by sequencing

Salmonella strains	Genome size (kb)			Other features	
	Mapping	Sequencing	Difference (%)	Mapping	Sequencing
Typhimurium LT2	4807; Liu *et al.*, 1993c	4857; McClelland *et al.*, 2001	50 kb, (1.00)	Four insertions located	All four insertions are prophages
Typhi Ty2	4780; Liu and Sanderson, 1995b	4792; Deng *et al.*, 2003	12 kb, (0.25)	Two inversions, 3 translocations, one large insertion	Two inversions, 3 translocations, SPI7 (134 kb)
Paratyphi A ATCC9150	4595; Liu and Sanderson, 1995a	4585; McClelland *et al.*, 2004	10 kb, (0.22)	Inversion in *rrnH/G*	Inversion in *rrnH/G*
				100 kb insertion	SPA prophages

rather than 'serovars' to refer to individual salmonellae, because a serovar may in fact contain genetically distinct pathogens, such as *Salmonella* Paratyphi B (human-adapted typhoid agent) and *Salmonella* Paratyphi B biovar Java (a host-generalist infecting a broad range of hosts causing salmonellosis but not typhoid), sharing the same antigenic formula 1,4,[5],12:b:1,2, or *Salmonella* Gallinarum biovar Pullorum (causing pullorum disease in fowl) and *Salmonella* Gallinarum biovar Gallinarum (causing typhoid in fowl), sharing 1,9,12:-:-. Most *Salmonella* lineages, such as *Salmonella* Typhimurium and *Salmonella* Enteritidis, infect a broad range of host species and cause self-limiting gastroenteritis in humans. On the other hand, some *Salmonella* lineages are limited to specific hosts, such as *Salmonella* Typhi infecting only humans and causing typhoid. *Salmonella* Typhimurium is interesting, because it causes different clinical situations in different host species, such as typhoid in mice, gastroenteritis in humans, and asymptomatic carrier state in adult chickens. Much research on *Salmonella* genomics has focused on the genetic basis of strict host adaptation and distinct pathogeneses, both requiring in depth comparative genomic analyses among the *Salmonella* lineages.

Core versus accessory genomes

Core genome refers to gene complement shared by a defined group of bacteria, such as *Salmonella* lineages. It includes mostly essential genes that are necessary for the bacterial cell to survive and grow, such as genes encoding enzymes which function in biosynthetic pathways. Virtually all identified genes on the linkage map of *Salmonella* Typhimurium LT2 (Fig. 3.1) are parts of the core genome, many of them are common to both *E. coli* and *Salmonella*, and they usually are highly conserved both in sequence identity and in gene order (synteny), as revealed initially by genetic maps, later by physical maps (Kohara *et al.*, 1987; Liu *et al.*, 1993c; Liu and Sanderson, 1992; Smith *et al.*, 1987; Wong and McClelland, 1992), and finally by sequencing (Blattner *et al.*, 1997; McClelland *et al.*, 2001; Parkhill *et al.*, 2001). The rest of the (hundreds of) genes are unique to a single *Salmonella* lineage or a small phylogenetic group of *Salmonella* lineages and are often called 'accessory genome' in comparison to the 'core genome'. Genes in the accessory genome are mainly in the following groups: genomic islands (GIs), including *Salmonella* pathogenicity islands (SPIs); prophages; insertion sequences (IS); etc. The definition of core or accessory genome is not absolute, however: genes may be defined to be in the core genome depending on what strains are compared. In pair-wise comparisons, the core genome includes genes that are common to the two members of the pair. *Salmonella* lineages may differ from one another in metabolism, host range, and type and severity of disease caused. It is reasonable to presume that the phenotypic differences among the distinct *Salmonella* lineages are primarily due to differences in the accessory genome.

Salmonella Typhi versus *Salmonella* Typhimurium: distinct pathogens but close relatives

Comparison of *Salmonella* Typhi and *Salmonella* Typhimurium was initially achieved by physical mapping using PFGE with the endonucleases I-*Ceu*I, *Xba*I, *Avr*II, and *Spe*I, which revealed multiple large scale genomic insertions/deletions and rearrangements (Liu and Sanderson, 1995b, d), and later by whole genome sequencing, which uncovered the genes shared by both and those possessed by one or the other (McClelland *et al.*, 2001; Parkhill *et al.*, 2001). This pair of *Salmonella* lineages was a very good choice for in depth comparative studies, especially at a time when sequencing of large bacterial genomes (~5 Mb) was still fairly expensive, because Typhi and Typhimurium are the two most important representatives of all *Salmonella* lineages, the former infecting only humans and causing only typhoid and the latter infecting a broad range of host species and causing different diseases in different hosts species. It was not until 2001 when researchers realized that the two pathogenically very different *Salmonella* lineages share as much as nearly 90% of their genes, with the remaining a little over 10% being unique to either of them.

Diverse *Salmonella* lineages with a common core genome

When representative strains of additional *Salmonella* lineages were sequenced, it became clear that distinct *Salmonella* lineages all share a common core genome. For example, the human typhoid agent *Salmonella* Paratyphi C strain RKS4594 and the swine pathogen *Salmonella* Choleraesuis strain SC-B67 have 4346 genes in common, spanning 96.66% of the Paratyphi C RKS4594 genome and 98.23% of the Choleraesuis SC-B67 genome. The remaining parts consist mainly of prophages or SPIs in their genomes. On the other hand, clinically similar *Salmonella* pathogens may not have as many genes in common, as demonstrated in the comparison of two human-adapted typhoid agents, Paratyphi C and Typhi: they have only 4008 genes (89.78% of the Paratyphi C RKS4594 genome and 90.23% of the Typhi CT18 genome) in common, a similar level of genomic divergence as between Typhimurium LT2 and Typhi CT18. Genomic differences between Paratyphi C RKS4594 and Typhi CT18 are mainly small gene clusters scattered throughout the genomes, an indication that these genomic differences are stochastic genomic modifications accumulated during the independent adaptation processes of the two close relatives to their own hosts; analysis of these small gene clusters may provide novel insights into host–pathogen interactions and, especially, molecular basis of host-adaptation.

Pan-genome of *Salmonella*

The term "pan-genome" describes all genes of a defined group of organisms, including the core genome and the accessory genomes of all members in the group. This can be defined for a species like *Bacillus anthracis* or *E. coli*, or a genus like *Salmonella*. Estimates by Lapierre and Gogarten for the total size of the pan-genome of the entire bacterial group, using 573 sequenced genomes, indicate that its size is infinite (Lapierre and Gogarten, 2009), which is supported by work of Medini *et al.* based on analysis of eight genomes of *Streptococcus agalactiae* that indicates that unique genes will continue to be identified even after hundreds or even thousands of genomes are sequenced (Medini *et al.*, 2005). Some bacterial species are reported to have a 'closed pan-genome', e.g. *Bacillus anthracis*, in which the total number of genes saturated when only a few strains were sequenced (Medini *et al.*, 2005). On the other hand, many bacteria are known to have an 'open pan-genome', so these species may never be fully described, because new genes will continue to be identified as new strains are sequenced. For bacteria in the 'open pan-genome' category, lateral gene transfer keeps bringing in new genes, facilitating rapid evolution and enabling an organism to adapt quickly as its environment changes.

The pan-genome of *E. coli* was estimated to be open based on analysis of 17 sequenced strains representing both pathogens and commensals, with 13,000 genes identified so far (Rasko *et al.*, 2008). As close relatives of *E. coli*, *Salmonella* also seem to have an open pan-genome, as suggested

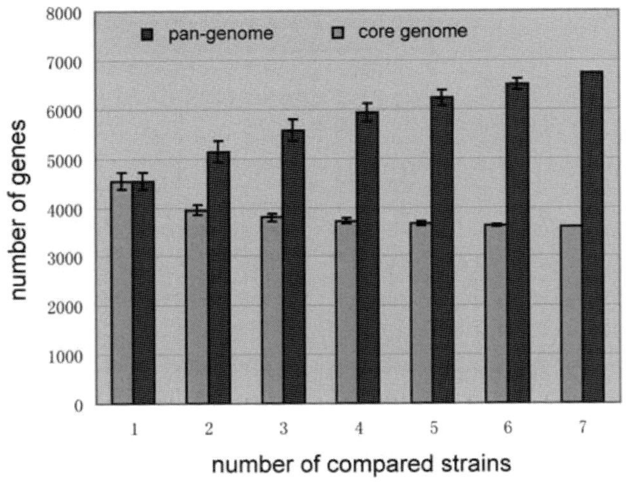

Figure 3.2. The pan-genome of *Salmonella* subspecies I lineages. The genomes of seven strains (SCS, SEN, SGA, SPAS, SPC, STYC, STM, listed in Table 3.1) were analyzed. The number of genomes included in the analysis is on the x-axis; the genomes are added to the analysis in 10 random orders. The numbers of genes common to the genomes (the core genome) are shown in light shading, and the total numbers of unique genes (the pan-genome) are in dark shading.

by analysis of seven sequenced strains, all of which belong to subgroup I of *Salmonella* (Fig. 3.2). The genome sizes of the seven strains are relatively conserved (Table 3.1). In a mathematical model, when only one genome is considered, the whole genome is core and the accessory genome is zero. As more sequences are added (in random order, up to seven), the core continues to decrease, but appears to be approaching a minimum; the core for all seven genomes was identified as 3598 genes. The total number of genes (composed of the core plus the accessory genes) continues to increase as more genomes are added in random order; at seven genomes it is 6732. The total number of genes in the pan-genome of *Salmonella* will likely continue to increase indefinitely. For example, if two *Salmonella* lineages share 90% of their ca. 4500 genes, the number of genes in the pan-genome of the two is

$$4500 \times 90\% + 4500 \times 10\% + 4500 \times 10\% = 4950$$

Assuming that all known 1400 *Salmonella* subspecies I lineages share an average of approximately 90% of their 4500 genes, we can estimate the number of genes in the pan-genome to be

$$4500 \times 90\% + 4500 \times 10\% \times 1400 = 634{,}050$$

Similarly, the number of genes in the pan-genome of all known 2500 *Salmonella* subspecies I–VI lineages would probably be over three million, considering the much greater divergence among them. Note that new *Salmonella* lineages continue to be isolated annually (Popoff *et al.*, 2004).

Genome degradation in *Salmonella*

While *Salmonella* diversify to become distinct pathogens with different host-ranges or pathogeneses by lateral gene transfer, individual lineages also shed no-longer needed genes, so genome degradation is essentially a natural part of the evolution process. Gene degradation results from either deletion or pseudogene formation and is much more common in host restricted pathogenic bacteria than in host generalists. For instance, the host-generalist *Salmonella* Typhimurium has as few as 40 pseudogenes in 4423 coding sequences (<1%), whereas the host-specialists Typhi and Paratyphi A both have over 200 pseudogenes (around 5% of coding sequences) (Table 3.1).

Comparisons on the role of pseudogenes between *Salmonella* Typhi and *Salmonella* Paratyphi A have led to new insights into evolution of these pathogens (Holt *et al.*, 2009). Sequence analysis reveals a mosaic structure of *Salmonella* Typhi and *Salmonella* Paratyphi A with about 20% of the genome having about 0.2% sequence divergence (unusually low for separate serovars of *Salmonella*) and 80% about 1.2% divergence, suggesting that the two lineages have undergone up to 100 different recombination events in the low divergence portion of the genome, postulated to have occurred when the two typhoid agents began to share the same host (humans) (Didelot *et al.*, 2007).

It is noteworthy that the two typhoid agents have diverged significantly in sequence, yet there is evidence of convergence on common phenotypes, especially common adaptation to humans. Analysis reveals that 66 pseudogenes are common to the four completely sequenced strains of these two typhoid agents, even with identical inactivating mutations in many of the shared pseudogenes. Twenty of the shared pseudogenes (54% of the non-phage/transposase shared pseudogenes) encode secreted or surface-exposed proteins, thus are likely to have contributed to convergence to a similar phenotype, namely host-restriction to humans. In addition, inactivation of different genes in the same pathway often results in similar loss of function and such examples have been reported (Liu *et al.*, 2009).

Holt *et al.* categorized the pseudogenes into the following three groups (Holt *et al.*, 2009): (i), ancestral (present prior to divergence of the compared lineages); (ii), shared (through recombination); and (iii), recent (arising by convergence after divergence of the compared lineages). The last category can also be seen between different members of the same *Salmonella* lineage, such as the two strains of SPA, which have as many as 22 different strain-specific pseudogenes, indicating that the process of pseudogene formation is continuing as the strains adapt to their specific host.

Comparison of pseudogenes between *Salmonella* Typhi and *Salmonella* Paratyphi A presents a very interesting case: following extensive recombination between their genomes, they have each accumulated lineage-specific pseudogenes. Some of the common pseudogenes between these two serovars appear to be associated with lack of selection on the gene function during adaptation of these different bacteria to the same niche, such as *sopD2*, which encodes an secreted effector protein required for virulence in *Salmonella* Typhimurium but may not really be needed by Typhi or Paratyphi A. Some other genes such as *shdA*, *ratB*, and *sivH* are involved in intestinal colonization and persistence, functions that are required in gastroenteritis but not in typhoid, so those genes are lost through pseudogene formation in Typhi and Paratyphi A.

Salmonella Typhi is now considered a recent lineage, having split from its last ancestor only 15,000–150,000 years ago (Kidgell *et al.*, 2002; Roumagnac *et al.*, 2006), and over 200 pseudogenes have been identified in this typhoid agent. If one assumes that the pseudogenes began to be formed in the process of adaptation of Typhi to humans, there should still be many superfluous genes that require more time to be inactivated. *Salmonella* Paratyphi A, another typhoid agent genetically distinct from Typhi, also has over 200 pseudogenes, but the two organisms have only 66 pseudogenes in common. Judging by the overall genetic similarity and common niche, one can reasonably anticipate a growing number of pseudogenes to be shared by Typhi and Paratyphi A as these lineages evolve in the future.

Divergence and evolution of the *Salmonella* genomes

Following the divergence of *Salmonella* and *E. coli* over 100 million years ago (Doolittle *et al.*, 1996; Ochman and Wilson, 1987), more than 2500 distinct lineages have developed from the *Salmonella* line. The divergence processes from a common ancestor to thousands of genetically and phenotypically different pathogens involve countless evolutionary events of the genomes, which can be summarized into three kinds: point mutations, indels (i.e. insertions by lateral gene transfer and deletions resulting from gene decay), and genomic rearrangements (which include inversions, translocations, and duplications). None of these kinds of evolutionary events could easily take place in a bacterial cell, due to the existence of mechanisms such as the mismatch repair (MMR) system, which strongly inhibits mutation and recombination. As MMR has been implicated in spontaneously regulating the bacterial mutability to inhibit or accept genomic changes (Gong *et al.*,

2007), we will briefly comment on the possible roles of MMR in the divergence and evolution of the *Salmonella* genomes at the end of this section.

Point mutations

Point mutations may be categorized into three classes: (i), single or multiple base insertions or deletions, which may cause frameshifts, abolishing a gene function; (ii), synonymous base substitutions, which may not cause obvious functional changes; and (iii), non-synonymous base substitutions, which may prematurely terminate a peptide chain by converting an amino acid codon into a stop codon or may change one amino acid codon into a different one, thereby changing the characteristics of the peptide. In cases of (i) and (iii), pseudogenes may be formed. It is possible that non-synonymous base substitutions may create novel protein products that are advantageous for the bacteria in adaptation to changed environment. In cases of (ii), protein products will largely be unaltered. Taken together, point mutations may bring about two main consequences: phylogenetic divergence among initially close relatives and adaptation of the organism to the niche.

Unless lethal, point mutations may accumulate over time, working like molecular clocks and reflecting levels of divergence between bacterial lineages. The phylogenetic relationships among the sequenced strains of representative subgroup I *Salmonella* lineages, with *E. coli* K-12 as a reference, are revealed by comparison of concatenated sequence of all orthologues shared by these strains (Fig. 3.3). As expected, strains of the same lineage (e.g. those of SPA or of STY) are very closely related to each other. Importantly, this phylogenetic tree shows no obvious link between phylogeny and host-range or pathogenicity: SGA is avian-restricted and causes fowl-typhoid while SEN is a host-generalist causing gastroenteritis in humans, but they are closely related (Thomson *et al.*, 2008); STY, SPA and SPC are human-restricted and cause enteric fever, yet are not more closely related among them than any of them to other *Salmonella* lineages; SCS has a narrow host-range, causing swine paratyphoid and occasionally causing invasive infections in humans, yet is most closely related to SPC.

The average nucleotide identity (%ANI) can be calculated for the orthologues shared by a given pair of genomes; it measures the conservation of base pairs for this pair of strains. In addition, the same pair of strains can be compared for the per cent of the genome that is shared, i.e. genes which are part of the core as per cent of the total genes of both strains. Fig. 3.4 shows the %ANI plotted against the per cent shared DNA for pairs of sequenced strains in *Salmonella* and in

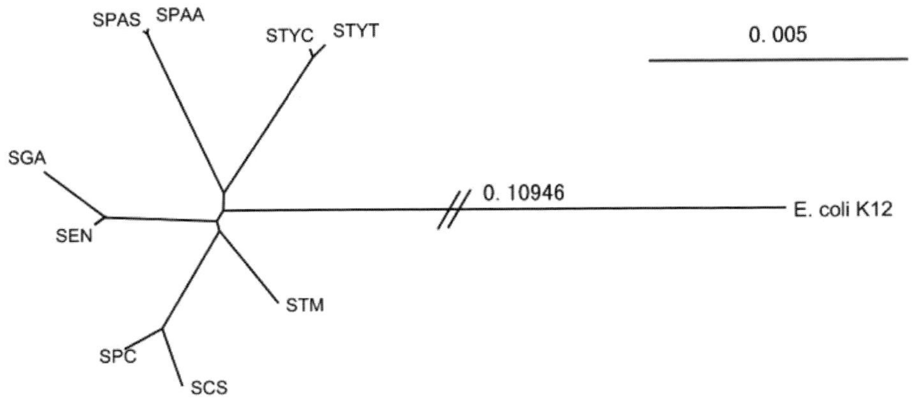

Figure 3.3. Phylogenetic relationships of *Salmonella* and *E. coli* strains based on concatenated sequences of all shared orthologues. Strain abbreviations are explained in Table 3.1. Nucleotide divergence was defined as the number of mismatched bases divided by that of comparable bases after pairwise alignment made by CLUSTALW.

two *E. coli* strains, K-12 (ECOK) and O157:H7 (ECOO). The points on the graph can be placed in seven groups. Group 1 includes four comparisons of very closely related *Salmonella* strains: a pair of SPA strains with ANI of 99.98% and shared DNA of 99.96%; a pair of STY strains with ANI of 99.94% and shared DNA of 99.07%; SEN and SGA with ANI of 99.83% and shared DNA of 97.8%; and SPC and SCS with ANI of 99.8% and shared DNA of 97.1%. Taking the SEN/SGA pair as an example, we can see approximately 0.2% base-pair divergence and 2% non-shared (accessory) DNA between them; the other points approximate a straight line. The second group includes all other pair-wise comparisons of the sequenced *Salmonella* subspecies I strains; these form a relatively continuous group from STM/SEN with ANI of 99.1% and shared DNA of 93.2% to SCS/STYT with ANI of 98.6% and shared DNA of 87%. Thus, group 2 has approximately 1% base pair divergence and 10% non-shared (accessory) DNA, a similar 1 to 10 ratio as in group 1. Group 3 is the only comparison presented between *E. coli* strains, ECOK and ECOO, with ANI of 98.4% and shared DNA of 81.1%. Group 4 shows all the pair-wise comparisons of SAR (in *Salmonella* subspecies IIIa) with strains in *Salmonella* subspecies I, with ANI close to 94% and shared DNA close to 77% (78.2% to 76.0). Thus, base pair divergence of 6% is associated with about 23% accessory DNA. Group 5 shows comparison of *S. bongori* with *Salmonella* subspecies I strains, with ANI of about 91% and shared DNA of about 76%. Group 6 is the comparison of SAR with *S. bongori*, with ANI of 91% and shared DNA of 74.7%. Group 7 compares *E. coli* and *Salmonella* strains, with ANI of about 85% and shared DNA of about 50%. Thus, base pair divergence of 15% is associated with 50% accessory DNA for a ratio of about 1 to 3.

Similar comparisons have been done among the genomes of 10 relatively divergent *Shewanella* strains, some of which showed as little as 70% ANI and only 35% shared DNA (Konstantinidis *et al.*, 2009). In these strains, most pair-wise comparisons fall on a straight line ($R^2 = 0.92$), indicating that strains with about 10% DNA divergence are associated with about 30% accessory DNA; thus, the ratio was about 1 to 3. The authors concluded that this linearity indicates that, although there is extensive genome fluidity, the genotypic similarity as indicated by the per

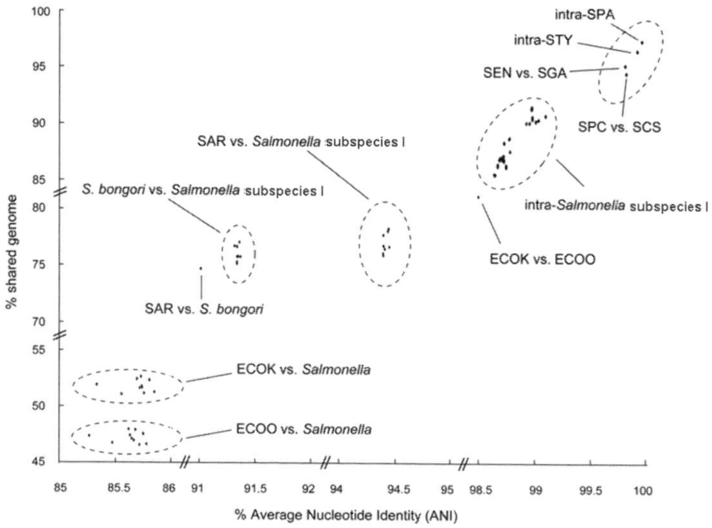

Figure 3.4. The average nucleotide identity (ANI%) of the core genome of pairs of strains of *Salmonella* and *E. coli* plotted against the per cent of shared (core) genome for the same pairs. The strains are listed in Table 3.1. The strain pairs represented in Groups 1 to 7 are discussed in the text.

cent shared DNA is predictable from their evolutionary relatedness as revealed by their ANI. For relatively less related strains in the *Salmonella* and *E. coli* group, our analysis reveals a similar conclusion: comparisons between *E. coli* and *Salmonella* and those among *Salmonella* lineages in different subspecies both show a ratio of divergence to accessory DNA of about 1 to 3. However, in comparisons between very closely related *Salmonella* strains, the ratio of divergence to accessory DNA is about 1 to 10 rather than 1 to 3. Therefore, it seems that *Salmonella* accumulate accessory DNA (mostly prophages) quickly in the short term but an average of over two thirds of the laterally acquired DNA will eventually be eliminated from the genome during long term evolution.

The other main consequence of point mutations, in the form of pseudogene formation, is the adaptation, or even confinement, of the organism to a particular niche. Human-adapted *Salmonella* lineages contain much greater numbers of pseudogenes than the *Salmonella* host-generalists, probably due to readily available nutrients in the protected niche (the human body), and hence lack of selection on those biosynthetic pathways and other related gene functions that are no longer needed. As a result, the no longer needed genes will gradually degrade, making the bacteria strictly adapted or even confined to the niche. Pseudogenes in bacterial genomes are mostly recognized by comparison with functional homologues. Generally, genes that contain frameshifts, insertions, deletions or mutations to stop codons are categorized as pseudogenes. Two factors affect the accuracy of pseudogene recognition. The first factor is proper choice of the reference sequence. As the sequence of most functional genes is defined by *in silico* prediction, reliability of its functional assignment is not guaranteed. For example, the similarity between reference and target sequences does not necessarily indicate its functional identity: a truncated sequence may be non-functional in one organism but functional in another. The second factor is how to set the criteria for defining pseudogenes. Not all frameshifts, deletions or stop mutations will inactivate genes; sometimes the shortened protein products may still carry out certain functions. Therefore, the predicted pseudogenes, though in most cases likely to be correct as annotated, eventually would require experimental confirmation.

Insertions and deletions (Indels)

Compared to the core genome, the remaining part of the gene complement, i.e. the 'accessory genome', consists of genes specific to a given bacterial lineage. Such genes are usually acquired by lateral transfer and are interspersed as large blocks or small clusters among the core genome. The main components of the accessory genomes are prophages and genomic islands (e.g. SPIs); many *Salmonella* lineages also contain IS elements.

A prophage is a phage genome inserted as part of the bacterial chromosome. It has become increasingly clear that lysogenic bacteriophages may provide essential virulence and fitness factors to their hosts, enhancing the capability of bacterial adhesion, invasiveness, exotoxin production, serum resistance, and/or resistance to antibiotics (Canchaya *et al.*, 2003). The gene layout of a typical prophage can be divided into several segments according to the encoded functions, including Head, Tail, Integration, Ea region, Recombination, Immunity, Replication, Nin region and Lysis. Additionally, a transfer RNA (tRNA) gene adjacent to the prophage is often present to provide the target site (attB) for phage integration, although the tRNA genes may disappear during the phage genome degradation process (e.g. remnants of phages are often not flanked by tRNA genes). The *Salmonella* Typhimurium LT2 genome carries four prophages: the P2-like Fels-2, and the lambda-like phages Fels-1, Gifsy-1 and Gifsy-2. The independent wild type Typhimurium strains SL1344 and 14028s both contain prophages Gifsy-1 and Gifsy-2, but not Fel-1 or Fels-2, and they each have one other prophage, SopEΦ and Gifsy-3, respectively (Frye *et al.*, 2005). Many of the *Salmonella* prophages are intact and inducible to produce free phage particles, which can in turn lysogenize *Salmonella* strains that do not carry the phage.

Similar to prophages, many genomic islands (e.g. SPIs) also occupy relatively large genomic regions ranging from 10–200 kb, and encode genes that contribute to the virulence of the bacterium, including adherence factors, toxins, iron uptake systems, invasion factors and secretion systems

(Schmidt and Hensel, 2004). SPIs are often flanked by direct repeats, insertion sequences or tRNA genes, which are sites for recombination into the bacterial genome. Site specific integration of SPIs is mediated by an integrase or recombinase. SPIs themselves are not able to 'jump' in or out of the bacterial genome independently; their horizontal transfer relies on a 'helper' phage. Before the whole genome sequencing era, five SPIs, i.e. SPIs 1–5, were identified in *Salmonella* Typhimurium; later SPI 6–10 were discovered in Typhi CT18. SPIs 1–5 are conserved in all sequenced *Salmonella* genomes; SPI6 is present also in all *Salmonella* genomes but the sequence is not as conserved in different *Salmonella* genomes as SPIs 1–5; SPI7 is present only in *Salmonella* Typhi, Paratyphi C and some strains of Dublin; SPI8 is present only in Typhi and Paratyphi A; SPI9 is present in all sequenced *Salmonella* genomes; and SPI10 is present in Typhi and Paratyphi A, with remnants being present in Enteritidis and Gallinarum. As of now, up to 18 SPIs have been identified (Vernikos and Parkhill, 2006), and more will likely be found.

Based on the structural characteristics of SPIs and prophages, some computational tools have been developed for identification of these elements, such as IslandPath for SPIs and other genomic islands and PhageFinder for prophages (Fouts, 2006; Hsiao *et al.*, 2003). However, the majority of the prophages are thought to be 'cryptic', i.e. the phage is defective, not being able to form plaques, and is stably integrated into the bacterial genome. Due to the accumulation of inactivating mutations and subsequent deletion events, some prophage remnants are even shorter than 5 kb, therefore the above tools cannot effectively find these short remnants. Some parameters of nucleotide composition, e.g. the GC content, oligonucleotide frequencies or codon usage, of the prophages and SPI elements, are in most cases different from the surrounding genome regions and can therefore often help to identify these laterally acquired genes.

In contrast to the inserted DNA such as prophages and SPIs discussed above, deletions have not been equally well documented. If two genomes are compared, deletion is relative: absence of a DNA segment might be due to deletion of the sequence in one of the two genomes or to insertion of the sequence in the other genome. Therefore, a deletion can be assumed to be real only when many genomes of related organisms are compared and the missing sequence is present in most of the genomes. Common loss of genes in the typhoid agents relative to the host-generalist *Salmonella* lineages such as *Salmonella* Typhimurium might be responsible for the confinement of these different bacteria to their common human host (Liu *et al.*, 2009).

Genomic rearrangements

Genomic rearrangements include inversions, translocations, and duplications (also called amplifications). When a strain of *Salmonella* Enteritidis was mapped in 1993, an 815 kb inversion relative to Typhimurium LT2, covering the terminus region of genomic replication, was found on the genome (Liu *et al.*, 1993b). Before that, a shorter inversion at that region was known in *E. coli* K12 relative to LT2, based on linkage data (Sanderson *et al.*, 1995). In fact, genomic inversion as a phenomenon had been extensively investigated by Hill and colleagues (Hill and Gray, 1988; Hill and Harnish, 1981) and the Roth laboratory (Mahan and Roth, 1988; Segall *et al.*, 1988), however its significance to the bacterial fitness was largely unknown. It was not until 1995 when genomic rearrangements, including large scale inversions and translocations, were associated with an over 100 kb insertion (later characterized as SPI7) in *Salmonella* Typhi Ty2 (Liu and Sanderson, 1995b, d), which forced genomic rearrangements for physical rebalancing between origin *oriC* and terminus *terC* of DNA replication (Liu and Sanderson, 1996).

In terms of overall genome organization and its stability, *Salmonella* lineages may be divided into three categories: (1) genome structure stable and not rearranged, relative to the reference strain *Salmonella* Typhimurium LT2 (Liu and Sanderson, 1992), which is representative of the majority of analysed *Salmonella* genomes (Liu *et al.*, 1993c; Liu and Sanderson, 1995c, 1998; Liu *et al.*, 1999); (2) genome structure rearranged but still stable, as represented by *Salmonella* Paratyphi A (Liu and Sanderson, 1995a); and (3) genome structure rearranged and plastic (unstable), as represented by *Salmonella* Typhi (Liu and Sanderson, 1996). The first category includes most *Salmonella* serotypes (Liu and Sanderson, 1998; Liu *et al.*, 1999); the second category includes only one known case (Paratyphi A has a 100 kb insertion region, which could have significantly disrupted the genome balance, but a 2400 kb inversion rebalanced the genome, making the genome highly stable) (Liu and Sanderson, 1995a; McClelland *et al.*, 2004); and the third category includes many host-adapted *Salmonella* serotypes, such as Typhi (Liu *et al.*, 2006; Liu and Sanderson, 1995d, 1996), Paratyphi C (both adapted to humans) (Liu *et al.*, 2009; Liu *et al.*, 2007), and the Gallinarum biovars Pullorum (Liu *et al.*, 2002) and Gallinarum (both adapted to fowl) (Wu *et al.*, 2005).

Most rearrangement events identified to date in *Salmonella* are mediated by genes encoding ribosomal RNA (*rrn* operons; Liu *et al.*, 1993c; Liu and Sanderson, 1998), some are due to recombination between prophages (Liu *et al.*, 2009) or IS elements (Alokam *et al.*, 2002). All *Salmonella* genomes contain seven *rrn* operons with nearly identical sequences, each being about 6.5 kb in size. An exemplary IS element is IS*200*, a 708 bp insertion sequence that encodes a transposase (Lam and Roth, 1983a, b). The number of IS*200* contained in the genome varies among the *Salmonella* lineages, from zero in Paratyphi C and Choleraesuis to six in Typhimurium and to as many as 25 in Typhi. It is reasonable to postulate that *Salmonella* Typhi harbours so many copies of IS*200* because it requires rebalancing of the genome by rearrangement through many repeat sequences. Paratyphi C also requires rebalancing of the genome by rearrangement, but it achieves this through other repeats like homologous sequences between prophages (Gifsy-1 and Gifsy-2; Liu *et al.*, 2009) in addition to the *rrn* operons. *Salmonella* Abortusovis contains multi-copies of IS1414 (Bacciu *et al.*, 2004), but whether this element mediates genomic rearrangements is not clear.

The importance of tandem duplication (amplification) has been emphasized in work from the Roth laboratory (Roth *et al.*, 1996). Duplications occur at high frequency in *Salmonella* Typhimurium; most segments of the genome are duplicated at frequencies of 10^{-2} to 10^{-5}, but these duplications are unstable and revert to haploid readily under non-selective conditions (Anderson and Roth, 1981). Cells with duplications may be selected during growth limitation due to reduced availability of carbon and energy sources (Sonti and Roth, 1989). Gene amplification, because of its high frequency, may have a major role in evolution of increased antibiotic resistance, especially where bacterial population size is too small for rare point mutations to occur. Evolution of increased β-lactam resistance was detected following exposure of *Salmonella* Typhimurium to cephalosporin (Sun *et al.*, 2009). The initial response to selection was due to very frequent amplification of the $bla_{\mathrm{TEM\text{-}1}}$ gene to copy numbers ranging from 2 to 40, which gave the cells a transient resistance to cephalosporins. This was followed by point mutations in other genes such as *envZ*, *cpxA*, or *nmpC*, which resulted in stable resistance to the antibiotic; the amplification was subsequently lost, but was a crucial step in the evolution of resistance. In addition, evidence has been presented that the phenomenon of 'adaptive mutation' in strains with a mutation in the *lac* operon when grown in selective growth conditions on lactose may not be due to an increase in mutation rate, but to the growth of rare cells with a duplication of the leaky *lac* operon, followed by a change in sequence among the increased copy numbers of the amplified gene (Kugelberg *et al.*, 2006).

Genetic switches to spontaneously regulate bacterial mutability in evolution

We have discussed genomic divergence of *Salmonella* from *E. coli* and further evolution into thousands of distinct lineages through two kinds of genomic evolutionary events, i.e. point mutation and lateral gene transfer (LGT). However, neither kind of events would be assumed to take place readily, because bacteria have powerful mechanisms like the mismatch repair (MMR) system that would strongly inhibit mutation and homologous recombination, preventing gene modification and LGT from happening. MMR maintains genetic stability by correcting DNA biosynthetic errors, participating in responses to

DNA damages and preventing homologous recombination (Harfe and Jinks-Robertson, 2000; Miller, 1998; Rayssiguier *et al.*, 1989; Schofield and Hsieh, 2003). Three of the MMR proteins, MutS, MutL and MutH, coordinate to initiate the mismatch repair process (Modrich and Lahue, 1996). Bacterial strains having defects in MMR are called mutators, which have elevated rates of spontaneous mutation (LeClerc *et al.*, 1996; Sniegowski *et al.*, 1997) and recombination (Rayssiguier *et al.*, 1989). It is thus reasonable to speculate that most genomic changes may take place only when MMR is not functioning. It then follows that once a bacterial cell under stress becomes a mutator and obtains beneficial traits by chance through mutation or LGT, it will become better fit than other cells under the same stress and grow more rapidly. This new line of cells will gradually establish a growing subpopulation, which may eventually take over the whole population. Some bacteria are known to have basal steady-state frequencies of defective MMR (LeClerc *et al.*, 1996; Matic *et al.*, 1997), suggesting that cells with defects in MMR may be important to a bacterial population for adaptation and survival under stressful conditions. However, as mutators tend to accumulate genetic changes, most of which would be deleterious, we need to understand how bacteria would maintain genetic stability and also allow genetic changes to occur, leading to genetically modified organisms.

Over 10,000 mutants derived from a mutator strain, *Salmonella* Typhimurium LT7, had been archived at University of Missouri at room temperature for over four decades. The genomes of a few representatives of these strains were mapped, and extensive genome diversification was observed, i.e. individual LT7-derived mutants kept changing the physical structure of the genome, reflected as frequent inversions, translocations, duplications, and nucleotide substitutions (Liu *et al.*, 2003). The sequences of MMR genes, including *mutL*, *mutS* and *mutH*, were analysed, and all screened Typhimurium LT7 archival mutants had intact *mutS* and *mutH* genes, but harboured an aberrant *mutL*. Their *mutL* version had a deletion of one of three tandem 6 bp repeats GCTGGC GCTGGC GCTGGC; the defective *mutL* allele was designated 6bpΔ*mutL* (Gong *et al.*, 2007). Similar repeats (G CTGGCG CTGGCG CTGGCG) had been reported in *E. coli* (Shaver and Sniegowski, 2003). The sequence (G) CTGGC GCTGGC GCTGGC C in *Salmonella* or (G) CTGGCG CTGGCG CTGGCG in *E. coli* codes for the amino acid sequence LALALA, which lies in a region of *mutL* that forms the lid of the ATP-binding pocket of the MutL protein (Ban *et al.*, 1999; Ban and Yang, 1998; Yang *et al.*, 2000). In the *Salmonella* Typhimurium LT7 mutants, deletion of the 6 bp repeat leads to a deletion of LA in the ATP lid structure of MutL and may thus impair the ATPase activity, eventually hampering the MMR process. As the structure of the three tandem repeats would easily lead to deletion or duplication of the 6 bp string via slipped-strand mispairing (Levinson and Gutman, 1987; Streisinger *et al.*, 1966), it is likely that MMR may transiently cease functioning when the 6 bp repeats decrease or increase copy numbers, and resume functioning when the copy number becomes three again.

The conversion between the wild type and defective *mutL* alleles has been shown to be spontaneous, so either one could be selected to increase in frequency in the bacterial population when it renders the bacteria better fitness. Therefore, *mutL* and other MMR genes may work like genetic switches spontaneously regulating bacterial mutability to maximize the chances of acquiring beneficial genetic traits, and in the meantime minimizing deleterious genetic changes (Gong *et al.*, 2007).

Remaining questions: answers to come from further genomic studies

Salmonella has been a well studied group of bacteria for over a century. Although *Salmonella* infections are essentially under control due to the use of a wide variety of antimicrobials, many questions remain to be answered, including the taxonomic status of individual *Salmonella* lineages. Additionally, questions about whether all typhoid agents use the same mechanisms to cause clinically similar disease and whether the human-adapted salmonellae all became unable to infect any host other than humans due to the same genetic changes remain unanswered and will require further genomic studies.

Salmonella: one or two species? Or thousands of species?

In this section, we will deviate from the current convention of *Salmonella* nomenclature, to illustrate the fluidity and ongoing debate with respect to classification of the different *Salmonella* lineages.

The first *Salmonella* pathogen was isolated in 1881 from a typhoid patient and was given many names by different authors, all meaning 'the bacteria that cause typhoid fever', such as *Typhusbacillen, Typhus abdominalis, Bacterium typhosum*, etc. The name *Salmonella typhi* was first used in 1929 and soon became the most widely used name. More *Salmonella* bacteria were isolated in the following years and many of them were given typical scientific names, each consisting of a genus name and a specific epithet like *Salmonella typhi*, such as *S. enteritidis* (Gaertner 1888), *S. typhimurium* (Loeffler 1892), *S. choleraesuis* (Smith 1894; Popoff and Le Minor, 2005), and so on. However, in the early 1980s, one *Salmonella* species was proposed for all over 2000 recognized serovars, under a new bacterial scientific name, *Salmonella enterica* (Le Minor and Popoff, 1987); later (in 1989), the one single *Salmonella* species, *Salmonella enterica*, was divided into two species, *Salmonella enterica* and *Salmonella bongori* (Reeves *et al.*, 1989). As a result, the two most important and most commonly familiar distinct pathogens *S. typhi* and *S. typhimurium* became members of the same species *Salmonella enterica* and of the same subspecies *Salmonella enterica* subsp. *enterica*, even though they have as much as 12–13% genetic differences (note that the genetic difference between man and chimpanzee is about 1% and that man and chimpanzee belong to different species, genera and families), dwell in different niches, and cause vastly different diseases in humans. Additionally, there seems an obvious inter–species barrier between *S. typhi* and *S. typhimurium* as judged by the drastically reduced frequencies of recombination from *S. typhi* to *S. typhimurium* or vice versa, compared to the frequencies of recombination from one strain to another within *S. typhi* or within *S. typhimurium*, even when the MMR function is inactivated (Zahrt *et al.*, 1994). At present, most scientific journals accept both nomenclature systems for *Salmonella*, reflecting a common opinion that it is not wise to change bacterial names, e.g. from *S. typhi* to *Salmonella enterica* subspecies *enterica* serovar Typhi, without solid scientific or practical reasons. It is therefore possible that many (thousands) individual *Salmonella* lineages may regain their once widely used scientific names (each consisting of the genus name and a specific epithet) to reflect their evolutionary positions in nature, taxonomic ranks in the groups of related bacteria, the host they primarily infect, and the disease they primarily cause.

Typhoid fever: common pathogenesis by different typhoid-causing agents?

Among the over 2500 known *Salmonella* lineages, roughly 1400 infect humans and most of them cause gastroenteritis, a mild self-limiting infection usually without the need of hospitalization or antimicrobial treatment. On the other hand, a few *Salmonella* lineages, such as Typhi, Paratyphi A, Paratyphi B and Paratyphi C, cause typhoid, a serious systemic infection that has high mortality if not properly treated. Many investigators have asked the question whether different typhoid agents use the same pathogenic traits in causing the clinically similar disease. Genomic comparisons among Typhi, Paratyphi A, Paratyphi B and Paratyphi C did not lead to a list of genes that are present only in these four typhoid agents but not in other sequenced *Salmonella* lineages, so it is still an open question.

Adaptation to humans: common genetic basis?

It is reasonable to presume that gene degradation has made the human-adapted *Salmonella* lineages unable to infect any other host species. *Salmonella* Typhi and Paratyphi A are strictly adapted to humans only and both have over 200 pseudogenes, but only a small number of these annotated pseudogenes are common to Typhi, Paratyphi A and Paratyphi C (Feng and Liu, unpublished work). It is possible that some of the shared pseudogenes are responsible for the limitation of these different bacteria to the human body.

Acknowledgements

We thank Nick Thomson of the Wellcome Trust Sanger Institute and Michael McClelland of the Vaccine Research Institute of San Diego for allowing us to analyse sequence data on *S. bongori* and on *S. enterica* subsp. *arizonae*.

References

Alokam, S., Liu, S.L., Said, K., and Sanderson, K.E. (2002). Inversions over the terminus region in *Salmonella* and *Escherichia coli*: IS200s as the sites of homologous recombination inverting the chromosome of *Salmonella enterica* serovar typhi. J. Bacteriol. *184*, 6190–6197.

Anderson, P., and Roth, J. (1981). Spontaneous tandem genetic duplications in *Salmonella* Typhimurium arise by unequal recombination between rRNA (rrn) cistrons. Proc. Natl. Acad. Sci. U.S.A. *78*, 3113–3117.

Andrews-Polymenis, H.L., Santiviago, C.A., and McClelland, M. (2009). Novel genetic tools for studying food-borne *Salmonella*. Curr. Opin. Biotechnol. *20*, 149–157.

Bacciu, D., Falchi, G., Spazziani, A., Bossi, L., Marogna, G., Leori, G.S., Rubino, S., and Uzzau, S. (2004). Transposition of the heat-stable toxin astA gene into a gifsy-2-related prophage of *Salmonella enterica* serovar Abortusovis. J. Bacteriol. *186*, 4568–4574.

Ban, C., Junop, M., and Yang, W. (1999). Transformation of MutL by ATP binding and hydrolysis: a switch in DNA mismatch repair. Cell *97*, 85–97.

Ban, C., and Yang, W. (1998). Crystal structure and ATPase activity of MutL: implications for DNA repair and mutagenesis. Cell *95*, 541–552.

Blattner, F.R., Plunkett, G., 3rd, Bloch, C.A., Perna, N.T., Burland, V., Riley, M., Collado-Vides, J., Glasner, J.D., Rode, C.K., Mayhew, G.F., *et al.* (1997). The complete genome sequence of *Escherichia coli* K-12. Science *277*, 1453–1474.

Canchaya, C., Proux, C., Fournous, G., Bruttin, A., and Brussow, H. (2003). Prophage genomics. Microbiol. Mol. Biol. Rev. *67*, 238–276, table of contents.

Chiu, C.H., Tang, P., Chu, C., Hu, S., Bao, Q., Yu, J., Chou, Y.Y., Wang, H.S., and Lee, Y.S. (2005). The genome sequence of *Salmonella enterica* serovar Choleraesuis, a highly invasive and resistant zoonotic pathogen. Nucleic. Acids. Res. *33*, 1690–1698.

Deng, W., Liou, S.R., Plunkett, G., 3rd, Mayhew, G.F., Rose, D.J., Burland, V., Kodoyianni, V., Schwartz, D.C., and Blattner, F.R. (2003). Comparative genomics of *Salmonella enterica* serovar Typhi strains Ty2 and CT18. J. Bacteriol. *185*, 2330–2337.

Didelot, X., Achtman, M., Parkhill, J., Thomson, N.R., and Falush, D. (2007). A bimodal pattern of relatedness between the *Salmonella* Paratyphi A and Typhi genomes: convergence or divergence by homologous recombination? Genome. Res. *17*, 61–68.

Doolittle, R.F., Feng, D.F., Tsang, S., Cho, G., and Little, E. (1996). Determining divergence times of the major kingdoms of living organisms with a protein clock. Science *271*, 470–477.

Fleischmann, R.D., Adams, M.D., White, O., Clayton, R.A., Kirkness, E.F., Kerlavage, A.R., Bult, C.J., Tomb, J.F., Dougherty, B.A., Merrick, J.M., *et al.* (1995). Whole-genome random sequencing and assembly of *Haemophilus influenzae* Rd. Science *269*, 496–512.

Fouts, D.E. (2006). Phage_Finder: automated identification and classification of prophage regions in complete bacterial genome sequences. Nucleic. Acids. Res. *34*, 5839–5851.

Frye, J.G., Porwollik, S., Blackmer, F., Cheng, P., and McClelland, M. (2005). Host gene expression changes and DNA amplification during temperate phage induction. J. Bacteriol. *187*, 1485–1492.

Gong, J., Liu, W.Q., Liu, G.R., Chen, F., Li, J.Q., Xu, G.M., Wang, L., Johnston, R.N., Eisenstark, A., and Liu, S.L. (2007). Spontaneous conversion between mutL and 6bpDeltamutL in *Salmonella* typhimurium LT7: association with genome diversification and possible roles in bacterial adaptation. Genomics *90*, 542–549.

Harfe, B.D., and Jinks-Robertson, S. (2000). DNA mismatch repair and genetic instability. Annu. Rev. Genet. *34*, 359–399.

Hill, C.W., and Gray, J.A. (1988). Effects of chromosomal inversion on cell fitness in *Escherichia coli* K-12. Genetics *119*, 771–778.

Hill, C.W., and Harnish, B.W. (1981). Inversions between ribosomal RNA genes of *Escherichia coli*. Proc. Natl. Acad. Sci. U.S.A. *78*, 7069–7072.

Holt, K.E., Parkhill, J., Mazzoni, C.J., Roumagnac, P., Weill, F.X., Goodhead, I., Rance, R., Baker, S., Maskell, D.J., Wain, J., *et al.* (2008). High-throughput sequencing provides insights into genome variation and evolution in *Salmonella* Typhi. Nat. Genet. *40*, 987–993.

Holt, K.E., Thomson, N.R., Wain, J., Langridge, G.C., Hasan, R., Bhutta, Z.A., Quail, M.A., Norbertczak, H., Walker, D., Simmonds, M., *et al.* (2009). Pseudogene accumulation in the evolutionary histories of *Salmonella enterica* serovars Paratyphi A and Typhi. BMC Genomics *10*, 36.

Hsiao, W., Wan, I., Jones, S.J., and Brinkman, F.S. (2003). IslandPath: aiding detection of genomic islands in prokaryotes. Bioinformatics *19*, 418–420.

Kauffmann, F., and Edwards, P.R. (1957). A revised, simplified Kauffmann-White schema. Acta. Pathol. Microbiol. Scand. *41*, 242–246.

Kidgell, C., Reichard, U., Wain, J., Linz, B., Torpdahl, M., Dougan, G., and Achtman, M. (2002). *Salmonella* Typhi, the causative agent of typhoid fever, is approximately 50,000 years old. Infect. Genet. Evol. *2*, 39–45.

Kohara, Y., Akiyama, K., and Isono, K. (1987). The physical map of the whole E. coli chromosome: application of a new strategy for rapid analysis and sorting of a large genomic library. Cell *50*, 495–508.

Konstantinidis, K.T., Serres, M.H., Romine, M.F., Rodrigues, J.L., Auchtung, J., McCue, L.A., Lipton,

M.S., Obraztsova, A., Giometti, C.S., Nealson, K.H., *et al.* (2009). Comparative systems biology across an evolutionary gradient within the *Shewanella* genus. Proc. Natl. Acad. Sci. U.S.A. *106,* 15909–15914.

Kothapalli, S., Nair, S., Alokam, S., Pang, T., Khakhria, R., Woodward, D., Johnson, W., Stocker, B.A., Sanderson, K.E., and Liu, S.L. (2005). Diversity of Genome Structure in *Salmonella enterica* Serovar Typhi Populations. J. Bacteriol. *187,* 2638–2650.

Kugelberg, E., Kofoid, E., Reams, A.B., Andersson, D.I., and Roth, J.R. (2006). Multiple pathways of selected gene amplification during adaptive mutation. Proc. Natl. Acad. Sci. U.S.A. *103,* 17319–17324.

Lam, S., and Roth, J.R. (1983a). Genetic mapping of IS200 copies in *Salmonella* Typhimurim strain LT2. Genetics *105,* 801–811.

Lam, S., and Roth, J.R. (1983b). IS200: a *Salmonella*-specific insertion sequence. Cell *34,* 951–960.

Lapierre, P., and Gogarten, J.P. (2009). Estimating the size of the bacterial pan-genome. Trends Genet. *25,* 107–110.

Le Minor, L., and Popoff, M.Y. (1987). Designation of *Salmonella enterica* sp. nov., nom. rev., as the type and only species of the genus *Salmonella.* Int. J. Syst. Bacteriol. *37,* 465–468.

LeClerc, J.E., Li, B., Payne, W.L., and Cebula, T.A. (1996). High mutation frequencies among *Escherichia coli* and *Salmonella* pathogens. Science *274,* 1208–1211.

Levinson, G., and Gutman, G.A. (1987). Slipped-strand mispairing: a major mechanism for DNA sequence evolution. Mol. Biol. Evol. *4,* 203–221.

Liu, G.R., Edwards, K., Eisenstark, A., Fu, Y.M., Liu, W.Q., Sanderson, K.E., Johnston, R.N., and Liu, S.L. (2003). Genomic diversification among archival strains of *Salmonella enterica* serovar typhimurium LT7. J. Bacteriol. *185,* 2131–2142.

Liu, G.R., Liu, W.Q., Johnston, R.N., Sanderson, K.E., Li, S.X., and Liu, S.L. (2006). Genome plasticity and ori-ter rebalancing in *Salmonella* typhi. Mol. Biol. Evol. *23,* 365–371.

Liu, G.R., Rahn, A., Liu, W.Q., Sanderson, K.E., Johnston, R.N., and Liu, S.L. (2002). The evolving genome of *Salmonella enterica* serovar Pullorum. J. Bacteriol. *184,* 2626–2633.

Liu, S.L., Hessel, A., and Sanderson, K.E. (1993a). Genomic mapping with I-Ceu I, an intron-encoded endonuclease specific for genes for ribosomal RNA, in *Salmonella* spp., *Escherichia coli,* and other bacteria. Proc. Natl. Acad. Sci. U.S.A. *90,* 6874–6878.

Liu, S.L., Hessel, A., and Sanderson, K.E. (1993b). The *XbaI-BlnI-CeuI* genomic cleavage map of *Salmonella enteritidis* shows an inversion relative to *Salmonella typhimurium* LT2. Mol. Microbiol. *10,* 655–664.

Liu, S.L., Hessel, A., and Sanderson, K.E. (1993c). The *XbaI-BlnI-CeuI* genomic cleavage map of *Salmonella typhimurium* LT2 determined by double digestion, end labelling, and pulsed-field gel electrophoresis. J. Bacteriol. *175,* 4104–4120.

Liu, S.L., and Sanderson, K.E. (1992). A physical map of the *Salmonella typhimurium* LT2 genome made by using *Xba*I analysis. J. Bacteriol. *174,* 1662–1672.

Liu, S.L., and Sanderson, K.E. (1995a). The chromosome of *Salmonella paratyphi A* is inverted by recombination between *rrnH* and *rrnG.* J. Bacteriol. *177,* 6585–6592.

Liu, S.L., and Sanderson, K.E. (1995b). Genomic cleavage map of *Salmonella* typhi Ty2. J. Bacteriol. *177,* 5099–5107.

Liu, S.L., and Sanderson, K.E. (1995c). I-CeuI reveals conservation of the genome of independent strains of *Salmonella* typhimurium. J. Bacteriol. *177,* 3355–3357.

Liu, S.L., and Sanderson, K.E. (1995d). Rearrangements in the genome of the bacterium *Salmonella* typhi. Proc. Natl. Acad. Sci. U.S.A. *92,* 1018–1022.

Liu, S.L., and Sanderson, K.E. (1996). Highly plastic chromosomal organization in *Salmonella* typhi. Proc. Natl. Acad. Sci. U.S.A. *93,* 10303–10308.

Liu, S.L., and Sanderson, K.E. (1998). Homologous recombination between rrn operons rearranges the chromosome in host-specialized species of *Salmonella.* FEMS Microbiol. Lett. *164,* 275–281.

Liu, S.L., Schryvers, A.B., Sanderson, K.E., and Johnston, R.N. (1999). Bacterial phylogenetic clusters revealed by genome structure. J. Bacteriol. *181,* 6747–6755.

Liu, W.Q., Feng, Y., Wang, Y., Zou, Q.H., Chen, F., Guo, J.T., Peng, Y.H., Jin, Y., Li, Y.G., Hu, S.N., *et al.* (2009). *Salmonella* paratyphi C: genetic divergence from *Salmonella choleraesuis* and pathogenic convergence with Salmonella typhi. PLoS ONE *4,* e4510.

Liu, W.Q., Liu, G.R., Li, J.Q., Xu, G.M., Qi, D., He, X.Y., Deng, J., Zhang, F.M., Johnston, R.N., and Liu, S.L. (2007). Diverse genome structures of *Salmonella* paratyphi C. BMC Genomics *8.*

Mahan, M.J., and Roth, J.R. (1988). Reciprocality of recombination events that rearrange the chromosome. Genetics *120,* 23–35.

Matic, I., Radman, M., Taddei, F., Picard, B., Doit, C., Bingen, E., Denamur, E., and Elion, J. (1997). Highly variable mutation rates in commensal and pathogenic *Escherichia coli.* Science *277,* 1833–1834.

McClelland, M., Sanderson, K.E., Clifton, S.W., Latreille, P., Porwollik, S., Sabo, A., Meyer, R., Bieri, T., Ozersky, P., McLellan, M., *et al.* (2004). Comparison of genome degradation in Paratyphi A and Typhi, human-restricted serovars of *Salmonella enterica* that cause typhoid. Nat. Genet. *36,* 1268–1274.

McClelland, M., Sanderson, K.E., Spieth, J., Clifton, S.W., Latreille, P., Courtney, L., Porwollik, S., Ali, J., Dante, M., Du, F., *et al.* (2001). Complete genome sequence of *Salmonella enterica* serovar Typhimurium LT2. Nature *413,* 852–856.

Medini, D., Donati, C., Tettelin, H., Masignani, V., and Rappuoli, R. (2005). The microbial pan-genome. Curr. Opin. Genet. Dev. *15,* 589–594.

Miller, J.H. (1998). Mutators in *Escherichia coli.* Mutat. Res. *409,* 99–106.

Modrich, P., and Lahue, R. (1996). Mismatch repair in replication fidelity, genetic recombination, and cancer biology. Annu. Rev. Biochem. *65,* 101–133.

Nair, S., Alokam, S., Kothapalli, S., Porwollik, S., Proctor, E., Choy, C., McClelland, M., Liu, S.L., and Sanderson, K.E. (2004). *Salmonella enterica* serovar Typhi strains from which SPI7, a 134-kilobase island with genes for

Vi exopolysaccharide and other functions, has been deleted. J. Bacteriol. *186*, 3214–3223.
Ochman, H., and Wilson, A.C. (1987). Evolution in bacteria: evidence for a universal substitution rate in cellular genomes. J. Mol. Evol. *26*, 74–86.
Parkhill, J., Dougan, G., James, K.D., Thomson, N.R., Pickard, D., Wain, J., Churcher, C., Mungall, K.L., Bentley, S.D., Holden, M.T., *et al.* (2001). Complete genome sequence of a multiple drug resistant *Salmonella enterica* serovar Typhi CT18. Nature *413*, 848–852.
Perna, N.T., Plunkett, G., 3rd, Burland, V., Mau, B., Glasner, J.D., Rose, D.J., Mayhew, G.F., Evans, P.S., Gregor, J., Kirkpatrick, H.A., *et al.* (2001). Genome sequence of enterohaemorrhagic *Escherichia coli* O157:H7. Nature *409*, 529–533.
Popoff, M.Y., Bockemuhl, J., and Gheesling, L.L. (2004). Supplement 2002 (no. 46) to the Kauffmann-White scheme. Res. Microbiol. *155*, 568–570.
Popoff, M.Y., and Le Minor, L.E. (2005). Genus XXXIII. Salmonella. In Bergey's Mannual of Systematic Bacteriology, Brenner, D.J., Krieg, N.R., and Stanley, J.T., eds (Springer), pp. 764–799.
Rasko, D.A., Rosovitz, M.J., Myers, G.S., Mongodin, E.F., Fricke, W.F., Gajer, P., Crabtree, J., Sebaihia, M., Thomson, N.R., Chaudhuri, R., *et al.* (2008). The pangenome structure of *Escherichia coli*: comparative genomic analysis of *E. coli* commensal and pathogenic isolates. J. Bacteriol. *190*, 6881–6893.
Rayssiguier, C., Thaler, D.S., and Radman, M. (1989). The barrier to recombination between *Escherichia coli* and *Salmonella* typhimurium is disrupted in mismatch-repair mutants. Nature *342*, 396–401.
Reeves, M.W., Evins, G.M., Heiba, A.A., Plikaytis, B.D., and Farmer, J.J., 3rd (1989). Clonal nature of *Salmonella* typhi and its genetic relatedness to other salmonellae as shown by multilocus enzyme electrophoresis, and proposal of *Salmonella bongori* comb. nov. J. Clin. Microbiol. *27*, 313–320.
Roth, J.R., Benson, N., Galitski, T., Haack, K., Lawrence, J.G., and Miesel, L. (1996). Rearrangements of the bacterial chromosome: formation and applications. In *Escherichia coli* and *Salmonella*: cellular and molecular biology, Neidhardt, F.C., ed. (ASM Press), pp. 2256–2276.
Roumagnac, P., Weill, F.-X., Dolecek, C., Baker, S., Brisse, S., Chinh, N.T., Le, T.A.H., Acosta, C.J., Farrar, J., Dougan, G., *et al.* (2006). Evolutionary History of *Salmonella* Typhi. Science *314*, 1301–1304.
Sanderson, K.E., and Demerec, M. (1965). The Linkage Map of *Salmonella* Typhimurium. Genetics *51*, 897–913.
Sanderson, K.E., Hessel, A., and Rudd, K.E. (1995). Genetic map of *Salmonella* typhimurium, edition VIII. Microbiol. Rev. *59*, 241–303.
Sanderson, K.E., and Roth, J.R. (1988). Linkage map of *Salmonella* typhimurium, edition VII. Microbiol. Rev. *52*, 485–532.
Schmidt, H., and Hensel, M. (2004). Pathogenicity islands in bacterial pathogenesis. Clin. Microbiol. Rev. *17*, 14–56.
Schofield, M.J., and Hsieh, P. (2003). DNA mismatch repair: molecular mechanisms and biological function. Annu. Rev. Microbiol. *57*, 579–608.
Segall, A., Mahan, M.J., and Roth, J.R. (1988). Rearrangement of the bacterial chromosome: forbidden inversions. Science *241*, 1314–1318.
Shaver, A.C., and Sniegowski, P.D. (2003). Spontaneously arising mutL mutators in evolving *Escherichia coli* populations are the result of changes in repeat length. J. Bacteriol. *185*, 6076–6082.
Smith, C.L., Econome, J.G., Schutt, A., Klco, S., and Cantor, C.R. (1987). A physical map of the *Escherichia coli* K12 genome. Science *236*, 1448–1453.
Sniegowski, P.D., Gerrish, P.J., and Lenski, R.E. (1997). Evolution of high mutation rates in experimental populations of *E. coli*. Nature *387*, 703–705.
Sonti, R.V., and Roth, J.R. (1989). Role of gene duplications in the adaptation of *Salmonella* typhimurium to growth on limiting carbon sources. Genetics *123*, 19–28.
Streisinger, G., Okada, Y., Emrich, J., Newton, J., Tsugita, A., Terzaghi, E., and Inouye, M. (1966). Frameshift mutations and the genetic code. This paper is dedicated to Professor Theodosius Dobzhansky on the occasion of his 66th birthday. Cold Spring Harb. Symp. Quant. Biol. *31*, 77–84.
Sun, S., Berg, O.G., Roth, J.R., and Andersson, D.I. (2009). Contribution of gene amplification to evolution of increased antibiotic resistance in *Salmonella* typhimurium. Genetics *182*, 1183–1195.
Thomson, N.R., Clayton, D.J., Windhorst, D., Vernikos, G., Davidson, S., Churcher, C., Quail, M.A., Stevens, M., Jones, M.A., Watson, M., *et al.* (2008). Comparative genome analysis of *Salmonella* Enteritidis PT4 and *Salmonella* Gallinarum 287/91 provides insights into evolutionary and host adaptation pathways. Genome. Res. *18*, 1624–1637.
Vernikas, G.S., and Parkhill, J. (2006). Interpolated variable order motifs for identification of horizontally acquired DNA: revisiting the *Salmonella* pathogenicity islands. Bioinformatics *22*, 2196–2203.
Wollman, E.L., Jacob, F., and Hayes, W. (1956). Conjugation and genetic recombination in *Escherichia coli* K-12. Cold Spring Harb. Symp. Quant. Biol. *21*, 141–162.
Wong, K.K., and McClelland, M. (1992). A BlnI restriction map of the *Salmonella* typhimurium LT2 genome. J. Bacteriol. *174*, 1656–1661.
Wu, K.Y., Liu, G.R., Liu, W.Q., Wang, A.Q., Zhan, S., Sanderson, K.E., Johnston, R.N., and Liu, S.L. (2005). The genome of *Salmonella enterica* serovar gallinarum: distinct insertions/deletions and rare rearrangements. J. Bacteriol. *187*, 4720–4727.
Yang, W., Junop, M.S., Ban, C., Obmolova, G., and Hsieh, P. (2000). DNA mismatch repair: from structure to mechanism. Cold Spring Harb. Symp. Quant. Biol. *65*, 225–232.
Zahrt, T.C., Mora, G.C., and Maloy, S. (1994). Inactivation of mismatch repair overcomes the barrier to transduction between *Salmonella* typhimurium and *Salmonella* typhi. J. Bacteriol. *176*, 1527–1529.

High-throughput Screening to Determine the Genetic Requirements for *Salmonella* Survival Under Different Growth Conditions

4

Mollie Megan Reynolds, Rocio Canals, Michael McClelland and Helene Andrews-Polymenis*

Abstract

Salmonella species are capable of survival in a wide range of niches, both in the environment and in an infected host. Genetic requirements for survival of *Salmonella* in different niches have traditionally been identified using gene expression and forward genetics. The availability of complete genome sequences, microarray technology, and cost-effective new sequencing capabilities enabled increasingly efficient high-throughput analyses of *Salmonella* genomes to identify elements that contribute to survival in these niches. In this chapter we describe many of the high-throughput tools that have been developed over the past two decades, and the genetic requirements for *Salmonella* survival that have been identified using these techniques.

Introduction

Salmonella are responsible for millions of cases of food-borne illness and typhoid fever worldwide, despite efforts to reduce transmission of these organisms (Pang *et al.*, 1995). The majority of cases of salmonellosis in the United States is the result of food-borne infection with broad host range serotypes Typhimurium and Enteritidis (Altekruse *et al.*, 1997; Mishu *et al.*, 1994; Rodrique *et al.*, 1990). *Salmonella* are able to survive in numerous niches, including in diverse animal hosts, on vegetables and fruits, and in the environment for livestock or vegetable production (farm, water, soil, etc.), an ability that has made transmission of this organism difficult to control.

Forward genetics combined with technologies developed in the wake of genomic sequencing are revolutionizing the identification of genes required for *Salmonella* survival in many environments. Identification of *Salmonella* genes necessary for growth in different niches has made rapid progress in recent years, as the molecular biological tools for their identification become more sophisticated and are applied to study this organism in an increasing number of environments. In this chapter, we detail the high throughput methods used to identify *Salmonella* genes required in various niches (summarized in Table 4.1).

Genomic sequencing

Comprehensive understanding of the *Salmonella* genome and its biology begins with sequencing. The first complete annotated *Salmonella* genome sequences were published in 2001 (McClelland *et al.*, 2001; Parkhill *et al.*, 2001), and recent reductions in cost along with improvements in sequencing technology have made it possible to do complete genomic content analysis by rapidly sequencing many *Salmonella* genomes (Holt *et al.*, 2008). The current status of genome sequencing of *Salmonellae* has recently been reviewed (Andrews-Polymenis *et al.*, 2009b). Over 100 genomes of *Salmonella* have been sequenced to date, including 23 genomes representing 15 different subspecies I serovars, one subspecies IIIa isolate and one *S. bongori* reported as completed. In addition to the annotated open reading frames, other features of these genomes continue to be identified, improving the annotation of these

*Corresponding author

Table 4.1 High throughput screens used to identify *Salmonella* genes necessary for survival, growth and virulence

Technique	Abbreviation	Description
In vivo expression technology	IVET	Identification of *Salmonella* promoters induced during infection using a reporter-based construct such as *purA-lacZY* or resistances such as Cm^R (*cat*)
Signature-tagged mutagenesis	STM	Negative selection strategy that utilizes a library of transposon mutants with unique barcode-like tags, enabling simultaneous infection and tracking of mutants in small pools during infection
Recombinase-based *in vivo* expression technology	RIVET	Identification of *Salmonella* promoters induced during infection using recombinase activity as a reporter
Differential fluorescence induction	DFI	Identification of *Salmonella* promoters induced during infection using GFP as a reporter
Transcriptional profiling by microarray		Identification of *Salmonella* transcripts that are differentially regulated by comparison of transcript levels under two different conditions
Selective capture of transcribed sequences	SCOTS	Hybridization and pull-down of *Salmonella* transcripts produced preferentially in infected cells or tissue
High-throughput sequencing	HTS	Methods using high throughput sequencing technology such as Illumina GS
2D gel analysis/mass spectrometry		Comparative high-throughput proteomic analysis
Chromatin immunoprecipitation	ChIP	Cross-linking of proteins to DNA *in vivo*, and capture of DNA–protein complexes: characterization of DNA targets of protein binding
Transposon-associated site hybridization	TraSH	Negative selection strategy using outward facing T7 promoter to generate a unique transcript from each mutant in a pool of mutants; abundance of these transcripts from two conditions is compared
Transposon-mediated differential hybridization	TMDH	Negative selection strategy utilizing transposon mutants containing outward facing T7 and SP6 promoters; abundance of each transcript is assessed using a tiling array
Transposon-directed insertion site sequencing	TraDIS	High-throughput sequencing for the identification of transposon insertion sites, and transposon mutants under selection in a given condition
Array-based analysis of cistrons under selection	ABACUS	High-throughput negative selection strategy using a library of targeted deletions

genomes, including small regulatory RNAs and very small membrane proteins (Alix and Blanc-Potard, 2009; Sittka *et al.*, 2009; see also Chapter 7 of this book).

The availability of complete genome sequences of *Salmonellae* has allowed the development of microarrays, and comparative genomic, transcriptomic, and forward genetic studies using these tools (Boyd *et al.*, 2003; Chan *et al.*, 2005; Lawley *et al.*, 2006; McClelland *et al.*, 2004; Porwollik and McClelland, 2007; Porwollik *et al.*, 2005; Porwollik, 2002; Reen *et al.*, 2005). Microarray technology has also recently been described as a method for the rapid molecular typing of *Salmonella* isolates (Andreoli, 2004; Scaria *et al.*, 2008; Wattiau *et al.*, 2008b, see also Chapter 1 of this book). One study using PremiTest, a commercial low density array, yielded results for 94.7% of samples tested over a nine month period, whereas traditional serotyping yielded results for 90.8% (Wattiau *et al.*, 2008a).

Sequencing and annotation are only the beginning of our understanding of the *Salmonella* genome, as in many cases the functions of particular genes must be determined experimentally. In the Typhimurium LT2 sequence, there are nearly 1000 genes out of 4500 that have not been assigned a function and are annotated only with a prediction of their subcellular location (McClelland *et al.*, 2001). The case for other genomes is similar. Typhi, for example, has approximately 200 hypothetical and 600 conserved hypothetical open reading frames (Parkhill *et al.*, 2001). It is vital to our understanding of virulence mechanisms and pathogenesis to assign well-characterized functions to all genes in the *Salmonella* genome.

Differences in genetic content between serovars

Genetic content, along with what is known about the niche preferences of different serotypes, may also provide information about the ability to survive within different hosts or environments. Comparison of the Typhimurium LT2 sequence with sequences of closely related serovars, such as Typhi, Enteritidis, Gallinarum, Choleraesuis and Dublin, has revealed interesting differences between serovars that infect a broad range of hosts and those that are more specialized and infect only specific hosts. Host specialist genomes, including Typhi, Choleraesuis and Gallinarum, typically contain a large number of pseudogenes (Typhi has approximately 200 (Parkhill *et al.*, 2001), Choleraesuis has 151 (Chiu *et al.*, 2005), and Gallinarum has 309 (Thomson *et al.*, 2008)), while the host generalist Typhimurium LT2 only has 39 (McClelland *et al.*, 2001). Typhimurium also has a large array of fimbrial operons, encoding up to 12, which may be necessary for survival within different hosts (McClelland *et al.*, 2001; Parkhill *et al.*, 2001; Reen *et al.*, 2005), whereas host specialized strains, such as Dublin and Typhi, often encode a much smaller repertoire of fimbria. In contrast, some *Salmonella* Pathogenicity Islands (SPI), such as SPI-1 and SPI-2, that encode systems to inject effector proteins into host cells and allow internalization, growth and survival within subcellular niches, appear to be rather stable across the serovars.

Comparative genomics also revealed that the genomes of strains within a serovar can differ more from each other than genomes from strains in different serovars (Porwollik *et al.*, 2004). These distinct and dramatically different genomic variants within serovars have been referred to as 'genovars' and might cause very different phenotypes for members of the same serovar.

Identification of genetic requirements across environments

In order to identify genes important in particular niches, two broad approaches have been employed. First, promoter-trapping strategies and microarray-based approaches have been used to evaluate gene expression under particular sets of conditions. Such studies allow a more comprehensive determination of gene expression changes that can be used to outline and analyse pathways necessary for a particular process. Second, forward genetic screens, and now functional genomics, are being used to identify genes required for survival and growth of *Salmonella* in particular niches. The generation of mutants using various methods, and the study of these mutants in various environments, remains a valid method for identification of necessary genes and elucidation of their function in the environment under study.

Identification of genetic requirements for growth and survival of *Salmonella* using gene expression

In vivo *gene expression*

Promoter trap strategies were first developed in the 1980s (Osbourn *et al.*, 1987), and these approaches enabled the identification of promoters that are up-regulated in particular environments, including infected tissue.

***In vivo* expression technology (IVET)** was originally described in 1993 as a method for identifying promoters specifically induced during *Salmonella* infection (*in vivo* induced genes, or *ivi*) (Mahan, 1993). In the original IVET screen, Typhimurium genomic DNA was cloned upstream of a promoterless *purA–lacZY* fusion in single copy on the bacterial chromosome, in a *purA* mutant background (Mahan, 1993). This approach identified genes involved in *de novo* pyrimidine synthesis (*carAB*) and LPS synthesis (*rfb*) as necessary during infection of BALB/c mice (Mahan, 1993). IVET was further modified to use an antibiotic resistance reporter, *cat*, instead of *purA* (Mahan *et al.*, 1995). Using this method, a gene required for fatty acid degradation, *fadB*, was identified, and expression of the promoter fusion was confirmed in the presence of oleate (Mahan *et al.*, 1995). This IVET strategy identified five *Salmonella* genes induced in RAW 264.7 macrophages (Mahan *et al.*, 1995). While these five specific fusions were not characterized further, in later studies they were shown to be PhoPQ regulated and/or iron responsive genes (Heithoff *et al.*, 1997; Heithoff *et al.*, 1999).

Genes required for global gene regulation (including *phoP, pmrB, cadC* and *vacBC*), during stationary phase (*spvB, cfa, otsA*), for metabolism (including *recD, hemA, entF, fhuA, cirA, mgtAB,* and *ndk*), and for adhesion and invasion were induced during infection in a large-scale IVET screen (Heithoff *et al.*, 1997).

IVET has also been used to identify promoters that are induced during infection of cultured murine hepatocytes (Janakiraman and Slauch, 2000). Genes identified in this screen included known virulence factors *sodA, pagJ* and *ssaE*, as well as an iron transport operon, *sitABCD* (Janakiraman and Slauch, 2000). The *sit* operon is located on *Salmonella* SPI-1 and was shown to be required for full virulence in BALB/c and C3H/HeN mice infected by the oral and intraperitoneal routes of infection, but it is not required for initial invasion of Henle 407 epithelial cells or hepatocytes (Janakiraman and Slauch, 2000).

Recombinase-based IVET (RIVET) (Camilli *et al.*, 1994) is a modified version of IVET that uses a highly sensitive recombinase method. A library of promoter fusions is generated upstream of a promoterless resolvase, *tnpR*, that induces excision of an antibiotic resistance cassette when it is expressed. The resulting loss of antibiotic resistance is a permanent marker of induction that occurred during the infection, enabling the identification of transient events. In *Salmonella* Typhimurium, RIVET was recently used to identify promoters expressed during infection in swine (Huang *et al.*, 2007). Genomic DNA fragments were cloned upstream of a promoterless Cre-LacZ cassette and the resulting library of plasmids was transformed into a strain harbouring a chromosomal cassette containing genes for sucrose sensitivity (*sacB*) and kanamycin resistance flanked by *loxP* sites. When genetic elements upstream of Cre are activated, the Cre recombinase is expressed and both the *sacB* and the kanamycin resistance genes are excised, resulting in sucrose resistance and kanamycin sensitivity. In conjunction with a modified signature tagged mutagenesis strategy (described later in this chapter), this method was used to identify 32 unique promoters that are induced during infection in swine (Huang *et al.*, 2007). Huang *et al.* also found expression of regions that were cloned in a reverse orientation, suggesting their function to be anti-sense regulators (Huang *et al.*, 2007). The recombinase plasmid used in this study encodes a promoterless *lacZ* reporter downstream of Cre, which was used to measure the level of induction of certain promoters by β-galactosidase activity under specific environmental conditions, such as high temperature and osmotic stress (Huang *et al.*, 2007).

Differential fluorescence induction (DFI) is an alternative promoter trap strategy that uses GFP as a reporter along with fluorescence-activated cell sorting (FACS) analysis to find promoters that are active during infection, independent from antibiotic susceptibility or metabolic requirements (Bumann, 2002; Valdivia and Falkow, 1996; Valdivia and Falkow, 1997). A library of *Salmonella* genomic DNA fragments is cloned upstream of a promoterless GFP, and expression of GFP can be examined under a variety of conditions. Initially, DFI was used to identify promoters induced upon exposure to acidic environments which might mimic the environment in the stomach, within macrophages and in intracellular vacuoles (Valdivia and Falkow, 1996). Promoters that were induced upon exposure to low pH, but repressed when exposed to high pH were termed acid inducible genes and included the promoter for cyclopropane fatty acyl phospholipid synthase (*cfa*) (Valdivia and Falkow, 1996), a protein necessary for modification of the *Salmonella* membrane in response to stress (Kim *et al.*, 2005). Promoters induced when *Salmonella* was associated with RAW 264.7 macrophages (Valdivia and Falkow, 1997) included those governing PhoPQ regulated and OmpR/EnvZ regulated genes, and members of SPI-2 (Valdivia and Falkow, 1997). Three expressed promoters were identified during oral infection of BALB/c mice. These promoters regulate genes that were previously known to be involved in virulence or biochemical pathways, and were identified using two-colour flow cytometry DFI, a method developed to eliminate confounding host cell autofluorescence (Bumann, 2002). The three promoters were those for *aroQ*, a chorismate mutase gene necessary for aromatic amino acid biosynthesis, *pipB* and *sifA*, both effectors translocated through the type three secretion system (TTSS) encoded by SPI-2 (Knodler *et al.*, 2002; Pfeifer *et al.*, 1999; Stein *et al.*, 1996).

The flow cytometry approach has also been used to identify *Salmonella* promoters that are up-regulated in tumours in mice, a tissue for which *Salmonella* has a specific tropism (Arrach *et al.*, 2008). Nude mice were infected with a library of 180,000 promoter–trap fusions to otherwise promoterless GFP, and clones were recovered from tumours and normal spleen tissues and sorted by FACS for GFP expression. Clones expressing GFP *in vitro* were removed, and the remaining clones were subjected to rounds of enrichment for tumour specific promoters. Promoters of genes required under hypoxic conditions, including *pflE* (pyruvate-formate-lyase activating enzyme), *flhB* (flagellar apparatus) and *ansB* (asparaginase), were identified in this study as specifically induced in tumour tissue (Arrach *et al.*, 2008). The identification of tumour specific promoters will be useful for targeting chemotherapeutic agents directly to tumours.

Transcriptional profiling

Transcriptional profiling by microarray is another high-throughput method to determine genes that have altered expression under a specific set of conditions. This method has been used to analyse global transcript changes in *Salmonella* in regulatory mutants, during infection, and during *in vitro* growth under infection relevant conditions, such as bile (reviewed in Thompson *et al.*, 2006). Identification of *Salmonella* genes expressed during infection presents a unique set of challenges because bacterial RNA represents a small minority of the RNA in such environments. Nevertheless, enriched bacterial RNA was used to obtain a global transcriptional profile of bacterial genes from bacteria grown in cultured macrophages (J774 cells) using microarray analysis (Eriksson *et al.*, 2003). This work showed that approximately one quarter of the Typhimurium genome (919 genes) displayed altered expression inside macrophages (Eriksson *et al.*, 2003). *Salmonella* Pathogenicity islands, along with other genes necessary for survival within the *Salmonella*-containing vacuole (SCV), were up-regulated, while many gene products expressed on the bacterial surface, including LPS, fimbria and flagellar genes, were down-regulated. Nearly half of the genes with altered expression inside macrophages are of unknown function. Two regions that display structural features of pathogenicity islands, *STM0854–0859* and *STM3117–3120*, were identified as expressed in macrophages (Eriksson *et al.*, 2003).

Transcriptional profiles of Typhimurium have also been determined in cultured human epithelial cells (HeLa cells), and compared to profiles in J774 murine macrophages (Hautefort *et al.*, 2008). While many genes show a similar pattern of expression in both cell types, including 128 genes of unknown function, differences in the transcriptome are also prominent. In epithelial cells, SPI-1 (encoding TTSS-1), which is necessary for invasion of normally non-phagocytic cells, was expressed throughout the infection. But in macrophages, a phagocytic cell type, the need for TTSS-1 appears to be less critical as SPI-1 transcription was down regulated (Hautefort *et al.*, 2008). While SPI-2 genes (encoding TTSS-2) were expressed in both cell types, they appear to be more highly transcribed in macrophages compared to epithelial cells (Hautefort *et al.*, 2008). In contrast, flagellar genes were up-regulated at later time points during infection of HeLa cells, which may indicate that motility at this stage is a necessary bacterial adaptation to prepare for lysis from dying epithelial cells (Hautefort *et al.*, 2008).

Selective capture of transcribed sequences (SCOTS) is a subtractive hybridization method used for preferentially isolating bacterial RNA from infected cells (Morrow *et al.*, 1999). In this method, *Salmonella* genomic DNA is first prepared and biotinylated. Eukaryotic cells are then infected with *Salmonella,* total RNA is isolated and cDNA is prepared. *Salmonella* cDNA from infected cells that hybridizes to the biotinylated *Salmonella* genomic DNA is subsequently isolated using streptavidin, enriching for the bacterial sequences transcribed during infection. Transcriptional differences between broad host and host-adapted *Salmonellae* (Typhimurium versus Typhi) have been examined using SCOTS. These studies identified a fimbrial operon (*stfACDEFG*) and a LysR type regulator *stmR* (*STM3121*), present in Typhimurium but not in Typhi, that may be necessary for survival in a wider range of hosts (Morrow *et al.*, 1999).

Recent studies using SCOTS have identified genes expressed in macrophages, utilizing microarray technology for higher throughput. Typhi genes that were expressed during infection of human monocytes (THP-1 cells) were identified using this combination of methods (Faucher *et al.*, 2006). The expression of approximately 117 genes, including genes of unknown function, was elevated throughout infection (Faucher *et al.*, 2006). The TTSSs were up-regulated early in infection. While expression of genes encoded on SPI-2 remained high throughout the infection, SPI-1 gene expression fell early in infection. Genes encoding proteins involved in antimicrobial peptide resistance were expressed at high levels at multiple times during infection, and genes involved in motility and iron transport were repressed throughout Typhi infection of macrophages.

Differences observed between this Typhi study and the results of earlier Typhimurium studies are perhaps a result of the host specificity of the infecting serovar (Eriksson *et al.*, 2003; Faucher *et al.*, 2006). For example, *phoP*, *slyA* and antimicrobial resistance genes appear to be differentially regulated when serotype Typhimurium is intracellular (Eriksson *et al.*, 2003; Hautefort *et al.*, 2008), while this is not the case for serotype Typhi (Faucher *et al.*, 2006). Such comparative analyses may yield further clues to the evolutionary basis of host preference.

The most recent methods for identification of differentially regulated transcripts use **high-throughput sequencing technology** (Croucher *et al.*, 2009; Perkins *et al.*, 2009). Illumina sequencing technology has been used as a tool to study the complete transcriptome of *Salmonella bongori,* among other prokaryotes, in a quantitative manner, using the proportionality between the number of sequencing reads and the level of transcription (Croucher *et al.*, 2009). This method uses single stranded cDNA and therefore maintains directionality of the transcript (Croucher *et al.*, 2009). Illumina sequencing has also been used to determine the transcriptome of serotype Typhi, using single stranded cDNA (Perkins *et al.*, 2009). This method has enabled the identification of novel non-coding sRNAs and resolved transcripts from overlapping DNA sequences (Perkins *et al.*, 2009). The authors also used Illumina sequencing for a comparative study between the parent Typhi strain and a Δ*ompR* strain in order to identify genes in the OmpR regulon (Perkins *et al.*, 2009). Results were compared to results from a parallel study using microarray technology. The authors found the technologies to be complementary to each other, but not necessarily overlapping (Perkins *et al.*, 2009).

Deep sequencing technology has also been employed to identify novel small regulatory RNAs (sRNA) and their regulatory targets (Sittka *et al.*, 2008; Sittka *et al.*, 2009). Using tagged Hfq protein, cDNAs that were targeted by this post-transcriptional regulator were identified in a pull down strategy (Sittka *et al.*, 2008). These cDNAs were both sequenced and studied by microarray analysis. To complement this data, a comparative microarray analysis was performed using transcripts from wild type and Δ*hfq* strains. In addition to type III secretion and flagellar biosynthesis machineries, several sRNAs were identified as targets of Hfq (Sittka *et al.*, 2008). The conservation of Hfq function was examined by expressing the Hfq protein of distant relatives in *Salmonella* in a related study

(Sittka *et al.*, 2009). Small RNAs, including novel ones discovered during this study, were identified as the major target of Hfq regulation (Sittka *et al.*, 2009). Deep sequencing technology will come into widespread use as its cost continues to fall, and the number and variability of its applications rises.

Identification of genetic requirements for growth and survival of Salmonella *using proteomics*

Expression at the protein level, if precisely quantified, is probably a more accurate reflection of the state of the cell than mRNA levels. One barrier to the wide use of protein levels in searches for important genes associated with a phenotype is the difficulty in getting highly quantitative and reproducible simultaneous measurements of levels of many proteins. Furthermore, proteins expressed at low levels can be difficult to assay in complex mixtures. Nevertheless, considerable progress has been made using a combination of **2D gel analysis and mass spectrometry (MS)**. Increasingly sensitive MS technologies have been used in comparative proteomic studies in order to better understand protein regulation and interaction during *Salmonella* infection.

One important area of study is the regulation of proteins under different environmental conditions. The proteomes of *Salmonella* Typhimurium LT2 and 14028 grown in conditions that mimic different infection conditions, such as rich media (mimicking invasion), stationary phase (mimicking systemic infection and growth in macrophages), and low magnesium/pH (mimicking conditions inside a *Salmonella*-containing vacuole, SCV) have been assayed (Adkins *et al.*, 2006). The importance of the ability to use propanediol as a carbon source during infection and growth within the SCV was inferred, as proteins encoded in the *pdu* operon were highly expressed under low magnesium conditions (Adkins *et al.*, 2006). Low magnesium conditions were further characterized in a comparative study using abrupt shift to low magnesium (MgM shock) versus gradual shift to low magnesium using dilution. The proteins encoded in the *pdu* operon were detected under shock conditions, along with DmsA and other proteins typically expressed under low oxygen conditions (Shi *et al.*, 2009). Low magnesium dilution conditions are considered to better mimic conditions inside the SCV, and SodA, SodCI, SpvBC, SitAB and other proteins of the TTSS-2 apparatus were more abundant under this condition (Shi *et al.*, 2009). Biotin and thiamin biosynthesis proteins were also identified under dilution conditions (Shi *et al.*, 2009). Conditions that favour the formation of biofilms were used to identify 14 proteins that were more abundantly expressed and 18 proteins that were down-regulated in biofilms compared to planktonic growth conditions (Mangalappalli-Illathu *et al.*, 2008). Protein requirements during infection were identified by comparing the proteome of the bacteria before and during infection. This method was used to identify a novel set of proteins (STM3117–3119) expressed upon infection of RAW264.7 macrophages with *Salmonella* (Shi *et al.*, 2006).

Comparative proteomics can also be used to identify differences among serotypes, for insights into host specificity or adaptation mechanisms. The proteomes of strains representing serovars Typhimurium, Enteritidis, Choleraesuis, Pullorum and Dublin were compared after growth on Columbia Blood Agar (Encheva *et al.*, 2007). Expression of a novel *sodA* (superoxide dismutase) isoform was identified in Typhimurium, while GabD (succinate semialdehyde dehydrogenase I) was shown to be specifically expressed in Gallinarum (Encheva *et al.*, 2007). A novel isoform of a lysine-arginine–ornithine transporter was identified in Pullorum and MglB (D-galactose-binding periplasmic protein precursor) was shown to be more highly expressed in Enteritidis (Encheva *et al.*, 2007).

The proteome of Typhi Ty2 was compared with that of Typhimurium LT2 under macrophage-like conditions (Ansong *et al.*, 2008). A subset of Typhi-unique proteins (CdtB, HlyE, proteins from genes *t0142, t1108, t1109, t1476* and *t1602*) were identified and therefore linked to host specificity (Ansong *et al.*, 2008). Proteins involved in host adaptation of serotypes Gallinarum and Enteritidis to poultry were compared using proteomics. Twenty-two proteins were shown to be expressed in Gallinarum, but not Enteritidis, under rich media conditions, including one with similarity to the cell invasion protein SipA (Osman *et al.*, 2009).

Identification of targets of global regulators of virulence

Chromatin immunoprecipitation (ChIP) combines proteomics with genomics and has been used for comprehensive identification of the targets of global regulators in *Salmonellae*. In this method, specific DNA-binding proteins, such as transcription factors, are cross-linked to genomic DNA at their recognition sites *in vivo*, the DNA is then sheared and captured using an epitope tag engineered into the protein or a specific antibody (Hayakawa *et al.*, 2004). The DNA fragments attached to the protein of interest are amplified and characterized by a variety of methods. The DNA binding sites of several of the master regulators of virulence in *Salmonella*, such as PhoP, HilA and the histone-like nucleoid structuring protein H-NS (Navarre *et al.*, 2006; Thijs *et al.*, 2007; Zwir *et al.*, 2005), were recently determined using ChIP-based approaches. PhoP was shown to bind to numerous promoters, including those for *mgtA*, *mgtC*, *pagC*, *mig-14*, and *ugd* (Zwir *et al.*, 2005). In studies to determine the DNA binding sites for HilA and H-NS, microarrays were used to identify the DNA targets bound by these proteins. HilA appeared to bind near and regulate expression of *sopB*, *sopE*, *sopA*, *siiA*, *invF*, *prgH*, *ssaH*, *ygiM*, *STM3598*, *gpmA*, *sseL*, *glpT*, *flhD*, *melA*, *miaB*, *atpF*, *lpp*, *wecC* and STM1533 (Thijs *et al.*, 2007). H-NS binding studies showed that H-NS primarily binds to and globally represses A+T rich regions, allowing for coordinate regulation of virulence even as the genome acquired new functions through lateral transfer of new virulence loci (Lucchini *et al.*, 2006; Navarre *et al.*, 2006). This approach is a powerful way to build relationships between regulators and gene expression.

Forward genetic screening

Large-scale genomic studies yield a broad overview of important aspects within the genomes of a variety of serovars, but they also create a platform for more focused studies using classical techniques, such as forward genetics. The power of forward genetics lies in the ability to examine the function of a gene by disrupting that gene and looking for a change in phenotype. For decades, researchers had only tools that would enable mutagenesis in the absence of sequence data, such as the use of chemical mutagens and radiation treatment. More recently, re-engineered, stable transposable elements have been used to create banks of mutants for phenotypic analysis (Bäumler *et al.*, 1994; Bowe *et al.*, 1998; Fields, 1986; Finlay, 1988; Gulig and Curtiss, 1987; Gulig and Curtiss, 1988; Kukral *et al.*, 1987; Lodge *et al.*, 1995; Miller *et al.*, 1989; Rhen *et al.*, 1989; Rubino *et al.*, 1993; Sizemore *et al.*, 1991; Turner *et al.*, 1998). Screening these mutants, however, continued to be slow and expensive, primarily because sequencing of individual candidate mutants with phenotypes was necessary to determine the location of transposons, and clean deletion mutants had to be constructed and re-tested. The development of negative selection techniques allowed screening for mutants that were not viable during infection and improved the throughput of transposon-based screens.

Identification of the genetic requirements for Salmonella *growth and virulence using signature-tagged mutagenesis*

In 1995, Hensel *et al.* described the first negative selection strategy used during infection with *Salmonella*, called **signature tagged mutagenesis (STM)**. STM combined transposon mutagenesis and tracking of each individual transposon mutant using short, exogenously introduced, unique sequence tags on the transposons themselves (Hensel *et al.*, 1995). Small pools of mutants were assembled and subjected to a genetic selection during intraperitoneal infection of BALB/c mice, an established model for typhoid fever (Hensel *et al.*, 1995). Mutants under-represented after infection were identified using Southern analysis (Hensel *et al.*, 1995). This study first identified *Salmonella* Pathogenicity Island 2 (SPI-2) (Shea *et al.*, 1996). Numerous STM studies in *Salmonella* followed, building upon the important work of this initial study, by determining genetic requirements for infection in numerous niches and for different serotypes of *Salmonella*.

STM has been used for comparative studies to determine the genetic requirements for *Salmonella* survival and growth during infection of different host species. A small pool of 260 Typhimurium STM mutants was used to identify genes necessary for infection in mice and calves (Tsolis *et al.*, 1999). Surprisingly, the largest category of *Salmonella* mutants that were identified in this study were necessary for survival in both mice and calves (Tsolis *et al.*, 1999), providing a basis for future work that initially identifies *Salmonella* gene requirements in mice and later tests these in more cumbersome and expensive livestock models. STM has also been employed to study genetic requirements for Typhimurium infection in calves and chicks, using a larger number of transposon mutants (Morgan *et al.*, 2004). SPI-4, a 27 Kb region (Wong *et al.*, 1998) encoding a type I secretion system (T1SS) and a large non-fimbrial adhesin SiiE (STM4261), was identified as an important determinant of virulence with a specific role in cattle (Gerlach *et al.*, 2007; Kiss *et al.*, 2007; Morgan *et al.*, 2007; Morgan *et al.*, 2004). The *siiABCDEF* operon is regulated by *sirA, hilA* and *rfaH* (Gerlach *et al.*, 2007; Nagy *et al.*, 2006; Thijs *et al.*, 2007), and mutants in *siiE* are attenuated for virulence in mice and cattle (Gerlach *et al.*, 2007; Morgan *et al.*, 2007; Morgan *et al.*, 2004). SPI-4 is co-regulated with SPI-1 and required for apical invasion of polarized epithelial cells, destruction of tight junctions, cytoskeletal rearrangement and effacement of the brush border (Gerlach *et al.*, 2008; Gerlach *et al.*, 2007). The large size of *siiE* is a requirement for its function, as it must reach past the lipopolysaccharide layer of the cell to interact with the host cell (Gerlach *et al.*, 2008).

Candidate genes necessary for virulence in swine have also been identified using STM (Carnell *et al.*, 2007). Mutants in genes in pathogenicity islands SPI-1 through 6 were under-represented after infection in swine. Under-representation of mutants in *Salmonella* atypical fimbria (*saf*) genes after infection of swine suggest a role for this adhesin during infection, and a subtle role for *saf* was confirmed in competitive oral infections. Mutants in a structural component of the TTSS-1 encoded on SPI-1, *prgH,* were also defective in colonization of cattle and pigs (Carnell *et al.*, 2007; Morgan *et al.*, 2004), and the efficacy of the *prgH* mutant as a vaccine candidate has been explored (Carnell *et al.*, 2007). Finally, over 40 genes encoding proteins of unknown or hypothetical function were identified after STM screening in swine, including STM0557 *(gtrC1),* a gene that has since been characterized as necessary for O12 antigen form variation and persistence of Typhimurium in the murine intestine (Bogomolnaya *et al.*, 2008). STM has also been used to identify Typhimurium genes in an *ex vivo* assay, with swine stomach contents as the selective mechanism (Bearson *et al.*, 2006). Several genes of putative function were identified from 34 pools (each containing 48 mutants) as defective in the ability to survive the harsh, acidic environment of the stomach.

STM has also been used to determine genes necessary during infection for host-specific serotypes Choleraesuis and Dublin (Bispham *et al.*, 2001; Lichtensteiger and Vimr, 2003). Choleraesuis mutants in the regulatory gene *hilA* (hyper-invasive locus A) were attenuated in swine after oral, but not intraperitoneal, inoculation (Lichtensteiger and Vimr, 2003), similar to Typhimurium *hilA* mutants (Galan and Curtiss, 1989; Murray and Lee, 2000; Tsolis *et al.*, 1999). After intranasal infection with Choleraesuis STM mutants, two genes required for virulence, STM1459 (putative oxidoreductase protein) and STM2626 (Gifsy-1 prophage replication protein) were identified as necessary (Ku *et al.*, 2005). *Salmonella* Dublin STM mutants were screened during infection of calves and BALB/c mice to determine genes necessary for virulence (Bispham *et al.*, 2001). Mutants in two genes encoding parts of the TTSS-2 machinery, *ssaT and sseD,* were defective in both host models (Bispham *et al.*, 2001). The *sseD* mutant was further characterized in oral and ileal infections in the calf. Consistent with previous conclusions that SPI-2 is not required for invasion of the intestinal epithelium, the *sseD* mutant and the wild type were recovered in equal numbers from ileal loop infections. However, *sseD* mutants were recovered in lower amounts from the faeces of calves after oral infection (Bispham *et al.*, 2001). These data are consistent with the hypothesis that *Salmonella* may proliferate in the intestinal mucosa, requiring the TTSS-2 encoded on SPI-2, and subsequently re-seed the intestinal lumen.

A defined, small-scale follow-up study to earlier work in cattle identified genes necessary for translocation from the intestine to systemic sites (Pullinger *et al.*, 2007). Thirty-six mutations were generated in serotype Dublin using the lambda red recombinase methods (Datsenko and Wanner, 2000), reproducing mutants previously characterized in Typhimurium (Morgan *et al.*, 2004; Pullinger *et al.*, 2007). These defined mutants were used to identify genes necessary for translocation of the bacteria from the gut to the mesenteric lymph node in cattle. This study showed that the SPI-4 T1SS, but not SPI-2 TTSS-2, is necessary for translocation from enteric to systemic sites.

As investigators continued to use STM, improvements have been made to the original technology. Alternate methods have been developed, such as the use of PCR to assay changes in representation of mutants (Lehoux *et al.*, 1999) in studies using mutant pools with reduced complexity. Using this strategy, the requirement for genes of the host-specific serotype Gallinarum for growth and survival in chicks was determined after oral infection (Shah *et al.*, 2005). Mutants in genes encoded in SPI-10 (*sefD*) and SPI-13 (STM3118-STM3120) were defective during this infection (Shah *et al.*, 2005).

Functional genomics

The use of unique sequence tags embedded within the transposon was revolutionary, but one drawback of the STM technology was the need to characterize each tag, making the construction of multiple libraries or large libraries quite laborious, and constraining the use of STM for comprehensive functional genomic studies. For large-scale genome-wide screening, two groups simultaneously developed methods to use microarray technology and the idea of signature tagging without exogenously added tags to study bacterial pathogens (Badarinarayana *et al.*, 2001; Sassetti *et al.*, 2001). Each group designed transposons bearing outwardly facing T7 promoters that could promote *in vitro* transcription from genomic DNA flanking each insertion site, generating a unique transcript for each mutant. The representation of these transcripts in a pool was evaluated by using ORF microarrays (Badarinarayana *et al.*, 2001; Sassetti *et al.*, 2001). This strategy was named **transposon site hybridization or TraSH** (Sassetti *et al.*, 2001), and made the use of additional exogenous sequence tags unnecessary. Although this method was developed for use in *Mycobacteria*, a similar approach was also used in *E. coli* (Badarinarayana *et al.*, 2001), and was subsequently adapted for use in *Salmonella*.

A variant of the TRaSH screening methodology was used to follow 50,000 pooled transposon mutants in Typhimurium during intraperitoneal infection of BALB/c mice of short duration, as well as infection of cultured RAW 264.7 macrophages (Chan *et al.*, 2005). Many of the genes identified in this study, including those in genes of SPI-2, were previously identified as important in this murine model using STM (Hensel *et al.*, 1995). These data verified that the T7-based TRaSH method variant worked in *Salmonella*. Unexpectedly however, genes encoded on SPI-1 were shown to play a role in replication and survival in macrophages, and *sirA* was identified as a potential regulator of SPI-1 function within macrophages (Chan *et al.*, 2005). Finally, the vast majority of genes required for virulence in these experiments are specific to *Salmonellae* (Chan *et al.*, 2005), suggesting that *Salmonella*-specific genes (Porwollik, 2002) are an important category of genes for further study.

The genetic requirements for Typhimurium persistence in systemic sites has also been studied using this T7-microarray-based method in *Salmonella*-resistant mouse models (Lawley *et al.*, 2006). *Salmonella* resistant mice were infected for up to 28 days, to investigate the genetic requirements for systemic persistence (Lawley *et al.*, 2006). Infected animals were sacrificed and the mutant pools were analysed at 7, 14, 21 and 28 days post infection, allowing specific identification of genes necessary for long-term intestinal persistence. Similar to short term systemic colonization, studies of long-term colonization of systemic sites identified a large complement of *Salmonella*-specific genes as necessary, including genes of SPI-2. Genes required for restructuring of the bacterial outer membrane, including the *rfa* and *rfb* clusters, also appear to be necessary for long term colonization of systemic sites by Typhimurium (Lawley *et al.*, 2006). Finally, mutants in SPI-1 effectors were underrepresented within 7 days post infection. These effectors, such as *sipCBD*, appear to be needed for establishment of infection, but also for a more long-term role that is poorly understood (Lawley *et al.*, 2006).

Until recently, pools of transposon mutants have been evaluated using primarily Open Reading Frame (ORF) microarrays. Transposon insertions in non-coding regions, or un-annotated coding regions, especially small genes, were not assayed. A new quantitative method termed **transposon-mediated differential hybridization (TMDH)** that uses a novel high-density tiling array containing 60-mer oligonucleotides spaced every 100 base pairs along both strands of the genome, alleviates these issues (Chaudhuri *et al.*, 2009). The transposons developed for generating mutants in this method, Tn5 and Mu, have an outward facing T7 promoter on one end and an outward facing SP6 promoter on the opposite end. Representation of transposon mutants in the pool was evaluated by hybridizing labelled transcripts to the tiling array. Use of this array allows determination of the location of each transposon insertion with much higher resolution than with other types of arrays.

Approximately 10,000 Typhimurium random transposon mutants were studied, pooled in groups of 480 or 384, and used to infect mice intravenously. Comparison of the representation of individual mutants in pools used to infect mice (input pools) to the pools recovered from mice (output pools) were evaluated, however, the input and output pools were not directly compared to each other on the same slide. A second minor limitation was that multiple transposons located very close to each other could not be distinguished from each other, limiting the pool size that can be tested by arrays. Using the TDMH method, Chauduri *et al.* identified the position of 8533 transposon mutations, with 6108 transposons unambiguously mapped to 2824 genes, roughly half of the *Salmonella* genome. A large data set of 'unbiased relative fitness' was generated from this study, and 47 genes were targeted for deletion and studied further using single infections in BALB/c mice. Mutants in *trxA*, *atpA*, and *tolA* were further tested in immunization and protection studies, and found to provide protection against intravenous challenge only (*trxA* and *atpA*) or against both intravenous and oral challenge (*tolA*).

Ultra high-throughput screening of extremely large transposon mutant pools has recently been developed using Illumina sequencing technology (Canals *et al.*, unpublished results; Langridge *et al.*, 2009). One method, called **transposon-directed insertion site sequencing or TraDIS**, uses the standard Illumina sequencing strategy of genome shearing and ligation of adaptors, with the innovation of the addition of a primer within the transposon for subsequent sequencing of tags adjacent from the gene (Langridge *et al.*, 2009). Using this method, 370,000 transposon insertions were mapped in the Typhi genome, an insertion every 13 base pairs on average across the genome (Langridge *et al.*, 2009). Exactly 356 genes were determined to be essential for growth of Typhi under standard laboratory conditions, of which 256 genes (or 70%) were also essential in *E. coli* (Baba *et al.*, 2006).

Our group has developed a slightly different high throughput sequencing (HTS) method for measuring tags (Canals *et al.*, unpublished results). This new HTS method employs strategies that we previously developed for microarrays (Santiviago *et al.*, 2009), and relies on shearing, polyadenylation, and nested PCR (outlined in Fig. 4.1), avoiding any ligation steps that are used in other methods. A specific code in the sequenced portion of each primer facilitates the inclusion of multiple samples in a single sequencing channel. We performed a multiplex of nine different samples in a single channel of an Illumina sequencing run, and mapped 22 million different tags. Using this protocol, differential fitness of transposon mutants after intraperitoneal inoculation of BALB/c mice was investigated, revealing many of the expected genes (Canals *et al.*, unpublished results).

Transposon mutagenesis is a valuable tool for genetic studies, but its use is limited in particular environments where members are lost at random from pools of mutants due to biological 'bottlenecks', where only a few independent clones clear a barrier in the environment or process under investigation, such as during infection of a host (Fig. 4.2). In order to generate a mutation in each

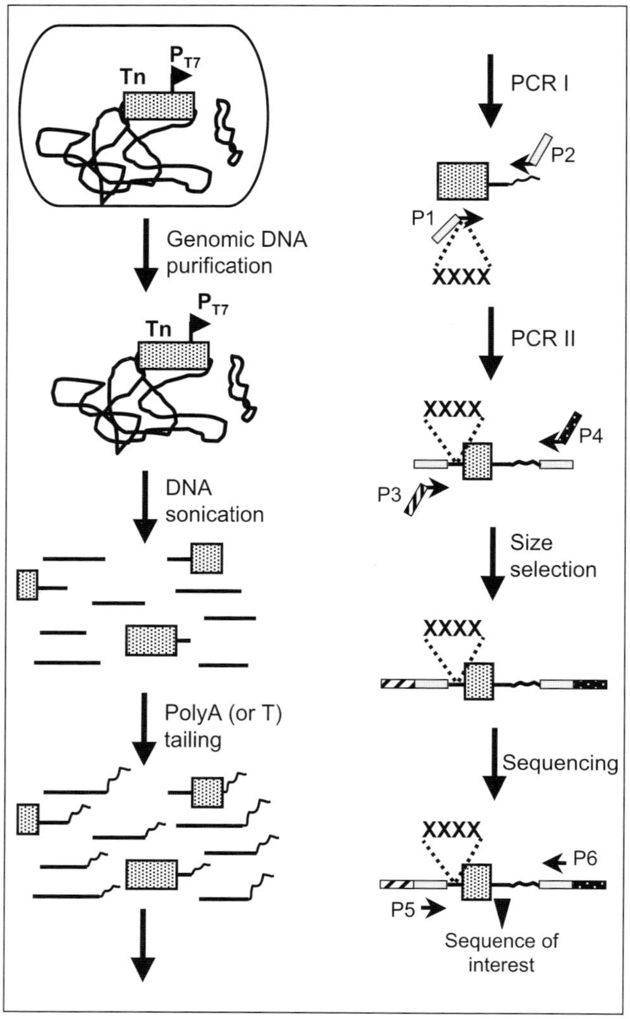

Figure 4.1 High-throughput sequencing of transposon libraries. Each clone of a transposon library contains a single transposon inserted at random in the genome. Genomic DNA is purified and fragmented by sonication, and DNA fragments are polyadenylated. PCR is used to amplify only the DNA fragments of interest, where primer P1 contains a 4-base signature code to identify the specific library used. A second PCR is carried out to add the sequences required for hybridization to the Illumina flow cell surface (indicated by patterns). PCR products of 250–450 base pairs are selected using an agarose gel. The sample is then used in a paired-end sequencing run using Illumina (GS) technology. Among the resulting sequences, those obtained using the P5 primer include the libraries' 4-base signature code, the last few bases of the 3′ end of the transposon to confirm non-arbitrarily amplified DNA, and the *Salmonella* sequence of interest adjacent to the transposon. Sequences obtained using the P6 primer contain polyT (or A) sequence and the site where the *Salmonella* genome was sheared.

gene in the *Salmonella* genome using random mutagenesis approaches, transposon libraries must be highly complex. A complexity of 5–10 times the number of genes in the genome of interest is necessary in order to have a high probability of generating a mutation in each non-essential gene in the genome. During genetic selections where many mutants are lost at random, such as an infection, representation of individual mutants in such complex pools can be affected by random effects, and not by biological fitness (Fig. 4.2). Therefore, smaller pools of mutants, such as in STM and in TMDH approaches used for *Salmonella,* are preferred during infection. Pool sizes that do not

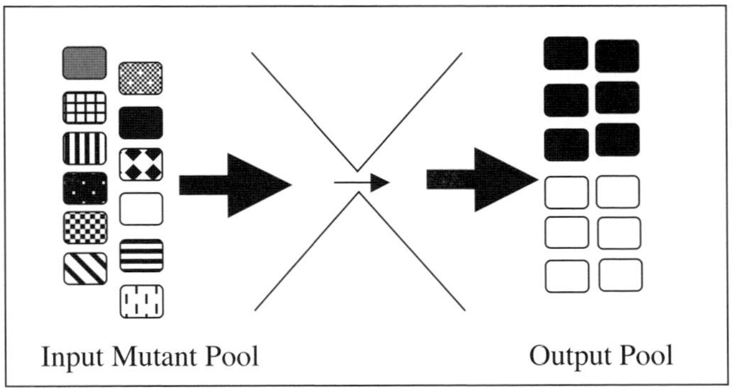

Figure 4.2 The complexity of a mixture of mutants can be reduced by passage through a bottleneck. In the example shown, only two founder clones pass a severe bottleneck, eliminating other, equally capable, clones from the output simply by chance.

fluctuate at random must be determined empirically for each route of infection and niche to be studied. However, use of small pool sizes does not allow comprehensive screening of the genome in non-traditional animal models, such as livestock, where it is prohibitively expensive to use many animals for screening. Additional drawbacks of transposon-based mutagenesis include the random location of the insertions themselves, and potential polar effects on downstream genetic elements. We discuss an approach using targeted mutants that we have developed to overcome these hurdles, in the next section.

Array-based analysis of targeted deletion mutants under selection

The rapid availability of complete genome sequence information, combined with straightforward techniques for generating targeted mutants in Enterobacteriaceae, allow the development of libraries of targeted deletion mutants, generating one mutant in each gene in the genome. The major advantage of using targeted deletion mutants over randomly generated mutants discussed earlier lies in the reduced total complexity of pools of mutants to be screened. Targeted deletion libraries containing mutations in each non-essential gene have been developed in *Saccharomyces cerevisae* (Kelly *et al.*, 2001), and *E. coli* (the Keio collection) (Baba *et al.*, 2006), and are under development for the fungal pathogen *Cryptococcus neoformans* (Loftus *et al.*, 2005). A similar collection is being developed in *Salmonella* Typhimurium, and currently contains mutants in all *Salmonella*-specific genes, a group of genes that contains most of the known virulence determinants in this pathogenic organism (Santiviago *et al.*, 2009). A combination of T7 and microarray-based technologies makes it possible to screen mutants from this *Salmonella* targeted deletion library in pools (Santiviago *et al.*, 2009), using a technique called **array-based analysis of cistrons under selection (ABACUS)**.

The targeted deletion library in Typhimurium is being constructed using lambda red recombination technology (Datsenko and Wanner, 2000). The original plasmids used for generation of the insert, pKD3 and pKD4, bearing antibiotic resistance cassettes (either Kan^R or Cm^R) to be inserted into the genome, have been modified to include an outwardly facing T7 promoter (GenBank Accession EU629213 and EU629214) (Santiviago *et al.*, 2009). The introduction of the T7 promoter in these targeted deletion mutants serves the same purpose as the insertion of outwardly facing promoters on transposons used for random mutagenesis described earlier. These promoters provide a unique transcript for each mutant, enabling determination of the representation of each mutant after selection. In addition, these modified plasmids maintain all of the favourable features of the original plasmids, including an in-frame stop codon and ribosome binding sites to minimize polar effects, and FRT sites flanking the antibiotic resistance markers to allow the generation of unmarked mutations.

Using the complete sequence for Typhimurium ATCC14028 (GenBank Accession# CP001363) (Jarvik *et al.*, 2010), 1052 genes specific to *Salmonella* and not found in *E. coli* and other closely related organisms, were targeted for deletion (Santiviago *et al.*, 2009). Positioning of each single gene deletion was standardized so that the authentic start and stop as well as the first ten and last nine amino acids of each targeted gene remain intact. After removal of the antibiotic resistance cassette, these coding regions and a small amount of inserted DNA encode a 39 amino acid mini-protein, so that polar effects on downstream open reading frames are minimized as much as possible.

The presence of an active T7 promoter in the correct location was simultaneously assayed in all of these mutants by production of directly labelled T7 transcripts from the pool, and subsequent assay of these transcripts on a tiling array (Santiviago *et al.*, 2009). This tiling array covers the entire Typhimurium ATCC14028 genome with a staggered 12 base overlap, allowing resolution between closely spaced inserted T7 promoters. At least 1023 of 1053 attempted mutants had correctly placed inserts and an active T7 promoter at the site targeted for deletion. These mutants are currently being tested both individually and as a pool for phenotypes (Santiviago *et al.*, 2009).

This novel Typhimurium deletion library has been pooled and used in a negative selection strategy, using a second inexpensive oligonucleotide microarray containing 65mers based on the 3′ and 5′ ends of each mutated gene. These oligos were originally used to produce the linear fragments for generating the targeted deletions. The oligos based on the 3′ end of the gene hybridize to the probe transcribed from the T7 promoter, while the 5′ oligos are included on the array as negative controls. We successfully obtained specific signal for 97% of all mutants, with minimal (0.7%) hybridization to negative controls.

Representation of each mutant in the pooled Typhimurium deletion library was tested during infection of BALB/c mice. A pool of targeted deletion mutants in Typhimurium was used to infect 6 mice intraperitoneally at a dose of 10^6 CFU (Input pool). 48 hours post-infection, infected mice were euthanized, and spleens and livers from infected animals were collected and homogenized. Output pools from the homogenized tissue were expanded overnight, and genomic DNA was prepared. Genomic DNA from both input and output pools was used for *in vitro* transcription from the T7 promoter with direct labelling of transcripts, and the resulting labelled transcripts were used as probes on our in-house microarray. A total of 12 array hybridizations were performed for statistical purposes, including dye swaps, and analysed using the online MIAME-compliant array data analysis suites WebArray (Xia *et al.*, 2005) and WebArrayDB (Xia *et al.*, 2009).

We identified 120 mutants undergoing selection during infection, including all genes present in the pool that had also been previously reported in similar models (Chan *et al.*, 2005; Hensel *et al.*, 1995; Lawley *et al.*, 2006). We also identified >40 previously unreported candidate novel mutant phenotypes at a greater than 2-fold threshold and a *p*-value of less than or equal to 0.0001, including many genes of putative or unknown function and six small RNAs (Santiviago *et al.*, 2009) (Table 4.2). Mutants in 6 small RNAs, and mutants in nine randomly chosen genes were chosen for verification after P22 transduction to a clean genetic background. Each single mutant was tested in competitive infection with wild type Typhimurium ATCC14028 during intraperitoneal infection in BALB/c mice. Ten of the fifteen mutants tested were statistically underrepresented compared to the wild type, while one was overrepresented, for a total of 11 confirmed phenotypes. Underrepresented mutants include small RNAs leuX, istR, and sroA, as well as mutants in STM2303 *(pmrM)*, STM2215 (*rtn*), STM1760 (TPR repeat protein), STM3120 *(citE)*, STM3121 *(stmR)*, STM2120 *(asmA)*, and STM1131. Mutation in the small RNA oxyS resulted in overrepresentation when compared to the wild type (Santiviago *et al.*, 2009). The antibiotic resistance cassette was removed from eight of these confirmed mutants, and the now unmarked, non-polar deletions were re-tested in mixed competitive infections with the wild type. The phenotype for each mutant tested was confirmed (Santiviago *et al.*, 2009). Finally, complementation analysis showed that the missing genes were indeed responsible for the defective phenotypes we observed. Four deletion

Table 4.2 Novel mutants with altered fitness during murine infection. Genes were identified during infection of BALB/c mice by the intraperitoneal route, using a pool of targeted mutants and ABACUS. Genes confirmed in individual competitive infections are underlined

Functional group	Gene
Regulator	STM0693 (Fur transcript repressor), STM1547 (putative *marR* family)
	STM2639 (*rseA*)
	<u>STM3121</u> (*stmR*)
	STM3687 (*mtlR*)
	STM4417
Small RNAs	<u>leuX</u>
	<u>sroA</u>
	sraA
	<u>istR</u>
	rybB
	<u>oxyS*</u>
Porin/transport	<u>STM1131</u>
	STM1258 (ABC transport)
	STM3763 (*mgtB,* magnesium transport)
	<u>STM3975</u> (*tatC*)
Scavenger	STM2777 (*iroN*)
Enzymatic	STM0719
	STM0857
	STM1151 (*mdoH*)
	STM3119 (*maoC, ripB* homologue)
	<u>STM3120</u> (*citE*-like)
	STM3339 (*nanA*)
	STM3846 (putative reverse transcriptase)
	STM4475 (*valS,* valine tRNA synthase)
	STM4489 (putative helicase)
PhoPQ related	STM1544 (*pqaA, phoQ* regulated)
	STM1601 (*ugtL*)*
LPS	<u>STM2120</u> (*asmA*)
Kinase	STM1670
	STM2116 (*wzc*)
c-di-GMP	STM0551
	STM1344 (*cdgR/ydiV*)
	STM1697
	<u>STM2215</u> (*rtn*)
	STM2672 (*yfiN*)*
Flagellar biosynthesis	STM1971 (*fliH*)*

Table 4.2 continued

Functional group	Gene
Unknown	STM0081
	STM0731
	STM1600
	STM1760 (TPR repeat)
	STM1948
	STM2303 (*pmrM, pbgE3*)
	STM2329
	STM3047 (*ygfY*)
	STM3089 (*yqgD*)
	STM3155
	STM3783
	STM4504*
	STM4529
	STM4599 (*yjjY*)

*Indicates mutant was positively selected during infection.

See text for details, modified from Santiviago *et al.* (2009).

mutants were complemented *in trans* (Andrews-Polymenis *et al.*, 2009a), which restored the ability of the complemented mutant strain to compete with the wild type during infection (Santiviago *et al.*, 2009).

The development of a targeted deletion library, in conjunction with the technological advancements of the protocols that allow screening of many mutants at the same time, will open the door to a wide range of possibilities. Note, for example, that changes in the representation of each mutant can be monitored by high-throughput sequencing, and 100 or more conditions could be sequenced in one channel of current sequencers and still provide enough data for monitoring 4000 mutants. This yield is likely to grow, supplanting arrays entirely for this purpose in the near future.

The Typhimurium targeted deletion library is currently being expanded to include deletions in all non-essential genes as well as deletions that cover many adjacent non-essential genes. The latter set of 'island' mutants will allow efficient hierarchical searches for mutants of interest in those circumstances where only one mutant can be screened at a time or where the 'bottleneck' that has to be overcome by a pool is very severe. An example of the former, where mutants must be used one at a time, is a screen for a host phenotype, such as diarrhea. An example of the latter, a severe bottleneck, is the transition of bacteria from the gut to the bloodstream of the host.

These targeted mutants can be subjected to any variety of selective environments *in vitro, in vivo*, or *ex vivo*, both as a pool and as individual mutants, to assay for genes important for numerous phenotypes. *In vivo* comparative studies are now possible, in different animal systems. Combining the use of this library with studies in knockout mice has great potential to lead to a better understanding of the interaction between *Salmonellae* and the diverse environments where this organism survives, grows, and causes disease.

Summary

Studies of differential gene expression have been heavily used in animal models and in cultured cells to provide lists of genes that may be critical during infection, or growth in a particular environment. However, genes that are expressed but not differentially regulated between the environments being compared may also be important in that niche. Furthermore, some genes are regulated as part of hierarchies, so the expression of a particular gene in a particular environment may not be necessary for survival in that environment. Thus, gene expression studies are most safely used as a platform for determining groups of genes that are coordinately regulated, in an effort to build an overview of the network of gene expression, and as a tool to determine and prioritize candidate genes for further genetic studies.

Methods to characterize differential expression of genes, and differential fitness of mutants, have greatly improved in the past few years. These improvements have a positive impact on our understanding of the *Salmonella* genome and the biology of this important organism, continually reducing costs and increasing the completeness and speed of analysis. Important hurdles that need new strategies remain, such as the need for methods that put a molecule in its regulatory context and methods to determine the function of molecules in a process.

References

Adkins, J.N., Mottaz, H.M., Norbeck, A.D., Gustin, J.K., Rue, J., Clauss, T.R., Purvine, S.O., Rodland, K.D., Heffron, F., and Smith, R.D. (2006). Analysis of the *Salmonella* Typhi*murium* proteome through environmental response toward infectious conditions. Mol. Cell Proteomics 5, 1450–1461.

Alix, E., and Blanc-Potard, A.B. (2009). Hydrophobic peptides: novel regulators within bacterial membrane. Mol. Microbiol. 72, 5–11.

Altekruse, S.F., Cohen, M.L., and Swerdlow, D.L. (1997). Emerging foodborne diseases. Emerg. Infect. Dis. 3, 285–293.

Andreoli, P.e.a. (2004). Fast method for detecting microorganisms in food samples. Patent WOZ004/106547.

Andrews-Polymenis, H., Reynolds, M.M., Santiviago, C., and McClelland, M. (2009a). (unpublished results)

Andrews-Polymenis, H.L., Santiviago, C.A., and McClelland, M. (2009b). Novel genetic tools for studying food-borne *Salmonella*. Curr. Opin. Biotechnol. *20*, 149–157.

Ansong, C., Yoon, H., Norbeck, A.D., Gustin, J.K., McDermott, J.E., Mottaz, H.M., Rue, J., Adkins, J.N., Heffron, F., and Smith, R.D. (2008). Proteomics analysis of the causative agent of typhoid fever. J. Proteome Res. 7, 546–557.

Arrach, N., Zhao, M., Porwollik, S., Hoffman, R.M., and McClelland, M. (2008). *Salmonella* promoters preferentially activated inside tumors. Cancer Res. *68*, 4827–4832.

Baba, T., Ara, T., Hasegawa, M., Takai, Y., Okumura, Y., Baba, M., Datsenko, K.A., Tomita, M., Wanner, B.L., and Mori, H. (2006). Construction of Escherichia coli K-12 in-frame, single-gene knockout mutants: the Keio collection. Mol. Syst. Biol. 2, 2006, 0008.

Badarinarayana, V., Estep, P.W., 3rd, Shendure, J., Edwards, J., Tavazoie, S., Lam, F., and Church, G.M. (2001). Selection analyses of insertional mutants using subgenic-resolution arrays. Nat. Biotechnol. *19*, 1060–1065.

Bäumler, A.J., Kusters, J.G., Stojiljkovic, I., and Heffron, F. (1994). *Salmonella* Typhimurium loci involved in survival within macrophages. Infect. Immun. *62*, 1623–1630.

Bearson, S.M., Bearson, B.L., and Rasmussen, M.A. (2006). Identification of *Salmonella enterica* serovar Typhimurium genes important for survival in the swine gastric environment. Appl. Environ. Microbiol. *72*, 2829–2836.

Bispham, J., Tripathi, B.N., Watson, P.R., and Wallis, T.S. (2001). *Salmonella* pathogenicity island 2 influences both systemic salmonellosis and *Salmonella*-induced enteritis in calves. Infect. Immun. *69*, 367–377.

Bogomolnaya, L.M., Santiviago, C.A., Yang, H.J., Baumler, A.J., and Andrews-Polymenis, H.L. (2008). 'Form variation' of the O12 antigen is critical for persistence of *Salmonella* Typhimurium in the murine intestine. Mol. Microbiol. *70*, 1105–1119.

Bowe, F., Lipps, C.J., Tsolis, R.M., Groisman, E.A., Kusters, J.G., and Heffron, F. (1998). At least four percent of the *Salmonella* Typhimurium genome is required for fatal infection of mice. Infect. Immun. *66*, 3372–3377.

Boyd, E.F., Porwollik, S., Blackmer, F., and McClelland, M. (2003). Differences in gene content among *Salmonella enterica* serovar typhi isolates. J. Clin. Microbiol. *41*, 3823–3828.

Bumann, D. (2002). Examination of *Salmonella* gene expression in an infected mammalian host using the green fluorescent protein and two-colour flow cytometry. Mol. Microbiol. *43*, 1269–1283.

Camilli, A., Beattie, D.T., and Mekalanos, J.J. (1994). Use of genetic recombination as a reporter of gene expression. Proc. Natl. Acad. Sci. U.S.A. *91*, 2634–2638.

Canals, R., Xia, X.-Q., Clifton, S.W., Fronick, C., Ahmer, B.M., Andrews-Polymenis, H., and McClelland, M. (unpublished results)

Carnell, S.C., Bowen, A., Morgan, E., Maskell, D.J., Wallis, T.S., and Stevens, M.P. (2007). Role in virulence and protective efficacy in pigs of *Salmonella enterica* serovar Typhimurium secreted components identified by signature-tagged mutagenesis. Microbiology *153*, 1940–1952.

Chan, K., Kim, C.C., and Falkow, S. (2005). Microarray-based detection of *Salmonella enterica* serovar Typhimurium transposon mutants that cannot survive in macrophages and mice. Infect. Immun. *73*, 5438–5449.

Chaudhuri, R.R., Peters, S.E., Pleasance, S.J., Northen, H., Willers, C., Paterson, G.K., Cone, D.B., Allen, A.G.,

Owen, P.J., Shalom, G., *et al.* (2009). Comprehensive identification of *Salmonella enterica* serovar Typhi*murium* genes required for infection of BALB/c mice. PLoS Pathog. *5*, e1000529.

Chiu, C.H., Tang, P., Chu, C., Hu, S., Bao, Q., Yu, J., Chou, Y.Y., Wang, H.S., and Lee, Y.S. (2005). The genome sequence of *Salmonella enterica* serovar Choleraesuis, a highly invasive and resistant zoonotic pathogen. Nucleic Acids Res *33*, 1690–1698.

Croucher, N.J., Fookes, M.C., Perkins, T.T., Turner, D.J., Marguerat, S.B., Keane, T., Quail, M.A., He, M., Assefa, S., Bahler, J., *et al.* (2009). A simple method for directional transcriptome sequencing using Illumina technology. Nucleic Acids Res *37*, e148.

Datsenko, K.A., and Wanner, B.L. (2000). One-step inactivation of chromosomal genes in Escherichia coli K-12 using PCR products. Proc. Natl. Acad. Sci. U.S.A. *97*, 6640–6645.

Encheva, V., Wait, R., Begum, S., Gharbia, S.E., and Shah, H.N. (2007). Protein expression diversity amongst serovars of *Salmonella enterica*. Microbiology *153*, 4183–4193.

Eriksson, S., Lucchini, S., Thompson, A., Rhen, M., and Hinton, J.C. (2003). Unravelling the biology of macrophage infection by gene expression profiling of intracellular *Salmonella enterica*. Mol. Microbiol. *47*, 103–118.

Faucher, S.P., Porwollik, S., Dozois, C.M., McClelland, M., and Daigle, F. (2006). Transcriptome of *Salmonella enterica* serovar Typhi within macrophages revealed through the selective capture of transcribed sequences. Proc. Natl. Acad. Sci. U.S.A. *103*, 1906–1911.

Fields, P.I., Swanson, R.V., Haidaris, C.G., and Heffron, F. (1986). Mutants of *Salmonella* Typhimurium that cannot survive within the macrophage are avirulent. Proc. Natl. Acad. Sci. U.S.A. *83*, 5189–5193.

Finlay, B.B., Starnbach, M.N., Francis, C.L., Stocker, B.A.D., Chatfield, S., Dougan, G., and Falkow, S. (1988). Identification and characterization of Tn*phoA* mutants of *Salmonella* that are unable to pass through a polarized MDCK epithelial cell monolayer. Mol. Microbiol. *2*, 757–766.

Galan, J.E., and Curtiss, R., 3rd (1989). Cloning and molecular characterization of genes whose products allow *Salmonella* Typhi*murium* to penetrate tissue culture cells. Proc. Natl. Acad. Sci. U.S.A. *86*, 6383–6387.

Gerlach, R.G., Claudio, N., Rohde, M., Jackel, D., Wagner, C., and Hensel, M. (2008). Cooperation of *Salmonella* pathogenicity islands 1 and 4 is required to breach epithelial barriers. Cell Microbiol. *10*, 2364–2376.

Gerlach, R.G., Jackel, D., Geymeier, N., and Hensel, M. (2007). *Salmonella* pathogenicity island 4-mediated adhesion is coregulated with invasion genes in *Salmonella enterica*. Infect. Immun. *75*, 4697–4709.

Gulig, P.A., and Curtiss, R. (1987). Plasmid-associated virulence of *Salmonella* Typhimurium. Infect. Immun. *1987*, 2891–2901.

Gulig, P.A., and Curtiss, R.D. (1988). Cloning and transposon insertion mutagenesis of virulence genes of the 100-kilobase plasmid of *Salmonella* Typhi*murium*. Infect. Immun. *56*, 3262–3271.

Hautefort, I., Thompson, A., Eriksson-Ygberg, S., Parker, M.L., Lucchini, S., Danino, V., Bongaerts, R.J., Ahmad, N., Rhen, M., and Hinton, J.C. (2008). During infection of epithelial cells *Salmonella enterica* serovar Typhimurium undergoes a time-dependent transcriptional adaptation that results in simultaneous expression of three type 3 secretion systems. Cell Microbiol. *10*, 958–984.

Hayakawa, J., Mittal, S., Wang, Y., Korkmaz, K.S., Adamson, E., English, C., Ohmichi, M., McClelland, M., and Mercola, D. (2004). Identification of promoters bound by c-Jun/ATF2 during rapid large-scale gene activation following genotoxic stress. Mol. Cell *16*, 521–535.

Heithoff, D.M., Conner, C.P., Hanna, P.C., Julio, S.M., Hentschel, U., and Mahan, M.J. (1997). Bacterial infection as assessed by *in vivo* gene expression. Proc. Natl. Acad. Sci. U.S.A. *94*, 934–939.

Heithoff, D.M., Conner, C.P., Hentschel, U., Govantes, F., Hanna, P.C., and Mahan, M.J. (1999). Coordinate intracellular expression of *Salmonella* genes induced during infection. J. Bacteriol. *181*, 799–807.

Hensel, M., Shea, J.E., Gleeson, C., Jones, M.D., Dalton, E., and Holden, D.W. (1995). Simultaneous identification of bacterial virulence genes by negative selection. Science *269*, 400–403.

Holt, K.E., Parkhill, J., Mazzoni, C.J., Roumagnac, P., Weill, F.X., Goodhead, I., Rance, R., Baker, S., Maskell, D.J., Wain, J., *et al.* (2008). High-throughput sequencing provides insights into genome variation and evolution in *Salmonella* Typhi. Nat. Genet. *40*, 987–993.

Huang, Y., Leming, C.L., Suyemoto, M., and Altier, C. (2007). Genome-wide screen of *Salmonella* genes expressed during infection in pigs, using *in vivo* expression technology. Appl. Environ. Microbiol. *73*, 7522–7530.

Janakiraman, A., and Slauch, J.M. (2000). The putative iron transport system SitABCD encoded on SPI1 is required for full virulence of *Salmonella* Typhi*murium*. Mol. Microbiol. 35, 1146–1155.

Jarvik, T., Smillie, C., Groisman, E.A., and Ochman, H. (2010). Short-term signatures of evolutionary change in the *Salmonella enterica* serovar Typhi*murium* 14028 genome. J. Bacteriol. *192*, 560–567.

Kelly, D.E., Lamb, D.C., and Kelly, S.L. (2001). Genome-wide generation of yeast gene deletion strains. Comp. Funct. Genomics *2*, 236–242.

Kim, B.H., Kim, S., Kim, H.G., Lee, J., Lee, I.S., and Park, Y.K. (2005). The formation of cyclopropane fatty acids in *Salmonella enterica* serovar Typhimurium. Microbiology *151*, 209–218.

Kiss, T., Morgan, E., and Nagy, G. (2007). Contribution of SPI-4 genes to the virulence of *Salmonella enterica*. FEMS Microbiol. Lett *275*, 153–159.

Knodler, L.A., Celli, J., Hardt, W.D., Vallance, B.A., Yip, C., and Finlay, B.B. (2002). *Salmonella* effectors within a single pathogenicity island are differentially expressed and translocated by separate type III secretion systems. Mol. Microbiol. *43*, 1089–1103.

Ku, Y.W., McDonough, S.P., Palaniappan, R.U., Chang, C.F., and Chang, Y.F. (2005). Novel attenuated

Salmonella enterica serovar Choleraesuis strains as live vaccine candidates generated by signature-tagged mutagenesis. Infect. Immun. *73*, 8194–8203.

Kukral, A.M., Strauch, K.L., Maurer, R.A., and Miller, C.G. (1987). Genetic analysis in *Salmonella* Typhi*murium* with a small collection of randomly spaced insertions of transposon Tn10 delta 16 delta 17. J. Bacteriol. *169*, 1787–1793.

Langridge, G.C., Phan, M.D., Turner, D.J., Perkins, T.T., Parts, L., Haase, J., Charles, I., Maskell, D.J., Peters, S.E., Dougan, G., *et al.* (2009). Simultaneous assay of every *Salmonella* Typhi gene using one million transposon mutants. Genome Res. *19*, 2308–2316.

Lawley, T.D., Chan, K., Thompson, L.J., Kim, C.C., Govoni, G.R., and Monack, D.M. (2006). Genome-wide screen for *Salmonella* genes required for long-term systemic infection of the mouse. PLoS Pathog. *2*, e11.

Lehoux, D.E., Sanschagrin, F., and Levesque, R.C. (1999). Defined oligonucleotide tag pools and PCR screening in signature-tagged mutagenesis of essential genes from bacteria. Biotechniques 26, 473–478, 480.

Lichtensteiger, C.A., and Vimr, E.R. (2003). Systemic and enteric colonization of pigs by a hilA signature-tagged mutant of *Salmonella* Choleraesuis. Microb Pathog *34*, 149–154.

Lodge, J., Douce, G.R., Amin, I.I., Bolton, A.J., Martin, G.D., Chatfield, S., Dougan, G., Brown, N.L., and Stephen, J. (1995). Biological and genetic characterization of TnphoA mutants of *Salmonella* Typhi*murium* TML in the context of gastroenteritis. Infect. Immun. *63*, 762–769.

Loftus, B.J., Fung, E., Roncaglia, P., Rowley, D., Amedeo, P., Bruno, D., Vamathevan, J., Miranda, M., Anderson, I.J., Fraser, J.A., *et al.* (2005). The genome of the basidiomycetous yeast and human pathogen Cryptococcus neoformans. Science *307*, 1321–1324.

Lucchini, S., Rowley, G., Goldberg, M.D., Hurd, D., Harrison, M., and Hinton, J.C. (2006). H-NS mediates the silencing of laterally acquired genes in bacteria. PLoS Pathog 2, e81.

Mahan, M.J., Slauch, J.M., and Mekalanos, J.J. (1993). Selection of bacterial virulence genes that are specifically induced in host tissues. Science *259*, 686–688.

Mahan, M.J., Tobias, J.W., Slauch, J.M., Hanna, P.C., Collier, R.J., and Mekalanos, J.J. (1995). Antibiotic-based selection for bacterial genes that are specifically induced during infection of a host. Proc. Natl. Acad. Sci. U.S.A. *92*, 669–673.

Mangalappalli-Illathu, A.K., Lawrence, J.R., Swerhone, G.D., and Korber, D.R. (2008). Architectural adaptation and protein expression patterns of *Salmonella enterica* serovar Enteritidis biofilms under laminar flow conditions. Int. J. Food Microbiol. *123*, 109–120.

McClelland, M., Sanderson, K.E., Clifton, S.W., Latreille, P., Porwollik, S., Sabo, A., Meyer, R., Bieri, T., Ozersky, P., McLellan, M., *et al.* (2004). Comparison of genome degradation in Paratyphi A and Typhi, human-restricted serovars of *Salmonella enterica* that cause typhoid. Nat. Genet. *36*, 1268–1274.

McClelland, M., Sanderson, K.E., Spieth, J., Clifton, S.W., Latreille, P., Courtney, L., Porwollik, S., Ali, J., Dante, M., Du, F., *et al.* (2001). Complete genome sequence of *Salmonella enterica* serovar Typhimurium LT2. Nature *413*, 852–856.

Miller, I., Maskell, D., Hormaeche, C., Johnson, K., Pickard, D., and Dougan, G. (1989). Isolation of orally attenuated *Salmonella* Typhi*murium* following TnphoA mutagenesis. Infect. Immun. *57*, 2758–2763.

Mishu, B., Koehler, J., Lee, L.A., Rodrigue, D., Brenner, F.H., Blake, P., and Tauxe, R.V. (1994). Outbreaks of *Salmonella* Enteritidis infections in the United States, 1985–1991. J. Infect Dis. *169*, 547–552.

Morgan, E., Bowen, A.J., Carnell, S.C., Wallis, T.S., and Stevens, M.P. (2007). SiiE is secreted by the *Salmonella enterica* serovar Typhimurium pathogenicity island 4-encoded secretion system and contributes to intestinal colonization in cattle. Infect. Immun. *75*, 1524–1533.

Morgan, E., Campbell, J.D., Rowe, S.C., Bispham, J., Stevens, M.P., Bowen, A.J., Barrow, P.A., Maskell, D., and Wallis, T.S. (2004). Identification of host-specific colonization factors of *Salmonella enterica* serovar Typhimurium. Mol. Microbiol. 53.

Morrow, B.J., Graham, J.E., and Curtiss, R., 3rd (1999). Genomic subtractive hybridization and selective capture of transcribed sequences identify a novel *Salmonella* Typhi*murium* fimbrial operon and putative transcriptional regulator that are absent from the *Salmonella* typhi genome. Infect. Immun. *67*, 5106–5116.

Murray, R.A., and Lee, C.A. (2000). Invasion genes are not required for *Salmonella enterica* serovar Typhi*murium* to breach the intestinal epithelium: evidence that *Salmonella* pathogenicity island 1 has alternative functions during infection. Infection and Immunity *68*, 5050–5055.

Nagy, G., Danino, V., Dobrindt, U., Pallen, M., Chaudhuri, R., Emody, L., Hinton, J.C., and Hacker, J. (2006). Down-regulation of key virulence factors makes the *Salmonella enterica* serovar Typhimurium rfaH mutant a promising live-attenuated vaccine candidate. Infect. Immun. *74*, 5914–5925.

Navarre, W.W., Porwollik, S., Wang, Y., McClelland, M., Rosen, H., Libby, S.J., and Fang, F.C. (2006). Selective silencing of foreign DNA with low GC content by the H-NS protein in *Salmonella*. Science *313*, 236–238.

Osbourn, A.E., Barber, C.E., and Daniels, M.J. (1987). Identification of plant-induced genes of the bacterial pathogen *Xanthomonas campestris* pathovar campestris using a promoter-probe plasmid. EMBO J. *6*, 23–28.

Osman, K.M., Ali, M.M., Radwan, M.I., Kim, H.K., and Han, J. (2009). Comparative proteomic analysis on *Salmonella Gallinarum* and *Salmonella* Enteritidis exploring proteins that may incorporate host adaptation in poultry. J. Proteomics *72*, 815–821.

Pang, T., Bhutta, Z.A., Finlay, B.B., and Altwegg, M. (1995). Typhoid fever and other salmonellosis: a continuing challenge. Trends Microbiol. *3*, 253–255.

Parkhill, J., Dougan, G., James, K.D., Thomson, N.R., Pickard, D., Wain, J., Churcher, C., Mungall, K.L., Bentley, S.D., Holden, M.T., *et al.* (2001). Complete genome sequence of a multiple drug resistant *Salmonella enterica* serovar Typhi CT18. Nature *413*, 848–852.

Perkins, T.T., Kingsley, R.A., Fookes, M.C., Gardner, P.P., James, K.D., Yu, L., Assefa, S.A., He, M., Croucher, N.J., Pickard, D.J., *et al.* (2009). A strand-specific RNA-Seq analysis of the transcriptome of the typhoid bacillus *Salmonella* typhi. PLoS Genet *5*, e1000569.

Pfeifer, C.G., Marcus, S.L., Steele-Mortimer, O., Knodler, L.A., and Finlay, B.B. (1999). *Salmonella* Typhi*murium* virulence genes are induced upon bacterial invasion into phagocytic and nonphagocytic cells. Infect. Immun. *67*, 5690–5698.

Porwollik, S., Boyd, E.F., Choy, C., Cheng, P., Florea, L., Proctor, E., and McClelland, M. (2004). Characterization of *Salmonella enterica* subspecies I genovars by use of microarrays. J. Bacteriol. *186*, 5883–5898.

Porwollik, S., and McClelland, M. (2007). Determination of the gene content of *Salmonella* genomes by microarray analysis. Methods Mol. Biol. *394*, 89–103.

Porwollik, S., Santiviago, C.A., Cheng, P., Florea, L., Jackson, S., and McClelland, M. (2005). Differences in gene content between *Salmonella enterica* serovar Enteritidis isolates and comparison to closely related serovars *Gallinarum* and Dublin. J. Bacteriol. *187*, 6545–6555.

Porwollik, S., Wong, R.M., and Michael McClelland (2002). Evolutionary genomics of *Salmonella*: Gene acquisitions revealed by microarray analysis. Proceedings of the National Academy of Sciences *99*, 8956–8961.

Pullinger, G.D., Paulin, S.M., Charleston, B., Watson, P.R., Bowen, A.J., Dziva, F., Morgan, E., Villarreal-Ramos, B., Wallis, T.S., and Stevens, M.P. (2007). Systemic translocation of *Salmonella enterica* serovar Dublin in cattle occurs predominantly via efferent lymphatics in a cell-free niche and requires type III secretion system 1 (T3SS-1) but not T3SS-2. Infect. Immun. *75*, 5191–5199.

Reen, F.J., Boyd, E.F., Porwollik, S., Murphy, B.P., Gilroy, D., Fanning, S., and McClelland, M. (2005). Genomic comparisons of *Salmonella enterica* serovar Dublin, Agona, and Typhimurium strains recently isolated from milk filters and bovine samples from Ireland, using a *Salmonella* microarray. Appl. Environ. Microbiol. *71*, 1616–1625.

Rhen, M., Virtanen, M., and Makela, P.H. (1989). Localization by insertion mutagenesis of a virulence-associated region on the *Salmonella* Typhi*murium* 96 kilobase pair plasmid. Microb Pathog *6*, 153–158.

Rodrique, D.C., Tauxe, R.V., and Rowe, B. (1990). International increase in *Salmonella* Enteritidis: a new pandemic? Epidemiol. Infect. *105*, 21–27.

Rubino, S., Leori, G., Rizzu, P., Erre, G., Colombo, M.M., Uzzau, S., Masala, G., and Cappuccinelli, P. (1993). TnphoA *Salmonella* abortusovis mutants unable to adhere to epithelial cells and with reduced virulence in mice. Infect. Immun. *61*, 1786–1792.

Santiviago, C.A., Reynolds, M.M., Porwollik, S., Choi, S.H., Long, F., Andrews-Polymenis, H.L., and McClelland, M. (2009). Analysis of pools of targeted *Salmonella* deletion mutants identifies novel genes affecting fitness during competitive infection in mice. PLoS Pathog. *5*, e1000477.

Sassetti, C.M., Boyd, D.H., and Rubin, E.J. (2001). Comprehensive identification of conditionally essential genes in mycobacteria. Proc. Natl. Acad. Sci. U.S.A. *98*, 12712–12717.

Scaria, J., Palaniappan, R.U., Chiu, D., Phan, J.A., Ponnala, L., McDonough, P., Grohn, Y.T., Porwollik, S., McClelland, M., Chiou, C.S., *et al.* (2008). Microarray for molecular typing of *Salmonella enterica* serovars. Mol. Cell Probes *22*, 238–243.

Shah, D.H., Lee, M.J., Park, J.H., Lee, J.H., Eo, S.K., Kwon, J.T., and Chae, J.S. (2005). Identification of *Salmonella gallinarum* virulence genes in a chicken infection model using PCR-based signature-tagged mutagenesis. Microbiology *151*, 3957–3968.

Shea, J.E., Hensel, M., Gleeson, C., and Holden, D.W. (1996). Identification of a virulence locus encoding a second type III secretion system in *Salmonella* Typhimurium. Proc. Natl. Acad. Sci. U.S.A. *93*, 2593–2597.

Shi, L., Adkins, J.N., Coleman, J.R., Schepmoes, A.A., Dohnkova, A., Mottaz, H.M., Norbeck, A.D., Purvine, S.O., Manes, N.P., Smallwood, H.S., *et al.* (2006). Proteomic analysis of *Salmonella enterica* serovar Typhi*murium* isolated from RAW 264.7 macrophages: identification of a novel protein that contributes to the replication of serovar Typhi*murium* inside macrophages. J. Biol. Chem. *281*, 29131–29140.

Shi, L., Ansong, C., Smallwood, H., Rommereim, L., McDermott, J.E., Brewer, H.M., Norbeck, A.D., Taylor, R.C., Gustin, J.K., Heffron, F., *et al.* (2009). Proteome of *Salmonella enterica* Serotype Typhimurium Grown in a Low Mg/pH Medium. J. Proteomics Bioinform. *2*, 388–397.

Sittka, A., Lucchini, S., Papenfort, K., Sharma, C.M., Rolle, K., Binnewies, T.T., Hinton, J.C., and Vogel, J. (2008). Deep sequencing analysis of small noncoding RNA and mRNA targets of the global post-transcriptional regulator, Hfq. PLoS Genet *4*, e1000163.

Sittka, A., Sharma, C.M., Rolle, K., and Vogel, J. (2009). Deep sequencing of *Salmonella* RNA associated with heterologous Hfq proteins *in vivo* reveals small RNAs as a major target class and identifies RNA processing phenotypes. RNA Biol. *6*, 266–275.

Sizemore, D.R., Fink, P.S., Ou, J.T., Baron, L., Kopecko, D.J., and Warren, R.L. (1991). Tn5 mutagenesis of the *Salmonella* Typhi*murium* 100 kb plasmid: definition of new virulence regions. Microb Pathog *10*, 493–499.

Stein, M.A., Leung, K.Y., Zwick, M., Garcia, del, P.F., and Finlay, B.B. (1996). Identification of a *Salmonella* virulence gene required for formation of filamentous structures containing lysosomal membrane glycoproteins within epithelial cells. Mol. Microbiol. *20*, 151–164.

Thijs, I.M., De Keersmaecker, S.C., Fadda, A., Engelen, K., Zhao, H., McClelland, M., Marchal, K., and Vanderleyden, J. (2007). Delineation of the *Salmonella enterica* serovar Typhimurium HilA regulon through genome-wide location and transcript analysis. J. Bacteriol. *189*, 4587–4596.

Thompson, A., Rowley, G., Alston, M., Danino, V., and Hinton, J.C. (2006). *Salmonella* transcriptomics: relating regulons, stimulons and regulatory networks to the process of infection. Curr. Opin Microbiol. *9*, 109–116.

Thomson, N.R., Clayton, D.J., Windhorst, D., Vernikos, G., Davidson, S., Churcher, C., Quail, M.A., Stevens, M., Jones, M.A., Watson, M., *et al.* (2008). Comparative genome analysis of *Salmonella* Enteritidis PT4 and *Salmonella Gallinarum* 287/91 provides insights into evolutionary and host adaptation pathways. Genome Res. *18*, 1624–1637.

Tsolis, R.M., Townsend, S.M., Miao, E.A., Miller, S.I., Ficht, T.A., Adams, L.G., and Baumler, A.J. (1999). Identification of a putative *Salmonella enterica* serotype Typhi*murium* host range factor with homology to IpaH and YopM by signature-tagged mutagenesis. Infect. Immun. *67*, 6385–6393.

Turner, A.K., Lovell, A., Hulme, S.D., Zhang-Barber, L., and Barrow, P.A. (1998). Identification of *Salmonella* Typhimurium genes required for colonization of the chicken alimentary tract and for virulence in newly hatched chicks. Infect. Immun. *66*, 2099–2106.

Valdivia, R.H., and Falkow, S. (1996). Bacterial genetics by flow cytometry: rapid isolation of *Salmonella* Typhi*murium* acid-inducible promoters by differential fluorescence induction. Mol. Microbiol. *22*, 367–378.

Valdivia, R.H., and Falkow, S. (1997). Fluorescence-based isolation of bacterial genes expressed within host cells. Science *277*, 2007–2011.

Wattiau, P., Van Hessche, M., Schlicker, C., Vander Veken, H., and Imberechts, H. (2008a). Comparison of classical serotyping and PremiTest assay for routine identification of common *Salmonella enterica* serovars. J. Clin. Microbiol. *46*, 4037–4040.

Wattiau, P., Weijers, T., Andreoli, P., Schliker, C., Veken, H.V., Maas, H.M., Verbruggen, A.J., Heck, M.E., Wannet, W.J., Imberechts, H., and Vos, P. (2008b). Evaluation of the Premi Test *Salmonella*, a commercial low-density DNA microarray system intended for routine identification and typing of *Salmonella enterica*. Int. J. Food Microbiol. *123*, 293–298.

Wong, K.K., McClelland, M., Stillwell, L.C., Sisk, E.C., Thurston, S.J., and Saffer, J.D. (1998). Identification and sequence analysis of a 27-kilobase chromosomal fragment containing a *Salmonella* pathogenicity island located at 92 minutes on the chromosome map of *Salmonella enterica* serovar Typhi*murium* LT2. Infect. Immun. *66*, 3365–3371.

Xia, X., McClelland, M., and Wang, Y. (2005). WebArray: an online platform for microarray data analysis. BMC Bioinformatics *6*, 306.

Xia, X.Q., McClelland, M., Porwollik, S., Song, W., Cong, X., and Wang, Y. (2009). WebArrayDB: Cross-platform microarray data analysis and public data repository. Bioinformatics *25*, 2425–2429.

Zwir, I., Shin, D., Kato, A., Nishino, K., Latifi, T., Solomon, F., Hare, J.M., Huang, H., and Groisman, E.A. (2005). Dissecting the PhoP regulatory network of Escherichia coli and *Salmonella enterica*. Proc. Natl. Acad. Sci. U.S.A. *102*, 2862–2867.

Evolutionary Trends Associated with Niche Specialization as Modelled by Whole Genome Analysis of Egg-contaminating *Salmonella enterica* Serovar Enteritidis

5

Jean Guard*, Devendra Shah, Cesar A. Morales and Doug Call

Abstract

The mosaic nature of the *Salmonella enterica* genome facilitates its access to multiple environments. Many large scale genomic events have been described that contribute to the combinatorial complexity of the pathogenic *Salmonellae*. However, the impact of small scale genetic change occurring at the level of single nucleotide polymorphism (SNP) on the emergence of niche specialization is just now becoming appreciated. This chapter describes concepts behind the evolution that culminated in the remarkable ability of *Salmonella enterica* serovar Enteritidis to contaminate and survive in the internal content of eggs produced by otherwise healthy hens. Evidence suggests that combinations of SNPs facilitate niche specialization by *Salmonella enterica*. However, few typing methods incorporate unbiased strategies for their detection. Selection of appropriate biological assays for ranking SNPs and combinations of SNPs for their impact on the ability of *Salmonella enterica* to propagate outbreaks, pandemics and disease will be a significant challenge to improve the safety of the food supply.

Introduction

The mosaic nature of the *Salmonella enterica* genome facilitates its access to multiple environments (Darling *et al.*, 2004; Doublet *et al.*, 2008; Groisman *et al.*, 1992; Hensel *et al.*, 1999; Li *et al.*, 1994; Martin, 1999; Mirold *et al.*, 2001). *Salmonella* colonizes food, water, soil, animals, insects, plants, surfaces and even air-borne dust by adjusting its repertoire of gene content and patterns of gene expression (Crump *et al.*, 2002; Davies and Wray, 1995a; Davies and Wray, 1995b; Guard-Petter *et al.*, 1997; Holt *et al.*, 2007; Mian *et al.*, 2002; Moro *et al.*, 2007; Olsen *et al.*, 2000; Singh *et al.*, 1980; Stocki *et al.*, 2007; Tavechio *et al.*, 2002; Valiente Moro *et al.*, 2007). The combinatorial complexity of the *Salmonella* genome would seem to prevent identification of genomic determinants useful for assessing the risk of illness from any one strain or its relative prevalence within the food production chain. However, only a handful of the 2500 serotypes of *Salmonella enterica* cause disease in humans or animals, and most of these belong to the approximately 1400 serotypes within subspecies I (Le Minor *et al.*, 1982). We suggest that sets of discrete genetic changes facilitate niche specialization by *Salmonella enterica,* but that few current methods incorporate unbiased strategies for their detection. The foundation for this hypothesis is that just a few serotypes emerge to be successful pathogens on a recurring basis. Serotype, as defined by the Kauffman–White lipopolysaccharide O-antigen and flagellar H-antigen scheme, was one of the few typing methods available for decades that could be used for tracking shifting patterns of *Salmonella* strains within human and animal populations (Brenner *et al.*, 2000; Hendriksen *et al.*, 2009). The Kauffman–White scheme has since been augmented by molecular approaches such as pulsed-field-gel-electrophoresis (PFGE)

*Corresponding author

that, unlike serotyping, provide some information about changes in genome content (Kerouanton *et al.*, 2007; Zhao *et al.*, 2006). PFGE detects infrequent restriction enzyme sites in the genome to produce a barcode-like scheme of classification that can be used to compare strains. However, PFGE measures the presence or absence of infinitesimally few nucleotide sequences relative to entire genomes and has no regard for their location in relationship to genes, regulatory regions or ribosomal sequences. There is evidence that PFGE analysis of *Salmonella* Enteritidis is dependent on sequence in ribosomal genes (Morales *et al.*, 2007). Other methods for typing of the *Salmonellae* evaluate single nucleotide polymorphisms that do not necessarily alter restriction enzyme sites and that are not stringently linked to phenotype (Fitzgerald *et al.*, 2007; Lindstedt *et al.*, 2004; Ross and Heuzenroeder, 2009; Yazdankhah and Lindstedt, 2007). What is more evident following application of emerging genomic tools is that combining several existing methods, such as serotyping, PFGE and analysis of randomly occurring SNPs, is still not enough to associate *Salmonella enterica* with environmental niches, foods, and animal hosts. Epidemiological analysis as currently practiced does not detect patterns of gene flow occurring within the *Salmonellae* that pose the highest risk to people and animals. The dominance of a few outer membrane epitopes useful for serotyping, infrequent variation in restriction enzyme sites and bias in selection of genomic targets introduces a skew in the search for those evolutionary determinants of highest consequence to the outbreak of food-borne disease. This statement is not a criticism of approaches being used today; indeed, the diligence and thoroughness with which they have been applied has advanced knowledge of the epidemiology of *Salmonella*. The issue is that methodology has only recently progressed to the point where multiple strains can be analysed for whole-genome content in a comparative manner to identify that most subtle of evolutionary event, the single nucleotide polymorphism (SNP) (Anonymous, 2007b; Matheson *et al.*, 2009). There is even less emphasis on screening SNPs for their linkage to phenotypes that impact epidemiology. Niche specialization in some cases can involve a subtle change in metabolic capability due to the presence of a fused gene, or a change in gene regulation or the substitution of a basic polar amino acid for one that is neutral and non-polar. Because subtle genomic change can cause major divergence in metabolic properties, virulence, invasiveness, colonization and survival it is impossible to discuss evolution that contributes to niche specialization without an understanding that technology has only just reached the point where analysis is possible.

The hand dealt by *Salmonella*

There are multiple combinations of genomic determinants that enable colonization of one niche over another by any one strain of *Salmonella enterica*. A useful analogy is to a deck of cards, where constant reshuffling and cutting of the deck eventually results in one individual drawing a combination of high value. Several hands have the potential to win the game, which in this case is the ability of any one serotype to cause a food-borne outbreak with some regularity that can be observed. The relative value of the hands matter, because even a fairly low value hand may win if all other hands present at the table are of lower value. For example, if an infrequently encountered serotype that does not have an optimal set of genes and serotype characteristics is still able to colonize a production facility that does not follow biosecurity guidelines (Anonymous, 1998; Anonymous, 2007a), even this 'suboptimal' isolate can cause food-borne outbreaks in people as it amplifies and is distributed through the food chain. The focus of this analogy is not so much on the innate ability of a strain to cause human gastrointestinal illness; in fact, the analogy assumes that any *Salmonellae* that can enter the game could cause gastrointestinal illness in humans under experimental settings. The focus of this discussion is on the ability of the pathogen to colonize and propagate through the food chain. The commonality between all winning hands is that certain patterns of gene inheritance add value over all the other hands in play. This situation accounts for the emergence of one serotype more than another; basically, the winner competes better than other circulating strains for available nutrients. In addition, the most frequent winner might also be better able to survive stressful conditions and to access new niches. In this evolutionary poker game, cutting the deck represents large genetic shifts such as those that occur with homologous recombination, transpositions or acquisition of extra-chromosomal DNA. Shuffling is analogous to genetic drift, which introduces genomic variation at the level of the single nucleotide polymorphism (SNP). Failure to consider the impact of both types of evolutionary events may influence a researcher to overemphasize one method of genomic change over another. Investigating niche specialization by the *Salmonellae* requires careful consideration of which methods and combinations of methods are applied to detect the set of genes that contribute to outbreak potential. Failure to overcome bias inherent in scientific methodology impedes detection of the flow of high-consequence genes, which in the right combination facilitates niche specialization and persistence in the myriad environments of the food supply. The job of the evolutionary biologist becomes clear; namely, identify recurring combinations of genes that have added value for contributing to food-borne disease.

Genetic shift and its importance in the evolution of *Salmonella enterica* as a pathogen

Advances in whole genome analysis have produced much information about the ability of the *Salmonella enterica* genome to rearrange and change gene content between serotypes and even with other bacteria (Falush *et al.*, 2006; Holt *et al.*, 2008; Porwollik *et al.*, 2005; Scaria *et al.*, 2008). Many examples of genetic shift, including homologous recombination, transposition, insertion events and acquisition of extra-chromosomal DNA, occur within the Enterobacteriaceae and between *Salmonella* serotypes, and have been well described (Baumler, 1997; Kalia *et al.*, 2004; Lecointre *et al.*, 1998; Liu and Sanderson, 1995). Substantial advances in knowledge about evolution within the *Salmonellae* have been made using *in vivo* expression technology or proteomics coupled with high throughput microarrays (Andrews-Polymenis *et al.*, 2009; Thijs *et al.*, 2007). However, the limit of detection of these methods remains approximately at the level of whole genes. These platforms can determine the presence or absence of specific genes and can be used to compare *Salmonella enterica* serotypes. The DNA–DNA microarray hybridization methodology has a limit of detection that could be made more stringent than the whole gene. For example, loss of 50% of a 3000 bp gene might be included in the design of the array. Nevertheless, it must be considered that overreliance on any single technique for genome characterization of the *Salmonellae* will bias the user to investigation of large or small events. In regards to *S. enterica* evolving niche specialization, the scale of event matters. If genetic shift and drift are evolutionary processes that both contribute to the pathogenic potential of the organism, then at least two approaches are required to document evolutionary trends, with perhaps a third applied to detect events that are at the interface of methodologies.

That which is seen and that which is not seen (Frederick Bastiat, 1850)

In this chapter, genetic drift is defined from the viewpoint of an epidemiologist as small insertions, deletions, and nucleotide substitutions that may or may not alter an open-reading frame. More commonly, the term genetic drift denotes what happens to the frequency of variant alleles across populations. All small-scale genetic events likely to escape detection by DNA–DNA microarray hybridization are referred to as single nucleotide polymorphism and a single SNP can involve multiple contiguous base pairs. The use of these terms is intended to make it clear to the reader that the two events, genetic shift and genetic drift, have distinctive characteristics related to scale.

Currently, much more is known about genetic shift than genetic drift within the *Salmonella enterica* genome. Completion of whole genome comparisons suggests that genetic drift can be defined as less than 250 contiguous base pairs. For example, the largest SNP discovered in PT13a *Salmonella* Enteritidis is a 215-bp deletion that removed the N-terminus and 80 bp upstream of the gene SEN4316. Also defining a SNP is the likelihood that it will be missed by DNA–DNA microarray hybridization (Morales *et al.*, 2005).

Microarray platforms have improved to where detection of at least small deletions is reliable, although uncertainty remains in regards to detection and identification of any one single base pair change (Dotsch *et al.*, 2009; Lin and Malanoski, 2009). Some epidemiological surveys have been designed to detect SNPs (Alix *et al.*, 2006; Esaki *et al.*, 2004; Holt *et al.*, 2009; Lermo *et al.*, 2007; Octavia and Lan, 2007). However, SNPs are not often selected for their linkage to phenotypic characteristics, with the exception of antibiotic resistance (Levy *et al.*, 2004). Limitations posed by methodology that is biased towards detection of genetic shift became clear while studying genomic variation that could be linked to virulence phenotypes in *Salmonella enterica* serovar Enteritidis (*Salmonella* Enteritidis). Basically, DNA–DNA microarrays did not detect any changes in the genome associated with variant virulence behaviour between strains (Morales *et al.*, 2005). Thus, *Salmonella* Enteritidis appeared to be an appropriate model organism for investigating evolutionary changes that escape characterization by any approach except whole genome comparative analysis. *Salmonella* Enteritidis may be an unusual case within the *Salmonellae* where pure genetic drift is driving the biology that impacts human health. A caveat to this statement is discussed later in this chapter when it is considered that even whole genome analysis has a limit of detection that arises yet again from genetic shift. An additional limit of detection is physiological variation that is introduced by epigenetic differences (Marinus and Casadesus, 2009).

The curious case of egg contamination by *Salmonella* Enteritidis

Salmonella Enteritidis is currently the world's leading cause of human food-borne salmonellosis. It is unusual among the *Salmonellae*, because it causes illness in people following the contamination of the internal contents of eggs produced by otherwise healthy hens (Anonymous, 1988; Guard-Petter, 2001; St Louis *et al.*, 1988). Its emergence worldwide was well documented in the early 1980s, but estimates suggest that it was circulating during the 1970s (Baumler *et al.*, 2000; Todd, 1997). Another egg-contaminating serovar, namely *Salmonella* Gallinarum, causes high mortality in chickens, but does not cause illness in people (Shivaprasad, 2000). Development of a narrow host range by the avian-restricted *Salmonellae* has been linked to the absence of a number of genes that are otherwise present in *Salmonella* Enteritidis. Whole genome comparisons comparing *Salmonella* Enteritidis to the host-restricted avian pathogen shows that the latter evolved from the former (Thomson *et al.*, 2008), although a set of genes may have undergone further positive selection within the avian-restricted serotypes (Eswarappa *et al.*, 2009). Host adaptation should be monitored, because newly evolved strains may periodically emerge from seemingly unimportant strains to cause high morbidity and mortality in the affected host. Host restriction may have unstable epidemiological aspects prior to emergence of a phenotype that is well-adapted to a particular niche. It is possible that the ebb and flow of serotypes between years are populations in flux.

Odd associations between phylogenetic patterns and classical serotype determinants are observed. This happens when homologous recombination between strains is layered on top of the longer evolutionary process resulting in host adaptation. For example, the host-restricted human pathogen *Salmonella* Typhi has the same O-antigen D1 serotype as that of *Salmonella* Enteritidis, but genomic analysis shows that the two pathogens have a greatly divergent evolutionary path (Bhan *et al.*, 2005; Boyd *et al.*, 1993; Morales *et al.*, 2006). In the case of egg contamination, all *Salmonellae* that are well-documented for their ability to contaminate the internal contents of eggs are within the *Salmonella* Enteritidis evolutionary lineage (Porwollik *et al.*, 2005; Thomson *et al.*, 2008). It has been suggested that clearing poultry flocks of *Salmonella* Gallinarum opened a favourable niche for the emergence of *Salmonella* Enteritidis, but no direct experimentation or testing of this theory has been done and some studies suggest co-habitation within regions (Baumler *et al.*, 2000; Kumar *et al.*, 2009). There have been reports of egg contamination propagated by *Salmonella* Heidelberg; however, it has never been a persistent association as has occurred with *Salmonella* Enteritidis (Chittick *et al.*, 2006).

Comparison of two prevalent broad-host range *Salmonellae*, *Salmonella* Enteritidis and *Salmonella* Typhimurium

Egg-contaminating *Salmonella* Enteritidis has retained a host range essentially as broad as that of *Salmonella* Typhimurium (Conner *et al.*, 1998; Ochman and Groisman, 1995). *Salmonella* Typhimurium is the second most frequently encountered serotype associated with human illness from contaminated foods (Anonymous, 2000). Prior to the emergence of *Salmonella* Enteritidis during the 1980s, *Salmonella* Typhimurium was the predominant serotype associated with food-borne illness as evidenced by descriptions recorded since the advent of a typing system in the 1920s and by yearly surveys. However, *Salmonella* Typhimurium does not contaminate the internal contents of eggs at any measurable frequency, even though experimental evidence shows that it can reach the ovum (Gantois *et al.*, 2008). Comparing the whole genomes of *Salmonella* Enteritidis to *Salmonella* Typhimurium would seem to provide an opportunity to find those determinants linked to the emergence of highly successful pathogens with different niches. However, there are so many genomic differences between these sequences that it is not possible to relate any particular difference to the ability to contaminate eggs. Direct comparison of outer membrane properties of these two serotypes revealed that strains of *Salmonella* Enteritidis associated with egg contamination produced a capsular-like lipopolysaccharide O-antigen region that mitigated signs of disease in hens (Parker *et al.*, 2002; Parker *et al.*, 2001). Other investigations have shown that *Salmonella* Enteritidis is especially capable of surviving within the internal contents of eggs (Gantois *et al.*, 2008). Further studies of survival of *Salmonella* Enteritidis in eggs identified the lipopolysaccharide gene *rfbH* as the only gene that is clearly up regulated (Gantois *et al.*, 2009). Analysis of SNPs did not reveal polymorphisms in *rfbH*, but this analysis did reveal evidence of genetic drift occurring in other genes within the same operon (Gantois *et al.*, 2009). Overall, these complementary approaches combining phenotypic and evolutionary analyses support the assertion that *Salmonella* Enteritidis is the best template available for evaluating how genetic drift impacts niche specialization when these changes do not also result in the ability to infect a broad range of hosts. The availability of the well-annotated genome sequence of *Salmonella* Typhimurium (McClelland *et al.*, 2001), as well as *Salmonella* Enteritidis and other *Salmonellae*, greatly facilitates characterization of SNPs. Because this information has been centralized, detailed genomic information for the many serotypes and of genomes from multiple strains of single serotypes should facilitate the search for patterns of gene flow linked to the emergence of egg contamination. How can associations between genes and virulence potential as defined by egg contamination, retention of a broad host range and the ability to propagate in the environment of the chicken be analysed?

Linking genetic drift to strain heterogeneity within *Salmonella* Enteritidis

The emergence of *Salmonella* Enteritidis as the world's leading cause of food-borne salmonellosis posed a logical conundrum while at the same time providing a rare opportunity for developing a strategy for analysis of genetic drift (Guard-Petter, 2001). Wherever it is isolated from around the world, *Salmonella* Enteritidis encompasses little genomic variation within its own serotype, in comparison to other serotypes (Hudson *et al.*, 2001; Kitamoto *et al.*, 2005; Liebana *et al.*, 2002; Malorny *et al.*, 2008; Pang *et al.*, 2005; Peters *et al.*, 2007; Soares and Machado, 2003). However, animal infection trials, cell invasion assays, survival assays and other biological approaches consistently report that there are significant differences in the ability of any one *Salmonella* Enteritidis strain to produce distinctive outer membrane structures and to grow, invade, colonize and survive within the internal contents of eggs (Guard-Bouldin *et al.*, 2007; Guard-Petter, 1998; Guard-Petter *et al.*, 1999; Humphrey *et al.*, 1998; Humphrey *et al.*, 1996). A turning point in understanding evolution of *Salmonella* Enteritidis occurred when it became evident that phenotypic heterogeneity within *Salmonella* Enteritidis of the same phage type (PT) lineage was as great as that that occurring between PT lineages (Morales *et al.*, 2005). Even more remarkably, variation occurring within a single PT appeared to be as great as that occurring between serotypes. Phage type was therefore at best partially linked to genomic determinants of egg contamination, and at worse, not linked at all. In addition, recurring patterns of metabolism could not be assumed to have a fixed relationship to serotype or phage type. In summary, metabolism and virulence varied as much within phage type as between phage types. This finding suggested that comparison of two near identical strains within the same phage type that nonetheless varied in phenotype was a shortcut to finding those polymorphisms most likely linked causally to the ability to contaminate eggs.

A comparison of whole genomes from strains within a single PT of *Salmonella* Enteritidis was done to find the minimal number of high consequence genomic polymorphisms that could be linked to the evolution of a remarkable niche, namely the ability to contaminate and survive in eggs. Concentrating on finding all SNPs that differentiated two phenotypic variants of the same PT increased the likelihood of finding SNPs of consequence without bias, because background genomic noise associated with random evolution would be minimized. Phylogenetic studies of the highly conserved gene *cyaA* and two adjacent ribosomal regions supported pursuing genetic drift as the main evolutionary force behind the emergence of pandemic egg contamination. Analysis of *cyaA* SNPs across serotypes showed that genetic noise was present between strains (Morales *et al.*, 2007). Conversely, phylogenetic analysis and complementary phenotype microarrays also confirmed that there were true clones in circulation, even if separated by source of isolation and geographic region (Olson *et al.*, 2007).

That which was not seen is now seen

Two PT13a strains that varied in their ability to contaminate eggs and to form biofilm were mutationally mapped and compared to the Sanger Institute reference PT4 *Salmonella* Enteritidis strain P125109 (GenBank AM933172). Strain PT13a 21027 formed biofilm, but did not contaminate eggs (EC-BF+). In contrast, PT13a 21046 contaminated eggs, but did not form biofilm (EC+BF-). The PT4 template strain, PT4 22079, contaminated eggs and formed biofilm (EC+BF+). A total of 247 evolutionary events were detected that distinguished the two PT13a strains from each other. The completed database of evolutionary events is available within a public database and will be released for searchable content in NCBI dbSNP Build 131 (http://www.ncbi.nlm.nih.gov/genomes/static/*Salmonella*_SNPS.html). None of these events that differentiated the genomes of the two PT13a strains were detectable by DNA–DNA hybridization microarray. Two ribosomal gene SNPs had been previously detected by pulsed-field gel electrophoresis (PFGE) and SNPs present within *cyaA* served as internal controls to assess the sensitivity of methodology. Of all the events, 82 (33.2%) were synonymous and thus did not alter predicted amino acid sequences of proteins, 38 (15.4%) occurred in intergenic regions, 99 (40.1%) introduced amino acid substitutions, 12 (4.9%) were in ribosomal genes and 15 (6.1%) SNPs were small deletions or early terminations in genes. The largest event was a 215 bp deletion, but most events were non-contiguous single nucleotide substitutions. PT13a 21046 (BF-EC+) had 3 gene mutations disrupting ORFs that differentiated it from the reference PT4 strain (BF+EC+); conversely, PT13a 21027 (BF+EC-) had 12 unique mutations that disrupted coding regions and differentiated it from the reference PT4 strain. Overall, evolution occurring within a single PT lineage of *Salmonella* Enteritidis appeared to be essentially an example of pure genetic drift that nonetheless had a profound impact on phenotype. Approximately 250 SNPs are thus linked to phenotypic differentiation occurring within phage type lineage and some of these are clearly of more consequence than others. It is possible that as little as 10% of these SNPs have a causal impact on phenotype. It is possible that there may be one signatory SNP that then requires rapid accumulation of compensatory SNPs due to introducing massive physiological changes within the cell or chemical changes within the chromosome. These data provide a basis for understanding how little genetic change is needed to impact niche specialization by the *Salmonellae*.

The misconception of the 'simple' bacterial genome

In comparison to a eukaryotic genome, the bacterial genome appears simple. But within that simplicity is the ability to colonize extreme environments ranging from spacecraft to extremes in weather to the internal contents of the egg (Miraglia *et al.*, 2009; Pierson, 2001; St Louis *et al.*, 1988). The argument has been put forth that a single bacterial spore has the ability to be the progenitor of all life forms on earth (Villarreal, 2009; Warmflash and Weiss, 2005). A typical food-borne pathogen like *Salmonella enterica* has about 5 million base pairs and 5000 genes, whereas the human genome is approximately 700-fold larger with 3.5 billion base pairs and nearly 25,000 genes. An accepted principle in genomics is that the complexity of a genome is not necessarily related to its size (Gregory, 2002; Petrov, 2001; Vellai and Vida, 1999). If the size of the microbial genome does not limit complexity, then bacteria must have a strategy for generating the degree of phenotypic variation that continually challenges our ability to protect the public health. Whole genome analysis of variation occurring within a single PT lineage of *Salmonella* Enteritidis provides some guidelines for appreciating the combinatorial complexity of the *Salmonellae*. The following notes derived from data are put forth for the sake of arguing that it is the sheer ability of *Salmonella enterica* to evolve strain heterogeneity in a nearly undetectable manner that has confounded our grasp of this pathogen (Baker *et al.*, 2008; Holt *et al.*, 2008). These comments are open for additional interpretations as information from massive sequencing projects accumulates and is processed (Achtman, 2008; Falush *et al.*, 2006; Kidgell *et al.*, 2002).

During analysis of mutational mapping occurring within a single PT lineage, data were gathered on how often SNPs specific to a phage type were encountered. Using the region located between base pairs 0.8 and 1.2 Mb of the reference PT4 genome, 41 SNPs were located that differentiated at least one of the PT13a strains from each other and from the PT4 reference strain. Of these, 22 (53%) differentiated both PT13a strains from PT4. Thus, the PT4 and PT13a lineages accumulated approximately twice as many genetic drift events in total between them as occurred within the PT13a lineage. Within this region, naturally occurring evolution present in the gene adenylate cyclase (*cyaA*) was investigated further (Morales *et al.*, 2007). The gene *cyaA* is a large (2547 bp) and highly conserved gene. It is the only adenylate cyclase of the *Salmonellae*. If *cyaA* is mutated, *Salmonella* is attenuated and loses serological and biochemical characteristics (Curtiss and Kelly, 1987). Specifically, *cyaA* mutants of *Salmonella enterica* are typed as *Hafnia alvei*, because biochemical reactions change and O-antigen reactivity is lost (Curtiss and Kelly, 1987; Romanowska *et al.*, 1999). Knowing that *Salmonella* Enteritidis has an adenylate cyclase that is accumulating naturally occurring genetic drift can be used as a point of reference for evaluating how much different taxonomic groupings of *Salmonella enterica* might vary in regards to SNP content. There was only a 2 bp difference between the *cyaA* genes of PT13a *Salmonella* Enteritidis, its close descendent *Salmonella* Gallinarum and the reference PT4 strain. Sometimes there are cases where homologous recombination and lack of evolutionary distance confuses the issue. For example, O-antigen group D1 *Salmonella* Dublin has a *cyaA* gene identical to that of *Salmonella* Enteritidis PT4 (Porwollik *et al.*, 2005). In contrast, there are 37 bp that differentiate the *cyaA* of *Salmonella* Enteritidis PT4 from that of *Salmonella* Javiana, which has the same O-antigen serogroup D1 as *Salmonella* Enteritidis. There are only 17 bp that differentiate *cyaA* of *Salmonella* Enteritidis from that of O-antigen serogroup B *Salmonella* Typhimurium. Comparison outside of *S. enterica* subspecies *enterica* to subspecies *arizonae* indicates that 116 bp vary. Comparison of *cyaA* of *Salmonella* Enteritidis PT4 to *E. coli* O157:H7 str. Sakei and *Shigella dysenteriae* showed a difference of 382 and 463 bp, respectively. These data suggest that SNP incidence increases 2 fold between phage type or virulence phenotype of closely related serotypes, approximately 10-fold between more distant serotypes and approximately another 10 fold between *Salmonella* species. Adenylate cyclase is just one gene in the *Salmonella* genome. Given approximately 4500 genes per serotype genome and a baseline SNP difference in the range of 10 SNPs per 1000 bp gene, *Salmonella* Enteritidis and *Salmonella* Typhimurium alone would be expected to have 45,000 SNPs between them. Since most of the *Salmonellae* do not have the clonal population structure that *Salmonella* Enteritidis has to *Salmonella* Gallinarum and *Salmonella* Dublin, the 45,000 SNPs differentiating the majority of serotypes extrapolates to a potential of 30,000,000 SNPs (1400 *S. enterica* subspecies I serotypes × 45,000 SNPs differentiating any two non-clonally related serotypes). This analysis suggests that comparing whole genomes of serotypes that are not clonally related will not be efficient for detection of those SNPs of high consequence for phenotype. Linkage of SNPs to phenotypic traits related to outbreak potential is more likely to be successful if analysis of critical gene content is made between strains within the same serotype that vary naturally in pathogenic potential. However, it must be recognized that some traits related to host preference or virulence attributes, such as O-antigen serotype, evolve by genetic shift across serotypes. An example, as previously mentioned, is *Salmonella* Typhi, which has the same O-antigen serotype group D1 as *Salmonella* Enteritidis. *Salmonella* Typhi has a *cyaA* gene that is more different from O-antigen serotype D1 *Salmonella* Enteritidis (28 bp) than it is from O-antigen serotype group B *Salmonella* Typhimurium (17 bp). Combinatorial complexity occurring within the *Salmonellae* is daunting.

Most times an SNP is just an SNP

The single most defensible statement that can be made from these data is that genetic drift is underappreciated for its impact on niche specialization by pathogenic bacteria. There is a need for further development of methods and approaches that are focused on evaluating genetic drift in comparison to genetic shift. Biological strategies will be crucial for determining causal links to phenotype, because SNPs of no consequence will be frequently encountered even when genetic noise has been lowered a great deal. Appropriate biological assays maintain selection pressure for the trait of interest. The concept that genetic drift is at least half of the evolutionary equation impacting the emergence of pathogenic bacteria can now be appreciated. Appreciating the impact of genetic drift on the emergence of niche specialization is to understand the sheer frequency with which it occurs. In the analogy to a game of poker, the dealer is shuffling and dealing out hands at a comically fast rate. Consider that there are approximately 2500 *Salmonella enterica* serotypes, with 4500 genes, and that one gene can harbour approximately 10 SNPs between serotypes. The prediction, therefore, is that there are upwards of 100 million genetic drift events across all the *Salmonellae*. However, finding that fewer than 250 SNPs are needed to generate niche specialization such as egg contamination suggests that only a tiny fraction of genetic drift has any causal role in facilitating food-borne outbreaks.

Using that which had been unseen to see more

Except for some notable outlying variants, *Salmonella* Enteritidis appears to have an exceptionally clonal population structure in comparison to other pathogenic *Salmonellae* (Porwollik *et al.*, 2005). Genetic drift appears to be the evolutionary force primarily responsible for the emergence of strain heterogeneity within the clonal population structure. Is there a role for genetic shift between these strains that are nearly identical yet differ greatly in virulence phenotype? The question requires a consideration of the limits of detection present within whole genome analysis. Sequencing alone will not detect genetic shift associated with a major exchange of genetic material between strains that are closely related, because the percentage of base pairs that differ before and after the exchange are too small to detect as anything other than genetic drift. An approach is required that evaluates differences in patterns of genetic drift across the genome. To overcome the inherent limit of detection imposed by genetic drift, non-parametric statistical analysis was applied to the analysis of the SNP database developed for *Salmonella* Enteritidis (http://www.ncbi.nlm.nih.gov/genomes/static/*Salmonella*_SNPS.html). To do this, the number of SNPs occurring in increments of 100,000 bp was tabulated for each strain being compared. For example, *Salmonella* Enteritidis strains PT13a 21027 and PT13a

21046 had 1 and 3 SNPs unique to them between 0 and 100,000 bp, respectively. The 4,685,848 bp PT4 reference genome has a total of 47 increments, with the last increment accepted as having less than a full 100,000 bp. Data in each set of 47 increments is graphed and curvilinear analysis using the 5th polynomial is applied. Results show that there is substantial divergence in the number of SNPs within the region of the reference genome spanning nucleotides 0.7 Mb to 2.9 MB (Fig. 5.1). The 2.2 Mb with divergence is 46.9% of the genome. This region contains the lysogenized phage ST64b of the PT4 bacteriophage lineage, which is located between 2.01 and 2.06 Mb in the reference genome.

Of note is the location of lysogenized Fels-2 bacteriophage, which defines the PT13a lineage. It is located at the latter border of divergence and is near base pair 2.9 Mb of the reference genome (Fig. 5.1). There were no insertion elements or phage remnants located near the edge of the first crossover in polynomial lines at around 0.7 Mb; thus, it is not clear that any exchange at this point was due to the presence of a particular genomic feature. One genetic feature near this site that is noteworthy is *Salmonella* Enteritidis gene SEN0712. It is a deoxycarbylase that forms pyruvate and CO_2 from oxaloacetate. SEN0712 has an exceptional number of SNPs in both strains of the PT13a lineage, namely 4 in PT13a 21027 and 5 in PT13a 21046. All of these SNPs are synonymous and thus none alter amino acid sequence. SEN0712 is the only protein-encoding gene that has this many SNPs. Because it is so unusual, SEN0712 is used to mark a site of potential crossover in homologous recombination between phage types of *Salmonella* Enteritidis, although no direct connection to the process of homologous recombination has been established. Curvilinear analysis can only approximate the location of genomic crossovers. The true location of a crossover within a graph is affected by the absence of ST64b bacteriophage in the PT13a lineage, which shifts nucleotide numbering. The fact that the Fels-2 bacteriophage has only been found in the PT8/PT13a lineage indicates that it is located outside of the divergent area rather than inside. Finding a lysogenized bacteriophage at the point of crossover of a polynomial line suggests that a homologous recombination event may have been initiated at some point during the infection of the host by the bacteriophage.

If the algorithm is applied in regards to homologous recombination occurring within the PT13a lineage, then ribosomal gene sequence appears to be important for maintaining identity to

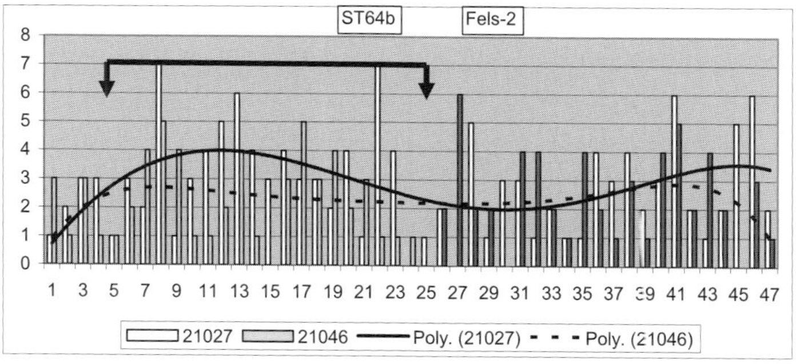

Figure 5.1 Curvilinear analysis of single nucleotide polymorphisms in *Salmonella enterica* serovar Enteritidis. The two strains compared are PT13a 21027 (white bar) and PT13a 21046 (patterned bar). The number of SNPs that differentiate strains were collated every 100,000 bp: thus for the 4,685,848 bp genome of the *Salmonella* Enteritidis there are 47 data points. Text in rectangles approximate the location of ST64b-like and Fels-2 bacteriophages, which are the phages defining PT4 and PT8/PT13a phage lineages, respectively. Note that ST64b is absent in the two strains being compared and that nucleotide locations are approximate due to shifts in numbering related to the presence of lysogenized phages. Bracketed is a region of curve divergence that suggests a homologous recombination event occurred between the two strains, because a non-parametric significantly different number of SNPs are present between the two strains. The 5th polynomial was applied to generate curves.

the originating strain. As noted in previous studies, the *Salmonella* Enteritidis ribosomal genes SEN_r012 and SEN_r015 (similar in location to *rrlC* and *rrlA* of *Salmonella* Typhimurium, respectively) have exceptional heterogeneity between subpopulations occurring within a single phage lineage of *Salmonella* Enteritidis (Morales *et al.*, 2007). In fact, these ribosomal genes encode the restriction enzyme sites identified by PFGE as differentiating strains. The region between *rrlC* and Fels-2, located at approximately 2.84 through 3.97 Mb of the PT4 reference genome, is the region most similar between PT13a 21027 and PT13a 21046. In the absence of evidence that homologous recombination occurred in this region, the SNPs in this region evolved spontaneously.

Niche specialization by *Salmonella* Enteritidis, aces high and four of a kind

The pathobiology of *Salmonella* Enteritidis enables the organism to contaminate and survive in the internal contents of eggs, to grow to high cell density, to produce a capsular-like LPS and to otherwise colonize a broad range of hosts. At this point, it is the serotype holding the highest value combination of genes of any *Salmonella* serotype in reference to its ability to cause outbreaks and human illness. By considering results from a variety of molecular typing approaches, it appears that this remarkable pathogen evolved to its present state through persistent and relentless genetic drift. However, it must be considered that genetic drift may have been coupled to a recent pinnacle shift in gene content (Baumler *et al.*, 2000). From this coupling of genetic drift and shift evolved a hybrid pathogen that spread worldwide in part because it had a remarkable new capacity, the ability to efficiently contaminate and survive in eggs. Genetic drift continues unabated before and after genetic shift; thus, strain heterogeneity in this model is to be an expected consequence, because it is driven in part by the pathogen encountering strong selection pressures within the new environments to which it had never previously had access. A theory for the emergence of a clonal strain of *Salmonella* Enteritidis that nonetheless has two evolved phage type lineages has been presented (Guard-Bouldin, 2006). The theory basically states that:

1. *Two strains of* Salmonella *Enteritidis were in circulation for hundreds of years.* Recent phylogenetic analyses suggest the strains were circulating since the Middle Ages. These strains had some host preferences, but occasionally overlapped within the environment of the different host animals (Miyamura *et al.*, 1986).
2. *In recent times, a large-scale homologous recombination event occurred between the two strains of* Salmonella *Enteritidis.* Application of an algorithm suggests this event took place about 4 decades ago, which is a time frame similar to a previous postulation (Baumler *et al.*, 2000).
3. *A hybrid strain emerged that had two bacteriophages lysogenized into the genome, namely the ST64b-like and Fels-2 bacteriophage.* The ST64b-like mosaic phage is found in PT4 strains of *Salmonella* Enteritidis and Fels-2 is found in strains of the PT8/PT13a lineage (Brown *et al.*, 1999; Kropinski *et al.*, 2007; Mmolawa *et al.*, 2003; Morales *et al.*, 2005; Porwollik *et al.*, 2005).
4. *The hybrid strain quickly diverged into the two major bacteriophage lineages in circulation today.* Harbouring two complete lysogenic bacteriophage is an inherently unstable situation within bacteria, because of phage incompatibility and competition for host resources (Villarreal, 2009; Wichman *et al.*, 2005).
5. *The hybrid strain had a new ability; namely, the ability to contaminate and survive in eggs produced by otherwise healthy hens.* Within the foundation parent strain, SNP differences were present in a different combination. It is the new combination of SNPs, and their impact on gene presence, absence and expression, that resulted in the ability of *Salmonella* Enteritidis to colonize and survive in a new niche, namely the internal contents of eggs.
6. *Genetic drift continued as the two new phage lineage strains grew and propagated throughout the world.* It is theorized that some

disequilibrium was present in both phage lineages, because of selection associated with accessing new environments in addition to eggs. Experimental evidence using bacteriophage genomes suggests that nucleotide substitution rates can be very frequent, and that it may occur more frequently during early phases of evolution (Wichman *et al.*, 2005).

7 *Both phage type lineages remain capable of causing egg contamination; however, the existence of subpopulations within a phage type suggests that additional niche specialization is in progress.* PT13a 21027 forms biofilm but does not contaminate eggs. PT13a 21046 does not form biofilm but contaminates eggs. PT4 22079 does both. Hen infection trials suggest that the two PT13a subpopulations and similar subpopulations of other phage types are more effective at egg contamination when used to simultaneously infect birds (Guard-Bouldin *et al.*, 2004). A strain such as PT4 22079, which served as a template for the reference PT4 genome, retained both the ability to form biofilm and to contaminate eggs and is thus an example of a dimorphic phenotype. Because the two PT13a strains have lost one of the two phenotypes of interest, they are examples of monomorphic subpopulations. Dimorphic and mixed monomorphic subpopulations of *Salmonella* Enteritidis may have slightly different patterns of infection within hens that allow the pathogen to cope with infection of the reproductive tract. Whenever reproductive tract tissue is the subject of analysis, cyclical changes in host cell immunobiology are expected to occur due to hormonal variation.

Application of data from analyses of both genetic drift and genetic shift

Epidemiology regarding analysis of food-borne bacteria is at a crossroads. The availability of genomic information is expanding rapidly. The bottleneck in producing new data is no longer the genome itself or the fact that genomes differ in size, but is instead the degree of detailed analysis needed to find information hidden in terabytes of data. To facilitate interpretation of raw genomic information, biological assays are needed that rapidly facilitate analysis of datasets for their linkage to food borne outbreaks. The case of *Salmonella* Enteritidis is an example where genetic drift, otherwise known as 'that which was not seen' was the driving force behind the emergence of a worldwide pathogen. It should be noted that it took a decade of detailed biological assays to select the strains most appropriate for comparison for studying genetic linkages to egg contamination. Another round of biological assays will be needed to decipher which of the 250 SNPs that differentiate phenotypes within a single phage lineage matter most, either alone or in combination with each other.

Transcriptome analysis: impact of SNPs on the biology of *S. enterica*

How can subtle introduction of SNPs in a genome affect the regulation of phenotype that in turn leads to adaptation of a particular niche? Comparative transcriptomics and proteomics provide complementary means to identify niche-specific regulatory changes; in addition, these methods can be followed by definitive mutational analysis and complementation studies. A classic experimental design involves growth of two strains, a test strain and a reference strain, under various conditions followed by analysis using a microarray to identify alteration in transcription of individual genes. SNPs can affect the expression of genes in several ways; most directly by a change in the transcriptional promoter sequence that either enhances or diminishes expression of a single gene or operon. Other effects include introduction of premature stop codons that will be apparent by microarray analysis if the gene probe is located downstream of the stop codon. Probably the greatest effect, however, would result if a SNP altered the behaviour of a global transcriptional regulator either in terms of altered regulator expression or alteration of the protein sequence that produces a regulator protein with different affinity for its original target promoter sequences. Under this scenario a strain would 'run' a different transcriptional programme and potentially produce a very different phenotype as a result of a single SNP. An analogous situation may occur when a SNP is selected by antibiotic selection (e.g. rifampicin) that alters the behaviour of a key protein (RNA polymerase in this case) and subsequently can cause global changes in transcriptional programs. SNPs can also introduce translational changes even when the SNP in question represents a synonymous change to the amino acid sequence. This can happen if the synonymous change encodes a rare codon for which the bacterium has few or no tRNAs available and this results in truncated proteins or altered expression of proteins (Kane, 1995). Proteomics tools can be used to analyse the effects of SNPs in a comparative experimental design. For example, if a SNP produces an altered transcriptional program we would expect a comparative proteomics experiment to reveal missing and novel proteins for the mutated strain compared with the reference strain. Recovery of sufficient quantities and purities of bacterial mRNA or protein for analysis is a significant challenge, although this can be partially circumvented using *in situ* approaches such as insertion of dialysis tubing within a niche compartment (e.g. peritoneum) where bacteria can grow under the influence of *in vivo* conditions, although clearly without a direct interaction with the host tissue. Such approaches can be very useful for identifying initial transcriptional changes that can be subsequently pursued with more sensitive analytical tools such as real-time PCR.

Eliminating *Salmonella* from the game and the need for a relevant biological assay

A challenge with high-throughput comparative genome analyses is determining the relative consequence of each SNP and the impact of combinations of SNPs on the biology of the pathogen and the host. Animal challenge experiments and epidemiological survey of circulating isolates are still needed, because the quantitative and additive nature of complex combinations occurring in the pathogen of concern defies predicting what will happen next. The issue of which biological approach is most suitable for identification of any one set of SNPs with a causal link to phenotype is pre-eminent. For example, it is unlikely that any studies with mice or chicks would have revealed the set of SNPs linked to the egg-contaminating genotype of *Salmonella* Enteritidis. As mentioned, transcriptomics can only focus on discrete environments; thus, specific niches within the animal or the environment or in humans that are likely to apply major selection pressure on the pathogen should be identified. The progress in understanding the contribution of genetic drift to the emergence of egg contamination by *Salmonella* Enteritidis was greatly facilitated by having the animal of concern, namely the hen, available for conducting challenge experiments. Only the mature hen has functioning ovaries, a mature oviduct and the hormonal cycle of a reproducing animal. Thus, the hen has a unique repertoire of niches impacting the evolution of *Salmonella* Enteritidis. High-throughput genomics is expected to reveal unexpected evolutionary trends within the *Salmonellae*, but applying information from analysis of terabytes of data to improve the safety of the food supply remains the ultimate challenge.

References

Achtman, M. (2008). Evolution, population structure, and phylogeography of genetically monomorphic bacterial pathogens. Annu. Rev. Microbiol. *62*, 53–70.

Alix, E., Godreuil, S., and Blanc-Potard, A.B. (2006). Identification of a Haarlem genotype-specific single nucleotide polymorphism in the *mgtC* virulence gene of *Mycobacterium tuberculosis*. J. Clin. Microbiol. *44*, 2093–2098.

Andrews-Polymenis, H.L., Santiviago, C.A., and McClelland, M. (2009). Novel genetic tools for studying food-borne *Salmonella*. Curr. Opin. Biotechnol. *20*, 149–157.

Anonymous (1988). *Salmonella* Enteritidis phage type 4: chicken and egg. Lancet *2*, 720–722.

Anonymous (1998). Hazard analysis and critical control point principles and application guidelines. National Advisory Committee on Microbiological Criteria for Foods. J. Food Prot. *61*, 762–775.

Anonymous (2000). Global Salm-Surv on Internet. Weekly Epidemiological Record 21, 75, 233–240.

Anonymous (2007a). Multistate outbreak of *Salmonella* serotype Tennessee infections associated with peanut butter–United States, 2006–2007. MMWR Morb Mortal Wkly Rep *56*, 521–524.

Anonymous (2007b). SNPs: Variations on a Theme, In Just the Facts: A Basic Introduction to the Science Underlying NCBI Resources, N.L.O. Medicine, ed. National Center for Biotechnology Information (http://www.ncbi.nlm.nih.gov/About/primer/snps.html).

Baker, S., Holt, K., van de Vosse, E., Roumagnac, P., Whitehead, S., King, E., Ewels, P., Keniry, A., Weill, F.X., Lightfoot, D., *et al.* (2008). High-throughput genotyping of *Salmonella enterica* serovar Typhi allowing geographical assignment of haplotypes and pathotypes within an urban District of Jakarta, Indonesia. J. Clin. Microbiol. *46*, 1741–1746.

Baumler, A.J. (1997). The record of horizontal gene transfer in *Salmonella*. Trends Microbiol. *5*, 318–322.

Baumler, A.J., Hargis, B.M., and Tsolis, R.M. (2000). Tracing the origins of *Salmonella* outbreaks. Science *287*, 50–52.

Bhan, M.K., Bahl, R., and Bhatnagar, S. (2005). Typhoid and paratyphoid fever. Lancet *366*, 749–762.

Boyd, E.F., Wang, F.S., Beltran, P., Plock, S.A., Nelson, K., and Selander, R.K. (1993). *Salmonella* reference collection B (SARB): strains of 37 serovars of subspecies I. J. Gen Microbiol. *139 Pt 6*, 1125–1132.

Brenner, F.W., Villar, R.G., Angulo, F.J., Tauxe, R., and Swaminathan, B. (2000). *Salmonella* nomenclature. J. Clin. Microbiol. *38*, 2465–2467.

Brown, D.J., Baggesen, D.L., Platt, D.J., and Olsen, J.E. (1999). Phage type conversion in *Salmonella enterica* serotype Enteritidis caused by the introduction of a resistance plasmid of incompatibility group X (IncX). Epidemiol. Infect. *122*, 19–22.

Chittick, P., Sulka, A., Tauxe, R.V., and Fry, A.M. (2006). A summary of national reports of foodborne outbreaks of *Salmonella* Heidelberg infections in the United States: clues for disease prevention. J. Food Prot. *69*, 1150–1153.

Conner, C.P., Heithoff, D.M., Julio, S.M., Sinsheimer, R.L., and Mahan, M.J. (1998). Differential patterns of acquired virulence genes distinguish *Salmonella* strains. Proc. Natl. Acad. Sci. U.S.A. *95*, 4641–4645.

Crump, J.A., Griffin, P.M., and Angulo, F.J. (2002). Bacterial contamination of animal feed and its relationship to human foodborne illness. Clin. Infect. Dis. *35*, 859–865.

Curtiss, R., 3rd, and Kelly, S.M. (1987). *Salmonella* Typhimurium deletion mutants lacking adenylate cyclase and cyclic AMP receptor protein are avirulent and immunogenic. Infect. Immun. *55*, 3035–3043.

Darling, A.C., Mau, B., Blattner, F.R., and Perna, N.T. (2004). Mauve: multiple alignment of conserved genomic sequence with rearrangements. Genome Res. *14*, 1394–1403.

Davies, R.H., and Wray, C. (1995a). Contribution of the lesser mealworm beetle (*Alphitobius diaperinus*) to carriage of *Salmonella* Enteritidis in poultry. Vet. Rec. *137*, 407–408.

Davies, R.H., and Wray, C. (1995b). Mice as carriers of *Salmonella* Enteritidis on persistently infected poultry units. Vet. Rec. *137*, 337–341.

Dotsch, A., Pommerenke, C., Bredenbruch, F., Geffers, R., and Haussler, S. (2009). Evaluation of a microarray-hybridization-based method applicable for discovery of single nucleotide polymorphisms (SNPs) in the *Pseudomonas aeruginosa* genome. BMC Genomics *10*, 29.

Doublet, B., Praud, K., Bertrand, S., Collard, J.M., Weill, F.X., and Cloeckaert, A. (2008). Novel insertion sequence- and transposon-mediated genetic rearrangements in genomic island SGI1 of *Salmonella enterica* serovar Kentucky. Antimicrob. Agents Chemother. *52*, 3745–3754.

Esaki, H., Noda, K., Otsuki, N., Kojima, A., Asai, T., Tamura, Y., and Takahashi, T. (2004). Rapid detection of quinolone-resistant *Salmonella* by real time SNP genotyping. J. Microbiol. Methods *58*, 131–134.

Eswarappa, S.M., Janice, J., Balasundaram, S.V., Dixit, N.M., and Chakravortty, D. (2009). Host-specificity of *Salmonella enterica* serovar *Gallinarum*: insights from comparative genomics. Infect. Genet. Evol. *9*, 468–473.

Falush, D., Torpdahl, M., Didelot, X., Conrad, D.F., Wilson, D.J., and Achtman, M. (2006). Mismatch induced speciation in *Salmonella*: model and data. Philos. Trans. R. Soc. Lond. B. Biol. Sci. *361*, 2045–2053.

Fitzgerald, C., Collins, M., van Duyne, S., Mikoleit, M., Brown, T., and Fields, P. (2007). Multiplex, bead-based suspension array for molecular determination of common *Salmonella* serogroups. J. Clin. Microbiol. *45*, 3323–3334.

Gantois, I., Ducatelle, R., Pasmans, F., Haesebrouck, F., and Van Immerseel, F. (2009). The *Salmonella* Enteritidis lipopolysaccharide biosynthesis gene rfbH is required for survival in egg albumen. Zoonoses Public Health 56, 145–149.

Gantois, I., Eeckhaut, V., Pasmans, F., Haesebrouck, F., Ducatelle, R., and Van Immerseel, F. (2008). A comparative study on the pathogenesis of egg contamination by different serotypes of *Salmonella*. Avian Pathol. *37*, 399–406.

Gregory, T.R. (2002). Genome size and developmental complexity. Genetica *115*, 131–146.

Groisman, E.A., Saier, M.H., Jr., and Ochman, H. (1992). Horizontal transfer of a phosphatase gene as evidence for mosaic structure of the *Salmonella* genome. EMBO J. *11*, 1309–1316.

Guard-Bouldin, J. (2006). Comparative Genome Sequencing of *Salmonella* Enteritidis Isolates that Vary in Virulence Characteristics, Paper presented at: 110th Animal, U.S. Health Association (Minneapolis, Minnesota: Skinner Printing, Montgomer, AL).

Guard-Bouldin, J., Gast, R.K., Humphrey, T.J., Henzler, D.J., Morales, C., and Coles, K. (2004). Subpopulation characteristics of egg-contaminating *Salmonella enterica* serovar Enteritidis as defined by the lipopolysaccharide O chain. Appl. Environ. Microbiol. *70*, 2756–2763.

Guard-Bouldin, J., Morales, C.A., Frye, J.G., Gast, R.K., and Musgrove, M. (2007). Detection of *Salmonella enterica* subpopulations by phenotype microarray antibiotic resistance patterns. Appl. Environ. Microbiol. *73*, 7753–7756.

Guard-Petter, J. (1998). Variants of smooth *Salmonella enterica* serovar Enteritidis that grow to higher cell density than the wild type are more virulent. Appl. Environ. Microbiol. *64*, 2166–2172.

Guard-Petter, J. (2001). The chicken, the egg and *Salmonella* Enteritidis. Environ. Microbiol. *3*, 421–430.

Guard-Petter, J., Henzler, D.J., Rahman, M.M., and Carlson, R.W. (1997). On-farm monitoring of mouse-invasive *Salmonella enterica* serovar Enteritidis and a model for its association with the production of contaminated eggs. Appl. Environ. Microbiol. *63*, 1588–1593.

Guard-Petter, J., Parker, C.T., Asokan, K., and Carlson, R.W. (1999). Clinical and veterinary isolates of *Salmonella enterica* serovar Enteritidis defective in lipopolysaccharide O-chain polymerization. Appl. Environ. Microbiol. *65*, 2195–2201.

Hendriksen, R.S., Mikoleit, M., Carlson, V.P., Karlsmose, S., Vieira, A.R., Jensen, A.B., Seyfarth, A.M., Delong, S.M., Weill, F.X., Lo Fo Wong, D.M., *et al.* (2009). WHO Global Salm-Surv External Quality Assurance System (EQAS) for serotyping of *Salmonella* isolates, 2000–2007. J. Clin. Microbiol. *47*, 2729–2736.

Hensel, M., Nikolaus, T., and Egelseer, C. (1999). Molecular and functional analysis indicates a mosaic structure of *Salmonella* pathogenicity island 2. Mol. Microbiol. *31*, 489–498.

Holt, K.E., Parkhill, J., Mazzoni, C.J., Roumagnac, P., Weill, F.X., Goodhead, I., Rance, R., Baker, S., Maskell, D.J., Wain, J., *et al.* (2008). High-throughput sequencing provides insights into genome variation and evolution in *Salmonella* Typhi. Nat. Genet. *40*, 987–993.

Holt, K.E., Teo, Y.Y., Li, H., Nair, S., Dougan, G., Wain, J., and Parkhill, J. (2009). Detecting SNPs and estimating allele frequencies in clonal bacterial populations by sequencing pooled DNA. Bioinformatics *25*, 2074–2075.

Holt, P.S., Geden, C.J., Moore, R.W., and Gast, R.K. (2007). Isolation of *Salmonella enterica* serovar Enteritidis from houseflies (*Musca domestica*) found in rooms containing *Salmonella* serovar Enteritidis-challenged hens. Appl. Environ. Microbiol. *73*, 6030–6035.

Hudson, C.R., Garcia, M., Gast, R.K., and Maurer, J.J. (2001). Determination of close genetic relatedness of the major *Salmonella* Enteritidis phage types by pulsed-field gel electrophoresis and DNA sequence analysis of several *Salmonella* virulence genes. Avian Dis. *45*, 875–886.

Humphrey, T.J., Williams, A., McAlpine, K., Jorgensen, F., and O'Byrne, C. (1998). Pathogenicity in isolates of *Salmonella enterica* serotype Enteritidis PT4 which differ in RpoS expression: effects of growth phase and low temperature. Epidemiol. Infect. *121*, 295–301.

Humphrey, T.J., Williams, A., McAlpine, K., Lever, M.S., Guard-Petter, J., and Cox, J.M. (1996). Isolates of *Salmonella enterica* Enteritidis PT4 with enhanced heat and acid tolerance are more virulent in mice and more invasive in chickens. Epidemiol. Infect. *117*, 79–88.

Kalia, A., Mukhopadhyay, A.K., Dailide, G., Ito, Y., Azuma, T., Wong, B.C., and Berg, D.E. (2004). Evolutionary dynamics of insertion sequences in *Helicobacter pylori*. J. Bacteriol. *186*, 7508–7520.

Kane, J.F. (1995). Effects of rare codon clusters on high-level expression of heterologous proteins in *Escherichia coli*. Curr. Opin. Biotechnol. *6*, 494–500.

Kerouanton, A., Marault, M., Lailler, R., Weill, F.X., Feurer, C., Espie, E., and Brisabois, A. (2007). Pulsed-field gel electrophoresis subtyping database for foodborne *Salmonella enterica* serotype discrimination. Foodborne Pathog. Dis. *4*, 293–303.

Kidgell, C., Reichard, U., Wain, J., Linz, B., Torpdahl, M., Dougan, G., and Achtman, M. (2002). *Salmonella* Typhi, the causative agent of typhoid fever, is approximately 50,000 years old. Infect. Genet. Evol. *2*, 39–45.

Kitamoto, N., Kato, Y., Hamada, K., Kanzya, S., Onaka, T., Yokota, M., and Tanaka, T. (2005). [Molecular epidemiology of *Salmonella* Enteritidis food poisoning by pulsed-field gel electrophoresis]. Kansenshogaku Zasshi *79*, 122–128.

Kropinski, A.M., Sulakvelidze, A., Konczy, P., and Poppe, C. (2007). *Salmonella* phages and prophages–genomics and practical aspects. Methods Mol. Biol. *394*, 133–175.

Kumar, Y., Sharma, A., Sehgal, R., and Kumar, S. (2009). Distribution trends of *Salmonella* serovars in India (2001–2005). Trans. R. Soc. Trop. Med. Hyg.*103*, 390–394.

Le Minor, L., Veron, M., and Popoff, M. (1982). [The taxonomy of *Salmonella*]. Ann Microbiol. (Paris) 133, 223–243.

Lecointre, G., Rachdi, L., Darlu, P., and Denamur, E. (1998). *Escherichia coli* molecular phylogeny using the incongruence length difference test. Mol. Biol. Evol *15*, 1685–1695.

Lermo, A., Campoy, S., Barbe, J., Hernandez, S., Alegret, S., and Pividori, M.I. (2007). In situ DNA amplification with magnetic primers for the electrochemical detection of food pathogens. Biosens. Bioelectron. *22*, 2010–2017.

Levy, D.D., Sharma, B., and Cebula, T.A. (2004). Single-nucleotide polymorphism mutation spectra and resistance to quinolones in *Salmonella enterica* serovar Enteritidis with a mutator phenotype. Antimicrob. Agents Chemother. *48*, 2355–2363.

Li, J., Nelson, K., McWhorter, A.C., Whittam, T.S., and Selander, R.K. (1994). Recombinational basis of serovar diversity in *Salmonella enterica*. Proc. Natl. Acad. Sci. U.S.A. *91*, 2552–2556.

Liebana, E., Garcia-Migura, L., Guard-Petter, J., McDowell, S.W., Rankin, S., Opitz, H.M., Clifton-Hadley, F.A., and Davies, R.H. (2002). *Salmonella enterica* serotype Enteritidis phage types 4, 7, 6, 8, 13a, 29 and 34: a comparative analysis of genomic fingerprints from geographically distant isolates. J. Appl. Microbiol. *92*, 196–209.

Lin, B., and Malanoski, A.P. (2009). Resequencing arrays for diagnostics of respiratory pathogens. Methods Mol. Biol. *529*, 231–257.

Lindstedt, B.A., Vardund, T., Aas, L., and Kapperud, G. (2004). Multiple-locus variable-number tandem-repeats analysis of *Salmonella enterica* subsp. *enterica* serovar Typhimurium using PCR multiplexing and multicolor capillary electrophoresis. J. Microbiol. Methods *59*, 163–172.

Liu, S.L., and Sanderson, K.E. (1995). Rearrangements in the genome of the bacterium *Salmonella* Typhi. Proc. Natl. Acad. Sci. U.S.A. *92*, 1018–1022.

Malorny, B., Junker, E., and Helmuth, R. (2008). Multilocus variable-number tandem repeat analysis for outbreak studies of *Salmonella enterica* serotype Enteritidis. BMC Microbiol. *8*, 84.

Marinus, M.G., and Casadesus, J. (2009). Roles of DNA adenine methylation in host–pathogen interactions: mismatch repair, transcriptional regulation, and more. FEMS Microbiol. Rev. *33*, 488–503.

Martin, W. (1999). Mosaic bacterial chromosomes: a challenge en route to a tree of genomes. Bioessays *21*, 99–104.

Matheson, N., Kingsley, R., Sturgess, K., Aliyu, S., Wain, J., Dougan, G., and Cooke, F.J. (2009). Ten Years Experience of *Salmonella* Infections in Cambridge, UK. J. Infect. *60*, 21–25.

McClelland, M., Sanderson, K.E., Spieth, J., Clifton, S.W., Latreille, P., Courtney, L., Porwollik, S., Ali, J., Dante, M., Du, F., *et al.* (2001). Complete genome sequence of *Salmonella enterica* serovar Typhimurium LT2. Nature *413*, 852–856.

Mian, L.S., Maag, H., and Tacal, J.V. (2002). Isolation of *Salmonella* from muscoid flies at commercial animal establishments in San Bernardino County, California. J. Vector Ecol *27*, 82–85.

Miraglia, M., Marvin, H.J., Kleter, G.A., Battilani, P., Brera, C., Coni, E., Cubadda, F., Croci, L., De Santis, B., Dekkers, S., *et al.* (2009). Climate Change and Food Safety: an Emerging Issue with Special Focus on Europe. Food Chem. Toxicol. *47*, 1009–1021.

Mirold, S., Ehrbar, K., Weissmuller, A., Prager, R., Tschape, H., Russmann, H., and Hardt, W.D. (2001). *Salmonella* host cell invasion emerged by acquisition of a mosaic of separate genetic elements, including *Salmonella* pathogenicity island 1 (SPI1), SPI5, and sopE2. J. Bacteriol. *183*, 2348–2358.

Miyamura, K., Tanimura, M., Takeda, N., Kono, R., and Yamazaki, S. (1986). Evolution of enterovirus 70 in nature: all isolates were recently derived from a common ancestor. Arch Virol *89*, 1–14.

Mmolawa, P.T., Schmieger, H., and Heuzenroeder, M.W. (2003). Bacteriophage ST64B, a genetic mosaic of genes from diverse sources isolated from *Salmonella enterica* serovar Typhimurium DT 64. J. Bacteriol. *185*, 6481–6485.

Morales, C.A., Gast, R., and Guard-Bouldin, J. (2006). Linkage of avian and reproductive tract tropism with sequence divergence adjacent to the 5S ribosomal subunit rrfH of *Salmonella enterica.* FEMS Microbiol. Lett. *264*, 48–58.

Morales, C.A., Musgrove, M., Humphrey, T.J., Cates, C., Gast, R., and Guard-Bouldin, J. (2007). Pathotyping of *Salmonella enterica* by analysis of single-nucleotide polymorphisms in cyaA and flanking 23S ribosomal sequences. Environ. Microbiol. *9*, 1047–1059.

Morales, C.A., Porwollik, S., Frye, J.G., Kinde, H., McClelland, M., and Guard-Bouldin, J. (2005). Correlation of phenotype with the genotype of egg-contaminating *Salmonella enterica* serovar Enteritidis. Appl. Environ. Microbiol. *71*, 4388–4399.

Moro, C.V., Desloire, S., Chauve, C., and Zenner, L. (2007). Detection of *Salmonella* sp. in *Dermanyssus gallinae* using an FTA filter-based polymerase chain reaction. Med. Vet. Entomol. *21*, 148–152.

Ochman, H., and Groisman, E.A. (1995). The evolution of invasion by enteric bacteria. Can J. Microbiol. *41*, 555–561.

Octavia, S., and Lan, R. (2007). Single-nucleotide-polymorphism typing and genetic relationships of *Salmonella enterica* serovar Typhi isolates. J. Clin. Microbiol. *45*, 3795–3801.

Olsen, S.J., MacKinnon, L.C., Goulding, J.S., Bean, N.H., and Slutsker, L. (2000). Surveillance for foodborne-disease outbreaks–United States, 1993–1997. MMWR CDC Surveill Summ *49*, 1–62.

Olson, A.B., Andrysiak, A.K., Tracz, D.M., Guard-Bouldin, J., Demczuk, W., Ng, L.K., Maki, A., Jamieson, F., and Gilmour, M.W. (2007). Limited genetic diversity in *Salmonella enterica* serovar Enteritidis PT13. BMC Microbiol. *7*, 87.

Pang, J.C., Chiu, T.H., Chiou, C.S., Schroeter, A., Guerra, B., Helmuth, R., and Tsen, H.Y. (2005). Pulsed-field gel electrophoresis, plasmid profiles and phage types for the human isolates of *Salmonella enterica* serovar Enteritidis obtained over 13 years in Taiwan. J. Appl. Microbiol. *99*, 1472–1483.

Parker, C.T., Harmon, B., and Guard-Petter, J. (2002). Mitigation of avian reproductive tract function by *Salmonella* Enteritidis producing high-molecular-mass lipopolysaccharide. Environ. Microbiol. *4*, 538–545.

Parker, C.T., Liebana, E., Henzler, D.J., and Guard-Petter, J. (2001). Lipopolysaccharide O-chain microheterogeneity of *Salmonella* serotypes Enteritidis and Typhimurium. Environ. Microbiol. *3*, 332–342.

Peters, T.M., Berghold, C., Brown, D., Coia, J., Dionisi, A.M., Echeita, A., Fisher, I.S., Gatto, A.J., Gill, N., Green, J., *et al.* (2007). Relationship of pulsed-field profiles with key phage types of *Salmonella enterica* serotype Enteritidis in Europe: results of an international multi-centre study. Epidemiol. Infect. *135*, 1274–1281.

Petrov, D.A. (2001). Evolution of genome size: new approaches to an old problem. Trends Genet *17*, 23–28.

Pierson, D.L. (2001). Microbial contamination of spacecraft. Gravit Space Biol. Bull *14*, 1–6.

Porwollik, S., Santiviago, C.A., Cheng, P., Florea, L., Jackson, S., and McClelland, M. (2005). Differences in gene content between *Salmonella enterica* serovar Enteritidis isolates and comparison to closely related serovars *Gallinarum* and Dublin. J. Bacteriol. *187*, 6545–6555.

Romanowska, E., Katzenellenbogen, E., Jachymek, W., Niedziela, T., Bogulska, M., and Lugowski, C. (1999). Non-typical lipopolysaccharide core regions of some *Hafnia alvei* strains: structural and serological studies. FEMS Immunol. Med. Microbiol. *24*, 63–71.

Ross, I.L., and Heuzenroeder, M.W. (2009). A comparison of two PCR-based typing methods with pulsed-field gel electrophoresis in *Salmonella enterica* serovar Enteritidis. Int. J. Med. Microbiol. *299,* 410–420.

Scaria, J., Palaniappan, R.U., Chiu, D., Phan, J.A., Ponnala, L., McDonough, P., Grohn, Y.T., Porwollik, S., McClelland, M., Chiou, C.S., *et al.* (2008). Microarray for molecular typing of *Salmonella enterica* serovars. Mol. Cell Probes *22,* 238–243.

Shivaprasad, H.L. (2000). Fowl typhoid and pullorum disease. Rev. Sci Tech *19,* 405–424.

Singh, S.P., Sethi, M.S., and Sharma, V.D. (1980). The occurrence of *Salmonellae* in rodent, shrew, cockroach and ant. Int. J. Zoonoses *7,* 58–61.

Soares, A.R., and Machado, J. (2003). Molecular epidemiology of *Salmonella* Enteritidis phage type 1b and 6a isolates in Portugal. Epidemiol. Infect. *131,* 607–612.

St Louis, M.E., Morse, D.L., Potter, M.E., DeMelfi, T.M., Guzewich, J.J., Tauxe, R.V., and. Blake, P.A. (1988). The emergence of grade A eggs as a major source of *Salmonella* Enteritidis infections. New implications for the control of salmonellosis. JAMA *259,* 2103–2107.

Stocki, S.L., Annett, C.B., Sibley, C.D., McLaws, M., Checkley, S.L., Singh, N., Surette, M.G., and White, A.P. (2007). Persistence of *Salmonella* on egg conveyor belts is dependent on the belt type but not on the rdar morphotype. Poult Sci *86,* 2375–2383.

Tavechio, A.T., Ghilardi, A.C., Peresi, J.T., Fuzihara, T.O., Yonamine, E.K., Jakabi, M., and Fernandes, S.A. (2002). *Salmonella* serotypes isolated from nonhuman sources in Sao Paulo, Brazil, from 1996 through 2000. J. Food Prot. 65, 1041–1044.

Thijs, I.M., De Keersmaecker, S.C., Fadda, A., Engelen, K., Zhao, H., McClelland, M., Marchal, K., and Vanderleyden, I. (2007). Combining omics data to unravel the regulatory network controlling *Salmonella* invasion of epithelial cells. Commun Agric Appl. Biol. Sci 72, 55–59.

Thomson, N.R., Clayton, D.J., Windhorst, D., Vernikos, G., Davidson, S., Churcher, C., Quail, M.A., Stevens, M., Jones, M.A., Watson, M., *et al.* (2008). Comparative genome analysis of *Salmonella* Enteritidis PT4 and *Salmonella Gallinarum* 287/91 provides insights into evolutionary and host adaptation pathways. Genome Res. *18,* 1624–1637.

Todd, E.C. (1997). Epidemiology of foodborne diseases: a worldwide review. World Health Stat Q 50, 30–50.

Valiente Moro, C., Chauve, C., and Zenner, L. (2007). Experimental infection of *Salmonella* Enteritidis by the poultry red mite, *Dermanyssus gallinae.* Vet Parasitol *146,* 329–336.

Vellai, T., and Vida, G. (1999). The origin of eukaryotes: the difference between prokaryotic and eukaryotic cells. Proc Biol. Sci *266,* 1571–1577.

Villarreal, L.P. (2009). Origin of Group Identity: Viruses, Addiction and Cooperation (New York: Springer).

Warmflash, D., and Weiss, B. (2005). Did life come from another world? Sci Am. *293,* 64–71.

Wichman, H.A., Millstein, J., and Bull, J.J. (2005). Adaptive molecular evolution for 13,000 phage generations: a possible arms race. Genetics *170,* 19–31.

Yazdankhah, S.P., and Lindstedt, B.A. (2007). Variable number tandem repeat typing of bacteria. Methods Mol. Biol. *396,* 395–405.

Zhao, S., McDermott, P.F., Friedman, S., Abbott, J., Ayers, S., Glenn, A., Hall-Robinson, E., Hubert, S.K., Harbottle, H., Walker, R.D., *et al.* (2006). Antimicrobial resistance and genetic relatedness among *Salmonella* from retail foods of animal origin: NARMS retail meat surveillance. Foodborne Pathog Dis. 3, 106–117.

Genomics and Pathogenesis of *Salmonella enterica* Serovars Typhi and Paratyphi A

6

Kathryn E. Holt, Tim T. Perkins, Gordon Dougan and Robert A. Kingsley*

Abstract

The genomics era has transformed the way that we can study bacterial pathogens. The availability of two complete and 17 draft genomes of *Salmonella* Typhi has made it possible to study the phylogenetic structure of this pathogen in unparalleled resolution, monitor gene flux, accumulation of pseudogenes, neutral mutations and loci under selective pressure. We describe the molecular basis of *Salmonella* Typhi pathogenesis, in particular where genomics has contributed to our understanding in the past decade. Potentially important *Salmonella* Typhi-specific virulence determinants include the Vi polysaccharide capsule, the type IV pilus, and a unique repertoire of fimbria. These may account for key differences in the disease outcome of this pathogen compared with non-typhoidal serotypes. Genome comparison with the closely related serotype *Salmonella* Paratyphi A identifies a core set of pseudogenes, some of which emerged independently, that may define important features of genome degradation associated with host restriction and pathogenesis of invasive disease. Geo-phylogenetics of *Salmonella* Typhi constructed from single nucleotide polymorphism data from high throughput draft genome sequences is now being applied to study molecular epidemiology in the field.

Introduction

Typhoid and paratyphoid fever, collectively referred to as enteric fever, are invasive diseases caused by *Salmonella enterica* serovar Typhi (*Salmonella* Typhi), and *S. enterica* serovars Paratyphi A (*Salmonella* Paratyphi A), B and C, respectively. The English physician Thomas Willis is frequently attributed with the first description of the syndrome in 1659. However, this disease is likely to have blighted human populations for many thousands of years. Until 1837, when William Wood Gerhard defined them as separate diseases, typhus, a disease caused by the unrelated bacterium of the Rickettsiae spp., and typhoid fever were frequently confused with one another. The typhoid bacillus was identified in 1880, by Carl Joseph Eberth, and in 1896 Georges Widal developed an agglutination test that simplified diagnosis. The Widal test is still widely used today.

Approximately 25 million cases of enteric fever occur worldwide each year, almost exclusively in developing countries (Crump *et al.*, 2004). The bacterial causative agents are transmitted by the faecal–oral route, predominantly following ingestion of faecal-contaminated water. These diseases claim the lives of more than 200,000 people each year, mostly in South-East Asia but also in Africa and South America (Crump *et al.*, 2004). Cases in resource-rich nations are almost exclusively associated with travel to the Asian subcontinent. In addition to causing febrile illness in the form of typhoid fever, *Salmonella* Typhi also causes asymptomatic infection of the human gall bladder resulting in periodic shedding in the faeces over months or years. Infamous cases of asymptomatic carriers of *Salmonella* Typhi include Mary Mallon, or 'Typhoid Mary', who infected at least 53 people, three of whom died of the disease (Anonymous, 1939). Paratyphoid fever is virtually indistinguishable from typhoid clinically, but is caused by an antigenically and phylogenetically

*Corresponding author

distinct *Salmonella enterica* serovar, Paratyphi A (Maskey *et al.*, 2006). Recent reports suggest that Paratyphi A can also be carried asymptomatically in the gall bladder (Khatri *et al.*, 2009). Effective vaccines against Typhi are available (Fraser *et al.*, 2007), but are mostly used by travellers rather than residents of endemic areas. Unfortunately, Typhi vaccines provide little protection against Paratyphi A, and a dedicated Paratyphi A vaccine has not yet been developed (Fangtham and Wilde, 2008).

Until recently, typhoid cases outnumbered paratyphoid cases worldwide by 10:1. However, the incidence of paratyphoid fever has been rising in some areas, particularly China (Ochiai *et al.*, 2005), Nepal (Maskey *et al.*, 2008; Woods *et al.*, 2006) and parts of India, indeed, cases of enteric fever caused by *Salmonella* Paratyphi A currently outnumber typhoid fever cases in China (Palit *et al.*, 2006). The reason for this rise in prominence is not clear, however, the absence of a *Salmonella* Paratyphi A vaccine, or cross protection from the *Salmonella* Typhi vaccines in use, may contribute. In addition, resistance to fluoroquinolones, the drug of choice for treatment of enteric fever in most of Asia, is high among Paratyphi A isolates (Gupta *et al.*, 2009). An IncHI1 plasmid (pAKU1, EMBL: AM412236), closely related to those found in *Salmonella* Typhi, is also responsible for multidrug resistance in Paratyphi A, particularly in isolates from Pakistan (Holt *et al.*, 2007a).

Approximately 2500 serovars of *Salmonella*, distinguished by O-antigen and flagella antigens, infect a variety of hosts including mammals, birds and reptiles, and generally cause self-limiting gastroenteritis (Aleksic *et al.*, 1996; Galanis *et al.*, 2006). *Salmonella* Typhi and *Salmonella* Paratyphi A on the other hand are human-restricted and cause indistinguishable systemic diseases in addition to chronic carriage, yet it is not clear why they share this highly specialized host-adapted lifestyle. This review highlights recent genomic studies of *Salmonella*, and the insights they offer regarding the virulence and host-adaptation of *Salmonella* Typhi and *Salmonella* Paratyphi A. These studies have opened the way for genomic epidemiology studies of invasive *Salmonella* disease, and the review outlines some novel technological approaches. Finally, we discuss the ways in which genomic epidemiology will be crucial for future studies of transmission, evolution of virulence and drug resistance in invasive *Salmonella* disease, potentially leading to drastic improvements in prevention and treatment strategies.

The *Salmonella* Typhi genome

The first reports of the complete genome sequence for *Salmonella* Typhi and *Salmonella* Typhimurium were published around the turn of the century (Boyd *et al.*, 2003; McClelland *et al.*, 2001; Parkhill *et al.*, 2001), paving the way for genomic studies in *Salmonella*. The *Salmonella* Typhi genome consists of a single circular chromosome of 4.8 Mbp, in which about 4600 protein-coding genes and seven ribosomal RNA (rRNA) operons were annotated (Parkhill *et al.*, 2001). Homologous recombination between rRNA operons is known to occur in *Salmonella* Typhi, causing genomic rearrangements (Deng *et al.*, 2003; Liu and Sanderson, 1995; Sanderson and Liu, 1998; see also Chapter 3 of this book). The genome sequences were initially employed to design DNA oligonucleotide arrays with which to study genome content and gene expression in *Salmonella* Typhi and other serovars (Boyd *et al.*, 2003; Porwollik *et al.*, 2003). The *Salmonella* Typhi genome shares about 90% of its genes with other *S. enterica* serovars including *Salmonella* Typhimurium, with an average of 98% nucleotide identity (McClelland *et al.*, 2001; Porwollik *et al.*, 2003). *E. coli* encodes orthologues of 70% of *Salmonella* genes with an average 80% identity at the nucleotide level and 90% identity at the amino acid level (McClelland *et al.*, 2001; Parkhill *et al.*, 2001). However, there are many *Salmonella*-specific regions, and several of these encode key functions associated with *Salmonella* pathogenesis. The most important of these are the *Salmonella* Pathogenicity Islands (SPIs), genetic islands encoding virulence genes that are probably acquired by horizontal transfer.

Virulence genes

Pathogenicity islands (SPIs)

A total of 18 SPIs have been described in *Salmonella* serotypes (SPI1–17 and CS54) (Kingsley *et al.*, 2003; Vernikos and Parkhill, 2006). SPI1 encodes a type III secretion system (TTSS) and is essential for invasion of non-phagocytic host cells such as enterocytes. Its acquisition, along with other factors such as the fimbriae *fim* and *lpf*, accompanied the divergence of *Salmonella* from *E. coli* approximately 120–160 million years ago (Bäumler, 1997). A second TTSS is encoded by SPI2, which was acquired following divergence of the *S. enterica* and *S. bongori* lineages.

SPI1 is 40 kb in size and consists of 25 genes encoding TTSS apparatus proteins, effector proteins and regulatory elements (Mills *et al.*, 1995). Induction of the TTSS *in vitro* occurs on a shift from acidic conditions to mildly alkaline, mimicking the conditions in the host as bacteria transit from the stomach to the small intestine (Daefler, 1999). Expression is regulated by a complex network including HilA, HilC and HilD, along with several other regulators encoded both within and outside SPI1 (Ellermeier *et al.*, 2005). The SPI1 locus mediates invasion of host cell enterocytes, a key step in the sequence of processes that eventually lead to diarrhoea in non-typhoidal serovars. In most cases invasion brings pathogen-associated molecular patterns present on the bacterial surface, notably LPS and flagella, into contact with toll-like receptors (TLRs) and Nods (Inohara *et al.*, 2002; Sieling and Modlin, 2002). These innate immune receptors signal a proinflammatory response that ultimately results in diarrhoea, and changes in the composition of the host microbiota (Stecher *et al.*, 2007; Zhang *et al.*, 2003). These features of *Salmonella* pathogenesis are both likely to be directly involved in transmission since diarrhoea affects the spreading of faecal matter while changes in the microbiota benefits *Salmonella* luminal persistence by depleting the competing microbiota. However, diarrhoea is not a major feature of typhoid fever, occurring in fewer than 1 in 3 patients, and then only later in infection following the onset of fever. The relative importance of the SPI1-encoded TTSS is not clear, particularly since alternative, SPI1-independent mechanisms of *trans*-epithelial transfer and systemic dissemination involving CD18-positive lymphocytes have been described, at least in *Salmonella* Typhimurium infections of mice (Vazquez-Torres and Fang, 2000; Vazquez-Torres *et al.*, 1999).

SPI2 consists of two segments that are likely to have distinct evolutionary histories; a 14.5 kb region present in both *S. bongori* and *S. enterica*, and a 25.3 kb region uniformly present in serovars of *S. enterica*, but absent from *S. bongori*. The latter segment encodes the apparatus, chaperones, effectors and regulatory elements for a second TTSS required for intracellular survival and replication and systemic infection in mice (Ochman *et al.*, 1996; Shea *et al.*, 1996). This TTSS translocates effector proteins into the cytosol of the host cell across the membrane of the *Salmonella*-containing vacuole. The effectors have disparate activities that affect host-cell functions including cytoskeletal structure and signalling pathways to modulate vesicular trafficking (reviewed in Abrahams and Hensel, 2006; see also Chapters 10 and 11 of this book). The two component systems PhoPQ, OmpR/EnvZ and the positive regulators SlyA and HilD are involved in overcoming the powerful silencing of the nucleoid-associated proteins H-NS, NAPs, Hha and YdgT (reviewed in Fass and Groisman, 2009). The major external signal in the up-regulation of SPI2 genes is magnesium and phosphate ion depletion sensed by the PhoPQ two component regulator (Deiwick and Hensel, 1999).

Determination of the complete genome sequence of *Salmonella* Typhi led to the characterization of *Salmonella* pathogenicity island 7 (SPI7), a 134 kb horizontally acquired region encoding multiple unrelated virulence genes (Pickard *et al.*, 2003). SPI7 encodes genes involved in the synthesis of Vi polysaccharide (*viaB* locus), the SopE phage and a type IVB pilus operon (Pickard, 2003). Vi polysaccharide is a 1,4 linked homopolymer of galactosaminuronic acid with variable O-acetylation at the C3 position. *Salmonella* Typhi isolates from typhoid fever patients typically express Vi (Wain *et al.*, 2005), even though SPI7 is genetically unstable on passage *in vitro* (Bueno *et al.*, 2004; Nair *et al.*, 2004), suggesting that it is important during infection. Indeed, *Salmonella* Typhi lacking the Vi capsule were avirulent in the mouse mucin model (Hone *et al.*, 1988), and infectivity of *Salmonella* Typhi was increased by the expression of Vi in human volunteer studies (Hornick *et al.*, 1970a,b). A major function of the Vi locus may be in shielding lipopolysaccharide (LPS) on the *Salmonella* Typhi surface from the innate immune receptor toll-like receptor-4 (TLR-4) (Wilson *et al.*, 2008). However, the *viaB* locus and Vi antigen are not encoded by *Salmonella* Paratyphi A, suggesting that alternative mechanisms of pathogenesis are used by this serovar. A vaccine based on purified Vi antigen is currently licensed, and provides as much protection against *Salmonella* Typhi as an alternative live attenuated vaccine strain of *Salmonella* Typhi (Fraser *et al.*, 2007). SPI7 also contains the SopE bacteriophage that encodes the SPI1 TTSS effector SopE (Mirold *et al.*, 1999), a G-nucleotide exchange factor for host Cdc42 and Rac1 that contributes to host cell invasion (Hardt *et al.*, 1998; Wood *et al.*, 1996).

CS54 is a 25-kb genetic island encoded by virtually all *S. enterica* subspecies I serovars, including *Salmonella* Typhi and *Salmonella* Paratyphi A (Kingsley *et al.*, 2003). While the *shdA* and *ratB* genes encoded on one half of the island are restricted to serovars of subspecies I, homologues of the *sivH, sivI* and *ratA* genes are additionally present in serovars of subspecies II, and surprisingly in *S. bongori* (Kingsley *et al.*, 2003). The three of the genes on CS54, *shdA, ratB* and *sivA,* have been implicated in persistence in the murine intestine albeit with subtly different phenotypes (Abd El Ghany *et al.*, 2007; Kingsley *et al.*, 2003; Kingsley *et al.*, 2000). The mechanism by which ShdA, RatB and SivH mediate intestinal persistence is currently unknown, although it is known that ShdA binds the host extracellular matrix protein fibronectin, by a mechanism involving molecular mimicry of the heparin (Kingsley *et al.*, 2004a; Kingsley *et al.*, 2004b). Significantly, each of these genes are present as pseudogenes in both *Salmonella* Typhi and *Salmonella* Paratyphi A (Holt *et al.*, 2009; McClelland *et al.*, 2004; Parkhill *et al.*, 2001), suggesting that this may at least in part explain the lack of intestinal persistence observed in enteric fever.

Flagella

Flagella are whip-like filament structures that are elaborated on the surface of most *S. enterica* serovars and are required for swimming motility and chemotaxis (see Chapter 9 of this book). Motility aids invasion of non-phagocytic host cells by *Salmonella* by increasing contact (Jones *et al.*, 1992). Furthermore, the monomer comprising the flagellum filament, flagellin, is an agonist of the toll-like receptor-5 (TLR-5), an innate immune receptor expressed on the basolateral surface of intestinal cells and on the surface of some intestinal dendritic cells (Evans and Evans, 1978; Gewirtz *et al.*, 2001a; Gewirtz *et al.*, 2001b). *Salmonella* Typhi evades detection by these innate immune receptors by down-regulating expression of flagellin, using a mechanism involving the SPI7-encoded TviA regulator (Winter *et al.*, 2008). Most *S. enterica* serovars are biphasic, in that they express one of two alternative flagellin subunits, FliC and FljB, expression of which is under phase variation. The phase-switch is controlled by the repressor (*hin*) encoded down-stream of the phase two gene (*fljB*). Both *Salmonella* Typhi and *Salmonella* Paratyphi A are monophasic as a result of independent mutations (McClelland *et al.*, 2004). In *Salmonella* Typhi, there is a deletion including the phase 2 gene and the *hin* gene, while a mutation in the *hin* gene of *Salmonella* Paratyphi A causes a frame-shift preventing phase switching, and consequently locks the bacterium in the phase 1 ('FliC') position. Some *Salmonella* Typhi isolates from typhoid fever cases in Indonesia are unusual in that they are expressing an alternative flagellar antigen, z66. This antigen is related to phase 2 flagella and is encoded by an *fljB* gene located on a 27 kbp linear plasmid (Baker *et al.*, 2007a). A *hin-like* repressor is encoded downstream of the *fljB*^{z66} gene that represses expression of *fliC,* with the result that FliC is not expressed except in the case of rare phase 1 revertants of previously z66 isolates that are known to occur due to deletion of part of the linear plasmid (Baker *et al.*, 2007b).

Toxins

Salmonella Typhi encodes at least two toxins. The first is encoded by the *cdtB* gene and has similarities to cytolethal-distending toxins from a number of pathogens including *Campylobacter jejuni* and some enteropathogenic *E. coli* isolates (Johnson and Lior, 1988a,b). In *C. jejuni,* CdtA and CdtC are components of the transport system required for the targeting of the toxin CdtB to the host cell nucleus. However, *Salmonella* Typhi lacks *cdtA* and *cdtC*, and the cytotoxic effects are dependent on internalization of the bacterium into the host cell vacuole. In *Salmonella* Typhi, CdtB is secreted from host cells by autocrine and paracrine pathways following formation of a complex with PltA and PltB, which are encoded by genes close to *cdtB* on a pathogenicity islet. PltA and PltB amino acid sequence have similarity to the pertussis toxin subunit A and B respectively (Spano *et al.*, 2008). The contribution, if any, of these toxins to typhoid disease is currently not known.

Salmonella Typhi and *Salmonella* Paratyphi A also encode a putative toxin with similarity to cytolysin A from certain *E. coli* pathogens, encoded by the *clyA* gene (Parkhill *et al.*, 2001; von Rhein *et al.*, 2009). ClyA has pore-forming activity in lipid bilayer assays and contributes to intracellular growth and/or survival in macrophage in tissue culture (Faucher *et al.*, 2009; von Rhein *et al.*, 2009).

Fimbriae

Fimbriae, proteinaceous hair like structures elaborated on the surface of bacteria, are thought to be critical for early interaction of the pathogen with the host mucosa leading to colonization. *Salmonella* Typhi encodes a single operon encoding fimbriae of the type IV family (Pil), a single operon encoding fimbriae of the nucleator family (Agf) and 13 operons with similarity to fimbriae of the usher-chaperone family (*sef, fim, saf, tcf, bcf, sta, stb, ste, std, stc, stg* and *sth*) (Townsend *et al.*, 2001, see also Chapter 8 of this book). The gene encoding type IVb fimbriae is encoded on SPI7, mediates bacterial self association (Pier *et al.*, 1998) and binds the cystic fibrosis conductance regulator (Tsui *et al.*, 2003). However, no functions for interaction with cultured epithelial cells was observed (Bishop *et al.*, 2008). Little is known about the expression and possible function of Agf or the usher-chaperone family fimbriae in *Salmonella* Typhi. Indeed, five of the usher-chaperone family fimbrial operons contain one or more pseudogenes, suggesting that these may not be elaborated on the surface. *Salmonella* Paratyphi A lacks the type IVb fimbriae and the *sta* and *stg* operons.

Pseudogenes

Pseudogenes are protein-coding genes that have been disrupted by a non-sense SNP, an insertion, a deletion or a rearrangement, resulting in truncation of the encoded protein. The two finished *Salmonella* Typhi genome sequences (CT18 and Ty2) each contain approximately 200 pseudogenes (Fig. 6.1A), representing 5% of

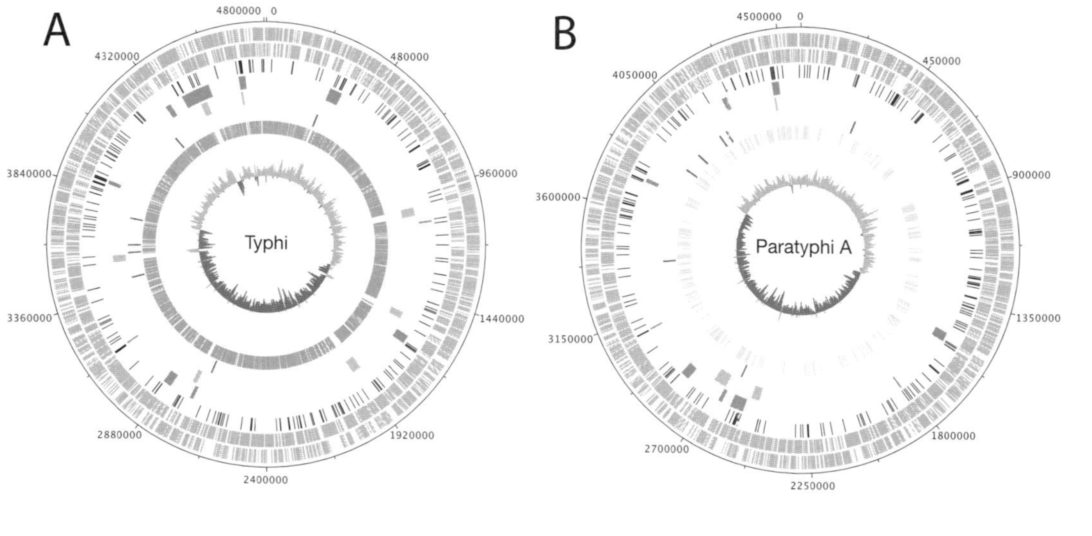

Figure 6.1 (A) *Salmonella* Typhi CT18 genome. (B) *Salmonella* Paratyphi A AKU_12601 genome. Rings, from outside: 1=genes on forward strand; 2=genes on reverse strand; 3=pseudogenes; 4=SPIs; 5=prophage; 6=rRNA operons; 7=SNPs (A) or genes recombined with Typhi (B). Inner plot shows GC skew (light=G>C, dark=G<C).

protein-coding sequences (Deng *et al.*, 2003; Holt *et al.*, 2009; Parkhill *et al.*, 2001), a greater proportion than present in *Salmonella* Typhimurium (1.1%) or *E. coli* K-12 (0.9%) (Andersson and Andersson, 1999; Cole *et al.*, 2001; McClelland *et al.*, 2001; Riley *et al.*, 2006). The accumulation of large numbers of pseudogenes results in genome degradation, a genomic feature of many host-restricted pathogenic bacteria when compared to related bacterial species with a broader host-range (Andersson and Andersson, 1999; Cole *et al.*, 2001; Parkhill *et al.*, 2003; Thomson *et al.*, 2006). Genome degradation probably results from a combination of adaptive selection for gene loss in a particular host niche, and genetic drift due to a sudden decrease in population size associated with entry into a new host-restricted niche. A recent analysis of 19 *Salmonella* Typhi genomes indicated that pseudogene formation is ongoing in *Salmonella* Typhi and different sublineages of *Salmonella* Typhi have accumulated different pseudogenes which may contribute to phenotypic variation within the population (Holt *et al.*, 2008b). Patterns of genome degradation due to pseudogene formation in pathogens that exhibit a distinct host range may provide insight into mechanisms of host-restriction and adaptation. For example, several genes involved in interactions between *Salmonella* and their mammalian hosts are pseudogenes in both *Salmonella* Typhi and *Salmonella* Paratyphi A. These include the secreted effectors *sopA, sopD2* and *slrP*, and intestinal persistence-related genes *ratB, shdA*, and *sivH* (Holt *et al.*, 2009; McClelland *et al.*, 2004). The loss of intestinal persistence genes may reflect the fact that enteric fever is largely a disease of the reticuloendothelial system (RES). Indeed, the period of intestinal colonization of an *Salmonella* Typhi vaccine strain was considerably shorter than that of an *Salmonella* Typhimurium vaccine strain in a human volunteer trial (Hindle *et al.*, 2002).

Plasmids

Several plasmids have been reported in *Salmonella* Typhi. The best characterized is the multidrug-resistance plasmid of the IncHI1 incompatibility type, exemplified by the 220 kbp sequenced plasmid pHCM1 (EMBL: AL513383) (Parkhill *et al.*, 2001; Wain *et al.*, 2003). This plasmid is self-transmissible, encoding all the genes required for conjugal transfer and mobilization between bacterial cells (Lawley *et al.*, 2002; Lawley *et al.*, 2003). The plasmid consists of a conserved backbone, including the conjugative transfer region, with insertions of various transposons and other mobile genetic elements carrying antibiotic resistance genes. Together, these genes encode resistance to all first- and second-line drugs used to treat typhoid fever, including chloramphenicol, ampicillin and co-trimoxazole (Parkhill *et al.*, 2001). The IncHI1 plasmid is thought to be responsible for the majority of multi-drug resistant typhoid observed in recent years, which in some areas accounts for up to 80% of typhoid cases (Chau *et al.*, 2007). Distinct lineages of IncHI1 plasmids have been identified in the *Salmonella* Typhi population and a recent study suggests the plasmids prevalent in Vietnam in the early 1990s have now been replaced by a plasmid of a distinct lineage (Phan *et al.*, 2009). Similar IncHI1 plasmids encoding multiple drug resistance have also been found in *Salmonella* Paratyphi A (Holt *et al.*, 2007b).

A 27 kbp linear plasmid, pBSSB1, was recently identified in *Salmonella* Typhi strains carrying the z66 flagella variant (Baker *et al.*, 2007a). The plasmid was shown to carry a flagellin gene ($fljB^{z66}$) encoding the z66 antigen (see above). The z66 flagella antigen has been observed exclusively in *Salmonella* Typhi isolates originating from typhoid fever cases in Indonesia (Guinée *et al.*, 1981; Song *et al.*, 1994; Tamura *et al.*, 1988). A recent study of flagellin antigens in *Salmonella* Typhi isolates from Jakarta, Indonesia found the plasmid in a single lineage of *Salmonella* Typhi (Baker *et al.*, 2008), suggesting it was acquired on a single occasion by just one *Salmonella* Typhi lineage, most likely in Indonesia, and has not spread to other lineages within the local *Salmonella* Typhi population since.

A 107 kbp plasmid of unknown function, pHCM2, was sequenced along with the first *Salmonella* Typhi genome (EMBL: AL513384) (Parkhill *et al.*, 2001). The plasmid exhibits some homology with the *Yersinia pestis* virulence plasmid pMT1, but its function in *Salmonella* Typhi has not been elucidated. The plasmid has been detected in *Salmonella* Typhi isolates from South-East Asia, India and Pakistan (Holt *et al.*, 2008b; Kidgell *et al.*, 2002a) and was present in two of 19 sequenced *Salmonella* Typhi genomes (Holt *et al.*, 2008b; Parkhill *et al.*, 2001). Other, often smaller, plasmids have been reported in *Salmonella* Typhi isolates, but their functions are as yet unknown.

Genome variation and population structure

A study of single nucleotide polymorphisms (SNPs) within a small fraction of the genome (1.85%) in a relatively large collection of 105 *Salmonella* Typhi strains identified 88 SNPs that provided the first insights into the population structure of *Salmonella* Typhi (Roumagnac *et al.*, 2006). This study indicated that there was no geographic or temporal distribution of *Salmonella* Typhi strains, with genetically distinct strains co-existing in the same time and geographical location. Identical strains were detected in multiple continents and in the same location over a period of more than 40 years.

Recent whole-genome sequence comparisons, as well as earlier genomic comparisons (Deng *et al.*, 2003) and DNA array work (Boyd *et al.*, 2003), have identified several regions of genomic variation in the *Salmonella* Typhi population (Holt *et al.*, 2008b). SNPs identified from sequence comparisons of 19 *Salmonella* Typhi genomes were used to assemble a rooted phylogenetic tree for *Salmonella* Typhi (Holt *et al.*, 2008b). The study identified fewer than 2000 SNPs in the 19 genomes (Fig. 6.1A), and provides a scaffold onto which genomic variation in the *Salmonella* Typhi can be overlaid. Twenty large deletion events (>100 bp) were identified and mapped to the tree, as were insertions of prophage sequences (Fig. 6.2). Several of these had been identified previously by microarray (Boyd *et al.*, 2003; Deng *et al.*, 2003). Three different variants of the recently defined SPI15 (Vernikos and Parkhill, 2006) were also identified among the 19 *Salmonella* Typhi genomes. Plasmids were not common among the sequenced strains, although IncHI1 multidrug resistance plasmids and pHCM2-like

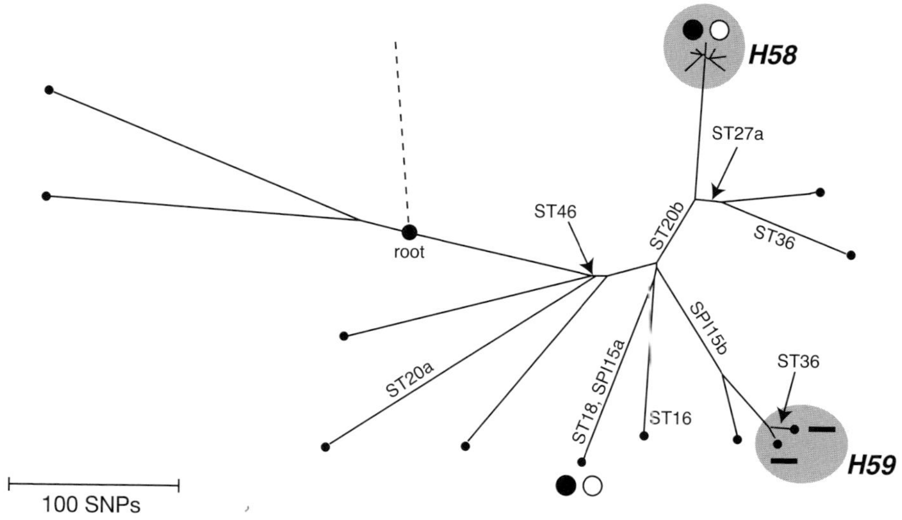

Figure 6.2 Phylogenetic tree of *Salmonella* Typhi. *Salmonella* Typhi phylogenetic tree based on 1787 SNPs determined from draft whole-genome sequences of 19 Typhi clinical isolates (Holt *et al.*, 2008a). Isolates and haplotypes are labelled in bold. Prophage (ST) and *Salmonella* pathogenicity island (SPI) insertions are labelled on the branches on which they occurred. H58 and H59 haplotypes, which were represented by multiple isolates, are highlighted in grey. Plasmid content is indicated with open circles (IncHI1 multidrug resistance plasmid), closed circles (cryptic plasmid pHCM2) and bars (linear plasmid pBSSB1 encoding the z66 flagellar antigen).

plasmids were found in two distinct lineages, suggesting multiple acquisition events (Fig. 6.2).

These studies point to a highly monomorphic and genetically isolated *Salmonella* Typhi population, with no evidence of recombination between *Salmonella* Typhi lineages or between particular *Salmonella* Typhi and other *S. enterica* serovars. Analysis of the SNP data revealed little evidence of stabilizing or diversifying selection within the *Salmonella* Typhi population but suggested the effective population size for *Salmonella* Typhi to be very small (Holt *et al.*, 2008b; Roumagnac *et al.*, 2006). This genetic isolation and small effective population size is consistent with a relatively small number of asymptomatic carriers being the main reservoir for *Salmonella* Typhi, implying that the bacteria responsible for long-term persistence in the human population are isolated in the gall bladders of individual human hosts with little opportunity for genetic exchange or even competition between strains. This may account for the lack of geographic or temporal distribution of *Salmonella* Typhi, as different carriers are likely to shed distinct variants of *Salmonella* Typhi persistently over a long period, and without competition between those variants there is no mechanism for long-term dominance of one strain over another in any given region.

The *Salmonella* Paratyphi A genome

The whole genome sequences of two *Salmonella* Paratyphi A isolates have been determined to date (Holt *et al.*, 2009; McClelland *et al.*, 2004), and these demonstrate several similarities with the *Salmonella* Typhi genome. Unlike *Salmonella* Typhi, *Salmonella* Paratyphi A is Vi-negative and does not carry SPI7. However, they do share several genes that are not present in other *Salmonella* serovars (shared 'rare genes'), including a fimbrial cluster (*sef*) and the *sopE* phage (McClelland *et al.*, 2004). The *Salmonella* Paratyphi A genomes accumulated pseudogenes to a similar level as the *Salmonella* Typhi genome (5%) (Fig. 6.1B), with independent inactivating mutations occurring in several of the same genes, as well as different genes within common functional pathways (Holt *et al.*, 2009; McClelland *et al.*, 2004). However, there is strong evidence to suggest that these shared genomic features are due to relatively recent convergent evolution, rather than divergence from a common human-restricted ancestor. A recent study reported that the *Salmonella* Typhi and *Salmonella* Paratyphi A genomes have undergone a relatively recent recombination event during which a quarter of genomic sequence was exchanged (Didelot *et al.*, 2007). The study suggested that this recombination was responsible for the majority of the 'rare genes' shared by the two serovars. Another recent study comparing the distribution of shared pseudogenes in recombined regions of the *Salmonella* Typhi and *Salmonella* Paratyphi A genomes suggested that extensive pseudogene accumulation began in both serovars only after the recombination event (Holt *et al.*, 2009). Thus, the hallmarks of host restriction and adaptation shared by *Salmonella* Typhi and *Salmonella* Paratyphi A (pseudogene accumulation, shared rare genes) appear to be the result of convergent evolution, arising after extensive recombination between the two serovars.

Genomic epidemiology: sequence-based approaches

A clear understanding of disease transmission and the response of a pathogenic bacterial population to environmental changes and human interventions requires epidemiological studies that consider variability within the pathogen population as well as clinical and environmental factors. High resolution typing of genomic variation within the pathogen population provides the tool to distinguish between closely related isolates, facilitating studies of disease transmission, as well as the evolution of novel traits (e.g. drug resistance, virulence phenotypes). Furthermore, information on genetic variation may provide direct mechanisms for variation in phenotype, analogous to genome–wide association studies routinely performed in human populations.

Established methods for typing in *Salmonella*, including phage typing, ribotyping and pulse-field gel electrophoresis (PFGE), are not ideal for elucidating phylogenetic relationships between isolates, and may not discriminate with enough resolution. Data are difficult to interpret and compare between laboratories, or even within a single laboratory over a period of time. The connection between phage type and chromosomal sequence type is particularly tenuous. Phage types and chromosomal haplotypes were not closely correlated in *Salmonella* Typhi (Roumagnac *et al.*, 2006). For example, for *Salmonella* Typhi phage ST36, identical phage can be acquired independently by different lineages and therefore do not reflect phylogeny (Fig. 6.2) (Holt *et al.*, 2008b). Sequence-based approaches, using next generation sequencing technology, are ideal for high resolution typing with a basis in phylogeny. Multi-locus sequence typing (MLST) has proven an invaluable tool in genomic epidemiology studies of many pathogenic bacteria (Maiden, 2006), but is not appropriate for low-diversity, monomorphic pathogens such as *Salmonella* Typhi and *Salmonella* Paratyphi A, which can be divided into only a few sequence types (Achtman, 2008; Kidgell *et al.*, 2002b). A better approach to genomic studies of *Salmonella* Typhi and *Salmonella* Paratyphi A is therefore likely to be SNP typing assays. However, it should be noted that SNP typing requires an initial SNP discovery phase before a useful typing regimen can be devised, which can bias the resulting view of population structure (Pearson *et al.*, 2004) and may therefore limit identification of novel variants as they arise.

Low-throughput approaches to SNP typing include allele-specific or real time PCR (Foster *et al.*, 2008; Ward *et al.*, 2008), which can be performed in most laboratories without the need for specialized SNP-typing hardware (Black and Vontas, 2007). However, dedicated technologies for high-throughput SNP typing allow the simultaneous typing of hundreds of SNP loci in hundreds of samples, ideal for large-scale genomic epidemiology studies. Current platforms suitable for bacterial SNP typing include the Sequenom platform (Tang *et al.*, 1999) and molecular inversion probes (Hardenbol *et al.*, 2003). Sequenom uses mass spectrometry to determine the nucleotide incorporated at a given SNP locus and can be multiplexed to type up to 35 SNPs in a single reaction. Additional SNPs can be assayed using additional multiplexes, for example 82 SNPs were targeted in a recent *Salmonella* Typhi study (Baker *et al.*, 2008). Molecular inversion probes (Hardenbol *et al.*, 2003) were recently used to type >1500 SNP loci in *Franciscella tularensis* (Vogler *et al.*, 2009), and the technology may be feasible for up to 20,000 SNPs.

Field applications of SNP typing in Typhi

SNPs currently identified in *Salmonella* Typhi include: (i) 450 SNPs identified between the two finished genome sequences (Deng *et al.*, 2003); (ii) 88 SNPs identified during analysis of 200 gene fragments (1.85% of the genome) in 105 isolates (Roumagnac *et al.*, 2006); and (iii) nearly 2000 SNPs identified in comparisons of draft genome sequences of 19 *Salmonella* Typhi isolates (Holt *et al.*, 2008b), which were targeted for sequencing based on their haplotypes according to SNP set (ii). To date, the only reported *Salmonella* Paratyphi A SNPs are those between the two finished genome sequences (Holt *et al.*, 2009).

A handful of genomic epidemiology studies have already utilized SNP typing in localized *Salmonella* Typhi populations (88 SNPs, set (ii) above). The first studies demonstrated that the apparent spread of fluoroquinolone resistant *Salmonella* Typhi in South-East Asia, in particular Vietnam, was due to the clonal expansion of a particular lineage or haplotype of *Salmonella* Typhi (designated H58) (Le *et al.*, 2007; Roumagnac *et al.*, 2006). The success of these studies demonstrates how SNP typing approaches can differentiate mechanisms of evolution of drug resistance. A highly localized study of 140 isolates collected in Jakarta, Indonesia, used Sequenom to type the 88 *Salmonella* Typhi SNPs (Baker *et al.*, 2008). The results demonstrated that the linear plasmid pBSSB1 carrying the z66 antigen was present in a single haplotype of *Salmonella* Typhi (known as H59), which was the most prevalent haplotype observed in the study. The study revealed a local population structure very different from that observed in the prior studies of Vietnamese populations, and reported GPS coordinates for the homes of typhoid patients from whom *Salmonella* Typhi was isolated, providing a first glimpse of the spatial distribution of *Salmonella* Typhi haplotypes within a local urban area. The Jakarta study is also a good example of SNP typing providing insight into the evolution of a particular phenotype. The z66 antigen was found to be common in Jakarta because it was present in the locally dominant H59 haplotype and not because

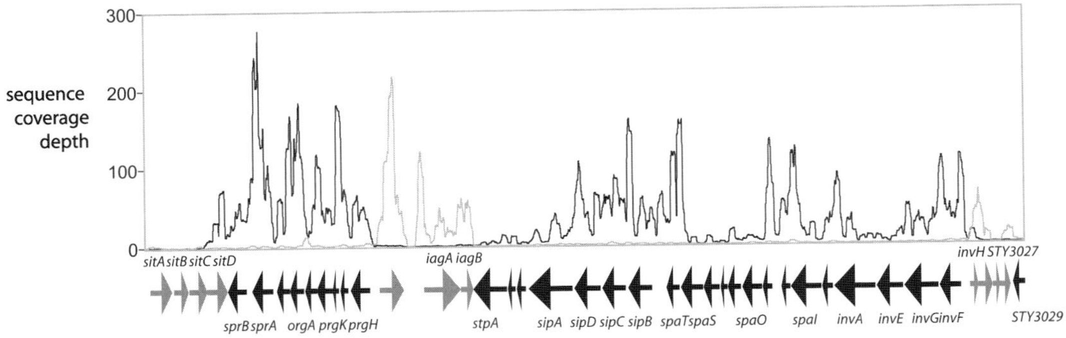

Figure 6.3 Strand-specific cDNA sequencing (ssRNA-seq) of the *Salmonella* Typhi SPI-1. Sequence coverage depth is indicated by a light (forward strand) or dark (reverse strand) line. The position of annotated genes is indicated by light (forward strand) or dark (reverse strand) arrows below. Select gene annotations are shown.

the plasmid carrying the *fljB*^{z66} gene was spread between Typhi isolates with a high frequency.

The *Salmonella* Typhi transcriptome

Classical annotation of the genome sequence of *Salmonella* Typhi has recently been complemented by strand-specific sequence analysis of the transcriptome, and whole cell proteome analysis (Perkins *et al.*, 2009). Strand-specific cDNA sequence (ssRNA-seq) was generated by direct sequencing of cDNA by Illumina next generation technology in which second strand synthesis was omitted, and peptide sequence was determined from whole cell lysates using liquid chromatography mass spectrometry (LCMS). SsRNA-seq provides a high resolution snap-shot of the coding and non-coding RNA content of the cell. For example, analysis of part of SPI1 identified transcriptionally silent regions, highly transcribed regions and 5′ *cis*-regulatory elements (Fig. 6.3). These approaches make high throughput annotation based on biological data possible, a significant improvement over classical annotation in which annotation of transcriptional units is largely based on conserved nucleotide sequence motifs and observations from homologous sequence in related organisms. High-throughput analyses made it possible to confirm the validity of previous annotations, provide re-annotation of a number of genes and validated the expression of previously hypothetical genes. Many non-coding RNA (ncRNA) species were identified including 55 previously unreported candidates. The ssRNA-seq approach was particularly useful for the identification of putative cargo genes on chromosomally encoded prophage elements (Perkins *et al.*, 2009). Although prophage elements are largely transcriptionally silent, they frequently encode 'cargo' genes that do not contribute to the phage lifecycle directly, but contribute to pathogenesis of the host bacterium, for example genes encoding TTSS effector proteins (Ehrbar and Hardt, 2005, see also Chapters 2 and 10 of this book). Several previously unreported transcriptionally active regions of prophage ST35, ST2–27 and ST46 were evident and may represent mobile genes that impact *Salmonella* Typhi virulence or transmission.

Future perspectives

Genomic studies to date form a solid foundation for future analysis of the genomic epidemiology of *Salmonella* Typhi. The expanded SNP set recently published for *Salmonella* Typhi (Holt *et al.*, 2008b) will form the basis for high-throughput studies of large collections of *Salmonella* Typhi isolates, contributing a valuable genomic component to epidemiological studies of typhoid fever. Future studies of *Salmonella* Paratyphi A should allow similar studies of paratyphoid fever, and potentially allow the two pathogen populations to be studied in parallel. Genomic epidemiology may be used to facilitate transmission studies, for example to detect clusters of related enteric fever cases that may be attributable to a common source, e.g. a contaminated water supply or an asymptomatic human carrier (Navarro *et al.*, 1996; Swaddiwudhipong and Kanlayanaphotporn, 2001; Usera *et al.*, 1995). Studies of the evolution of drug resistance in *Salmonella* Typhi and *Salmonella* Paratyphi A will continue to develop, potentially at higher resolution than has been achieved so far and incorporating co-evolution of IncHI1 drug resistance plasmids with their host strains (Phan *et al.*, 2009). A further application of genomic epidemiology in *Salmonella* Typhi and *Salmonella* Paratyphi A populations will be to study the impact of public health intervention strategies, including novel vaccines, antibiotics and improved water supply or hygiene practices, on the bacterial populations they target. For example, it may be possible to distinguish whether particular sub-populations are able to escape immunity provided by a novel vaccine, enabling improvement of vaccine design.

References

Abd El Ghany, M., Jansen, A., Clare, S., Hall, L., Pickard, D., Kingsley, R.A., and Dougan, G. (2007). Candidate live, attenuated *Salmonella enterica* serotype Typhimurium vaccines with reduced fecal shedding are immunogenic and effective oral vaccines. Infect. Immun. *75*, 1835–1842.

Abrahams, G.L., and Hensel, M. (2006). Manipulating cellular transport and immune responses: dynamic interactions between intracellular *Salmonella enterica* and its host cells. Cell Microbiol. *8*, 728–737.

Achtman, M. (2008). Evolution, Population Structure, and Phylogeography of Genetically Monomorphic Bacterial Pathogens. Annual Review of Microbiology *62*, 53–70.

Aleksic, S., Heinzerling, F., and Bockemühl, J. (1996). Human infection caused by *Salmonellae* of subspecies II to VI in Germany, 1977–1992. Zbl Bakt *283*, 391–398.

Andersson, J.O., and Andersson, S.G.E. (1999). Genome degradation is an ongoing process in Rickettsia. Molecular Biology and Evolution *16*, 1178–1191.

Anonymous (1939). Mary Mallon (Typhoid Mary). Am. J. Public Health Nations Health *29*, 66–68.

Baker, S., Hardy, J., Sanderson, K.E., Quail, M., Goodhead, I., Kingsley, R.A., Parkhill, J., Stocker, B., and Dougan, G. (2007a). A Novel Linear Plasmid Mediates Flagellar Variation in *Salmonella* Typhi. PLoS Pathog 3.

Baker, S., Holt, K., van de Vosse, E., Roumagnac, P., Whitehead, S., King, E., Ewels, P., Keniry, A., Weill, F.X., Lightfoot, D., *et al.* (2008). High-throughput genotyping of *Salmonella enterica* serovar Typhi allowing geographical assignment of haplotypes and pathotypes within an urban District of Jakarta, Indonesia. J. Clin. Microbiol. *46*, 1741–1746.

Baker, S., Holt, K., Whitehead, S., Goodhead, I., Perkins, T., Stocker, B., Hardy, J., and Dougan, G. (2007b). A linear plasmid truncation induces unidirectional flagellar phase change in H:z66 positive *Salmonella* Typhi. Mol. Microbiol. *66*, 1207–1218.

Bäumler, A.J. (1997). The record of horizontal gene transfer in *Salmonella*. Trends in Microbiol. *5*, 318–322.

Bishop, A., House, D., Perkins, T., Baker, S., Kingsley, R.A., and Dougan, G. (2008). Interaction of *Salmonella enterica* serovar Typhi with cultured epithelial cells: roles of surface structures in adhesion and invasion. Microbiology *154*, 1914–1926.

Black, W.C.t., and Vontas, J.G. (2007). Affordable assays for genotyping single nucleotide polymorphisms in insects. Insect Mol. Biol. *16*, 377–387.

Boyd, E.F., Porwollik, S., Blackmer, F., and McClelland, M. (2003). Differences in gene content among *Salmonella enterica* serovar Typhi isolates. Journal of Clinical Microbiology *41*, 3823–3828.

Bueno, S.M., Santiviago, C.A., Murillo, A.A., Fuentes, J.A., Trombert, A.N., Rodas, P.I., Youderian, P., and Mora, G.C. (2004). Precise excision of the large pathogenicity island, SPI7, in *Salmonella enterica* serovar Typhi. J. Bacteriol. *186*, 3202–3213.

Chau, T.T., Campbell, J.I., Galindo, C.M., Van, M., Diep, T.S., Nga, T.T., Van, V., Tuan, P.Q., Page, A.L., Ochiai, R.L., *et al.* (2007). Antimicrobial Drug Resistance of *Salmonella enterica* Serovar Typhi in Asia and Molecular Mechanism of Reduced Susceptibility to the Fluoroquinolones. Antimicrobial Agents and Chemotherapy *51*, 4315–4323.

Cole, S.T., Eiglmeier, K., Parkhill, J., James, K.D., Thomson, N.R., Wheeler, P.R., Honore, N., Garnier, T., Churcher, C., Harris, D., *et al.* (2001). Massive gene decay in the leprosy bacillus. Nature *409*, 1007–1011.

Crump, J.A., Luby, S.P., and Mintz, E.D. (2004). The global burden of typhoid fever. Bull World Health Organ *82*, 346–353.

Daefler, S. (1999). Type III secretion by *Salmonella* Typhimurium does not require contact with a eukaryotic host. Molecular microbiology *31*, 45–51.

Deiwick, J., and Hensel, M. (1999). Regulation of virulence genes by environmental signals in *Salmonella* Typhimurium. Electrophoresis *20*, 813–817.

Deng, W., Liou, S.R., rd, P., Mayhew, G.F., Rose, D.J., Burland, V., Kodoyianni, V., Schwartz, D.C., and Blattner, F.R. (2003). Comparative genomics of *Salmonella enterica* serovar Typhi strains Ty2 and CT18. J. Bacteriol. *185*, 2330–2337.

Didelot, X., Achtman, M., Parkhill, J., Thomson, N.R., and Falush, D. (2007). A bimodal pattern of relatedness between the *Salmonella* Paratyphi A and Typhi genomes: convergence or divergence by homologous recombination? Genome Research *17*, 61–68.

Ehrbar, K., and Hardt, W.D. (2005). Bacteriophage-encoded type III effectors in *Salmonella enterica* subspecies 1 serovar Typhimurium. Infect. Genet. Evol. *5*, 1–9.

Ellermeier, C.D., Ellermeier, J.R., and Slauch, J.M. (2005). HilD, HilC and RtsA constitute a feed forward loop that controls expression of the SPI1 type three secretion system regulator hilA in *Salmonella enterica* serovar Typhimurium. Molecular microbiology *57*, 691–705.

Evans, D.G., and Evans, D.J. (1978). New surface-associated heat-labile colonization factor antigen (CFA/II) produced by enterotoxic *Escherichia coli* of serogroups O6and O8. Infect. Immun. *21*, 638–647.

Fangtham, M., and Wilde, H. (2008). Emergence of *Salmonella* Paratyphi A as a major cause of enteric fever: need for early detection, preventive measures, and effective vaccines. Journal of travel medicine: official publication of the International Society of Travel Medicine and the Asia Pacific Travel Health Association *15*, 344–350.

Fass, E., and Groisman, E.A. (2009). Control of *Salmonella* pathogenicity island-2 gene expression. Current opinion in microbiology *12*, 199–204.

Faucher, S.P., Forest, C., Beland, M., and Daigle, F. (2009). A novel PhoP-regulated locus encoding the cytolysin ClyA and the secreted invasin TaiA of *Salmonella enterica* serovar Typhi is involved in virulence. Microbiology *155*, 477–488.

Foster, J., Okinaka, R., Svensson, R., Shaw, K., De, B., Robison, R., Probert, W., Kenefic, L., Brown, W., and Keim, P. (2008). Real-Time PCR Assays of Single-Nucleotide Polymorphisms Defining the Major Brucella Clades. J. Clin. Microbiol. *46*, 296–301.

Fraser, A., Goldberg, E., Acosta, C.J., Paul, M., and Leibovici, L. (2007). Vaccines for preventing typhoid fever. Cochrane database of systematic reviews (Online).

Galanis, E., Lo Fo Wong, D.M., Patrick, M.E., Binsztein, N., Cieslik, A., Chalermchikit, T., Aidara-Kane, A., Ellis, A., Angulo, F.J., and Wegener, H.C. (2006). Web-based surveillance and global *Salmonella* distribution, 2000–2002. Emerging infectious diseases *12*, 381–388.

Gewirtz, A.T., Navas, T.A., Lyons, S., Godowski, P.J., and Madara, J.L. (2001a). Cutting edge: bacterial flagellin activates basolaterally expressed TLR5 to induce epithelial proinflammatory gene expression. J. Immunol. *167*, 1882–1885.

Gewirtz, A.T., Simon, P.O., Jr., Schmitt, C.K., Taylor, L.J., Hagedorn, C.H., O'Brien, A.D., Neish, A.S., and Madara, J.L. (2001b). *Salmonella* Typhimurium translocates flagellin across intestinal epithelia, inducing a proinflammatory response. J. Clin. Invest. *107*, 99–109.

Guinée, P.A., Jansen, W.H., Maas, H.M., Le Minor, L., and Beaud, R. (1981). An unusual H antigen (Z66) in strains of *Salmonella* Typhi. Annales de microbiologie *132*, 331–334.

Gupta, V., Kaur, J., and Chander, J. (2009). An increase in enteric fever cases due to *Salmonella* Paratyphi A in and around Chandigarh. The Indian journal of medical research *129*, 95–98.

Hardenbol, P., Baner, J., Jain, M., Nilsson, M., Namsaraev, E., Karlin-Neumann, G., Fakhrai-Rad, H., Ronaghi, M., Willis, T., Landegren, U., *et al.* (2003). Multiplexed genotyping with sequence-tagged molecular inversion probes. Nat. Biotech *21*, 673–678.

Hardt, W.D., Chen, L.M., Schuebel, K.E., Bustelo, X.R., and Galan, J.E. (1998). S. Typhimurium encodes an activator of Rho GTPases that induces membrane ruffling and nuclear responses in host cells. Cell *93*, 815–826.

Hindle, Z., Chatfield, S.N., Phillimore, J., Bentley, M., Johnson, J., Cosgrove, C.A., Ghaem-Maghami, M., Sexton, A., Khan, M., Brennan, F.R., *et al.* (2002). Characterization of *Salmonella enterica* derivatives harboring defined aroC and *Salmonella* pathogenicity island 2 type III secretion system (ssaV) mutations by immunization of healthy volunteers. Infect. Immun. *70*, 3457–3467.

Holt, K., Parkhill, J., Mazzoni, C., Roumagnac, P., Weill, F.o.-X., Goodhead, I., Rance, R., Baker, S., Maskell, D., Wain, J., *et al.* (2008a). High-throughput sequencing provides insights into genome variation and evolution in *Salmonella* Typhi. Nature Genetics *40*, 987–993.

Holt, K.E., Parkhill, J., Mazzoni, C.J., Roumagnac, P., Weill, F.X., Goodhead, I., Rance, R., Baker, S., Maskell, D.J., Wain, J., *et al.* (2008b). High-throughput sequencing provides insights into genome variation and evolution in *Salmonella* Typhi. Nat. Genet. *40*, 987–993.

Holt, K.E., Thomson, N.R., Wain, J., Langridge, G.C., Hasan, R., Bhutta, Z.A., Quail, M.A., Norbertczak, H., Walker, D., Simmonds, M., *et al.* (2009). Pseudogene accumulation in the evolutionary histories of *Salmonella enterica* serovars Paratyphi A and Typhi. BMC Genomics *10*, 36.

Holt, K.E., Thomson, N.R., Wain, J., Minh, D.P., Nair, S., Hasan, R., Bhutta, Z.A., Quail, M.A., Norbertczak, H., Walker, D., *et al.* (2007a). Multidrug-resistant *Salmonella enterica* serovar Paratyphi A harbors IncHI1 plasmids similar to those found in serovar Typhi. J. Bacteriol. *189*, 4257–4264.

Holt, K.E., Thomson, N.R., Wain, J., Phan, M.D., Nair, S., Hasan, R., Bhutta, Z.A., Quail, M.A., Norbertczak, H., Walker, D., *et al.* (2007b). Multidrug-resistant *Salmonella enterica* serovar Paratyphi A harbors IncHI1 plasmids similar to those found in serovar Typhi. J. Bacteriol. *189*, 4257–4264.

Hone, D.M., Attridge, S.R., Forrest, B., Morona, R., Daniels, D., LaBrooy, J.T., Bartholomeusz, R.C., Shearman, D.J., and Hackett, J. (1988). A galE via (Vi antigen-negative) mutant of *Salmonella* Typhi Ty2 retains virulence in humans. Infect. Immun. *56*, 1326–1333.

Hornick, R.B., Greisman, S.E., Woodward, T.E., DuPont, H.L., Dawkins, A.T., and Snyder, M.J. (1970a). Typhoid fever: pathogenesis and immunologic control. N. Engl. J. Med. *283*, 686–691.

Hornick, R.B., Greisman, S.E., Woodward, T.E., DuPont, H.L., Dawkins, A.T., and Snyder, M.J. (1970b). Typhoid fever: pathogenesis and immunologic control. 2. N. Engl. J. Med. *283*, 739–746.

Inohara, N., Ogura, Y., and Nunez, G. (2002). Nods: a family of cytosolic proteins that regulate the host response to pathogens. Curr. Opin Microbiol. *5*, 76–80.

Johnson, W.M., and Lior, H. (1988a). A new heat-labile cytolethal distending toxin (CLDT) produced by Campylobacter spp. Microb Pathog *4*, 115–126.

Johnson, W.M., and Lior, H. (1988b). A new heat-labile cytolethal distending toxin (CLDT) produced by Escherichia coli isolates from clinical material. Microb Pathog *4*, 103–113.

Jones, B.D., Lee, C.A., and Falkow, S. (1992). Invasion of *Salmonella* Typhimurium is affected by the direction of flagellar rotation. Infect. Immun. *60*, 2475–2480.

Khatri, N.S., Maskey, P., Poudel, S., Jaiswal, V.K., Karkey, A., Koirala, S., Shakya, N., Agrawal, K., Arjyal, A., Basnyat, B., *et al.* (2009). Gallbladder carriage of *Salmonella* Paratyphi A may be an important factor in the increasing incidence of this infection in South Asia. Annals of internal medicine *150*, 567–568.

Kidgell, C., Pickard, D., Wain, J., James, K., Nga, D., Diep, T.S., Levine, M.M., O'Gaora, P., Prentice, M.B., Parkhill, J., *et al.* (2002a). Characterisation and distribution of a cryptic *Salmonella* Typhi plasmid pHCM2. Plasmid *47*, 159–171.

Kidgell, C., Reichard, U., Wain, J., Linz, B., Torpdahl, M., Dougan, G., and Achtman, M. (2002b). *Salmonella* Typhi, the causative agent of typhoid fever, is approximately 50,000 years old. Infect. Genet. Evol. *2*, 39–45.

Kingsley, R.A., Abi Ghanem, D., Puebla-Osorio, N., Keestra, A.M., Berghman, L., and Baumler, A.J. (2004a). Fibronectin binding to the *Salmonella enterica* serotype Typhimurium ShdA autotransporter protein is inhibited by a monoclonal antibody recognizing the A3 repeat. J. Bacteriol. *186*, 4931–4939.

Kingsley, R.A., Humphries, A.D., Weening, E.H., De Zoete, M.R., Winter, S., Papaconstantinopoulou, A., Dougan, G., and Baumler, A.J. (2003). Molecular and phenotypic analysis of the CS54 island of *Salmonella enterica* serotype Typhimurium: identification of intestinal colonization and persistence determinants. Infect. Immun. *71*, 629–640.

Kingsley, R.A., Keestra, A.M., de Zoete, M.R., and Baumler, A.J. (2004b). The ShdA adhesin binds to the cationic cradle of the fibronectin 13FnIII repeat module: evidence for molecular mimicry of heparin binding. Mol. Microbiol. *52*, 345–355.

Kingsley, R.A., van Amsterdam, K., Kramer, N., and Baumler, A.J. (2000). The shdA gene is restricted to serotypes of *Salmonella enterica* subspecies I and contributes to efficient and prolonged fecal shedding. Infect. Immun. *68*, 2720–2727.

Lawley, T.D., Gilmour, M.W., Gunton, J.E., Standeven, L.J., and Taylor, D.E. (2002). Functional and mutational analysis of conjugative transfer region 1 (Tra1) from the IncHI1 plasmid R27. J. Bacteriol. *184*, 2173–2180.

Lawley, T.D., Gilmour, M.W., Gunton, J.E., Tracz, D.M., and Taylor, D.E. (2003). Functional and mutational analysis of conjugative transfer region 2 (Tra2) from the IncHI1 plasmid R27. J. Bacteriol. *185*, 581–591.

Le, T.A., Fabre, L., Roumagnac, P., Grimont, P.A., Scavizzi, M.R., and Weill, F.X. (2007). Clonal Expansion and Microevolution of Quinolone-Resistant *Salmonella enterica* Serotype Typhi in Vietnam from 1996 to 2004. J. Clin. Microbiol. *45*, 3485–3492.

Liu, S.-L., and Sanderson, K.E. (1995). Rearrangements in the genome of the bacterium *Salmonella* Typhi. cad Sci USA *92*, 1018–1022.

Maiden, M.C. (2006). Multilocus sequence typing of bacteria. Annual review of microbiology *60*, 561–588.

Maskey, A.P., Basnyat, B., Thwaites, G.E., Campbell, J.I., Farrar, J.J., and Zimmerman, M.D. (2008). Emerging trends in enteric fever in Nepal: 9124 cases confirmed by blood culture 1993–2003. Transactions of the Royal Society of Tropical Medicine and Hygiene *102*, 91–95.

Maskey, A.P., Day, J.N., Phung, Q.T., Thwaites, G.E., Campbell, J.I., Zimmerman, M., Farrar, J.J., and Basnyat, B. (2006). *Salmonella enterica* serovar Paratyphi A and *S. enterica* serovar Typhi cause indistinguishable clinical syndromes in Kathmandu, Nepal. Clinical infectious diseases: an official publication of the Infectious Diseases Society of America *42*, 1247–1253.

McClelland, M., Sanderson, K.E., Clifton, S.W., Latreille, P., Porwollik, S., Sabo, A., Meyer, R., Bieri, T., Ozersky, P., McLellan, M., *et al.* (2004). Comparison of genome degradation in Paratyphi A and Typhi, human-restricted serovars of *Salmonella enterica* that cause typhoid. Nat. Genet. *36*, 1268–1274.

McClelland, M., Sanderson, K.E., Spieth, J., Clifton, S.W., Latreille, P., Courtney, L., Porwollik, S., Ali, J., Dante, M., Du, F., *et al.* (2001). Complete genome sequence of *Salmonella enterica* serovar Typhimurium LT2. Nature *413*, 852–856.

Mills, D.M., Bajaj, V., and Lee, C.A. (1995). A 40 kb chromosomal fragment encoding *Salmonella* Typhimurium invasion genes is absent from the corresponding region of the *Escherichia coli* K-12 chromosome. Mol. Microbiol. *15*, 749–759.

Mirold, S., Rabsch, W., Rohde, M., Stender, S., Tschäpe, H., Rüssmann, H., H., Igwe, E., and Hardt, W.-D. (1999). Isolation of a temperate bacteriophage encoding the type III effector protein SopE from an epidemic *Salmonella* Typhimurium strain. Proceedings of the National Academy of Sciences of the United States of America *96*, 9845–9850.

Nair, S., Alokam, S., Kothapalli, S., Porwollik, S., Proctor, E., Choy, C., McClelland, M., Liu, S.L., and Sanderson, K.E. (2004). *Salmonella enterica* serovar Typhi strains from which SPI7, a 134-kilobase island with genes for Vi exopolysaccharide and other functions, has been deleted. J. Bacteriol. *186*, 3214–3223.

Navarro, F., Llovet, T., Echeita, M.A., Coll, P., Aladuena, A., Usera, M.A., and Prats, G. (1996). Molecular typing of *Salmonella enterica* serovar Typhi. J. Clin. Microbiol. *34*, 2831–2834.

Ochiai, R.L., Wang, X., von Seidlein, L., Yang, J., Bhutta, Z.A., Bhattacharya, S.K., Agtini, M., Deen, J.L., Wain, J., Kim, D.R., *et al.* (2005). *Salmonella* Paratyphi A rates, Asia. Emerging infectious diseases (Print) 11, 1764–1766.

Ochman, H., Soncini, F.C., Solomon, F., and Groisman, E.A. (1996). Identification of a pathogenicity island for *Salmonella* survival in host cells. Proc. Natl. Acad. Sci. U.S.A. *93*, 7800–7804.

Palit, A., Ghosh, S., Dutta, S., Sur, D., Bhattacharya, M.K., and Bhattacharya, S.K. (2006). Increasing prevalence of *Salmonella enterica* serotype Paratyphi-A in patients with enteric fever in a periurban slum setting of Kolkata, India. International journal of environmental health research *16*, 455–459.

Parkhill, J., Dougan, G., James, K.D., Thomson, N.R., Pickard, D., Wain, J., Churcher, C., Mungall, K.L., Bentley, S.D., Holden, M.T., *et al.* (2001). Complete genome sequence of a multiple drug resistant *Salmonella enterica* serovar Typhi CT18. Nature *413*, 848–852.

Parkhill, J., Sebaihia, M., Preston, A., Murphy, L.D., Thomson, N., Harris, D.E., Holden, M.T., Churcher, C.M., Bentley, S.D., Mungall, K.L., *et al.* (2003). Comparative analysis of the genome sequences of Bordetella pertussis, Bordetella parapertussis and Bordetella bronchiseptica. Nature genetics 35, 32–40.

Pearson, T., Busch, J.D., Ravel, J., Read, T.D., Rhoton, S.D., U'Ren, J.M., Simonson, T.S., Kachur, S.M., Leadem, R.R., Cardon, M.L., *et al.* (2004). Phylogenetic discovery bias in Bacillus anthracis using single-nucleotide polymorphisms from whole-genome sequencing. Proceedings of the National Academy of Sciences of the United States of America *101*, 13536–13541.

Perkins, T.T., Kingsley, R.A., Fookes, M.C., Gardner, P.P., James, K.D., Yu, L., Assefa, S.A., He, M., Croucher, N.J., Pickard, D.J., *et al.* (2009). A strand-specific RNA-Seq analysis of the transcriptome of the typhoid bacillus *Salmonella* Typhi. PLoS Genet 5, e1000569.

Phan, M.D., Kidgell, C., Nair, S., Holt, K.E., Turner, A.K., Hinds, J., Butcher, P., Cooke, F.J., Thomson, N.R., Titball, R., *et al.* (2009). Variation in *Salmonella enterica* serovar Typhi IncHI1 plasmids during the global spread of resistant typhoid fever. Antimicrob. Agents Chemother. *53*, 716–727.

Pickard, D., Wain, J., Baker, S., Line, A., Chohan, S., Fookes, M., Barron, A., Gaora, P.O., Chabalgoity, J.A., Thanky, N., *et al.* (2003). Composition, acquisition, and distribution of the Vi exopolysaccharide-encoding *Salmonella enterica* pathogenicity island SPI-7. J. Bacteriol. *185*, 5055–5065.

Pier, G.B., Grout, M., Zaidi, T., Meluleni, G., Mueschenborn, S.S., Banting, G., Ratcliff, R., Evans, M.J., and Colledge, W.H. (1998). *Salmonella* Typhi uses CFTR to enter intestinal epithelial cells. Nature *393*, 79–82.

Porwollik, S., Frye, J., Florea, L.D., Blackmer, F., and McClelland, M. (2003). A non-redundant microarray of genes for two related bacteria. Nucleic acids research *31*, 1869–1876.

Riley, M., Abe, T., Arnaud, M.B., Berlyn, M.K., Blattner, F.R., Chaudhuri, R.R., Glasner, J.D., Horiuchi, T., Keseler, I.M., Kosuge, T., *et al.* (2006). Escherichia coli K-12: a cooperatively developed annotation snapshot--2005. Nucleic acids research *34*, 1–9.

Roumagnac, P., Weill, F.X., Dolecek, C., Baker, S., Brisse, S., Chinh, N.T., Le, T.A., Acosta, C.J., Farrar, J., Dougan, G., *et al.* (2006). Evolutionary history of *Salmonella* Typhi. Science *314*, 1301–1304.

Sanderson, K.E., and Liu, S.L. (1998). Chromosomal rearrangements in enteric bacteria. Electrophoresis *19*, 569–572.

Shea, J.E., Hensel, M., Gleeson, C., and Holden, D.W. (1996). Identification of a virulence locus encoding a second type III secretion system in *Salmonella* Typhimurium. Proc. Natl. Acad. Sci. U.S.A. *93*, 2593–2597.

Sieling, P.A., and Modlin, R.L. (2002). Toll-like receptors: mammalian 'taste receptors' for a smorgasbord of microbial invaders. Curr. Opin Microbiol. *5*, 70–75.

Song, J.H., Cho, H., Park, M.Y., Kim, Y.S., Moon, H.B., Kim, Y.K., and Pai, C.H. (1994). Detection of the H1-j strain of *Salmonella* Typhi among Korean isolates by the polymerase chain reaction. The American journal of tropical medicine and hygiene *50*, 608–611.

Spano, S., Ugalde, J.E., and Galan, J.E. (2008). Delivery of a *Salmonella* Typhi exotoxin from a host intracellular compartment. Cell Host Microbe 3, 30–38.

Stecher, B., Robbiani, R., Walker, A.W., Westendorf, A.M., Barthel, M., Kremer, M., Chaffron, S., Macpherson, A.J., Buer, J., Parkhill, J., *et al.* (2007). *Salmonella enterica* serovar Typhimurium exploits inflammation to compete with the intestinal microbiota. PLoS Biol. *5*, 2177–2189.

Swaddiwudhipong, W., and Kanlayanaphotporn, J. (2001). A common-source water-borne outbreak of multidrug-resistant typhoid fever in a rural Thai community. Journal of the Medical Association of Thailand = Chotmaihet thangphaet *84*, 1513–1517.

Tamura, K., Sakazaki, R., Kuramochi, S., and Nakamura, A. (1988). Occurrence of H-antigen Z66 of R phase in cultures of *Salmonella* serovar Typhi originated from Indonesia. Epidemiol. Infect. *101*, 311–314.

Tang, K., Fu, D.J., Julien, D., Braun, A., Cantor, C.R., and Koster, H. (1999). Chip-based genotyping by mass spectrometry. Proc. Natl. Acad. Sci. U.S.A. *96*, 10016–10020.

Thomson, N.R., Howard, S., Wren, B.W., Holden, M.T., Crossman, L., Challis, G.L., Churcher, C., Mungall, K., Brooks, K., Chillingworth, T., *et al.* (2006). The complete genome sequence and comparative genome analysis of the high pathogenicity Yersinia enterocolitica strain 8081. PLoS genetics 2.

Townsend, S., Kramer, N., Edwards, R., Baker, S., Hamlin, N., Simmonds, M., Stevens, K., Maloy, S., Parkhill, J., Dougan, G., *et al.* (2001). *Salmonella enterica* Serovar Typhi Possesses a Unique Repertoire of Fimbrial Gene Sequences. Infection and Immunity *69*, 2894–2901.

Tsui, I.S., Yip, C.M., Hackett, J., and Morris, C. (2003). The type IVB pili of *Salmonella enterica* serovar Typhi bind to the cystic fibrosis transmembrane conductance regulator. Infect. Immun. *71*, 6049–6050.

Usera, M.A., Echeita, A., Aladueña, A., Alvarez, J., Carreño, C., Orcau, A., and Planas, C. (1995). [Investigation of an outbreak of water-borne typhoid fever in Catalonia in 1994]. Enferm Infecc Microbiol Clin. *13*, 450–454.

Vazquez-Torres, A., and Fang, F.C. (2000). Cellular routes of invasion by enteropathogens. Curr. Opin Microbiol. *3*, 54–59.

Vazquez-Torres, A., Jones-Carson, J., Baumler, A.J., Falkow, S., Valdivia, R., Brown, W., Le, M., Berggren, R., Parks, W.T., and Fang, F.C. (1999). Extraintestinal dissemination of *Salmonella* by CD18-expressing phagocytes. Nature *401*, 804–808.

Vernikos, G.S., and Parkhill, J. (2006). Interpolated variable order motifs for identification of horizontally acquired DNA: revisiting the *Salmonella* pathogenicity islands. Bioinformatics (Oxford, England); Bioinformatics (Oxford, England) 22, 2196–2203.

Vogler, A., Birdsell, D., Price, L., Bowers, J., Beckstrom-Sternberg, S., Auerbach, R., Beckstrom-Sternberg, J., Johansson, A., Clare, A., Buchhagen, J., *et al.* (2009). Phylogeography of Francisella tularensis: Global Expansion of a Highly Fit Clone. J. Bacteriol. *191*, 2474–2484.

von Rhein, C., Bauer, S., Lopez Sanjurjo, E.J., Benz, R., Goebel, W., and Ludwig, A. (2009). ClyA cytolysin from *Salmonella*: distribution within the genus, regulation of expression by SlyA, and pore-forming characteristics. Int. J. Med. Microbiol. *299*, 21–35.

Wain, J., House, D., Zafar, A., Baker, S., Nair, S., Kidgell, C., Bhutta, Z., Dougan, G., and Hasan, R. (2005). Vi antigen expression in *Salmonella enterica* serovar Typhi clinical isolates from Pakistan. J. Clin. Microbiol. *43*, 1158–1165.

Wain, J., Nga, D., Kidgell, C., James, K., Fortune, S., Diep, S., Ali, T., Gaora, Parry, C., Parkhill, J., *et al.* (2003). Molecular analysis of incHI1 antimicrobial resistance plasmids from *Salmonella* serovar Typhi strains associated with typhoid fever. Antimicrobial Agents and Chemotherapy *47*, 2732–2739.

Ward, T., Ducey, T., Usgaard, T., Dunn, K., and Bielawski, J. (2008). Multilocus Genotyping Assays for Single Nucleotide Polymorphism-Based Subtyping of Listeria monocytogenes Isolates. Appl. Environ. Microbiol. *74*, 7629–7642.

Wilson, R.P., Raffatellu, M., Chessa, D., Winter, S.E., Tukel, C., and Baumler, A.J. (2008). The Vi-capsule prevents Toll-like receptor 4 recognition of *Salmonella*. Cell Microbiol. *10*, 876–890.

Winter, S.E., Raffatellu, M., Wilson, R.P., Russmann, H., and Baumler, A.J. (2008). The *Salmonella enterica* serotype Typhi regulator TviA reduces interleukin-8 production in intestinal epithelial cells by repressing flagellin secretion. Cell Microbiol. *10*, 247–261.

Wood, M.W., Rosqvist, R., Mullan, P.B., Edwards, M.H., and Galyov, E.E. (1996). SopE, a secreted protein of *Salmonella dublin*, is translocated into the target eukaryotic cell via a *sip*-dependent mechanism and promotes bacterial entry. Mol. Microbiol. *22*, 327–338.

Woods, C.W., Murdoch, D.R., Zimmerman, M.D., Glover, W.A., Basnyat, B., Wolf, L., Belbase, R.H., and Reller, L.B. (2006). Emergence of *Salmonella enterica* serotype Paratyphi A as a major cause of enteric fever in Kathmandu, Nepal. Transactions of the Royal Society of Tropical Medicine and Hygiene *100*, 1063–1067.

Zhang, S., Kingsley, R.A., Santos, R.L., Andrews-Polymenis, H., Raffatellu, M., Figueiredo, J., Nunes, J., Tsolis, R.M., Adams, L.G., and Baumler, A.J. (2003). Molecular pathogenesis of *Salmonella enterica* serotype Typhimurium-induced diarrhea. Infect. Immun. *71*, 1–12.

The Small RNAs of *Salmonella*

7

Sridhar Javayel, Kai Papenfort and Jörg Vogel*

Abstract

To date, close to one hundred distinct small non-coding RNAs (sRNAs) have been identified in *Salmonella* by a variety of biocomputational or wet-lab approaches including RNA sequencing. The function of more than twenty of these sRNAs is known from studies in *Salmonella* itself or can be inferred from conserved homologues in *E. coli*. Many of these sRNAs act in conjunction with the RNA-chaperone Hfq to post-transcriptionally repress or activate *trans*-encoded target genes, but *cis*-antisense RNAs and regulators of protein activity are also abundantly present. In addition to a large number of sRNAs conserved in other enteric bacteria, *Salmonella* also expresses a set of sRNAs specific to this genus. Interestingly, such regulators have been shown to control the expression of conserved genes encoded on the 'core' *Salmonella* genome. Conversely, conserved sRNA can act as regulators of recently acquired *Salmonella*-specific genes, indicating significant cross-talk of conserved and horizontally acquired elements at the RNA level. The present chapter covers strategies for the identification of sRNAs as well as their characterized functional roles in *Salmonella*.

Introduction

Bacterial adaptation to changing environmental conditions requires dynamic gene expression changes within specific regulatory circuits. Transcriptional activation or repression by DNA-binding proteins has long been considered the most relevant way to switch on and off bacterial genes. However, post-transcriptional control at the mRNA level is now well accepted to play a pivotal role in bacterial gene expression, too. To date, small RNAs (sRNAs) constitute the largest class of post-transcriptional regulators in bacteria, and those known in *Salmonella* will be covered in this chapter.

Most enteric sRNAs range from 50 to 250 nucleotides (nt) in length, lack open reading frames, and are encoded by freestanding genes in the inter-ORF regions of the chromosome. Disregarding the ubiquitous tRNA and rRNAs, sRNAs fall into four major groups (Fig. 7.1). The first group comprises three ubiquitous housekeeping RNAs that control fundamental cellular processes, namely the catalytic subunit of the tRNA-processing RNase P (Guerrier-Takada *et al.*, 1983); transfer-messenger RNA (tmRNA) which functions in quality control of protein synthesis (Keiler, 2008); and 4.5S RNA which serves in the Signal Recognition Particle (SRP) by targeting proteins to the membrane (Grudnik *et al.*, 2009).

Members of the second class of sRNAs bind to proteins and thereby alter their activity (Waters and Storz, 2009). *Salmonella* is currently known to express three sRNAs that act by protein-binding: 6S RNA targets RNA polymerase (RNAP) and negatively modulates its affinity to sigma factors (Wassarman, 2007), whilst the CsrB and CsrC sRNAs antagonize the globally acting RNA-binding protein CsrA (Altier *et al.*, 2000b; Fortune *et al.*, 2006; Viegas *et al.*, 2007).

The third group of sRNAs acts by base-pairing at the level of mRNA translation or stability, and this is by far the largest group of *Salmonella* sRNAs. Depending on their location relative to targets, these base-pairing sRNAs are referred

*Corresponding author

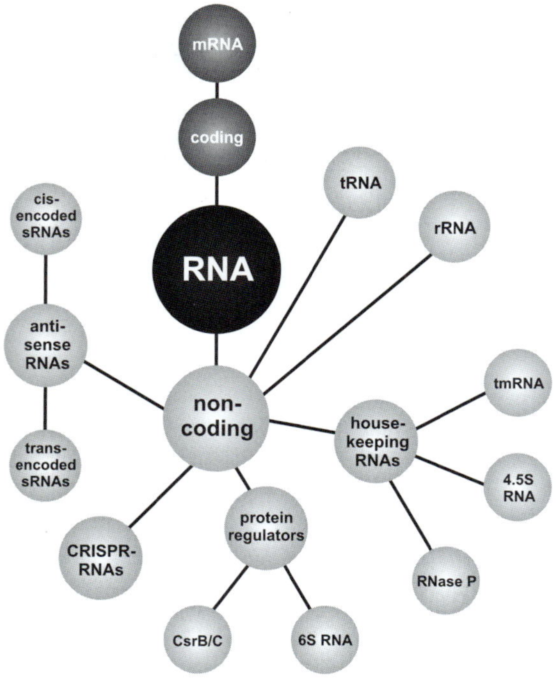

Figure 7.1 Prokaryotic RNA organigram. Classes of RNA regulators present in bacterial organisms with special emphasis on non-coding RNA molecules. These RNAs were further divided into six major groups: tRNAs, rRNAs, housekeeping RNAs, protein regulators, CRISPR RNAs and antisense RNAs.

to as '*cis*-antisense' if they overlap with the 5′ or 3′ region of a target gene, or '*trans*-antisense' if the target mRNA(s) is encoded elsewhere in the genome. *Cis*-encoded sRNAs typically possess full or extensive complementary sequence with targets, although full duplex-formation rarely is a prerequisite for efficient target regulation (Wagner *et al.*, 2002). Notably, some *cis*-antisense sRNAs might not act by direct base-pairing but instead affect the transcription of oppositely located genes at the DNA level (Brantl, 2007). *Trans*-antisense sRNAs (exemplified in Fig. 7.2) seem to constitute the largest class of regulatory sRNAs in *Salmonella*. Unlike their *cis*-encoded counterparts, each of them might act on several target mRNAs in parallel. Base-pairing by these sRNAs is usually short (10 to 25 base pairs) and imperfect, and involves the 5′ UTR (untranslated region) or early CDS (coding sequence) of target mRNAs. The outcome of the regulation is most often negative, leading to translational repression and/or decreased stability of the target (Waters and Storz, 2009), but several cases of translational activation by 'anti–antisense pairing' have been reported as well (Fröhlich and Vogel, 2009). Most of the *trans*-encoded sRNAs require the RNA chaperone Hfq for both productive annealing to targets (Møller *et al.*, 2002a; Zhang *et al.*, 2002a) and their own intracellular stability (Folichon *et al.*, 2003; Moll *et al.*, 2003).

The fourth group of sRNAs is a very recent addition: these are the processed ~60 nt transcripts that originate from CRISPR (Clustered, Regularly Interspaced Short Palindromic Repeats) loci belonging to a widespread prokaryotic immunity system present in ~40% of all Eubacteria and almost all Archaea (Sorek *et al.*, 2008). CRISPR loci are arrays of a variable number of direct repeats interspersed with non-repetitive spacers, which together with the adjacently located *cas* (CRISPR associated) genes form the functional CRISPR system (Sorek *et al.*, 2008). Following precursor processing, the CRISPR sRNAs of Eubacteria interfere with invading bacteriophages or plasmids, presumably by targeting the foreign DNA (Marraffini and Sontheimer, 2009).

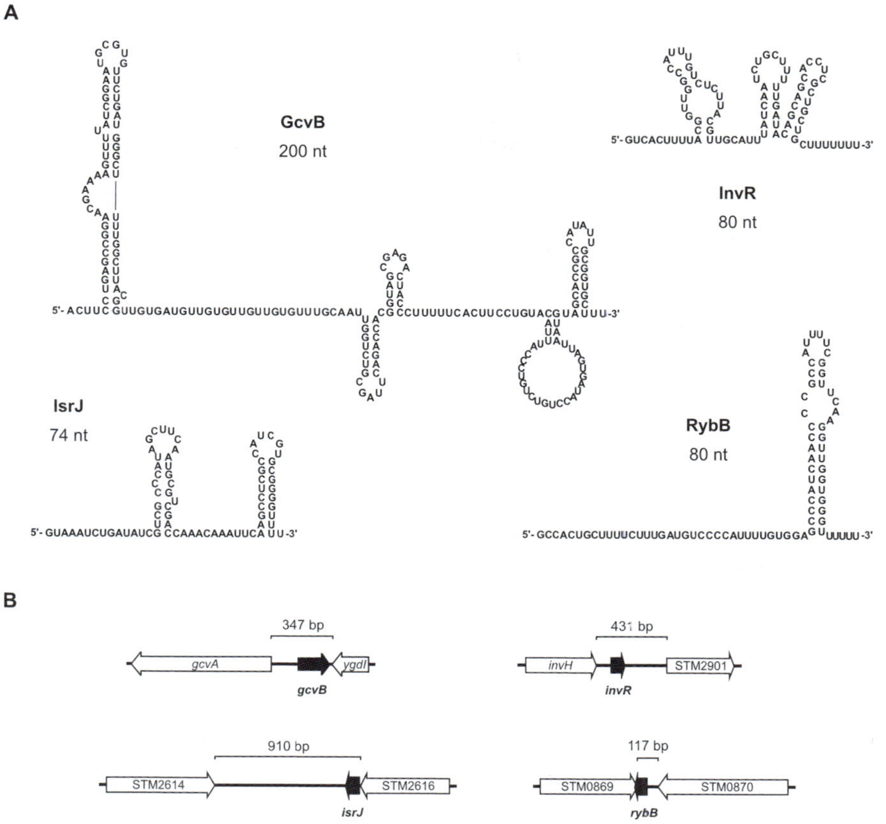

Figure 7.2 Examples of *Salmonella* sRNAs and their genomic location. (A) Secondary structures of *Salmonella* sRNA obtained by chemical probing (GcvB, InvR, RybB) or *in silico* structure prediction (IsrJ). (B) Location of these four sRNA genes in the intergenic regions (IGRs) of the *Salmonella* chromosome. Neighbouring genes and IGR lengths are given for orientation.

Whilst most of the *Salmonella* sRNAs are encoded by free-standing genes, UTR-derived transcripts of mRNAs constitute another rich source of short transcripts. Specifically, *cis*-regulatory 'riboswitches', i.e. 5′ UTR borne structured RNA elements sensing tRNAs or metabolites (Roth and Breaker, 2009), can produce stable small transcripts such as the SroA and SroH sRNAs (Vogel *et al.*, 2003). Small transcripts originating from S-adenosylmethionine-binding riboswitches in *Listeria monocytogens* were recently proposed to function *in trans* as base-pairing sRNAs to regulate the expression of virulence genes (Loh *et al.*, 2009).

The functional importance of sRNAs in *Salmonella* has become evident not only from studies of individual sRNA genes (as detailed below) but also from analysis of mutant strains that lack the conserved and ubiquitous RNA-chaperone, Hfq. Deletion of the *hfq* gene attenuates *Salmonella* virulence in mice and in cell-culture models, and causes strong phenotypes with relevance to infection including the loss of motility and type 3 secretion system (T3SS)-mediated cell invasion, and impaired adherence and survival in cultured epithelial and macrophage-like cells (Ansong *et al.*, 2009; Sittka *et al.*, 2007). In addition, *Salmonella* Typhimurium has been the paradigm of a pathogen in which Hfq acts as a truly global regulator (Chao and Vogel, 2010). Several gene expression studies including one performed during space flight, and involving genome-wide *lacZ* reporter screens, microarrays, 1D and 2D protein gel analyses, or sample-matched transcriptomics and gel-free proteomics measurements, have established that Hfq might directly or indirectly regulate at least 20% of all protein-coding genes of *Salmonella* (Ansong *et al.*, 2009; Figueroa-Bossi *et al.*, 2006; Sittka *et al.*, 2008; Sittka *et al.*, 2007; Wilson *et al.*, 2007). Given the key function of Hfq as an sRNA chaperone, the strong phenotypes and gene expression alterations observed in *hfq* mutants might reflect the collective loss of functions of many individual *Salmonella* sRNAs.

The present chapter will summarize our current knowledge of sRNAs and their functions determined in *Salmonella* itself or inferred from work in closely related *E. coli*.

Small RNA discovery in *Salmonella*

The availability of complete genome sequences in the late 1990s prompted systematic screens for sRNAs in bacteria. An extensive annotation of sRNA genes was initiated in *E. coli* K12 and involved a number of molecular and biocomputational techniques (Vogel and Sharma, 2005). Searches for sequence conservation and 'orphan' promoter/terminator pairs within the 'empty' intergenic regions (IGRs) predicted a large number of new *E. coli* sRNA genes (Argaman *et al.*, 2001; Chen *et al.*, 2002; Wassarman *et al.*, 2001), many of which would also be conserved in *Salmonella* and other enteric bacteria (Hershberg *et al.*, 2003). As part of these studies, a number of computational prediction tools such as QRNA (Rivas *et al.*, 2001), RNAGENiE (Carter *et al.*, 2001), RNAz (Washietl *et al.*, 2005) and sRNAPredict2 (Livny *et al.*, 2006) have become available that allow *in silico* predictions of sRNA genes in any available genome sequence.

More recently, two biocomputational studies focussed on finding *Salmonella*-specific sRNAs. Reminiscent of the early sRNA-searches in *E. coli*, these screens identified sRNA candidate genes by prediction of σ^{70}-type promoters and ρ-independent transcriptional terminator pairs in *Salmonella* IGRs. In the first study (Pfeiffer *et al.*, 2007), searches in the IGRs of the *S. enterica* Typhimurium LT2 genome predicted 46 sRNA genes, which were denoted STnc (Salmonella Typhimurium non-coding RNA). Most of these STnc candidates are absent from *E. coli* K12, yet a considerable number of them is conserved in the early branching *S. bongori*, suggesting that these loci were acquired shortly after *E. coli* and *Salmonella* shared a common ancestor. The candidates included the *invR* (STnc270) gene located in the *Salmonella* pathogenicity island 1 (SPI-1), which was shown to express the first abundant Hfq-dependent sRNA from an enterobacterial pathogenicity island (Pfeiffer *et al.*, 2007).

The second study (Padalon-Brauch *et al.*, 2008) focussed on IGRs extracted from the horizontally acquired genetic islands of *Salmonella* (again, discriminating against *E. coli*); of the 28 sRNAs predicted, expression of 19 was confirmed by northern blot analysis, and one of them was implicated in SPI-1-mediated invasion and effector secretion. These sRNAs are referred to as IsrA-Q, for island-encoded sRNAs, and include many *cis*-antisense RNAs (Padalon-Brauch *et al.*, 2008).

A further biocomputational study used the sRNAPredict2 algorithm to predict novel sRNA genes in *Salmonella* Typhi CT18 (Livny *et al.*, 2006). Again, this approach was based on the identification of intergenic ρ-independent terminators, however, added an additional parameter by scoring conservation of adjacent sequences. The algorithm detected 27 previously annotated sRNA genes, and predicted 572 new candidates, albeit without further evaluation.

Wet-lab-based discovery of novel sRNA genes has included immunoprecipitation (IP) of an epitope-tagged Hfq protein in *Salmonella* lysates, followed by deep sequencing of the associated RNA (Sittka *et al.*, 2008). Analysing either endogenous or ectopically expressed 'foreign' Hfq proteins, close to 30 previously unknown sRNAs were detected at the level of expression (Sittka *et al.*, 2008; Sittka *et al.*, 2009). These experiments were limited to a single stage of growth (early stationary phase) in rich medium. However, the extension of this approach to systematic profiling of growth and stress conditions has already detected additional abundant Hfq-associated sRNAs, including species that originate from mRNAs (Y. Chao and J. Vogel, unpublished). In addition, deep sequencing-based transcriptome profiling has become a powerful tool to detect the expression of known or new sRNAs in an unbiased fashion (Sharma and Vogel, 2009); such analysis of the *Salmonella* Typhi transcriptome revealed an additional set of 40 novel candidate sRNAs in this organism which await independent validation (Perkins *et al.*, 2009).

An updated list of sRNAs whose expression are experimentally validated in *Salmonella* Typhimurium or whose genes can be deemed functional because of conservation in other enteric bacteria is provided in Table 7.1 at the end of this chapter. There is currently no common nomenclature for *Salmonella* sRNA genes or candidates thereof, although the experimental discovery via Hfq co-IP continued to use the STnc system until functional characterization (Pfeiffer *et al.*, 2007) unless a function can be inferred from related work in *E. coli* or other bacteria.

As it is true for most of the *Salmonella* proteins, the accumulation patterns of sRNAs during growth are complex, often reflecting control by dedicated stress or virulence transcription factors as demonstrated for the *invR* gene and its regulation by the SPI-1 master transcription factor, HilD (Pfeiffer *et al.*, 2007). The most comprehensive profiling so far has included expression analysis of the *isr* sRNA loci in J774 macrophages in addition to several *in vitro* conditions mimicking environments of *Salmonella* during infection (Padalon-Brauch *et al.*, 2008). Nonetheless, a catalogue of sRNA expression clearly remains to be established, since their genes are not commonly covered by the currently available microarray data.

Getting to sRNA functions

Whilst the above listed genome-wide screens have drastically increased the number of known or candidate sRNAs in *Salmonella* species, their functional characterization has been lagging behind. Regarding cellular targets of base-pairing regulators, the Hfq-associated 'core' sRNAs that are widely conserved in enteric bacteria are understood best. The study of these conserved sRNAs has revealed new biology, often so with immediate relevance for *E. coli* and other model species. Recent examples include the discovery of feedback control of the σ^E-controlled envelope stress response by the conserved RybB sRNA (Papenfort *et al.*, 2006), multiple ABC transporter mRNA regulation by an ultra-conserved RNA domain in the GcvB sRNA (Sharma *et al.*, 2007), or the controlled decay of a regulatory sRNA by a 'pseudo-target' mRNA (Figueroa-Bossi *et al.*, 2009).

To facilitate systematic analyses, libraries of sRNA deletion strains have been generated to successfully reveal the CRP-cAMP controlled CyaR sRNA as the previously missing Hfq-dependent repressor of OmpX porin synthesis (Papenfort *et al.*, 2008), or phenotypes of IstR, OxyS and SroA sRNA mutant strains in competitive infection experiments in mice (Santiviago *et al.*, 2009). Experimental approaches to identify target mRNAs of base-pairing sRNAs at the genome-wide scale (Sharma and Vogel, 2009) have successfully been established in *Salmonella*. Typically, the sRNA gene of interest is cloned in a plasmid under tight control of a heterologous arabinose-inducible P_{BAD} promoter, and 'pulse-expressed' in early stationary phase (Papenfort *et al.*, 2006). The logic behind this approach is to transiently overexpress sRNAs (for 10–15 min), long enough to induce the decay of direct target mRNAs, while keeping the expression pulse short enough to avoid significant alterations at the protein level which might cause secondary effects. Combined with microarray-based transcriptome analysis, this approach revealed that the σ^E-activated RybB sRNA directly targeted the mRNAs of at least ten major and minor outer membrane proteins (OMPs) (Papenfort *et al.*, 2006). Although pulse-expression has become the method of choice, it requires the targets to be transcribed under the assay condition. This also applies to microarray-based profiling of sRNA deletion strains, which successfully predicted targets of the ArcZ sRNA (Papenfort *et al.*, 2009). Note, however, that the deregulation of target genes observed in sRNA deletion strains is usually not as dramatic as seen with pulse-expression; it might also be growth-dependent (Papenfort *et al.*, 2008; Papenfort *et al.*, 2009) or impossible to score if the regulatory sRNA regulates the target at the level of translation without effects on mRNA stability, as described for Spot42 sRNA repressing GalK production in *E. coli* (Møller *et al.*, 2002b). Regarding the prediction and validation of sRNA–mRNA pairing, all biocomputational, biochemical and genetic tools available for *E. coli* surely work for *Salmonella* as well (Sharma and Vogel, 2009).

To identify associated proteins of sRNAs, be it a very specific sRNA target or an ancillary factor such as Hfq, protein co-purification with *in vivo*-expressed, aptamer-tagged sRNAs has been established in *Salmonella* (Said *et al.*, 2009).

Functionally characterized sRNAs in *Salmonella*

The next section will summarize sRNA functions that were discovered and/or investigated in *Salmonella*. We will include references to work in other species if conservation is strong enough to be predictive of a similar function in *Salmonella*. Unless specified otherwise, '*Salmonella*' sRNAs and their flanking genes will be referred to as known in *Salmonella* Typhimurium.

Housekeeping RNAs

As expected, all three ubiquitous sRNAs with specialized housekeeping functions (RNase P, 4.5S RNA and tmRNA) have been observed to be well-expressed in *Salmonella* Typhimurium (Sittka *et al.*, 2008), but only tmRNA encoded by the *ssrA* gene has been experimentally studied (Julio *et al.*, 2000). Note that this gene is not to be confused with the SsrAB two-component system that regulates SPI-2 gene expression.

The *ssrA*-encoded tmRNA (a.k.a. 10Sa RNA) is well-established to serve as both a tRNA and an mRNA together with SmpB protein in tagging partially synthesized proteins for degradation, thereby aiding the release of stalled ribosomes from 3′ truncated messengers (Keiler, 2008). The *ssrA* locus was selected in a genetic screen by scoring *Salmonella* Typhimurium fitness within the host; a deletion of *ssrA* resulted in a 200-fold virulence defect in mice. In addition, *ssrA* was shown to impact on the expression of several genes induced during infection, to serve as the integration site for a large region of *Salmonella*-specific sequence, and to be required for bacteriophage P22 at both the levels of growth and induction of lysogens (Julio *et al.*, 2000).

Prior to the analysis of *ssrA*, a *Salmonella* Typhimurium *smpB* mutant was found to exhibit defects in intra-macrophage survival (Baumler *et al.*, 1994). More recently, global systems-level analysis using sample-matched transcriptomics and proteomics measurements of four different growth conditions reported that SmpB, and thus likely tmRNA as well, directly or indirectly modulates >4% of all annotated *Salmonella* proteins (Ansong *et al.*, 2009). However, mechanisms by which the tmRNA/SmpB system might function as a bona fide regulator are yet to be determined.

Regulators of protein activity

The Csr post-transcriptional system is widespread in bacterial pathogens, and consists of the RNA-binding protein, CsrA, which acts globally to repress the translation of target mRNAs, and two or several CsrB-like sRNAs that sequester CsrA and thereby control its activity. The genes of the CsrB (360 nt) and CsrC (240 nt) sRNAs are located in the *yqcC/syd* and *yihA/yihI* IGRs, respectively, and are controlled by the BarA/SirA two-component system (Altier *et al.*, 2000a; Fortune *et al.*, 2006). CsrA regulates a large set (~8%) of *Salmonella* genes including SPI-1, flagellar synthesis operons and many metabolically relevant genes (Altier *et al.*, 2000a; Lawhon *et al.*, 2003). Specifically, CsrA post-transcriptionally influences production of the ubiquitin-like AvrA effector protein in *Salmonella* Typhimurium and *Salmonella* Paratyphi B (Kerrinnes *et al.*, 2009) and affects the expression of at least eight genes involved in the production of (3′–5′)-cyclic-diguanosine monophosphate (ci-di-GMP), a second messenger controlling the switch between the sessile and motile lifestyles of *Salmonella* (Jonas *et al.*, 2009).

Deletion mutants of either CsrB or CsrC display only mild effects on the *Salmonella* virulence, but the combination of both mutants strongly impairs host cell invasion, suggesting that these sRNAs act redundantly to control CsrA activity in the cell (Altier *et al.*, 2000a; Fortune *et al.*, 2006).

Both the CsrB and CsrC sRNAs were reported to be stabilized in a *Salmonella* mutant strain deficient in assembly of the degradosome, a multi–protein complex that associates with the essential RNase E. The 3′ to 5′ exonuclease PNPase, one of four members of the degradosome, was shown to be critical for CsrB/C breakdown. Additionally, the major endoribonuclease RNase III accelerates decay of CsrB without impacting on CsrC levels (Viegas *et al.*, 2007).

The ubiquitous and abundant 6S RNA (183 nt) is encoded in the *ygfE/ygfA* IGR (Papenfort *et al.*, 2008) and well-expressed in both *Salmonella* Typhimurium and *Salmonella* Typhi (Perkins *et al.*, 2009; Sittka *et al.*, 2008). The regulatory activity of 6S RNA has not been addressed or studied in *Salmonella*; however, because of almost identical sequence identity, it is safe to predict from reports in *E. coli* (Wassarman, 2007) that it acts as a regulator of RNAP by direct binding, and serves as template for the production of 14–20 nt RNA transcripts (pRNAs) in the dissociation phase of the regulatory 6S RNA/RNAP cycle.

A large network of Hfq-dependent sRNAs regulates cell envelope composition

Regulation of the composition of the bacterial envelope including the synthesis of OMPs has clearly emerged as the major activity of enteric sRNAs (Fig. 7.3; Guillier *et al.*, 2006; Vogel and Papenfort, 2006). All of the OMP-regulatory *Salmonella* sRNAs described below associate with Hfq (Sittka *et al.*, 2008) and act as mRNA repressors; their majority is conserved in many other species. The prototypical **MicF** sRNA (93 nt) is encoded between the *ompC/yojN* genes, detectable under standard growth conditions (Papenfort *et al.*, 2006), and induced by a number of antimicrobial compounds, including salicylate, paraquat and decanoate (Hartog *et al.*, 2008). As in *E. coli*, MicF acts as repressor of the synthesis of OmpF in *Salmonella* but also has additional targets (D. Podkaminski and J. Vogel, unpublished results).

MicC sRNA is ~109 nt in length and encoded in the *nifJ/ynaF* IGR. Following its identification in an *in silico* screen for new *E. coli* sRNAs (Chen *et al.*, 2002), MicC was shown to repress the *ompC* mRNA in *E. coli* (Chen *et al.*, 2004). Both the MicC–*ompC* pairing and the mechanism of sequestration of the *ompC* RBS to inhibit 30S ribosome binding (Chen *et al.*, 2004) have been experimentally shown to be conserved in *Salmonella* (Pfeiffer *et al.*, 2009). Importantly, MicC also represses the abundant OmpD porin in *Salmonella*, yet different from *ompC*, targets the *ompD* mRNA at codons 23–26. Genetic and molecular analyses have established MicC as a paradigm for mRNA repression in the coding sequence (CDS), and led to the proposal of an alternate gene silencing pathway downstream of translational initiation by which Hfq-dependent sRNA might act primarily through target mRNA destruction, potentially increasing the local concentration of RNase E. *Salmonella* MicC accumulates towards stationary phase (in rich medium) and under SPI-1 inducing conditions (high salt, low oxygen), and, similar to MicF, its expression is regulated by the EnvZ-OmpR two component system (Pfeiffer *et al.*, 2009).

In contrast to the above sRNAs regulating one or two porins, the ~80 nt **RybB** sRNA from the STM0869/STM0870 IGR (Fig. 7.2) is a rather global regulator of *Salmonella omp* mRNAs. Pulse-expression of RybB revealed a total of eleven mRNAs as targets, including those of OmpA, OmpC, OmpD, and OmpF, i.e. the four most abundant porins constituting 25% of the protein content of the *Salmonella* outer membrane, and the *fadL, pagC, ompN, ompW, ompS, tsx,* and *yaeC* mRNAs encoding less abundant *Salmonella* OMPs (Papenfort *et al.*, 2006). In terms of mechanisms, analysis of RybB pairing with the *ompN* mRNA, which occurs exclusively downstream of the start codon, defined a 'five codon window' for translational repression in the CDS (Bouvier *et al.*, 2008). RybB is one of two *Salmonella* sRNAs, the other being **MicA** (74 nt; encoded in the *luxS/gshA* IGR), whose transcription is directly activated by the alternative sigma factor, σ^E. Both sRNAs are rapidly induced upon envelope stress, for example, when misfolded OMPs accumulate in the periplasm. Through accelerating the decay of important *omp* mRNAs, MicA and RybB lower the folding load in the periplasm and achieve auto-regulation of the σ^E response. Of the two sRNAs, RybB seems to be the major player since only deletion of *rybB* perturbs envelope homeostasis and induces σ^E in otherwise unstressed cells (Papenfort *et al.*, 2006). This might reflect the fact that *Salmonella* MicA has only two target genes, i.e. the messengers of OmpA (Papenfort *et al.*, 2006) and LamB (Bossi and Figueroa-Bossi, 2007).

CyaR sRNA (~86 nt) is encoded between *yegQ* and STM2137, and is transcriptionally induced by the metabolic signal molecule, cAMP, in conjunction with CRP protein (Papenfort *et al.*,

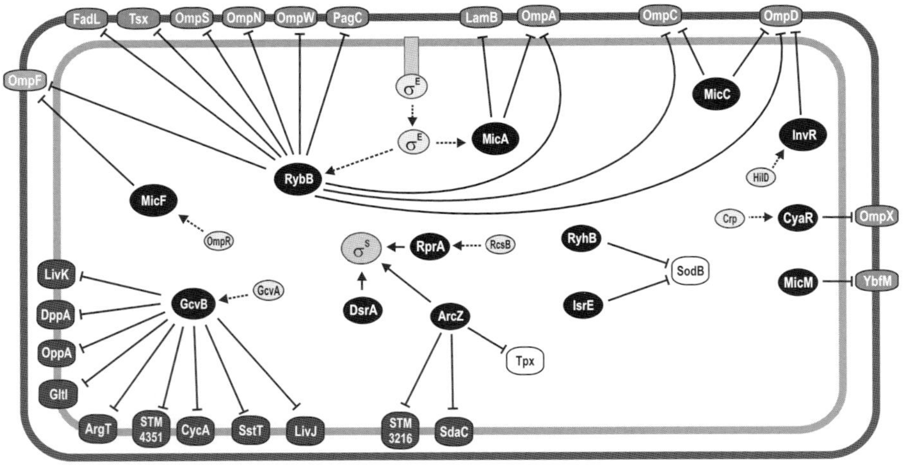

Figure 7.3 Network of sRNA regulators in *Salmonella*. Regulatory sRNAs (black) controlling the expression of outer membrane proteins (grey), periplasmic and inner membrane proteins (dark grey) or cytoplasmic proteins (white) are depicted with their relevant transcriptional regulators (light grey).

2008). A reverse genetic approach employing a *Salmonella* sRNA mutant library as well as global transcriptome analysis identified CyaR as a direct repressor of OmpX synthesis (Papenfort *et al.*, 2008). The CyaR-*ompX* RNA duplex is structurally interesting since a conserved C-rich apical loop at the 3′ end of CyaR is used to sequester the *ompX* RBS (Papenfort *et al.*, 2008). Work in *E. coli* has suggested additional CyaR targets, i.e. the *luxS* mRNA, whose control alongside that of *ompX* would suggest a link between porin control and bacterial group behaviour (De Lay and Gottesman, 2009).

From a mechanistic point of view, **MicM** (a.k.a. ChiX, SroB, or RybC) is one of the most fascinating OMP regulators. This ~80 nt sRNA is encoded in the *ybaK/ybaP* IGR, and almost constitutively expressed (Figueroa-Bossi *et al.*, 2009; Rasmussen *et al.*, 2009; Vogel *et al.*, 2003). MicM acts to repress the synthesis of the YbfM (a.k.a. ChiP) porin required for the uptake of chitosugars. However, MicM also pairs with a fragment of the *chb* (chitobiose) operon mRNA upon induction of the latter by the presence of chitosugars in the environment. Unlike the MicM–*ybfM* interaction, pairing with the *chb* fragment results in degradation of MicM through destabilization of its terminal hairpin structure (Figueroa-Bossi *et al.*, 2009). Thus, the *chb* mRNA acts as an RNA decoy to trap and inactivate MicM in order to derepress the synthesis of the functionally related YbfM porin. This novel mechanism was also reported in *E. coli*, in which MicM has at least one additional target (Vogel, 2009).

In contrast to the above 'core' sRNAs, **InvR** (80 nt; Fig. 7.2) is not conserved outside the genus of *Salmonella*. The *invR* gene was originally discovered as candidate STnc270 at the right-hand border of the SPI-1 invasion gene island (between *invH* and STM2901) in a biocomputational screen for *Salmonella*-specific sRNAs. InvR does not seem to regulate SPI-1 functions (Pfeiffer *et al.*, 2007) but is the third sRNA to repress *ompD* mRNA in addition to MicC and RybB (see above). Interestingly, for reasons unknown, all of these sRNAs bind *ompD* in the CDS (Pfeiffer *et al.*, 2009 and unpublished results). Unlike MicC and RybB, which are accumulating in late stationary phase, InvR is transiently expressed in early stationary phase and highly induced under SPI-1 activating conditions. This expression pattern reflects tight control of the *invR* promoter by the SPI-1 master transcription factor, HilD (Pfeiffer *et al.*, 2007). InvR has been the paradigm of an sRNA that is encoded by a horizontally acquired virulence region but acts through Hfq to regulate a gene of the *Salmonella* core genome (Fig. 7.4).

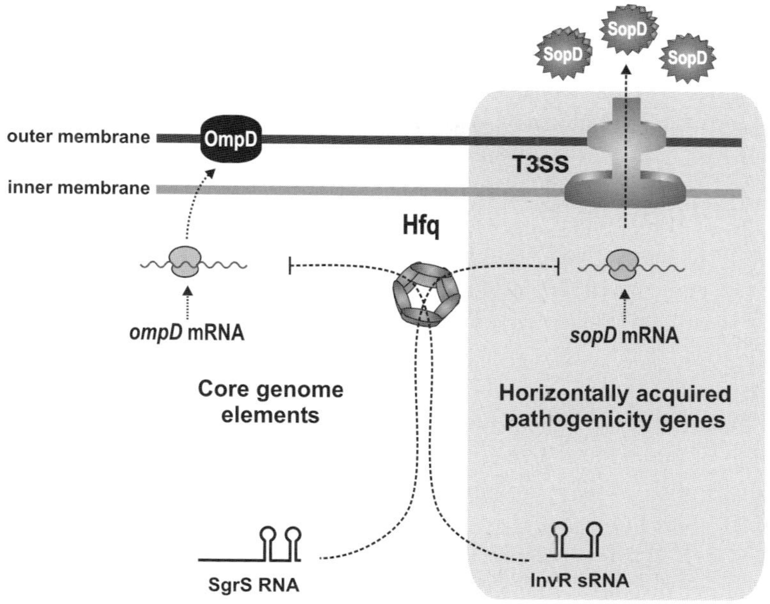

Figure 7.4 Cross-regulatory activity of two Hfq-dependent sRNAs originating from the *Salmonella* core genome (SgrS) or the horizontally acquired elements (InvR). Left: SgrS sRNA is a highly conserved sRNA that represses the synthesis of the horizontally acquired *sopD* locus, encoding a translocated effector protein. Right: InvR is encoded on the *Salmonella* pathogenicity island 1 and post-transcriptionally limits the expression of the core-genome encoded OmpD porin.

Given that many sRNAs remain to be characterized, we assume that the list of OMP regulators will grow. In *E. coli*, the highly similar **OmrA** (88 nt) and **OmrB** (82 nt) sRNAs, whose genes are found in tandem orientation in the *aas/galR* IGR in *Salmonella*, repress the outer membrane protease, OmpT, and the specific gated channels for iron–siderophore complexes, CirA, FecA, and FepA (Guillier and Gottesman, 2006). In addition, OmrAB repress their own transcriptional regulator, OmpR (Guillier and Gottesman, 2006, 2008). Whilst this latter feedback control mechanism is conserved (K. Papenfort and J. Vogel, unpublished), other OmrAB targets remain to be identified in *Salmonella*.

Of the new *Salmonella* sRNA identified by co-immunoprecipitation with Hfq (Sittka *et al.*, 2008), STnc560 (~98 nt) has lately been investigated and renamed to MgrR in *E. coli* (Moon and Gottesman, 2009). **MgrR** expression was shown to depend on magnesium concentrations and the PhoQ/P two-component system in both *E. coli* and *Salmonella*. This Hfq-associated sRNA takes regulation of envelope composition to the very outer level: MgrR represses two mRNAs (*eptB*, *ygdQ*), where the product of *eptB* (*pmrC* in *Salmonella*) regulates the net anion charge and decreases electrostatic repulsion of LPS molecules. In *E. coli*, deletion of *mgrR* confers a 10-fold increase in Polymyxin B resistance, indicating a probable change in the LPS composition (Moon and Gottesman, 2009). We expect the MgrR-mediated repression to be conserved in *Salmonella* because *eptB* is one of the most highly deregulated genes in an *hfq* deletion strain of *Salmonella* Typhimurium (Figueroa-Bossi *et al.*, 2006).

Regulation of ABC transporters and amino acid metabolism

Salmonella express several sRNAs with relevance to transport and metabolism of amino acids. The **GcvB** sRNA was originally reported in *E. coli* as a novel component of the GcvA-controlled glycine cleavage system (Urbanowski *et al.*, 2000); both the *gcvB* gene itself and its genomic location adjacent to *gcvA* are exceptionally well-conserved in many Gram-negative bacteria (Sharma *et al.*, 2007). The combined data of *Salmonella* and *E. coli* show that GcvB acts globally to repress the mRNAs of major ABC transporters of small peptides and single amino acids (*argT, dppA, gltI, livK, livJ, oppA, sstT* and STM4351; Fig. 7.3), including the *cycA* mRNA of the major glycine transporter, indicating feedback control of amino acid uptake through glycine signalling (Pulvermacher *et al.*, 2009a,b; Sharma *et al.*, 2007; Urbanowski *et al.*, 2000). The activity of GcvB depends on Hfq, and a loss of GcvB function might explain the aberrant accumulation of periplasmic proteins in an *hfq* deletion mutant of *Salmonella* Typhimurium (Sharma *et al.*, 2007; Sittka *et al.*, 2007)

The ~200 nt GcvB RNA is structured and contains two ultra-conserved single-stranded domains: a ~30 nucleotide stretch composed of G and U, and a shorter (9 nt) region resembling an anti-SD sequence (Fig. 7.2A). The G/U-rich domain interacts with all characterized targets, which are marked by C/A-rich sequence elements (acting as translational enhancers) located in the 5′ UTR, enabling GcvB to repress translation even far upstream of the SD sequence and start codon of target mRNAs (Sharma *et al.*, 2007).

GcvB is only detectable in exponential phase of *Salmonella* Typhimurium grown in rich media, suggesting that it acts as a global repressor of amino acid uptake under conditions of feast (Sharma *et al.*, 2007). Whether this is the only physiological function of GcvB remains to be seen; a recent screen in *E. coli* showed that GcvB enhances the ability of *E. coli* to survive low pH (Jin *et al.*, 2009), a function that might be relevant for the intracellular survival of *Salmonella* as well.

The dicistronic *sdaCB* operon encodes both a specific transporter and a deaminase relevant for the catabolic usage of serine; the *sdaCB* mRNA is not a target of GcvB, but of **ArcZ** (arc-associated sRNA Z), a conserved sRNA encoded in the *yhbL/arcB* IGR (Papenfort *et al.*, 2009). ArcZ is transcribed as a longer primary transcript (~120 nt) that undergoes 5′ processing to accumulate as an abundant ~50 nt species, constituting the shortest active Hfq-associated sRNA known to date (Papenfort *et al.*, 2009; Sittka *et al.*, 2008). ArcZ represses additional targets, including the *tpx* mRNA, which encodes an important thiol-peroxidase required for resistance against oxidative compounds, and the STM3216 mRNA, which encodes a predicted methyl-accepting protein with documented functions in chemotaxis and motility. Interestingly, whilst ArcZ is expressed throughout growth, it regulates its individual targets in different growth phases. In addition, ArcZ has been noted for its exceptionally pleitropic effects; if slightly overexpressed, this sRNA deregulates nearly a fifth of all *Salmonella* genes (Papenfort *et al.*, 2009). Some of these effects might be attributed to regulation of another target, i.e. the *rpoS* mRNA (see below).

Preliminary data from *E. coli* predict a third conserved sRNA, RydC (64 nt), to regulate ABC transporter expression, i.e. the putative one encoded by the *yejABEF* operon (Antal *et al.*, 2005). In *Salmonella*, **RydC** is encoded in the STM1638/*cybB* IGR, and its putative target was shown to be required for counteracting antimicrobial peptides and for full virulence of *Salmonella* in mice (Eswarappa *et al.*, 2008). For the time being, neither *yejABEF* nor other potential targets of RydC have been firmly validated in *E. coli* or *Salmonella*. Nonetheless, its strong association with Hfq (Sittka *et al.*, 2008) argues that RydC directly regulates mRNAs.

Iron starvation, sugar synthesis and metabolic switches

Salmonella expresses a number of Hfq-dependent sRNAs predicted to help adjust the fluxes and storage of nutrients as well as metabolic pathways. A primary example is the conserved iron-responsive **RyhB** (95 nt) sRNA, which is encoded in the *yhhX/yhhY* IGR and under negative control by the Fur transcriptional regulator (Massé and Gottesman, 2002). Under iron starvation conditions, this sRNA accumulates to regulate many messengers of iron homeostasis-related functions or of dispensable iron-containing proteins, including *sodB, iscSUA* and *shiA* (references in Desnoyers *et al.*, 2009). Moreover, RyhB feedback regulates its own transcriptional regulator by translational inhibition of a small upstream ORF to which expression of *fur* is coupled (Vecerek *et al.*, 2007). Most of these regulations that were originally observed in *E. coli* have been confirmed by pulse-expression and microarray analysis in *Salmonella* (K. Papenfort and J. Vogel, unpublished results).

Salmonella express an additional Fur-controlled RyhB-like sRNA from the STM1273/*yeaQ* IGR, which is currently referred to as **IsrE** (Padalon-Brauch *et al.*, 2008), RyhB-2 (Sittka *et al.*, 2008; Vogel and Sharma, 2005), or RfrB (Ellermeier and Slauch, 2008). IsrE shares with RyhB its own regulation by Fur and its repressor function of *sodB* mRNA (Ellermeier and Slauch, 2008); however, these two similar sRNAs show quite different expression patterns during growth and non-redundant activity in the iron stress response of *Salmonella* (Padalon-Brauch *et al.*, 2008).

There are several well-expressed Hfq-dependent sRNAs (Sittka *et al.*, 2008) whose documented roles in metabolism of *E. coli* are known or are likely to be conserved in *Salmonella*. First, **Spot 42** (109 nt; encoded by *spf*) from the *polA/yihA* IGR is likely to regulate differential translation of the *galETKM* mRNA as it does in *E. coli* (Møller *et al.*, 2002b). Second, **GlmY** (184nt) and **GlmZ** (207nt) which are encoded in the *yfhK/purG* and *yifK/hemY* IGRs, respectively, can be expected to form the same regulatory cascade as in *E. coli*, activating translation of the *glmS* mRNA with consequences for amino-sugar synthesis (Görke and Vogel, 2008; Urban and Vogel, 2008). Third, STnc580 sRNA (~100 nt), which is expressed from the *zntB/dbpA* IGR in *Salmonella* (Sittka *et al.*, 2008), should now be referred to as **FnrS**, following its functional characterization in *E. coli* (Boysen *et al.*, 2010; Durand and Storz, 2010). In *E. coli*, the *fnrS* gene is strongly regulated by FNR (fumarate and nitrate reduction), and to some extent by ArcA and CRP; collectively, these factors ensure high induction of the sRNA upon switch from aerobic to anaerobic growth. *E. coli* FnrS down-regulates more than thirty genes of metabolic enzymes; almost half of them are part of the FNR regulon, and several ones (*cydC*, *folE/X*, *gpmA*, *maeA*, *metE*, *sodA*, *sodB*, *yobA*) are confirmed to be direct targets of FnrS. It thus appears that FnrS functions to help reprogram metabolism during aerobic to anaerobic transition (Boysen *et al.*, 2010; Durand and Storz, 2010). The fact that FnrS was detected only under SPI-1 inducing conditions, i.e. low oxygen (Sittka *et al.*, 2008), argues for the same physiological role in *Salmonella*.

Lastly, there is evidence in *Salmonella* that the bifunctional **SgrS** RNA (227 nt), whose gene is found between *yabN* (*sgrR*) and *leuD*, affects glucose uptake and metabolism (Wadler and Vanderpool, 2009). In *E. coli*, SgrS is induced by glucose-phosphate stress to negatively regulate the *ptsG* mRNA encoding a major glucose transporter by direct base-pairing (Vanderpool, 2007). SgrS carrying the small *sgrT* ORF peptide in its 5′ region is also an mRNA; the corresponding SgrT peptide (40 amino acids) is both highly expressed and required for combating glucose-phosphate stress in *Salmonella* (Wadler and Vanderpool, 2009). Work on the 3′ located base-pairing region has shown that *Salmonella* SgrS regulates additional targets as *trans*-antisense RNA (K. Papenfort and J. Vogel, in preparation), which includes the translational activation of an enzyme required for the degradation of the stress-responsible phosphorylated sugars. Moreover, SgrS acts to repress the *sopD* mRNA of a horizontally acquired effector protein that is secreted via the SPI-1 T3SS (see below).

Oxidative stress and DNA damage

The oxidative stress response includes at least one sRNA, **OxyS** (110 nt), whose expression is induced about 35-fold in the early phase of infection after *Salmonella* enters macrophages (Padalon-Brauch *et al.*, 2008). OxyS, which is encoded next to the gene of, and transcriptionally regulated by, OxyR, was originally described in *E. coli* where it has an anti-mutator function (Altuvia *et al.*, 1997). *E. coli* OxyS directly represses the *fhlA* and *yobF* mRNAs (Argaman and Altuvia, 2000; Hemm *et al.*, 2010) and reduces *rpoS* expression by an as yet unclear mechanism (Altuvia *et al.*, 1997; Zhang *et al.*, 1998). These or other potential targets have not been investigated in *Salmonella*. However, competitive infections experiments of mice found an *oxyS* deletion mutant of *Salmonella* Typhimurium to exhibit increased survival in the host (Santiviago *et al.*, 2009), arguing that OxyS is somehow involved in *Salmonella* pathogenesis.

The SOS response to DNA damage includes a small RNA producing locus, *istR-tisAB* (Vogel *et al.*, 2004), of which *istR* is required for full virulence of *Salmonella* in mice (Santiviago *et al.*, 2009). The locus was originally identified between the *ilvB* and *emrD* genes in *E. coli* K12 (Vogel *et al.*, 2004), and its expression subsequently confirmed in *Salmonella* Typhimurium (Sittka *et al.*, 2008). The stable ~350 nt *tisAB* mRNA encodes the short hydrophobic TisB peptide that targets the inner membrane. The partially overlapping **IstR-1** (75 nt) and **IstR-2** (140 nt) sRNAs are transcribed upstream (from the opposite strand) from two different promoters with a shared transcriptional terminator (Vogel *et al.*, 2004), and do not seem to associate with Hfq (Sittka *et al.*, 2008). *E. coli* IstR-1 sRNA functions as an antisense repressor of TisB synthesis, by both inhibiting standby ribosomes on and inducing RNase III cleavage of the *tisAB* mRNA (Darfeuille *et al.*, 2007; Vogel *et al.*, 2004). The conservation of these mechanisms remains to be confirmed in *Salmonella*.

Small RNAs and RpoS

The *rpoS* mRNA encoding the major stress sigma factor σ^S has been a classic target of Hfq-dependent post-transcriptional regulation in both *E. coli* and *Salmonella* (Brown and Elliott, 1996; Muffler *et al.*, 1996). Moreover, we now know of three conserved Hfq-dependent sRNAs that directly activate the translation of *rpoS* mRNA in *E. coli*, namely **DsrA** (84 nt), **RprA** (107nt) and the aforementioned ArcZ sRNA (Papenfort *et al.*, 2009, S. Gottesman, personal communication). All three sRNAs use an 'anti-antisense mechanism' to increase RpoS synthesis, i.e. their binding competes with the formation of an inhibitory secondary structure in the 5′ UTR that limits ribosome association of *rpoS* mRNA (Fröhlich and Vogel, 2009). Irrespective of the mechanistic overlap, the three sRNAs belong to different transcriptional regulons and thus exert their activator functions under distinct stress or growth conditions (Papenfort *et al.*, 2009; Repoila *et al.*, 2003).

In *Salmonella*, the *rpoS* activator function has only been shown for ArcZ (Papenfort *et al.*, 2009). In contrast, the DsrA and RprA did not significantly impact on *rpoS* mRNA translation in *Salmonella* Typhimurium LT2, despite strong sequence conservation of both the sRNA and *rpoS* leader sequences between *Salmonella* and *E. coli* (Jones *et al.*, 2006). However, results from our laboratory clearly show that both DsrA and RprA activate RpoS synthesis in the virulent *Salmonella* Typhimurium strain SL1344, and that either sRNA regulates additional mRNAs in an Hfq-dependent fashion (K. Papenfort and J. Vogel, unpublished), as previously shown for DsrA-mediated repression of the nucleoid-associated protein H-NS in *E. coli* (Lease *et al.*, 1998; Sledjeski and Gottesman, 1995).

There is preliminary evidence to suggest that sRNAs might act downstream in the σ^S -mediated stress response. Specifically, unpublished results from our laboratory have shown that transcription of the ~100 nt, stationary phase-specific **SdsR** (formerly RyeB) sRNA ultimately depends on σ^S. SdsR is encoded between the *Salmonella* STM1871 and STM1872 genes, and opposite from the much longer, ~270 nt **SraC** (RyeA) sRNA, which affects processing and accumulation of SdsR (Vogel *et al.*, 2003; Wassarman *et al.*, 2001). Both sRNAs associate with Hfq (Sittka *et al.*, 2008; Wassarman *et al.*, 2001), suggesting antisense activity, but targets are currently unknown. Interestingly, the *sdsR/sraC* locus was also observed to be one of eleven hotspots for the integration of 'novel' genetic pools in *E. coli*/*Salmonella* (Sridhar and Rafi, 2007); specifically, it seems to serve as an attachment site for ΦW104 and other present-day bacteriophages in *Salmonella* (Balbontin *et al.*, 2008). In addition to SdsR, the accumulation of the iron starvation-related IsrE and RyhB sRNAs in stationary phase *Salmonella* depends on a functional *rpoS* gene, and there are several additional Isr sRNAs whose upregulation at a later stage of growth makes them good candidates for being members of the σ^S regulon (Padalon-Brauch *et al.*, 2008).

Evidence for virulence-associated functions of sRNAs

Given that *Salmonella* is one of the most intensely studied model pathogens, and that Hfq clearly impacts on its virulence properties, the organism lends itself to address whether sRNAs directly regulate virulence genes. There has been good indication of such virulence-related function in the case of **IsrJ**, a ~74 nt sRNA (Fig. 7.2) that is expressed under low oxygen and low magnesium conditions in *Salmonella* and whose transcription directly or indirectly depends on the SPI-1 transcription factor HilA. Furthermore, IsrJ is required for efficient invasion of *Salmonella* into non-phagocytic cells, and effector translocation by the SPI-1 T3SS (Padalon-Brauch *et al.*, 2008). It remains to be determined, however, at which level the SPI-1 genes are controlled by IsrJ.

Whilst IsrJ is a 'foreign' sRNA harboured by the Gifsy-1 prophage region of *Salmonella* Typhimurium (Padalon-Brauch *et al.*, 2008), core sRNAs might also be recruited to regulate horizontally acquired virulence genes post-transcriptionally, just as transcription factors are at the DNA level. We have shown that SgrS RNA of the phosphosugar stress response (see above) was recruited to directly repress the *sopD* mRNA (Fig. 7.4; K. Papenfort and J. Vogel, in preparation). SopD is an abundant *Salmonella* virulence protein secreted by the T3SSs encoded by both SPI-1 and SPI-2, and plays a significant role in promoting enteropathogenesis (Brumell *et al.*, 2003; Jiang *et al.*, 2004; Jones *et al.*, 1998; Wallis *et al.*, 1999; Zhang *et al.*, 2002b; see also Chapter 10 of this book). The monocistronic *sopD* gene is inserted between *ycgB* and *cysH* (Jones *et al.*, 1998), a locus otherwise co-linear in *Salmonella* and *E. coli*, and is expressed independently of the major SPI-1 and SPI-2 virulence regions (Brumell *et al.*, 2003). Intriguingly, the highly similar *sopD2* mRNA, which is assumed to have arisen from a duplication of *sopD* during evolution of the *Salmonella* clade, is not regulated by SgrS; a single G:U instead of a G:C pair in the SgrS-target duplex spares it from repression (K. Papenfort and J. Vogel, in preparation). Whilst the biological meaning of this post-transcriptional link between phosphosugar stress and effector secretion remains to be understood, this example illustrates that *Salmonella* selectively employs existing Hfq-dependent sRNAs of general stress responses to tame the expression of laterally acquired virulence genes. The ArcZ-mediated regulation of STM3216 encoding a horizontally acquired methyl-accepting chemotaxis protein (MCP) is another example of interplay between the core and variable genome, mediated by sRNAs (Papenfort *et al.*, 2009).

Small RNAs with *cis*-regulatory functions

Based on work in *E. coli*, there are several small RNA species that originate from riboswitches, namely **SroA** or **SroG** from the thiamine and riboflavin riboswitches upstream of the *thiB* or *ribB* genes, respectively (Vogel *et al.*, 2003), and **SraF** that is produced by pH-responsive transcriptional termination upstream of the *alx* (*ygjT*) ORF (Nechooshtan *et al.*, 2009). Interestingly, a deletion of the *sroA* region was shown to reduce the virulence of *Salmonella* Typhimurium in mice; however, it remains to be tested whether this phenotype is simply due to altered expression of the *thi* operon (Santiviago *et al.*, 2009). Whether these *cis*-regulatory RNAs in the range of 100–150 nt can also function *in trans* as recently suggested for riboswitch-derived sRNAs in *Listeria* (Loh *et al.*, 2009) is yet to be determined. We note that SraF is highly enriched in Hfq co-immunoprecipitations in *Salmonella* (Sittka *et al.*, 2008), rendering it a candidate for a *trans*-antisense sRNA.

Besides the well-known antisense functions in plasmid and transposon replication, and in toxin-antitoxin systems (Brantl, 2007), *cis*-activity has also been predicted for sRNAs from the *Salmonella* chromosome. Padalon-Brauch *et al.* (2008) described five sRNAs (IsrA, IsrG, IsrH, IsrC, IsrN) that most likely influence the expression of overlapping genes. Importantly, such expression control is not limited to mRNA repression, but can also activate flanking genes as shown for IsrN and STM2765 in *Salmonella* (Padalon-Brauch *et al.*, 2008), and GadY and *gadX* in *E. coli* (Opdyke *et al.*, 2004).

CRISPR encoding crRNAs in *Salmonella*

The group of bacterial RNA regulators has recently been expanded by the discovery and characterization of the CRISPR (Clustered, Regularly Interspaced Short Palindromic Repeats) loci, a widespread prokaryotic immunity system present in about 40% of all bacteria and almost all Archaea (Sorek *et al.*, 2008). These loci consist of a variant number of inversely repeated DNA sequences of 24 to 47 bp in length. When transcribed, the repeats fold into short hairpin structures interrupted by spacer sequences of 26 to 72 nt length, which often display sequence homology with and confer resistance to matching phage- or plasmid-encoded sequences. Activity of CRISPR RNAs is mediated by conserved Cas (CRISPR associated) proteins which assemble into multi–protein complexes to guide the cleavage of full-length CRISPR transcripts. The mechanism by which eubacterial CRISPR RNAs control the propagation of foreign DNA sequences is still not entirely clear; it appears to involve recognition of DNA rather than RNA (Horvath and Barrangou, 2010; van der Oost *et al.*, 2009).

In silico analysis of the *Salmonella* Typhimurium LT2 (NC_003197) genome using CRISPRFinder (Grissa *et al.*, 2007) has indicated the existence of three CRISPR loci at chromosomal positions 3076611–3077006, 3077080–3078147, and 3094279–3096260 carrying 6, 17 and 32 spacers, respectively. Comparison of CRISPR spacer sequences (Mojica *et al.*, 2005) of *Salmonella* species with available sequences in prophage database (Srividhya *et al.*, 2007) indicates similarity of spacers with several known phages. For example, the second spacer of the CRISPR locus 2 (TGGACCGATGGGGCCAACATCGCCGAACGTGG) of *Salmonella* Typhimurium LT2 shows high overall similarity with the Sti1 prophage that is specific to *Salmonella* Typhi CT18 (Srividhya *et al.*, 2006), suggesting that *Salmonella* Typhimurium LT2 might have obtained immunity towards the Sti1 phage due to incoorperation of this spacer sequence. Interestingly, the CRISPR-mediated immunity systems in *Salmonella* have not received much attention, despite the fact that this type of riboregulation might involve important regulatory capacities that contributed to the genomic organization of *Salmonella* and its close relatives.

Conclusion

Recent functional and computational analyses have revealed many sRNA-based regulations in *Salmonella,* the sum of which can be expected to significantly contribute to the saprophytic and virulent lifestyles of this pathogen. Part of the attraction of *Salmonella* for the study of sRNAs stems from its close relationship with *E. coli* K12, the long-standing model organism for bacterial RNA research. Specifically, the core machinery for RNA processing and degradation is highly conserved between these organisms. Therefore, *Salmonella* work can readily make use of a wealth of knowledge and mutant strains available for *E. coli.* Nonetheless, more than one-quarter of the total genetic material has been horizontally acquired since *Salmonella* and *E. coli* shared a common ancestor (Porwollik and McClelland, 2003). Thus, the two organisms lend themselves for comparative analysis in order to understand the evolution of sRNA functions and mechanisms. Such analysis will help us understand what is needed for a conserved sRNA to be recruited as a regulator of newly acquired genes, and vice versa, what it takes for an sRNA of a mobile element to establish itself as a regulator in a new host organism.

It is fair to say that *Salmonella* now rivals *E. coli,* and surpasses any other known pathogen, in terms of the number of experimentally validated sRNAs. Moreover, the catalogues of candidate mRNA targets obtained by deep sequencing-based analysis of Hfq-bound RNA (Sittka *et al.*, 2008; Sittka *et al.*, 2009), as well as the depth of characterization of phenotypes and affected pathways of *hfq* mutants (Chao and Vogel, 2010), have been unique to *Salmonella.* Most importantly, the recent work in *Salmonella* Typhimurium has produced new mechanistic concepts of sRNA-mediated gene regulation such as the multiple targeting by a conserved RNA domain (Sharma *et al.*, 2007), mRNA regulation in the coding sequence (Bouvier *et al.*, 2008; Pfeiffer *et al.*, 2009), or controlled sRNA decay by a pseudo-target (Figueroa-Bossi *et al.*, 2009), all of which have been relevant for other organisms.

Nonetheless, the majority of sRNAs are as yet to be fully characterized in *Salmonella.* Their systematic analysis will help us understand how the often subtle phenotypes of individual sRNA deletions sum up to the drastic virulence and fitness defects of *Salmonella hfq* or *csrA* mutants

(Ansong *et al.*, 2009; Lawhon *et al.*, 2003; Sittka *et al.*, 2007). The *cis*-encoded functions which were predicted for most of the Isr sRNAs (Padalon-Brauch *et al.*, 2008) also deserve more investigation, because this class of bacterial riboregulators has received comparatively little attention over the last few years (Brantl, 2007). In conclusion, now that *Salmonella* is firmly established as a pathogenic model organism of bacterial RNA research, we need a systems analysis approach to understand when and where its sRNAs are expressed, and which proteins, RNAs or other cellular components are targeted.

Acknowledgements

Work in the Vogel lab is supported by project grants SPP1258 *Sensory and Regulatory RNAs in Prokaryotes* (DFG; German Research Council) and *RNomics of Infectious Diseases* (BMBF; German Ministry of Education and Research).

References

Altier, C., Suyemoto, M., and Lawhon, S.D. (2000a). Regulation of *Salmonella enterica* serovar Typhimurium invasion genes by csrA. Infect. Immun. *68*, 6790–6797.

Altier, C., Suyemoto, M., Ruiz, A.I., Burnham, K.D., and Maurer, R. (2000b). Characterization of two novel regulatory genes affecting *Salmonella* invasion gene expression. Mol. Microbiol. *35*, 635–646.

Altuvia, S., Weinstein-Fischer, D., Zhang, A., Postow, L., and Storz, G. (1997). A small, stable RNA induced by oxidative stress: role as a pleiotropic regulator and antimutator. Cell *90*, 43–53.

Ansong, C., Yoon, H., Porwollik, S., Mottaz-Brewer, H., Petritis, B.O., Jaitly, N., Adkins, J.N., McClelland, M., Heffron, F., and Smith, R.D. (2009). Global systems-level analysis of Hfq and SmpB deletion mutants in *Salmonella*: implications for virulence and global protein translation. PLoS ONE *4*, e4809.

Antal, M., Bordeau, V., Douchin, V., and Felden, B. (2005). A small bacterial RNA regulates a putative ABC transporter. Biol. Chem. *280*, 7901–7908.

Argaman, L., and Altuvia, S. (2000). fhlA repression by OxyS RNA: kissing complex formation at two sites results in a stable antisense-target RNA complex. J. Mol. Biol. *300*, 1101–1112.

Argaman, L., Hershberg, R., Vogel, J., Bejerano, G., Wagner, E.G., Margalit, H., and Altuvia, S. (2001). Novel small RNA-encoding genes in the intergenic regions of Escherichia coli. Curr. Biol. *11*, 941–950.

Balbontin, R., Figueroa-Bossi, N., Casadesus, J., and Bossi, L. (2008). Insertion hot spot for horizontally acquired DNA within a bidirectional small-RNA locus in *Salmonella enterica*. J. Bacteriol. *190*, 4075–4078.

Baumler, A.J., Kusters, J.G., Stojiljkovic, I., and Heffron, F. (1994). *Salmonella* Typhimurium loci involved in survival within macrophages. Infect. Immun. *62*, 1623–1630.

Bossi, L., and Figueroa-Bossi, N. (2007). A small RNA downregulates LamB maltoporin in *Salmonella*. Mol. Microbiol. *65*, 799–810.

Bouvier, M., Sharma, C.M., Mika, F., Nierhaus, K.H., and Vogel, J. (2008). Small RNA binding to 5′ mRNA coding region inhibits translational initiation. Mol. Cell *32*, 827–837.

Boysen, A., Moller-Jensen, J., Kallipolitis, B.H., Valentin-Hansen, P., and Overgaard, M. (2010). Translational regulation of gene expression by an anaerobically induced small non-coding RNA in Escherichia coli. J. Biol. Chem. *285*, 10690–10702.

Brantl, S. (2007). Regulatory mechanisms employed by cis-encoded antisense RNAs. Curr. Opin Microbiol. *10*, 102–109.

Brown, L., and Elliott, T. (1996). Efficient translation of the RpoS sigma factor in *Salmonella* Typhimurium requires host factor I, an RNA-binding protein encoded by the hfq gene. J. Bacteriol. *178*, 3763–3770.

Brumell, J.H., Kujat-Choy, S., Brown, N.F., Vallance, B.A., Knodler, L.A., and Finlay, B.B. (2003). SopD2 is a novel type III secreted effector of *Salmonella* Typhimurium that targets late endocytic compartments upon delivery into host cells. Traffic (Copenhagen, Denmark) 4, 36–48.

Carter, R.J., Dubchak, I., and Holbrook, S.R. (2001). A computational approach to identify genes for functional RNAs in genomic sequences. Nucleic Acids Res. *29*, 3928–3938.

Chao, Y., and Vogel, J. (2010). The role of Hfq in bacterial pathogens. Curr. Opin Microbiol. *13*, 24–33.

Chen, S., Lesnik, E.A., Hall, T.A., Sampath, R., Griffey, R.H., Ecker, D.J., and Blyn, L.B. (2002). A bioinformatics-based approach to discover small RNA genes in the Escherichia coli genome. Biosystems *65*, 157–177.

Chen, S., Zhang, A., Blyn, L.B., and Storz, G. (2004). MicC, a second small-RNA regulator of Omp protein expression in Escherichia coli. J. Bacteriol. *186*, 6689–6697.

Darfeuille, F., Unoson, C., Vogel, J., and Wagner, E.G. (2007). An antisense RNA inhibits translation by competing with standby ribosomes. Mol. Cell *26*, 381–392.

De Lay, N., and Gottesman, S. (2009). The Crp-activated small noncoding regulatory RNA CyaR (RyeE) links nutritional status to group behavior. J. Bacteriol. *191*, 461–476.

Desnoyers, G., Morissette, A., Prevost, K., and Masse, E. (2009). Small RNA-induced differential degradation of the polycistronic mRNA iscRSUA. EMBO J. *28*, 1551–1561.

Douchin, V., Bohn, C., and Bouloc, P. (2006). Down-regulation of porins by a small RNA bypasses the essentiality of the regulated intramembrane proteolysis protease RseP in Escherichia coli. J. Biol. Chem. *281*, 12253–12259.

Durand, S., and Storz, G. (2010). Reprogramming of Anaerobic Metabolism by the FnrS Small RNA. Mol. Microbiol. *75*, 1215–1231.

Ellermeier, J.R., and Slauch, J.M. (2008). Fur regulates expression of the *Salmonella* pathogenicity island 1 type III secretion system through HilD. J. Bacteriol. *190*, 476–486.

Eswarappa, S.M., Panguluri, K.K., Hensel, M., and Chakravortty, D. (2008). The yejABEF operon of *Salmonella* confers resistance to antimicrobial peptides and contributes to its virulence. Microbiology (Reading, England) 154, 666–678.

Figueroa-Bossi, N., Lemire, S., Maloriol, D., Balbontin, R., Casadesus, J., and Bossi, L. (2006). Loss of Hfq activates the sigma-dependent envelope stress response in *Salmonella enterica*. Mol. Microbiol. *62*, 838–852.

Figueroa-Bossi, N., Valentini, M., Malleret, L., and Bossi, L. (2009). Caught at its own game: regulatory small RNA inactivated by an inducible transcript mimicking its target. Genes Dev. *23*, 2004–2015.

Folichon, M., Arluison, V., Pellegrini, O., Huntzinger, E., Regnier, P., and Hajnsdorf, E. (2003). The poly(A) binding protein Hfq protects RNA from RNase E and exoribonucleolytic degradation. Nucleic Acids Res. *31*, 7302–7310.

Fortune, D.R., Suyemoto, M., and Altier, C. (2006). Identification of CsrC and characterization of its role in epithelial cell invasion in *Salmonella enterica* serovar Typhimurium. Infect. Immun. *74*, 331–339.

Fröhlich, K.S., and Vogel, J. (2009). Activation of gene expression by small RNA. Curr. Opin. Microbiol. *12*, 674–682.

Görke, B., and Vogel, J. (2008). Noncoding RNA control of the making and breaking of sugars. Genes Dev. *22*, 2914–2925.

Grissa, I., Vergnaud, G., and Pourcel, C. (2007). CRISPRFinder: a web tool to identify clustered regularly interspaced short palindromic repeats. Nucleic Acids Res. *35*, W52–57.

Grudnik, P., Bange, G., and Sinning, I. (2009). Protein targeting by the signal recognition particle. J. Biol. Chem. *390*, 775–782.

Guerrier-Takada, C., Gardiner, K., Marsh, T., Pace, N., and Altman, S. (1983). The RNA moiety of ribonuclease P is the catalytic subunit of the enzyme. Cell *35*, 849–857.

Guillier, M., and Gottesman, S. (2006). Remodelling of the Escherichia coli outer membrane by two small regulatory RNAs. Mol. Microbiol. *59*, 231–247.

Guillier, M., and Gottesman, S. (2008). The 5′ end of two redundant sRNAs is involved in the regulation of multiple targets, including their own regulator. Nucleic Acids Res. *36*, 6781–6794.

Guillier, M., Gottesman, S., and Storz, G. (2006). Modulating the outer membrane with small RNAs. Genes Dev. *20*, 2338–2348.

Hartog, E., Ben-Shalom, L., Shachar, D., Matthews, K.R., and Yaron, S. (2008). Regulation of marA, soxS, rob, acrAB and micF in *Salmonella enterica* serovar Typhimurium. Microbiol. Immunol. *52*, 565–574.

Hemm, M.R., Paul, B.J., Miranda-Rios, J., Zhang, A., Soltanzad, N., and Storz, G. (2010). Small stress response proteins in Escherichia coli: proteins missed by classical proteomic studies. J. Bacteriol. *192*, 46–58.

Hershberg, R., Altuvia, S., and Margalit, H. (2003). A survey of small RNA-encoding genes in Escherichia coli. Nucleic Acids Res. *31*, 1813–1820.

Horvath, P., and Barrangou, R. (2010). CRISPR/Cas, the immune system of bacteria and archaea. Science *327*, 167–170.

Jiang, X., Rossanese, O.W., Brown, N.F., Kujat-Choy, S., Galan, J.E., Finlay, B.B., and Brumell, J.H. (2004). The related effector proteins SopD and SopD2 from *Salmonella enterica* serovar Typhimurium contribute to virulence during systemic infection of mice. Mol. Microbiol. *54*, 1186–1198.

Jin, Y., Watt, R.M., Danchin, A., and Huang, J.D. (2009). Small noncoding RNA GcvB is a novel regulator of acid resistance in Escherichia coli. BMC Genomics *10*, 165.

Johansen, J., Eriksen, M., Kallipolitis, B., and Valentin-Hansen, P. (2008). Down-regulation of outer membrane proteins by noncoding RNAs: unraveling the cAMP-CRP- and sigmaE-dependent CyaR-ompX regulatory case. J. Mol. Biol. *383*, 1–9.

Jonas, K., Edwards, A.N., Ahmad, I., Romeo, T., Romling, U., and Melefors, O. (2009). Complex regulatory network encompassing the Csr, c-di-GMP and motility systems of *Salmonella* Typhimurium. Environ. Microbiol. *12*, 524–540.

Jones, A.M., Goodwill, A., and Elliott, T. (2006). Limited role for the DsrA and RprA regulatory RNAs in rpoS regulation in *Salmonella enterica*. J. Bacteriol. *188*, 5077–5088.

Jones, M.A., Wood, M.W., Mullan, P.B., Watson, P.R., Wallis, T.S., and Galyov, E.E. (1998). Secreted effector proteins of *Salmonella* dublin act in concert to induce enteritis. Infect. Immun. *66*, 5799–5804.

Julio, S.M., Heithoff, D.M., and Mahan, M.J. (2000). ssrA (tmRNA) plays a role in *Salmonella enterica* serovar Typhimurium pathogenesis. J. Bacteriol. *182*, 1558–1563.

Keiler, K.C. (2008). Biology of trans-translation. Annu. Rev. Microbiol. *62*, 133–151.

Kerrinnes, T., Zelas, Z.B., Streckel, W., Faber, F., Tietze, E., Tschape, H., and Yaron, S. (2009). CsrA and CsrB are required for the post-transcriptional control of the virulence-associated effector protein AvrA of *Salmonella enterica*. Int. J. Med. Microbiol. *299*, 333–341.

Lawhon, S.D., Frye, J.G., Suyemoto, M., Porwollik, S., McClelland, M., and Altier, C. (2003). Global regulation by CsrA in *Salmonella* Typhimurium. Mol. Microbiol. *48*, 1633–1645.

Lease, R.A., Cusick, M.E., and Belfort, M. (1998). Riboregulation in Escherichia coli: DsrA RNA acts by RNA:RNA interactions at multiple loci. Proc. Natl. Acad. Sci. U.S.A. *95*, 12456–12461.

Livny, J., Brencic, A., Lory, S., and Waldor, M.K. (2006). Identification of 17 Pseudomonas aeruginosa sRNAs and prediction of sRNA-encoding genes in 10 diverse pathogens using the bioinformatic tool sRNAPredict2. Nucleic Acids Res. *34*, 3484–3493.

Loh, E., Dussurget, O., Gripenland, J., Vaitkevicius, K., Tiensuu, T., Mandin, P., Repoila, F., Buchrieser, C., Cossart, P., and Johansson, J. (2009). A trans-acting riboswitch controls expression of the virulence regulator PrfA in Listeria monocytogenes. Cell *139*, 770–779.

Majdalani, N., Chen, S., Murrow, J., St John, K., and Gottesman, S. (2001). Regulation of RpoS by a novel small RNA: the characterization of RprA. Mol. Microbiol. *39*, 1382–1394.

Marraffini, L.A., and Sontheimer, E.J. (2009). Invasive DNA, chopped and in the CRISPR. Structure *17*, 786–788.

Massé, E., and Gottesman, S. (2002). A small RNA regulates the expression of genes involved in iron metabolism in *Escherichia coli*. Proc. Natl. Acad. Sci. U.S.A. *99*, 4620–4625.

Mizuno, T., Chou, M.Y., and Inouye, M. (1984). A unique mechanism regulating gene expression: translational inhibition by a complementary RNA transcript (micRNA). Proc. Natl. Acad. Sci. U.S.A. *81*, 1966–1970.

Mojica, F.J., Diez-Villasenor, C., Garcia-Martinez, J., and Soria, E. (2005). Intervening sequences of regularly spaced prokaryotic repeats derive from foreign genetic elements. J. Mol. Evol. *60*, 174–182.

Moll, I., Afonyushkin, T., Vytvytska, O., Kaberdin, V.R., and Blasi, U. (2003). Coincident Hfq binding and RNase E cleavage sites on mRNA and small regulatory RNAs. RNA9, 1308–1314.

Møller, T., Franch, T., Hojrup, P., Keene, D.R., Bachinger, H.P., Brennan, R.G., and Valentin-Hansen, P. (2002a). Hfq: a bacterial Sm-like protein that mediates RNA–RNA interaction. Mol. Cell *9*, 23–30.

Møller, T., Franch, T., Udesen, C., Gerdes, K., and Valentin-Hansen, P. (2002b). Spot 42 RNA mediates discoordinate expression of the *E. coli* galactose operon. Genes Dev. *16*, 1696–1706.

Moon, K., and Gottesman, S. (2009). A PhoQ/P-regulated small RNA regulates sensitivity of Escherichia coli to antimicrobial peptides. Mol. Microbiol. *74*, 1314–1330.

Muffler, A., Fischer, D., and Hengge-Aronis, R. (1996). The RNA-binding protein HF-I, known as a host factor for phage Qbeta RNA replication, is essential for rpoS translation in Escherichia coli. Genes Dev. *10*, 1143–1151.

Nechooshtan, G., Elgrably-Weiss, M., Sheaffer, A., Westhof, E., and Altuvia, S. (2009). A pH-responsive riboregulator. Genes Dev. *23*, 2650–2662.

Opdyke, J.A., Kang, J.G., and Storz, G. (2004). GadY, a small-RNA regulator of acid response genes in Escherichia coli. J. Bacteriol. *186*, 6698–6705.

Padalon-Brauch, G., Hershberg, R., Elgrably-Weiss, M., Baruch, K., Rosenshine, I., Margalit, H., and Altuvia, S. (2008). Small RNAs encoded within genetic islands of *Salmonella* Typhimurium show host-induced expression and role in virulence. Nucleic Acids Res. *36*, 1913–1927.

Papenfort, K., Pfeiffer, V., Lucchini, S., Sonawane, A., Hinton, J.C., and Vogel, J. (2008). Systematic deletion of *Salmonella* small RNA genes identifies CyaR, a conserved CRP-dependent riboregulator of OmpX synthesis. Mol. Microbiol. *68*, 890–906.

Papenfort, K., Pfeiffer, V., Mika, F., Lucchini, S., Hinton, J.C., and Vogel, J. (2006). sigma(E)-dependent small RNAs of *Salmonella* respond to membrane stress by accelerating global omp mRNA decay. Mol. Microbiol. *62*, 1674–1688.

Papenfort, K., Said, N., Welsink, T., Lucchini, S., Hinton, J.C., and Vogel, J. (2009). Specific and pleiotropic patterns of mRNA regulation by ArcZ, a conserved, Hfq-dependent small RNA. Mol. Microbiol. *74*, 139–158.

Perkins, T.T., Kingsley, R.A., Fookes, M.C., Gardner, P.P., James, K.D., Yu, L., Assefa, S.A., He, M., Croucher, N.J., Pickard, D.J., *et al.* (2009). A strand-specific RNA-Seq analysis of the transcriptome of the typhoid bacillus *Salmonella* Typhi. PLoS Genet *5*, e1000569.

Pfeiffer, V., Papenfort, K., Lucchini, S., Hinton, J.C., and Vogel, J. (2009). Coding sequence targeting by MicC RNA reveals bacterial mRNA silencing downstream of translational initiation Nat. Struct. Mol. Biol. *16*, 840–846.

Pfeiffer, V., Sittka, A., Tomer, R., Tedin, K., Brinkmann, V., and Vogel, J. (2007). A small non-coding RNA of the invasion gene island (SPI-1) represses outer membrane protein synthesis from the *Salmonella* core genome. Mol. Microbiol. *66*, 1174–1191.

Porwollik, S., and McClelland, M. (2003). Lateral gene transfer in *Salmonella*. Microbes Infect. *5*, 977–989.

Pulvermacher, S.C., Stauffer, L.T., and Stauffer, G.V. (2009a). Role of the sRNA GcvB in regulation of cycA in Escherichia coli. Microbiology (Reading, England) 155, 106–114.

Pulvermacher, S.C., Stauffer, L.T., and Stauffer, G.V. (2009b). The small RNA GcvB regulates sstT mRNA expression in Escherichia coli. J. Bacteriol. *191*, 238–248.

Rasmussen, A.A., Eriksen, M., Gilany, K., Udesen, C., Franch, T., Petersen, C., and Valentin-Hansen, P. (2005). Regulation of ompA mRNA stability: the role of a small regulatory RNA in growth phase-dependent control. Mol. Microbiol. *58*, 1421–1429.

Rasmussen, A.A., Johansen, J., Nielsen, J.S., Overgaard, M., Kallipolitis, B., and Valentin-Hansen, P. (2009). A conserved small RNA promotes silencing of the outer membrane protein YbfM. Mol. Microbiol. *72*, 566–577.

Repoila, F., Majdalani, N., and Gottesman, S. (2003). Small non-coding RNAs, co-ordinators of adaptation processes in Escherichia coli: the RpoS paradigm. Mol. Microbiol. *48*, 855–861.

Rivas, E., Klein, R.J., Jones, T.A., and Eddy, S.R. (2001). Computational identification of noncoding RNAs in *E. coli* by comparative genomics. Curr. Biol. *11*, 1369–1373.

Roth, A., and Breaker, R.R. (2009). The structural and functional diversity of metabolite-binding riboswitches. Annu. Rev. Biochem. *78*, 305–334.

Said, N., Rieder, R., Hurwitz, R., Deckert, J., Urlaub, H., and Vogel, J. (2009). *In vivo* expression and purification of aptamer-tagged small RNA regulators. Nucleic Acids Res. *37*, e133.

Santiviago, C.A., Reynolds, M.M., Porwollik, S., Choi, S.H., Long, F., Andrews-Polymenis, H.L., and McClelland, M. (2009). Analysis of pools of targeted *Salmonella* deletion mutants identifies novel genes affecting fitness during competitive infection in mice. PLoS Pathogens *5*, e1000477.

Sharma, C.M., Darfeuille, F., Plantinga, T.H., and Vogel, J. (2007). A small RNA regulates multiple ABC transporter mRNAs by targeting C/A-rich elements inside and upstream of ribosome-binding sites. Genes Dev. *21*, 2804–2817.

Sharma, C.M., and Vogel, J. (2009). Experimental approaches for the discovery and characterization of regulatory small RNA. Curr. Opin. Microbiol. *12*, 536–546.

Sittka, A., Lucchini, S., Papenfort, K., Sharma, C.M., Rolle, K., Binnewies, T.T., Hinton, J.C., and Vogel, J. (2008). Deep sequencing analysis of small noncoding RNA and mRNA targets of the global post-transcriptional regulator, Hfq. PLoS Genet *4*, e1000163.

Sittka, A., Pfeiffer, V., Tedin, K., and Vogel, J. (2007). The RNA chaperone Hfq is essential for the virulence of *Salmonella* Typhimurium. Mol. Microbiol. *63*, 193–217.

Sittka, A., Sharma, C.M., Rolle, K., and Vogel, J. (2009). Deep sequencing of *Salmonella* RNA associated with heterologous Hfq proteins *in vivo* reveals small RNAs as a major target class and identifies RNA processing phenotypes. RNA Biol. *6*, 266–275.

Sledjeski, D., and Gottesman, S. (1995). A small RNA acts as an antisilencer of the H-NS-silenced rcsA gene of Escherichia coli. Proc. Natl. Acad. Sci. U.S.A. *92*, 2003–2007.

Sorek, R., Kunin, V., and Hugenholtz, P. (2008). CRISPR–a widespread system that provides acquired resistance against phages in bacteria and archaea. Nat. Rev. Microbiol. *6*, 181–186.

Sridhar, J., and Rafi, Z.A. (2007). Identification of novel genomic islands associated with small RNAs. In Silico Biol. 7, 601–611.

Srividhya, K.V., Alaguraj, V., Poornima, G., Kumar, D., Singh, G.P., Raghavenderan, L., Katta, A.V., Mehta, P., and Krishnaswamy, S. (2007). Identification of prophages in bacterial genomes by dinucleotide relative abundance difference. PLoS ONE *2*, e1193.

Srividhya, K.V., Rao, G.V., Raghavenderan, L., Mehta, P., Prilusky, J., Manicka, S., Sussman, J.L., and Krishnaswamy, S. (2006). Database and Comparative Identification of Prophages. LNCIS *344*, 863–868.

Tjaden, B., Saxena, R.M., Stolyar, S., Haynor, D.R., Kolker, E., and Rosenow, C. (2002). Transcriptome analysis of Escherichia coli using high-density oligonucleotide probe arrays. Nucleic Acids Res. *30*, 3732–3738.

Udekwu, K.I., Darfeuille, F., Vogel, J., Reimegard, J., Holmqvist, E., and Wagner, E.G. (2005). Hfq-dependent regulation of OmpA synthesis is mediated by an antisense RNA. Genes Dev. *19*, 2355–2366.

Urban, J.H., and Vogel, J. (2008). Two seemingly homologous noncoding RNAs act hierarchically to activate glmS mRNA translation. PLoS Biol. *6*, e64.

Urbanowski, M.L., Stauffer, L.T., and Stauffer, G.V. (2000). The gcvB gene encodes a small untranslated RNA involved in expression of the dipeptide and oligopeptide transport systems in Escherichia coli. Mol. Microbiol. *37*, 856–868.

van der Oost, J., Jore, M.M., Westra, E.R., Lundgren, M., and Brouns, S.J. (2009). CRISPR-based adaptive and heritable immunity in prokaryotes. Trends Biochem. Sci. *34*, 401–407.

Vanderpool, C.K. (2007). Physiological consequences of small RNA-mediated regulation of glucose-phosphate stress. Curr. Opin. Microbiol. *10*, 146–151.

Vecerek, B., Moll, I., and Blasi, U. (2007). Control of Fur synthesis by the non-coding RNA RyhB and iron-responsive decoding. EMBO J. *26*, 965–975.

Viegas, S.C., Pfeiffer, V., Sittka, A., Silva, I.J., Vogel, J., and Arraiano, C.M. (2007). Characterization of the role of ribonucleases in *Salmonella* small RNA decay. Nucleic Acids Res. *35*, 7651–7664.

Vogel, J. (2009). An RNA trap helps bacteria get the most out of chitosugars. Mol. Microbiol. *73*, 737–741.

Vogel, J., Argaman, L., Wagner, E.G., and Altuvia, S. (2004). The small RNA IstR inhibits synthesis of an SOS-induced toxic peptide. Curr. Biol. *14*, 2271–2276.

Vogel, J., Bartels, V., Tang, T.H., Churakov, G., Slagter-Jager, J.G., Hüttenhofer, A., and Wagner, E.G. (2003). RNomics in Escherichia coli detects new sRNA species and indicates parallel transcriptional output in bacteria. Nucleic Acids Res. *31*, 6435–6443.

Vogel, J., and Papenfort, K. (2006). Small non-coding RNAs and the bacterial outer membrane. Curr. Opin. Microbiol. *9*, 605–611.

Vogel, J., and Sharma, C.S. (2005). How to find small non-coding RNAs in bacteria. Biol. Chem. *386*, 1219–1238.

Wadler, C.S., and Vanderpool, C.K. (2009). Characterization of homologs of the small RNA SgrS reveals diversity in function. Nucleic Acids Res. *37*, 5477–5485.

Wagner, E.G., Altuvia, S., and Romby, P. (2002). Antisense RNAs in bacteria and their genetic elements. Adv. Genet. *46*, 361–398.

Wallis, T.S., Wood, M., Watson, P., Paulin, S., Jones, M., and Galyov, E. (1999). Sips, Sops, and SPIs but not stn influence *Salmonella* enteropathogenesis. Adv. Exp. Med. Biol. *473*, 275–280.

Washietl, S., Hofacker, I.L., and Stadler, P.F. (2005). Fast and reliable prediction of noncoding RNAs. Proc. Natl. Acad. Sci. U.S.A. *102*, 2454–2459.

Wassarman, K.M. (2007). 6S RNA: a regulator of transcription. Mol. Microbiol. *65*, 1425–1431.

Wassarman, K.M., Repoila, F., Rosenow, C., Storz, G., and Gottesman, S. (2001). Identification of novel small RNAs using comparative genomics and microarrays. Genes Dev. *15*, 1637–1651.

Waters, L.S., and Storz, G. (2009). Regulatory RNAs in bacteria. Cell *136*, 615–628.

Wilson, J.W., Ott, C.M., Zu Bentrup, K.H., Ramamurthy, R., Quick, L., Porwollik, S., Cheng, P., McClelland,

M., Tsaprailis, G., Radabaugh, T., *et al.* (2007). Space flight alters bacterial gene expression and virulence and reveals a role for global regulator Hfq. Proc. Natl. Acad. Sci. U.S.A. *104*, 16299–16304.

Zhang, A., Altuvia, S., Tiwari, A., Argaman, L., Hengge-Aronis, R., and Storz, G. (1998). The OxyS regulatory RNA represses rpoS translation and binds the Hfq (HF-I) protein. EMBO J. *17*, 6061–6068.

Zhang, A., Wassarman, K.M., Ortega, J., Steven, A.C., and Storz, G. (2002a). The Sm-like Hfq protein increases OxyS RNA interaction with target mRNAs. Mol. Cell *9*, 11–22.

Zhang, A., Wassarman, K.M., Rosenow, C., Tjaden, B.C., Storz, G., and Gottesman, S. (2003). Global analysis of small RNA and mRNA targets of Hfq. Mol. Microbiol. *50*, 1111–1124.

Zhang, S., Santos, R.L., Tsolis, R.M., Stender, S., Hardt, W.D., Baumler, A.J., and Adams, L.G. (2002b). The *Salmonella enterica* serotype Typhimurium effector proteins SipA, SopA, SopB, SopD, and SopE2 act in concert to induce diarrhea in calves. Infect. Immun. *70*, 3843–3855.

Table 7.1 sRNA loci of *Salmonella* Typhimurium

Name[a]	Alternative IDs[b]	Adjacent genes[c]	Orientation[d]	Flanking gene IDs	5′ end[e]	3′ end[e]	Comment[f]	Hfq coIP[g]	Reference[h]
SroA	tpe79	tbpA/yabN	←→←	STM0108/STM0109	126707	126796	riboswitch element of the *thiBPQ* operon		Santiviago *et al.* (2009), Vogel *et al.* (2003)
SgrS	ryaA	yabN/leuD	←→←	STM0109/STM0110	128574	128812	represses *ptsG* and *sopD* mRNAs	Yes	Wadler and Vanderpool, (2009)
Tp2	–	pdhR/aceE	→←→	STM0151/STM0152	176238	176082	–		Rivas *et al.*, (2001)
T44	–	map/rpsB	←→→	STM0215/STM0216	254130	254268	–		Tjaden *et al.* (2002)
IsrA	–	STM0294.In/ STM0295	→→→	STM0294.In/STM0295	339338	339760	*cis*-antisense regulation of STM0294.In		Padalon-Brauch *et al.* (2008)
SraA	psrA/t15	clpX/lon	→←→	STM0449/STM0450	505439	505384	constitutively expressed under all tested conditions		Argaman *et al.* (2001)
MicM/ ChiX	rybC/sroB	ybaK/ybaP	←→←	STM0495/STM0496	556005	556085	represses *chiP* mRNA	Yes	Figueroa-Bossi *et al.* (2009), Vogel *et al.* (2003)
SroC	–	gltJ/gltI	←←←	STM0664/STM0665	728913	728762	–	Yes	Vogel *et al.* (2003)
RybA	–	ybiP/MntR	←←→	STM0834/STM0835	902128	902040	–		Wassarman *et al.* (2001)
RybB	p25	STM0869/ STM0870	→←←	STM0869/STM0870	942632	942554	Repressor of multiple *omp* mRNAs	Yes	Papenfort *et al.* (2006)
STnc490	–	clpA/tnpA	→←→	STM0945/STM0946	1025165	1024975	antisense to IS200 element	Yes	Sittka *et al.* (2008)
IsrB-1	IS092	sbcA/ STM1010	←→←	STM1009/STM1010	1104179	1104266	induced in minimal media and macrophage infection	Yes	Padalon-Brauch *et al.* (2008)
STnc670	–	put/phoH	→→→	STM1125/STM1126	1213489	1214315	–	Yes	Sittka *et al.* (2009)

STnc500	–	STM1127/ STM1128	←←←	STM1127/STM1128	1216440	1216157	–	Yes	Sittka *et al.* (2008)
SraB	pke2	*STM1189/ yceD*	←→→	STM1189/STM1190	1275071	1275236	–		Argaman *et al.* (2001)
STnc150	–	*icdA/STM1239*	→←→	STM1238/STM1239	1325914	1325649	–	Yes	Sittka *et al.* (2008)
IsrC	*IS102*	*envF/msgA*	←→←	STM1240/STM1241	1329145	1329432	*cis* control of the downstream *mgsA* mRNA		Padalon-Brauch *et al.* (2008)
STnc520	–	*STM1248/ STM1249*	→←←	STM1248/STM1249	1334044	1332809	–	Yes	Sittka *et al.* (2008)
IsrD	–	*STM1261/ STM1263*	→←→	STM1261/STM1263	1345788	1345738	induced in stat. phase		Padalon-Brauch *et al.* (2008)
IsrE	*ryhB*-2, *rfrB*	STM1273/ *yeaQ*	→←→	STM1273/STM1274	1352972	1352875	represses *sodB* mRNA, induced during macrophage infection	Yes	Ellermeier and Slauch (2008), Padalon-Brauch *et al.* (2008), Vogel and Sharma (2005)
STnc540	–	*ihfA/btuC*	→→→	STM1339/STM1340	1419369	1419570	–	Yes	Sittka *et al.* (2008)
RprA	*IS083*	*ydiL/ydiK*	←←←	STM1362/STM1364	1444938	1444832	activates *rpoS* mRNA	Yes	Jones *et al.* (2006), Majdalani *et al.* (2001)
RydB	*tpe7, IS082*	*ydiH*/STM1368	→→←	STM1367/STM1368	1450426	1450520	–	Yes	Wassarman *et al.* (2001)
MgrR	*STnc560*	*ydeI/ydeE*	→→←	STM1515/STM1516	1593723	1594413	resistance against antimicrobial peptides	Yes	Moon and Gottesman (2009), Sittka *et al.* (2008)
STnc570	*yneM ORF*	*ydeI/ydeE*	→←←	STM1515/STM1516	1594413	1593723		Yes	Sittka *et al.* (2008)
IsrF	–	*STM1552/ STM1554*	→←←	STM1552/STM1554	1630160	1629871	expressed in stationary phase and upon osmotic shock		Padalon-Brauch *et al.* (2008)
RydC	*IS067*	STM1638/ *cybB*	→→←	STM1638/STM1639	1729673	1729735	controls the *yejABEF* operon	Yes	Antal *et al.* (2005), Zhang *et al.* (2003)
MicC	*IS063, tke8*	*nifJ/ynaF*	→←→	STM1651/STM1652	1745786	1745678	represses *ompC* and *ompD* mRNAs	Yes	Chen *et al.* (2004), Pfeiffer *et al.* (2009)

Table 7.1 continued

Name[a]	Alternative IDs[b]	Adjacent genes[c]	Orientation[d]	Flanking gene IDs	5′ end[e]	3′ end[e]	Comment[f]	Hfq coIP[g]	Reference[h]
FnrS	*STnc580*	*dbpA/zntB*	←←←	STM1655/STM1656	1750147	1749662	multiple repressor of oxygen regulated genes	Yes	Boysen *et al.* (2010), Durand and Storz (2010)
SraC	*ryeA*	STM1871/ STM1872	→→→	STM1871/STM1872	1968059	1968196	hotspot for integration of foreign DNA sequences	Yes	Balbontin *et al.* (2008)
SdsR	*ryeB*, *tpke79*	STM1871/ STM1872	→←←	STM1871/STM1872	1968156	1968053	hotspot for integration of foreign DNA sequences	Yes	Vogel *et al.* (2003), Balbontin *et al.* (2008)
DsrA	–	*yodD/yedP*	→←→	STM1984/STM1986	2068736	2068649	activates *rpoS* mRNA	Yes	Majdalani *et al.* (2001)
RseX	–	*STM1994/* ompS	←→→	STM1994/STM1995	2077136	2077269	represses *ompA* and *ompC* mRNAs	Yes	Douchin *et al.* (2006)
RyeC	*tp11*	*yegD/* STM2126	→→→	STM2125/STM2126	2213871	2214016	–	Yes	Sittka *et al.* (2008), Wassarman *et al.* (2001)
CyaR	*ryeE*	*yegQ/* STM2137	→→→	STM2136/STM2137	2231131	2231216	represses *ompX* synthesis	Yes	De Lay and Gottesman (2009), Johansen *et al.* (2008), Papenfort *et al.* (2008)
STnc740	–	*STM2148/stcD*	→←←	STM2148/STM2149	2242570	2242831	–	Yes	Sittka *et al.* (2009)
IsrG	–	*STM2243/ STM2244*	←→→	STM2243/STM2244	2344732	2345013	*cis*-antisense regulator of STM2243		Padalon-Brauch *et al.* (2008)
MicF	–	*ompC/yojN*	←→→	STM2267/STM2269	2366913	2367005	represses *ompF* mRNA	Yes	Hartog *et al.* (2008), Mizuno *et al.* (1984)

IsrH-1	–	*glpC/sseL*	→←→	STM2286/STM2287	2394753	2394303	*cis*-antisense regulator of STM2287, induced during macrophage infection		Padalon-Brauch *et al.* (2008)
RyfA	*tp1*	STM2534/*sseB*	→→←	STM2534/STM2535	2674934	2675228	–	Yes	Wassarman *et al.* (2001)
GlmY	*tke1*, *sroF*	*yfhK/purG*	←←←	STM2564/STM2565	2707847	2707664	prevents inactivation of GlmZ	Yes	Urban and Vogel (2008)
IsrI	–	*STM2614/STM2616*	→←←	STM2614/STM2616	2761576	2761329	expressed in stationary phase and upon osmotic shock	Yes	Padalon-Brauch *et al.* (2008)
IsrJ	–	*STM2614/STM2616*	→←←	STM2614/STM2616	2762030	2761957	control of effector protein production		Padalon-Brauch *et al.* (2008)
IsrK	–	*STM2616/STM2617*	←←←	STM2616/STM2617	2762867	2762791	induced in minimal media and stat. phase		Padalon-Brauch *et al.* (2008)
IsrB-2	–	*STM2631/sbcA*	→←→	STM2631/STM2632	2770965	2770872	induced in minimal media and macrophage infection		Padalon-Brauch *et al.* (2008)
IsrL	–	*smpB/STM2690*	→←→	STM2688/STM2690	2839399	2839055	induced in minimal media		Padalon-Brauch *et al.* (2008)
IsrM	–	*STM2762/STM2763*	←→→	STM2762/STM2763	2905050	2905378	induced upon cold shock		Padalon-Brauch *et al.* (2008)
IsrN	–	*STM2764/STM2765*	←→←	STM2764/STM2765	2906925	2907067	increases STM2765 mRNA, induced in early phases of macrophage infection		Padalon-Brauch *et al.* (2008)
MicA	*sraD*	*luxS/gshA*	←→←	STM2817/STM2818	2966853	2966926	represses *ompA* and *lamB* mRNAs	Yes	Papenfort *et al.* (2006), Rasmussen *et al.* (2005), Udekwu *et al.* (2005)

Table 7.1 continued

Name[a]	Alternative IDs[b]	Adjacent genes[c]	Orientation[d]	Flanking gene IDs	5′ end[e]	3′ end[e]	Comment[f]	Hfq coIP[g]	Reference[h]
C0664	–	*ygbD/hypF*	→→←	STM2841/STM2842	2987638	2987745	–		Tjaden *et al.* (2002)
InvR	*STnc270*	*invH*/STM2901	→→→	STM2900/STM2901	3044924	3045014	represses *ompD* mRNA	Yes	Pfeiffer *et al.* (2007)
CsrB	–	*yqcC/syd*	←←←	STM2965/STM2967	3117059	3116697	antagonizes the CsrA regulatory protein		Fortune *et al.* (2006)
GcvB	*IS145*	*gcvA/ygdI*	←→←	STM2982/STM2983	3135313	3135522	represses multiple mRNAs encoding substrate binding proteins	Yes	Sharma *et al.* (2007)
OmrA/B	*t59, rygA/B, sraE*	*aas/galR*	←←→	STM3010/STM3011	3170408	3170122	represses multiple *omp* mRNAs and *ompR*	Yes	Guillier and Gottesman (2008)
STnc290	–	*tnpA4/ STM3033*	←←←	STM3032/STM3033	3194996	3194914	–	Yes	Sittka *et al.* (2008)
IsrO	–	*STM3038/ STM3039*	←→→	STM3038/STM3039	3198380	3198580	induced upon cold shock		Padalon-Brauch *et al.* (2008)
6S	*ssrS*	*ygfE/ygfA*	→→→	STM3060/STM3061	3222098	3222280	represses polymerase activity in late stationary phase growing cells	Yes	Wassarman (2007)
RygC	*t27*	*ygfA/serA*	→→←	STM3060/STM3061	3222913	3223065	–		Wassarman *et al.* (2001), Zhang *et al.* (2003)
SroG	–	*ribB/yqiC*	→→←	STM3195/STM3196.S	3358708	3358608	regulates *rib* gene expression		Vogel *et al.* (2003)
RygD	*tp8, C0730*	*yqiK/rfaE*	→←←	STM3060/STM3061	3362474	3362327	–	Yes	Wassarman *et al.* (2001)

SraF	*tpk1*, *IS160*	*ygjR*/*ygjT*	→→→	STM3223/STM3224	3392069	3392261	pH responsive riboregulator	Yes	Nechooshtan *et al.* (2009)
SraG	–	*pnp*/*rpsO*	→←→	STM3282/STM3283	3451437	3451607	–		Argaman *et al.* (2001)
ArcZ	*sraH*/*ryhA*	*yhbL*/*arcB*	←→←	STM3327/STM3328	3490383	3490500	represses *sdaC*, STM3216 and *tpx* mRNAs	Yes	Papenfort *et al.* (2009)
STnc760	–	*yhgF*/*feoA*	→←→	STM3504/STM3505	3663454	3663889	–	Yes	Sittka *et al.* (2009)
RyhB	*sraI*, *IS176*, *RfrA*	*yhhX*/*yhhY*	←←→	STM3545/STM3546	3715495	3715401	represses *sodB* mRNA and controls multiple other mRNAs involved in iron starvation	Yes	Ellermeier and Slauch (2008), Padalon-Brauch *et al.* (2008)
STnc710	–	*yhjU*/ *STM3624A*	→→→	STM3624/STM3624A	3808100	3808672	–	Yes	Sittka *et al.* (2009)
IstR	–	*ilvB*/*emrD*	←←→	STM3796/STM3798	3998147	3998018	Negative regulator of TisB toxin production		Sittka *et al.* (2008), Vogel *et al.* (2004)
STnc400	–	*STM3844*/ *STM3845*	→→→	STM3844/STM3845	4051145	4051340	–	Yes	Sittka *et al.* (2008)
GlmZ	*k19*, *ryiA*, *sraJ*	*yifK*/*hemY*	→→←	STM3930/STM3935	4141647	4141854	activates *glmS* mRNA	Yes	Görke and Vogel (2008), Urban and Vogel (2008)
Spot42	*spf*	*polA*/*yihA*	→→←	STM3999/STM4001	4209066	4209175	represses *galK* mRNA fragment	Yes	Møller *et al.* (2002b), Sittka *et al.* (2008)
CsrC	*sraK*, *ryiB*, *tpk2*	*yihA*/*yihI*	←→→	STM4001/STM4003	4210157	4210400	antagonizes the CsrA regulatory protein		Fortune *et al.* (2006)
IsrP	–	*STM4097*/ *STM4098*	←→←	STM4097/STM4098	4306719	4306866	expressed under invasion conditions	Yes	Padalon-Brauch *et al.* (2008)
OxyS	–	*argH*/*oxyR*	→←→	STM4123/STM4125	4342986	4342866	represses *fhlA* and *rpoS* expression	Yes	Altuvia *et al.*, (1997), Padalon-Brauch *et al.*, (2008)

Table 7.1 continued

Name[a]	Alternative IDs[b]	Adjacent genes[c]	Orientation[d]	Flanking gene IDs	5′ end[e]	3′ end[e]	Comment[f]	Hfq coIP[g]	Reference[h]
SraL	*ryjA*	*soxR*/ STM4267	→←→	STM4266/STM4267	4505010	4504870	–		Argaman *et al.* (2001)
STnc440	–	*STM4310/ STM4311*	→→→	STM4310/STM4311	4559193	4559277	–	Yes	Sittka *et al.* (2008)
IsrQ	–	*STM4508/ STM4509.S*	←→→	STM4508/STM4509.S	4762997	4763158	induced upon cold shock		Padalon-Brauch *et al.* (2008)

[a]sRNA name and [b]alternative sRNA gene IDs are provided based on a summary of predicted or confirmed sRNAs.

[c]Flanking genes of the intergenic region in which the sRNA candidate is located. The sRNA specific conserved flanking genes are shown with their functions, and strain specific flanking genes are indicated by their gene IDs.

[d]Orientation of sRNA candidate (middle) and flanking genes (→ and ← denote location of a gene on the plus and minus strand of the *Salmonella* chromosome, respectively).

[e]Genomic location of sRNA candidate gene according to the *Salmonella* Typhimurium LT2 genome.

[f]Functional roles of the sRNAs originally discovered in *Salmonella* or inferred from *E. coli.*

[g]Hfq–sRNA interactions reported from these studies: Sittka *et al.* (2008, 2009).

[h]Most relevant reference for sRNA function.

Fimbrial Signature Arrangements in *Salmonella*

8

Sean-Paul Nuccio, Nicholas R. Thomson, Maria C. Fookes and Andreas J. Bäumler*

Abstract

The complement of fimbrial operons held within a genome represents one of the key differentiating features of the sequenced *Salmonella* serovars and one of the single largest sources of genetic diversity. Generically described as filamentous non-flagellar surface appendages, fimbriae (also known as pili) typically imbue an adhesive trait to the cells expressing them. While much is known about the general biology of fimbrial assembly mechanisms, the role of these structures in *Salmonella* pathogenesis remains poorly characterized. Here we present fimbrial operon data gathered from the seventeen completed *Salmonella* genome sequences and discuss its implications in *Salmonella* pathogenesis and dissemination.

Introduction

The current classification of *Salmonella* is extremely complex; based on DNA–DNA hybridization the genus *Salmonella* is divided into two species: *Salmonella enterica* and *Salmonella bongori*. *S. enterica* can then be subdivided into the subspecies *enterica, salamae, arizonae, diarizonae, houtenae* and *indica* based on biochemical differences (Brenner *et al.*, 2000). The genus is then further divided by serology into over 2500 serovars using the Kauffman and White Scheme (Grimont and Weill, 2007), which defines the serogroup according to expression of somatic lipopolysaccharide O antigens and the serovar by expression of flagellar H antigens.

To date, genome sequences have been completed for one representative of *S. bongori*, one representative of *S. enterica* subspecies *arizonae*, and 15 representatives of *S. enterica* subspecies *enterica* (Fig. 8.1). The chromosomes of these pathogens reveal a high degree of synteny, which is interrupted only by large inversions around the terminus or origin of replication (Thomson *et al.*, 2008). The core genome is interspersed with the accessory or variable genome, which has been shown to consist of either single genes or gene groups up to large genomic islands of greater than 100 kb, such as *Salmonella* pathogenicity island (SPI)–7 (see below).

Fimbriae have been shown to be important for host–pathogen interactions in a wide range of human and animal pathogenic bacteria (Kline *et al.*, 2009); *Salmonella* is no exception. Whole genome analysis has shown that there are >2500 genes that are considered core and shared by most of the sequenced isolates. However, *Salmonella* also possesses genes that are restricted in their distribution and are present only in a small subset of *Salmonella* serovars. Among these accessory elements, fimbrial operons exhibit perhaps the greatest variability besides phages in phylogenetic distribution amongst *Salmonella* serovars. Fimbriae (also known as pili) are filamentous, non-flagellar proteinaceous surface appendages involved in adhesion and are classified based on their mechanism of assembly (Fronzes *et al.*, 2008). The genus *Salmonella* harbours fimbriae belonging to the nucleation/precipitation pathway, the chaperone/usher pathway and type IV fimbriae. Here we will review the fimbrial complements and their genomic arrangements

*Corresponding author

Salmonella	Subspecies	Serovar	Strain	Genbank Accession	bcf	fim	sth	lpf	1 peg	saf	stb	1 stc	std	ste	stf	sti	pef	2 sba	sbb	sbc	3 sbd	sef	2 sta	3 stg	stj	stk	tcf	N1	N2	N3	N4	N5	csg	R64-like pil	sbe
				C/U Clade:	γ1	γ1	γ1	γ1	γ4	γ3	γ4	γ4	π	π	π	γ1	κ	γ4	π	κ	γ1	γ3	γ4	γ1	β	γ4	α	γ1	σ	γ4	κ	γ3	Curli	Type IVb	
bongori	-	-	12419	*Sanger Institute*	■		■	■	■									■	■	■	■												■		■
enterica	*arizonae*	62:z4,z23:--	RSK2980	CP000880.1	■	■			■																			■	■				□		
	enterica	Agona	SL483	CP001138.1	■	■	■	■		■	■	■	□	■	■	■									■					■			■		
		Choleraesuis	SC-B67	AE017220.1	□	□	■	□		■	□	■	■	■	■	■	■^{P}										□						■		
		Dublin	CT_02021853	CP001144.1	■	□	■	□	■	□	■		■	■	■	■						■											■		
		Enteritidis	P125109	AM933172.1	■	■	■	■	■	■	■		■	■	■	■						■											■		
		Gallinarum	287/91	AM933173.1	■	■	□	□	□	□	□			■	□	□						□											■		
		Heidelberg	SL476	CP001120.1	■	■	■	■			■	■	■	■	■	■									■	■	■						■	■^{P}	
		Newport	SL254	CP001113.1	■	■	■	■	■	■	■		■	■	■	■									■							□	■		
		Paratyphi A	AKU_12601	FM200053.1	□	■	■		■	□	■		■	■	□							□				■	■						□		
		Paratyphi A	ATCC 9150	CP000026.1	□	■	■		■	□	■		■	■	□							□				■	■						□		
		Paratyphi B	SPB7	CP000886.1	■	■	■	□		■	■	■	■		■	■							■										■	■	
		Paratyphi C	RKS4594	CP000857.1	□	■	■	■		□	■	□	■	■	□	■	■^{P}																■	□	
		Schwarzengrund	CVM19633	CP001127.1	■	■	■		■	■	■		■			■							■				■	■			■		■		
		Typhi	CT18	AL513382.1	□	□	□			□	■	■	■	□								□	■	□			■						■	□	
		Typhi	Ty2	AE014613.1	□	■	□			□	□	□	■	□								□	■	□			■						■	□	
		Typhimurium	LT2	AE006468.1	■	■	■	■		■	■	■	■		■	■	■^{P}								■								■		
					Core (>90%)			Conserved (>45%)									Accessory (<45%)																		

Figure 8.1 *Salmonella* fimbrial repertoire composition. Fimbrial operon data was gathered from the genomes of each completed *Salmonella* strain utilizing the National Center for Biotechnology Information conserved domain database search (Marchler-Bauer and Bryant, 2004), assigned to clades based on the classification scheme of Nuccio and Bäumler (2007), and assignments were confirmed by fimbrial usher primary amino acid sequence comparisons with known members of each clade. All currently known fimbrial operons are listed in the column headers, with five newly identified operons labelled N1 through N5. The fimbrial clade (C/U fimbriae) or general assembly class (Curli and Type IV fimbriae) is shown. A number above an operon name indicates insertion at the same relative position: 1, between *thiM* and *mrp*; 2, between *glmS* and *pstS*; 3, between *panB* and *folK*. A black box within a column indicates the operon is present and intact in a given organism, while a white box signifies the operon is present, but with at least one pseudogene in a chaperone, usher or subunit. A superscript 'P' denotes the operon in question is carried on an episome in that organism.

revealed by whole genome sequencing of *Salmonella* serovars, and explore selective forces that may have given rise to the observed diversity.

Diversity and nature of bacterial fimbrial operons

Curli

The *csgDEFG csgAB* gene cluster (Fig. 8.2) encodes the formation of extracellular filaments known as curli (Olsen *et al.*, 1989), thin curled fimbriae (Grund and Weber, 1988) or thin aggregative fimbriae (Collinson *et al.*, 1991) through the nucleation/precipitation pathway. The major subunit of curli, CsgA, contains a standard signal peptide directing transport across the cytoplasmic membrane and is secreted through the outer membrane by CsgG (Robinson *et al.*, 2006). Once secreted, CsgA monomers fold into an insoluble cross β sheet structure (Shewmaker *et al.*, 2009), resulting in the precipitation of individual subunits into fibrils resembling amyloids. The precipitation process is initiated by a second secreted protein, CsgB, which serves as a nucleator for amyloid fibril formation by CsgA (Chapman *et al.*, 2002). Curli fibrils mediate attachment to the host extracellular matrix protein fibronectin (Collinson *et al.*, 1993; Olsen *et al.*, 1989), which may play a role in host pathogen interaction (Gophna *et al.*, 2002). Host-derived amyloid fibrils also exhibit fibronectin binding activity (de Beer *et al.*, 1981; Rostagno *et al.*, 1986) and this property thus may be a common feature of amyloids.

Curli are distinct from other fimbriae not only in their assembly pathway, but also in several other aspects of their biology. First, unlike other

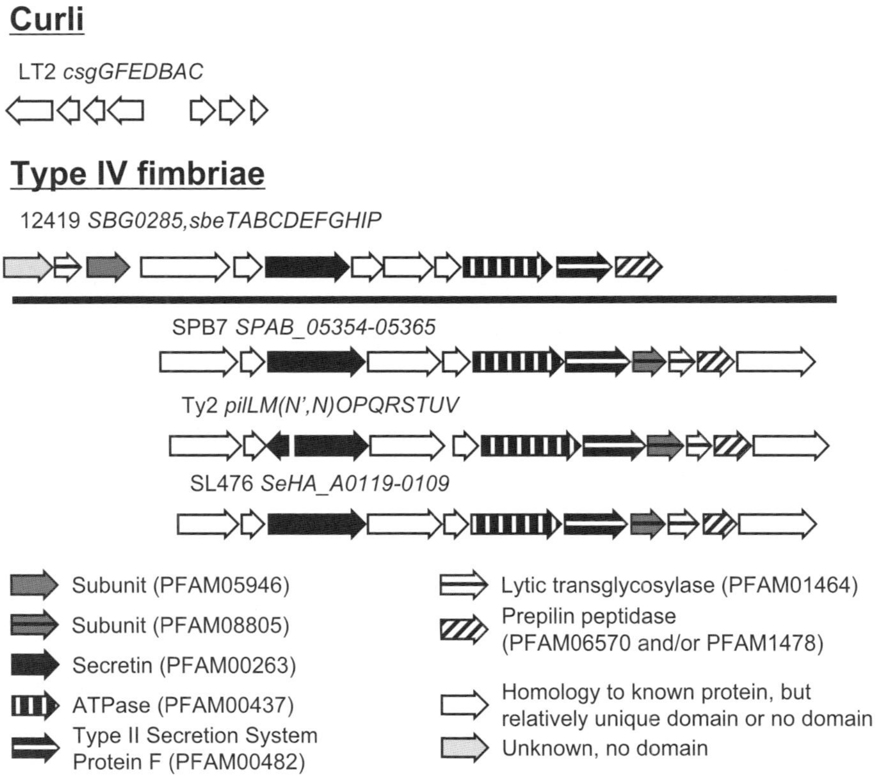

Figure 8.2 Curli and type IV fimbrial operon organization. A representative curli operon and several type IV fimbrial operons listed in Fig. 8.1 are displayed. The strain and gene names associated with an operon are listed, and strain classification can be gleaned from Fig. 8.1. Genes are represented by arrows; their shades correspond to the protein domain (Protein Family, PFAM) and associated function.

fimbrial gene clusters, the *csgDEFG* and *csgAB* gene clusters are highly conserved among a group of related genera within the Enterobacteriaceae, including *Citrobacter, Enterobacter, Escherichia, Klebsiella, Salmonella* and *Shigella* (Zogaj *et al.*, 2003). Secondly, curli do not form individual filaments protruding from the bacterial surface but have been shown in *E. coli* and *Salmonella* to form part of an extracellular biofilm matrix (Prigent-Combaret *et al.*, 2000; Romling *et al.*, 2003; Vidal *et al.*, 1998). Thirdly, in addition to mediating attachment, curli play an important role during host pathogen interaction by serving as a pathogen-associated molecular pattern (PAMP) recognized by the innate immune system through Toll-like receptor (TLR) 2 (Tukel *et al.*, 2005; Tukel *et al.*, 2009). It has recently been demonstrated that the mechanism of innate immune recognition through TLR2 involves shared structural features of amyloid fibrils of both host and bacterial origin (Tukel *et al.*, 2009). Recognition of curli fibrils following bacterial invasion contributes to proinflammatory cytokine production and neutrophil recruitment in a bovine ligated ileal loop model of *S. enterica* serovar Typhimurium-induced gastroenteritis (Tukel *et al.*, 2005).

Type IV fimbriae

Subunits of type IV fimbriae are assembled into fimbrial filaments during their transport across the cytoplasmic membrane, which involves a specialized prepilin peptidase cleaving a type IV prepilin signal peptide (Nunn and Lory, 1991; Strom and Lory, 1992; Strom *et al.*, 1991). The fimbrial filament extends from the cytoplasmic membrane to the bacterial cell surface through a multimeric outer membrane channel formed by a protein known as secretin (Bitter, 2003; Collins *et al.*, 2001, 2004; Frye *et al.*, 2006; Wolfgang *et al.*, 2000). Several components of type IV fimbrial assembly (such as secretin) share extensive homology with proteins used in Gram-negative Type II secretion systems (note that the 'Type IV' moniker reflects historical placement within the fimbrial classification scheme of Ottow (1975), and not relation to type IV secretion systems). Energy for adding subunits to or subtracting subunits from the fimbrial filament is provided by ATPases located in the cytoplasmic membrane (Jakovljevic *et al.*, 2008; Sakai *et al.*, 2001; Turner *et al.*, 1993). This assembly mechanism allows for active extension and retraction of fimbrial filaments, thereby providing a means of bacterial locomotion through twitching motility in some organisms (Merz *et al.*, 2000; Wolfgang *et al.*, 1998; Wu *et al.*, 1997).

Type IV fimbriae in general are currently the most widespread assembly class with representatives throughout the Eubacteria and Archaebacteria (Albers and Pohlschroder, 2009; Pelicic, 2008). The type IV fimbrial assembly class is currently subdivided into two groups, type IVa and IVb, based on homology of major fimbrial subunits. Amongst the sequenced *Salmonellae*, only a limited number of serovars possess type IVb fimbrial operons. The *S. bongori* chromosome contains the *sbeTABCDEFGHIP* operon, which remains uncharacterized though bears homology to the colonization factor antigen III gene cluster of enterotoxigenic *E. coli* (Taniguchi *et al.*, 2001; our unpublished data). *S. enterica* serovars Paratyphi B and Heidelberg each carry a gene cluster encoding type IV fimbriae in the chromosome and an episome, respectively (Fig. 8.2) which share homology with the plasmid R64 thin pilus (Yoshida *et al.*, 1999). Finally, the R64 thin pilus-like *pilM(N',N)OPQRSTUV* operon of *S. enterica* serovar Typhi is located on SPI7, a 134 kb DNA region which also carries the *tviABCDEvexABCDE* operon encoding the Vi-capsular antigen (Parkhill *et al.*, 2001). SPI7 is absent from other *Salmonella* serovars, except *S. enterica* serovar Paratyphi C and some isolates of *S. enterica* serovar Dublin. The *pilM(N',N)OPQRSTUV* operon encodes type IV fimbriae which bind the cystic fibrosis transmembrane conductance regulator (Tsui *et al.*, 2003), thereby enhancing *S. enterica* serovar Typhi invasion of epithelial cell lines (Zhang *et al.*, 2000).

Chaperone/usher fimbriae

The most common fimbrial gene sequences found in the genomes of *Salmonella* serovars belong to the chaperone/usher (C/U) assembly class. Structural proteins (i.e. fimbrial subunits) and assembly proteins (i.e. chaperone and usher) are predominantly encoded by discrete operons and standard signal peptides direct their transport across the cytoplasmic membrane by the general secretory system. Fimbrial subunits possess an incomplete immunoglobulin (Ig)-like fold, missing the final antiparallel β strand (Choudhury *et al.*, 1999; Sauer *et al.*, 1999). In the periplasmic space, each fimbrial subunit is bound by a chaperone which inserts a donor β strand in the parallel orientation, thereby completing the Ig-like fold of the bound subunit in a process termed donor-strand complementation (Waksman and Hultgren, 2009). The usher protein forms monofunctional dimers in the outer membrane that transport subunits to the cell surface (Huang *et al.*, 2009; Li *et al.*, 2004). This process is energized through a donor-strand exchange reaction that replaces the parallel donor β strand of the chaperone with the more energetically favourable antiparallel donor β strand (N-terminal extension) of another fimbrial subunit, thereby connecting fimbrial subunits into a filament that is anchored to the cell surface via the usher (Waksman and Hultgren, 2009). In many cases fimbrial filaments are capped by a tip adhesin that mediates attachment to host surfaces (Buts *et al.*, 2003; Dodson *et al.*, 2001; Haslam *et al.*, 1994; Hung *et al.*, 2002). Within the genus *Salmonella*, the binding specificity is known for only a small number of fimbriae belonging to the chaperone/usher assembly class. In each case, the receptor is a terminal structure of a host carbohydrate-chain. Type-1 fimbriae encoded by the *fim* operon bind terminal mannose residues (Duguid *et al.*, 1966). Plasmid encoded fimbriae encoded by the *pef* operon bind the Lewis X (Le^x) blood group antigen, a trisaccharide composed of galactose β1–4(fucose α1–3) N-acetyl-glucosamine (Chessa *et al.*, 2008). Fimbriae encoded by the *std* operon bind terminal α1–2 fucose residues (Chessa *et al.*, 2009).

Based on sequence alignments of usher proteins, the chaperone/usher assembly pathway can be subdivided into six major phylogenetic clades termed σ-fimbriae (or archaic chaperone/usher family), α-fimbriae (or alternate chaperone/usher family), β-fimbriae, γ-fimbriae, κ-fimbriae and π-fimbriae (Nuccio and Baumler, 2007). *Salmonella* serovars contain gene clusters representing each of these fimbrial types (Fig. 8.3). Serovars belonging to *S. enterica* subspecies *enterica* carry between 11 and 13 fimbrial operons of the chaperone/usher assembly class (Fig. 8.1). The diversity is lower for *S. enterica* subspecies *arizonae* (5 operons) and *S. bongori* (8 operons), while *E. coli* K-12 carries a similar number of fimbrial gene sequences (12 operons) as *S. enterica* subspecies *enterica* serovars. The role this multitude of fimbrial gene sequences plays in host–pathogen interactions remains incompletely understood.

In a few instances, chaperone/usher fimbriae have been implicated in the systemic phase of infection (Clayton *et al.*, 2008; Edwards *et al.*, 2000). However, the majority of studies support a role during intestinal colonization. Fimbriae can mediate attachment of *Salmonella* serovars to cultured cell lines (Bäumler *et al.*, 1996; Chessa *et al.*, 2009; Ernst *et al.*, 1990; Forest *et al.*, 2007; Wilson *et al.*, 2000) and contribute to biofilm formation on this surface (Boddicker *et al.*, 2002; Ledeboer *et al.*, 2006). It is not clear from these data whether attachment *in vivo* results in colonization of the apical surface of epithelial cells or in binding to mucin carbohydrates, thereby promoting colonization of the mucus layer. During gastroenteritis, diarrhea removes intestinal contents, thereby making mucus carbohydrates one of the few remaining high-energy carbon sources. The ability to chemotax towards carbohydrate gradients originating from the mucus layer has been shown to be important for colonizing the inflamed gut (Stecher *et al.*, 2008). Fimbriae encoded by the *std* operon bind mucus carbohydrates *in situ*, suggesting a possible role in colonizing this niche (Chessa *et al.*, 2009). Inactivation of an individual fimbrial operon may result in no defect in recovery from intestinal contents (Althouse *et al.*, 2003) or produce a defect in competition with wild type bacteria to colonize this niche (Weening *et al.*, 2005). These results suggest that functional redundancy complicates an analysis of the 11 to 13 fimbrial C/U gene clusters present in *S. enterica* subspecies *enterica* serovars.

Fimbrial operon distribution

A major unresolved question raised by whole genome sequencing remains why *S. enterica* subspecies *enterica* serovars require such a large repertoire of fimbrial operons. Nearly every

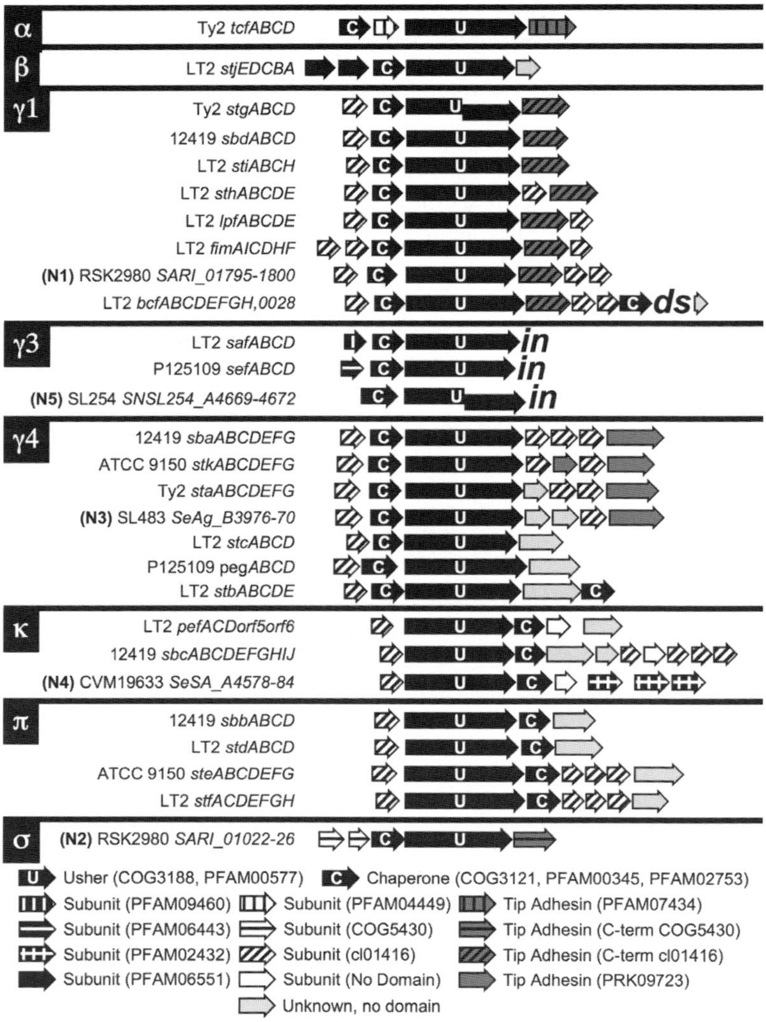

Figure 8.3 Chaperone/Usher fimbrial operon organization. A representative C/U operon from each group listed in Fig. 8.1 is displayed with the operons sorted by C/U fimbrial clade (white lettering on black relief). Operons are preceded by the strain (see Fig. 8.1 for full details) and associated gene names. Genes are represented by arrows, the shades correspond to the protein domain (Protein Family, PFAM; Cluster of Orthologous Genes, COG; Protein Clusters, PRK; Superfamily Cluster, cl) and associated function. Light grey arrows have no homology to a protein domain or characterized protein, while broken arrows indicate pseudogenes. Abbreviations: C-term, carboxy-terminal; *ds*, homology to thiol-disulfide isomerase domain (PFAM01323); *in*, homology to invasin domain (PFAM05775).

sequenced *Salmonella* serovar possesses a unique repertoire of chaperone/usher fimbriae (Fig. 8.1), employing combinations of 28 common fimbrial operons that encompass all of the known fimbrial clades (Fig. 8.3). In spite of this heterogeneity in fimbrial repertoire composition, orthologous fimbrial operons are always located in the same genomic context (Fig. 8.4), the relative positions only vary due to an occasional large chromosomal inversion. Even when distinct operons emerge from a common precursor operon (i.e. *peg/stc*, *sba/sta*, *sbd/stg*), their distribution amongst the *Salmonellae* becomes mutually exclusive (Fig. 8.1) as the operons continue to reside between the same genetic loci. This stability notwithstanding, no clear patterns of fimbrial operon acquisition and vertical inheritance are evident when comparing fimbrial repertoire content to serovar

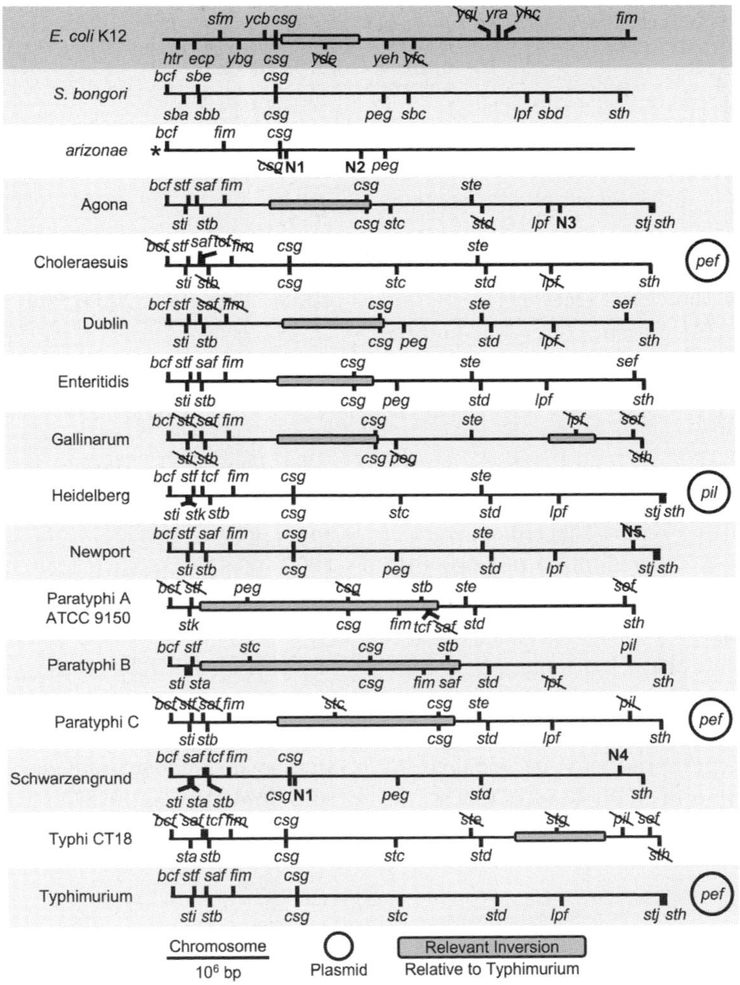

Figure 8.4 The chromosomal and episomal locations of the fimbrial operons from *E. coli* K12 and the sequenced *Salmonella* serovars. Additional information about each serovar can be found in Fig. 8.1; the *E. coli* K12 genome GenBank accession number is U00096.2 (Blattner *et al.*, 1997). Chromosomes are displayed to scale in a linear fashion with the 5′ end set at the thrA gene (* The *Salmonella enterica* subsp. *arizonae* genome is not annotated with *thrA* in this fashion, so we have arbitrarily set it as such). Operon labels above a chromosome indicate that its coding sequences are located on the Watson strand, while labels beneath the chromosome indicate coding sequences are on the Crick strand. Non-scale circles represent plasmids and the name within each corresponds to the fimbrial operon present. A crossed out operon name indicates that a pseudogene is present. Chromosome inversions relative to *Salmonella* Typhimurium LT2 (McClelland *et al.*, 2001) that affect fimbrial operon locations are indicated by grey rectangles and were identified utilizing the Artemis Comparison Tool (Carver *et al.*, 2005).

genealogy (unpublished observations), suggesting that once a fimbrial operon is introduced into the *Salmonellae*, further distribution is mediated both by vertical inheritance and by horizontal transfer processes involving homologous recombination (transduction, conjugation). However, some patterns are evident in the repertoires of the sequenced *Salmonellae*.

A subset of three C/U fimbrial operons (*bcf*, *fim* and *sth*), is highly conserved within the genus *Salmonella*, while a second subset of nine operons (*lpf*, *peg*, *saf*, *stb*, *stc*, *std*, *ste*, *stf*, and *sti*) is present in more than 45% of sequenced genomes (Fig. 8.1). However, the remaining 16 fimbrial operons found in *Salmonella* serovars are present in less than half of the completed genomes. The presence of a large number of fimbrial operons with limited distribution gives rise to fimbrial signature arrangements that seem to be unique for most, if not all *Salmonella* serovars (Townsend *et al.*, 2001). One possible reason for the presence of unique fimbrial signature arrangements in each *Salmonella* serovar may be the evasion of cross-immunity during circulation of two pathogens in the same host reservoir (Norris and Baumler, 1999). The presence of unique fimbrial signature arrangements may thus fulfil a similar function as proposed for O-antigen polymorphism among *Salmonella* serovars (Kingsley and Baumler, 2000). As reptiles appear to not possess immunological memory (Nahm *et al.*, 1992), the observation that reptile-adapted serovars of *Salmonella* possess fewer fimbrial operons than other isolates with broad host range specificity (5–8 operons versus 12–13, Fig. 8.5) is in line with the conjecture that cross-immunity evasion is a driving force in fimbrial repertoire expansion and diversification. It can also be speculated that large fimbrial repertoires may be required to ensure colonization of multiple host species, or to deal with the variable glycosylation patterns observed within an individual host species. While additional empirical studies must be performed to shed light on these processes, further insight into the potential roles fimbriae play in infection can be drawn from the pathogenesis of *Salmonella* serovars which have lost substantial portions of their fimbrial repertoires (Fig. 8.5) (Chiu *et al.*, 2005; Deng *et al.*, 2003; Holt *et al.*, 2009; Liu *et al.*, 2009; McClelland *et al.*, 2004; Parkhill *et al.*, 2001; Thomson *et al.*, 2008).

Whole genome sequencing of the strictly human adapted *S. enterica* serovar Typhi provides some further insight into this question. *S. enterica* serovar Typhi is a clone that emerged approximately 50,000 years ago (Kidgell *et al.*, 2002). Unlike closely related *Salmonella* serovars, which cause a localized gastroenteritis, *S. enterica* serovar Typhi causes a systemic infection, leading to chronic gall bladder carriage in a fraction of patients (Kingsley and Baumler, 2000). By adopting chronic gall bladder carriage as an important new strategy for transmission, genes for using intestinal colonization as the main transmission strategy

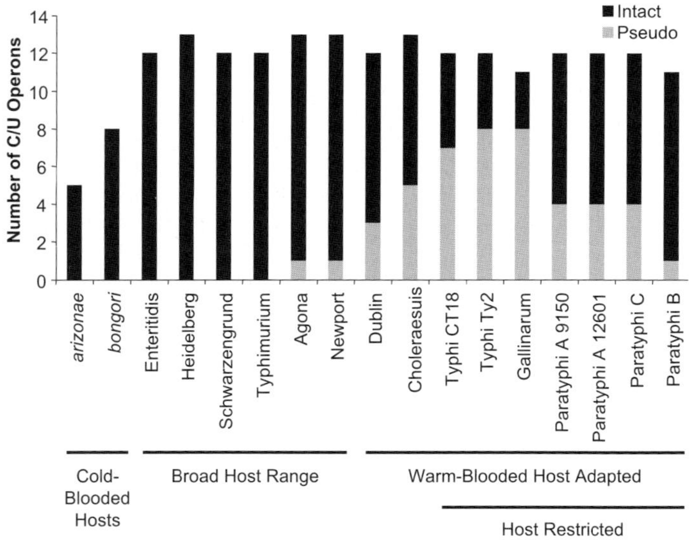

Figure 8.5 Chaperone/usher fimbrial repertoire pseudogene content amongst the analysed *Salmonella* serovars. The total number of C/U fimbrial operons present in each *Salmonella* serovar is represented by the height of a bar. The black portion of the bar denotes the number of intact operons, while the grey portion indicates the number of operons containing pseudogenes (Pseudo). The generic host range characteristic of each serovar is indicated.

became obsolete in *S. enterica* serovar Typhi. This recent change in its mode of transmission is visible in the genomic degradation of obsolete genes, which was revealed by whole genome sequencing (Parkhill *et al.*, 2001). For instance, *S. enterica* serovar Typhi strain CT18 contains pseudogenes in 7 of its 11 C/U fimbrial operons and strain Ty2 contains pseudogenes in 8 of 11 C/U operons (Fig. 8.5).

Genomic degradation of fimbrial operons is a trend observed in the genomes of other *Salmonella* serovars associated with systemic infections. Since these serovars generally exhibit a narrow host range, it is also possible that host restriction is responsible for fimbrial operons becoming obsolete. However, with 10 of its 11 fimbrial operons still intact, *S. enterica* serovar Paratyphi B represents a notable exception to this trend (Fig. 8.5). A potential explanation for this inconsistency can be found in the observation that Paratyphi B is a relatively 'young' host-restricted serovar (Selander *et al.*, 1990), implying substantial pseudogene accumulation has yet to occur according to current models of genome degradation following host restriction (Holt *et al.*, 2009). Assuming there is a common theme in host restriction amongst the various host-restricted serovars (which in and of themselves do not form a monophyletic group), the intact fimbrial repertoire of Paratyphi B may indicate that host restriction occurs through mutations at other loci, followed by fimbrial gene degradation. It should be noted, however, that pseudogene presence within a fimbrial operon might not always result in the loss of that structure. For example, the usher within the *stg* operon of *S. enterica* serovar Typhi is a pseudogene, yet expression of the whole *stg* operon still yields an adherence phenotype (Forest *et al.*, 2007), presumably through the relatively rare instance of fimbrial subunits reaching the cell surface via the usher of another operon (Klemm *et al.*, 1995; Ruer *et al.*, 2007; Saulino *et al.*, 1998).

A hypothesis for a mechanism common to both the transmission mode and host restriction postulates for fimbrial gene degradation can be formulated when one considers the observation that fimbrial subunits can serve as antigens during *S. enterica* serovar Typhimurium infections (Humphries *et al.*, 2005). With the predominantly systemic lifestyles of host-restricted serovars, the prospect exists that fimbrial loss is generally driven by selective pressures to rid the cell of unnecessary immunogenic surface appendages, which are vestiges of their previous enteric lifestyle.

Conclusions

Fimbriae represent one of the greatest single sources of diversity amongst the *Salmonellae*, however, their roles in infection and transmission remain largely an enigma. These structures serve as likely candidates for enabling colonization, persistence and dissemination, while their diversity in structural proteins and binding specificities may be the products of immune evasion and the need to cope with altered mucus and epithelial surface glycosylation profiles of different hosts. It is clear that there is not one fimbrial type per host, tissue or niche and that a deeper characterization of the full fimbrial complement and their combined functions will probably lead to a better understanding of their role in pathogenesis. Certainly, a key finding when looking at functional gene loss in host-restricted *Salmonellae* is that there is a strong selection for fimbrial loss if there is a change in niche. However, the cause and consequence of fimbrial loss remains an open question.

Through a combination of the aforementioned processes, and probably others we have not considered, each *Salmonella* serovar has acquired a signature fimbrial repertoire. It remains to be seen whether this trend is maintained as more *Salmonella* serovars are sequenced and of what benefit fimbrial diversity and quantity are to this genus.

Acknowledgements

Work in AJB's laboratory was supported by Public Health Service grants AI040124, AI044170, AI073120, AI076246 and AI079173. SPN was supported by Public Health Service grant AI060555 and the Floyd and Mary Schwall Fellowship in Medical Research. We thank The Wellcome Trust for its support of The Sanger Institute Pathogen Genomics group.

References

Albers, S.V., and Pohlschroder, M. (2009). Diversity of archaeal type IV pilin-like structures. Extremophiles *13*, 403–410.

Althouse, C., Patterson, S., Fedorka-Cray, P., and Isaacson, R.E. (2003). Type 1 fimbriae of *Salmonella enterica* serovar Typhimurium bind to enterocytes and contribute to colonization of swine *in vivo*. Infect. Immun. *71*, 6446–6452.

Bäumler, A.J., Tsolis, R.M., and Heffron, F. (1996). Contribution of fimbrial operons to attachment to and invasion of epithelial cell lines by *Salmonella* Typhimurium. Infect. Immun. *64*, 1862–1865.

Bitter, W. (2003). Secretins of *Pseudomonas aeruginosa*: large holes in the outer membrane. Arch. Microbiol. *179*, 307–314.

Blattner, F.R., Plunkett, G., 3rd, Bloch, C.A., Perna, N.T., Burland, V., Riley, M., Collado-Vides, J., Glasner, J.D., Rode, C.K., Mayhew, G.F., *et al.* (1997). The complete genome sequence of Escherichia coli K-12. Science *277*, 1453–1462.

Boddicker, J.D., Ledeboer, N.A., Jagnow, J., Jones, B.D., and Clegg, S. (2002). Differential binding to and biofilm formation on, HEp-2 cells by *Salmonella enterica* serovar Typhimurium is dependent upon allelic variation in the *fimH* gene of the fim gene cluster. Mol. Microbiol. *45*, 1255–1265.

Brenner, F.W., Villar, R.G., Angulo, F.J., Tauxe, R., and Swaminathan, B. (2000). *Salmonella* nomenclature. J. Clin. Microbiol. *38*, 2465–2467.

Buts, L., Bouckaert, J., De Genst, E., Loris, R., Oscarson, S., Lahmann, M., Messens, J., Brosens, E., Wyns, L., and De Greve, H. (2003). The fimbrial adhesin F17-G of enterotoxigenic Escherichia coli has an immunoglobulin-like lectin domain that binds N-acetylglucosamine. Mol. Microbiol. *49*, 705–715.

Carver, T.J., Rutherford, K.M., Berriman, M., Rajandream, M.A., Barrell, B.G., and Parkhill, J. (2005). ACT: the Artemis Comparison Tool. Bioinformatics *21*, 3422–3423.

Chapman, M.R., Robinson, L.S., Pinkner, J.S., Roth, R., Heuser, J., Hammar, M., Normark, S., and Hultgren, S.J. (2002). Role of Escherichia coli curli operons in directing amyloid fiber formation. Science *295*, 851–855.

Chessa, D., Dorsey, C.W., Winter, M., and Baumler, A.J. (2008). Binding specificity of *Salmonella* plasmid-encoded fimbriae assessed by glycomics. J. Biol. Chem. *283*, 8118–8124.

Chessa, D., Winter, M.G., Jakomin, M., and Baumler, A.J. (2009). *Salmonella enterica* serotype Typhimurium Std fimbriae bind terminal alpha(1,2)fucose residues in the cecal mucosa. Mol. Microbiol. *71*, 864–875.

Chiu, C.H., Tang, P., Chu, C., Hu, S., Bao, Q., Yu, J., Chou, Y.Y., Wang, H.S., and Lee, Y.S. (2005). The genome sequence of *Salmonella enterica* serovar Choleraesuis, a highly invasive and resistant zoonotic pathogen. Nucleic Acids Res *33*, 1690–1698.

Choudhury, D., Thompson, A., Stojanoff, V., Langermann, S., Pinkner, J., Hultgren, S.J., and Knight, S.D. (1999). X-ray structure of the FimC-FimH chaperone–adhesin complex from uropathogenic *Escherichia coli*. Science *285*, 1061–1066.

Clayton, D.J., Bowen, A.J., Hulme, S.D., Buckley, A.M., Deacon, V.L., Thomson, N.R., Barrow, P.A., Morgan, E., Jones, M.A., Watson, M., *et al.* (2008). Analysis of the role of 13 major fimbrial subunits in colonisation of the chicken intestines by *Salmonella enterica* serovar Enteritidis reveals a role for a novel locus. BMC Microbiol. *8*, 228.

Collins, R.F., Davidsen, L., Derrick, J.P., Ford, R.C., and Tonjum, T. (2001). Analysis of the PilQ secretin from *Neisseria meningitidis* by transmission electron microscopy reveals a dodecameric quaternary structure. J. Bacteriol. *183*, 3825–3832.

Collins, R.F., Frye, S.A., Kitmitto, A., Ford, R.C., Tonjum, T., and Derrick, J.P. (2004). Structure of the Neisseria meningitidis outer membrane PilQ secretin complex at 12 A resolution. J. Biol. Chem. *279*, 39750–39756.

Collinson, S.K., Doig, P.C., Doran, J.L., Clouthier, S., Trust, T.J., and Kay, W.W. (1993). Thin, aggregative fimbriae mediate binding of *Salmonella* Enteritidis to fibronectin. J. Bacteriol. *175*, 12–18.

Collinson, S.K., Emody, L., Muller, K.H., Trust, T.J., and Kay, W.W. (1991). Purification and characterization of thin, aggregative fimbriae from *Salmonella* Enteritidis. J. Bacteriol. *173*, 4773–4781.

de Beer, F.C., Baltz, M.L., Holford, S., Feinstein, A., and Pepys, M.B. (1981). Fibronectin and C4-binding protein are selectively bound by aggregated amyloid P component. J. Exp. Med. *154*, 1134–1139.

Deng, W., Liou, S.R., Plunkett, G., 3rd, Mayhew, G.F., Rose, D.J., Burland, V., Kodoyianni, V., Schwartz, D.C., and Blattner, F.R. (2003). Comparative genomics of *Salmonella enterica* serovar Typhi strains Ty2 and CT18. J. Bacteriol. *185*, 2330–2337.

Dodson, K.W., Pinkner, J.S., Rose, T., Magnusson, G., Hultgren, S.J., and Waksman, G.J. (2001). Structural basis of the Interaction of the pyelonephritic *E. coli* adhesin to Its human kidney receptor. Cell *105*, 733–743.

Duguid, J.P., Anderson, E.S., and Campbell, I. (1966). Fimbriae and adhesive properties in *Salmonellae*. J. Pathol. Bacteriol. *92*, 107–138.

Edwards, R.A., Schifferli, D.M., and Maloy, S.R. (2000). A role for *Salmonella* fimbriae in intraperitoneal infections. Proc. Natl. Acad. Sci. U.S.A. *97*, 1258–1262.

Ernst, R.K., Dombroski, D.M., and Merrick, J.M. (1990). Anaerobiosis, type 1 fimbriae, and growth phase are factors that affect invasion of HEp-2 cells by *Salmonella* Typhimurium. Infect. Immun. *58*, 2014–2016.

Forest, C., Faucher, S.P., Poirier, K., Houle, S., Dozois, C.M., and Daigle, F. (2007). Contribution of the stg fimbrial operon of *Salmonella enterica* serovar Typhi during interaction with human cells. Infect. Immun. *75*, 5264–5271.

Fronzes, R., Remaut, H., and Waksman, G. (2008). Architectures and biogenesis of non-flagellar protein appendages in Gram-negative bacteria. EMBO J. *27*, 2271–2280.

Frye, S.A., Assalkhou, R., Collins, R.F., Ford, R.C., Petersson, C., Derrick, J.P., and Tonjum, T. (2006). Topology of the outer-membrane secretin PilQ from Neisseria meningitidis. Microbiology *152*, 3751–3764.

Gophna, U., Oelschlaeger, T.A., Hacker, J., and Ron, E.Z. (2002). Role of fibronectin in curli-mediated internalization. FEMS Microbiol. Lett *212*, 55–58.

Grimont, P.A.D., and Weill, F.-X. (2007). Antigenic formulae of the *Salmonella* serovars, 9th edn (Paris, WHO Collaborating Centre for Reference and Research on *Salmonella*).

Grund, S., and Weber, A. (1988). A new type of fimbriae on *Salmonella* Typhimurium. Zentralbl Veterinarmed B *35*, 779–782.

Haslam, D.B., Boren, T., Falk, P., Ilver, D., Chou, A., Xu, Z., and Normark, S. (1994). The amino-terminal domain of the P-pilus adhesin determines receptor specificity. Mol. Microbiol. *14*, 399–409.

Holt, K.E., Thomson, N.R., Wain, J., Langridge, G.C., Hasan, R., Bhutta, Z.A., Quail, M.A., Norbertczak, H., Walker, D., Simmonds, M., *et al.* (2009). Pseudogene accumulation in the evolutionary histories of *Salmonella enterica* serovars Paratyphi A and Typhi. BMC Genomics *10*, 36.

Huang, Y., Smith, B.S., Chen, L.X., Baxter, R.H., and Deisenhofer, J. (2009). Insights into pilus assembly and secretion from the structure and functional characterization of usher PapC. Proc. Natl. Acad. Sci. U.S.A. *106*, 7403–7407.

Humphries, A., Deridder, S., and Baumler, A.J. (2005). *Salmonella enterica* serotype Typhimurium fimbrial proteins serve as antigens during infection of mice. Infect. Immun. *73*, 5329–5338.

Hung, C.S., Bouckaert, J., Hung, D., Pinkner, J., Widberg, C., DeFusco, A., Auguste, C.G., Strouse, R., Langermann, S., Waksman, G., *et al.* (2002). Structural basis of tropism of Escherichia coli to the bladder during urinary tract infection. Mol. Microbiol. *44*, 903–915.

Jakovljevic, V., Leonardy, S., Hoppert, M., and Sogaard-Andersen, L. (2008). PilB and PilT are ATPases acting antagonistically in type IV pilus function in *Myxococcus xanthus*. J. Bacteriol. *190*, 2411–2421.

Kidgell, C., Reichard, U., Wain, J., Linz, B., Torpdahl, M., Dougan, G., and Achtman, M. (2002). *Salmonella* Typhi, the causative agent of typhoid fever, is approximately 50,000 years old. Infect. Genet. Evol. *2*, 39–45.

Kingsley, R.A., and Baumler, A.J. (2000). Host adaptation and the emergence of infectious disease: the *Salmonella* paradigm. Mol. Microbiol. *36*, 1006–1014.

Klemm, P., Jorgensen, B.J., Kreft, B., and Christiansen, G. (1995). The export systems of type 1 and F1C fimbriae are interchangeable but work in parental pairs. J. Bacteriol. *177*, 621–627.

Kline, K.A., Falker, S., Dahlberg, S., Normark, S., and Henriques-Normark, B. (2009). Bacterial adhesins in host–microbe interactions. Cell Host Microbe *5*, 580–592.

Ledeboer, N.A., Frye, J.G., McClelland, M., and Jones, B D. (2006). *Salmonella enterica* serovar Typhimurium requires the Lpf, Pef, and Tafi fimbriae for biofilm formation on HEp-2 tissue culture cells and chicken intestinal epithelium. Infect. Immun. *74*, 3156–3169.

Li, H., Qian, L., Chen, Z., Thibault, D., Liu, G., Liu, T., and Thanassi, D.G. (2004). The outer membrane usher forms a twin-pore secretion complex. J. Mol. Biol. *344*, 1397–1407.

Liu, W.Q., Feng, Y., Wang, Y., Zou, Q.H., Chen, F., Guo, J.T., Peng, Y.H., Jin, Y., Li, Y.G., Hu, S.N., *et al.* (2009). *Salmonella* Paratyphi C: genetic divergence from *Salmonella* Choleraesuis and pathogenic convergence with *Salmonella* Typhi. PLoS One *4*, e4510.

Marchler-Bauer, A., and Bryant, S.H. (2004). CD-Search: protein domain annotations on the fly. Nucleic Acids Res *32*, W327–331.

McClelland, M., Sanderson, K.E., Clifton, S.W., Latreille, P., Porwollik, S., Sabo, A., Meyer, R., Bieri, T., Ozersky, P., McLellan, M., *et al.* (2004). Comparison of genome degradation in Paratyphi A and Typhi, human-restricted serovars of *Salmonella enterica* that cause typhoid. Nat. Genet. *36*, 1268–1274.

McClelland, M., Sanderson, K.E., Spieth, J., Clifton, S.W., Latreille, P., Courtney, L., Porwollik, S., Ali, J., Dante, M., Du, F., *et al.* (2001). Complete genome sequence of *Salmonella enterica* serovar Typhimurium LT2. Nature *413*, 852–856.

Merz, A.J., So, M., and Sheetz, M.P. (2000). Pilus retraction powers bacterial twitching motility. Nature *407*, 98–102.

Nahm, M.H., Kroese, F.G., and Hoffmann, J.W. (1992). The evolution of immune memory and germinal centers. Immunol. Today *13*, 438–441.

Norris, T.L., and Baumler, A.J. (1999). Phase variation of the lpf operon is a mechanism to evade cross-immunity between *Salmonella* serotypes. Proc. Natl. Acad. Sci. U.S.A. *96*, 13393–13398.

Nuccio, S.P., and Baumler, A.J. (2007). Evolution of the chaperone/usher assembly pathway: fimbrial classification goes Greek. Microbiol. Mol. Biol. Rev. *71*, 551–575.

Nunn, D.N., and Lory, S. (1991). Product of the *Pseudomonas aeruginosa* gene pilD is a prepilin leader peptidase. Proc. Natl. Acad. Sci. U.S.A. *88*, 3281–3285.

Olsen, A., Jonsson, A., and Normark, S. (1989). Fibronectin binding mediated by a novel class of surface organelles on Escherichia coli. Nature *338*, 652–655.

Ottow, J.C. (1975). Ecology, physiology, and genetics of fimbriae and pili. Annu. Rev. Microbiol. *29*, 79–108.

Parkhill, J., Dougan, G., James, K.D., Thomson, N.R., Pickard, D., Wain, J., Churcher, C., Mungall, K.L., Bentley, S.D., Holden, M.T., *et al.* (2001). Complete genome sequence of a multiple drug resistant *Salmonella enterica* serovar Typhi CT18. Nature *413*, 848–852.

Pelicic, V. (2008). Type IV pili: e pluribus unum? Mol. Microbiol. *68*, 827–837.

Prigent-Combaret, C., Prensier, G., Le Thi, T.T., Vidal, O., Lejeune, P., and Dorel, C. (2000). Developmental pathway for biofilm formation in curli-producing Escherichia coli strains: role of flagella, curli and colanic acid. Environ. Microbiol. *2*, 450–464.

Robinson, L.S., Ashman, E.M., Hultgren, S.J., and Chapman, M.R. (2006). Secretion of curli fibre subunits is mediated by the outer membrane-localized CsgG protein. Mol. Microbiol. *59*, 870–881.

Romling, U., Bokranz, W., Rabsch, W., Zogaj, X., Nimtz, M., and Tschape, H. (2003). Occurrence and regulation of the multicellular morphotype in *Salmonella* serovars important in human disease. Int. J. Med. Microbiol. *293*, 273–285.

Rostagno, A., Frangione, B., Pearlstein, E., and Garcia-Pardo, A. (1986). Fibronectin binds to amyloid P component. Localization of the binding site to the 31,000 dalton C-terminal domain. Biochem. Biophys. Res. Commun. *140*, 12–20.

Ruer, S., Stender, S., Filloux, A., and de Bentzmann, S. (2007). Assembly of fimbrial structures in Pseudomonas aeruginosa: functionality and specificity of chaperone-usher machineries. J. Bacteriol. *189*, 3547–3555.

Sakai, D., Horiuchi, T., and Komano, T. (2001). ATPase activity and multimer formation of Pilq protein are required for thin pilus biogenesis in plasmid R64. J. Biol. Chem. *276*, 17968–17975.

Sauer, F.G., Fütterer, K., Pinkner, J.S., Dodson, K.W., Hultgren, S.J., and Waksman, G. (1999). Structural basis of chaperone function and pilus biogenesis. Science *285*, 1058–1061.

Saulino, E.T., Thanassi, D.G., Pinkner, J.S., and Hultgren, S.J. (1998). Ramifications of kinetic partitioning on usher-mediated pilus biogenesis. EMBO J. *17*, 2177–2185.

Selander, R.K., Beltran, P., Smith, N.H., Helmuth, R., Rubin, F.A., Kopecko, D.J., Ferris, K., Tall, B.D., Cravioto, A., and Musser, J.M. (1990). Evolutionary genetic relationships of clones of *Salmonella* serovars that cause human typhoid and other enteric fevers. Infect. Immun. *58*, 2262–2275.

Shewmaker, F., McGlinchey, R.P., Thurber, K.R., McPhie, P., Dyda, F., Tycko, R., and Wickner, R.B. (2009). The functional curli amyloid is not based on in-register parallel beta-sheet structure. J. Biol. Chem. *284*, 25065–25076.

Stecher, B., Barthel, M., Schlumberger, M.C., Haberli, L., Rabsch, W., Kremer, M., and Hardt, W.D. (2008). Motility allows S. Typhimurium to benefit from the mucosal defence. Cell Microbiol. *10*, 1166–1180.

Strom, M.S., and Lory, S. (1992). Kinetics and sequence specificity of processing of prepilin by PilD, the type IV leader peptidase of Pseudomonas aeruginosa. J. Bacteriol. *174*, 7345–7351.

Strom, M.S., Nunn, D., and Lory, S. (1991). Multiple roles of the pilus biogenesis protein pilD: involvement of pilD in excretion of enzymes from Pseudomonas aeruginosa. J. Bacteriol. *173*, 1175–1180.

Taniguchi, T., Akeda, Y., Haba, A., Yasuda, Y., Yamamoto, K., Honda, T., and Tochikubo, K. (2001). Gene cluster for assembly of pilus colonization factor antigen III of enterotoxigenic Escherichia coli. Infect. Immun. *69*, 5864–5873.

Thomson, N.R., Clayton, D.J., Windhorst, D., Vernikos, G., Davidson, S., Churcher, C., Quail, M.A., Stevens, M., Jones, M.A., Watson, M., *et al.* (2008). Comparative genome analysis of *Salmonella* Enteritidis PT4 and *Salmonella Gallinarum* 287/91 provides insights into evolutionary and host adaptation pathways. Genome Res. *18*, 1624–1637.

Townsend, S.M., Kramer, N.E., Edwards, R., Baker, S., Hamlin, N., Simmonds, M., Stevens, K., Maloy, S., Parkhill, J., Dougan, G., *et al.* (2001). *Salmonella enterica* serovar Typhi possesses a unique repertoire of fimbrial gene sequences. Infect. Immun. *69*, 2894–2901.

Tsui, I.S., Yip, C.M., Hackett, J., and Morris, C. (2003). The type IVB pili of *Salmonella enterica* serovar Typhi bind to the cystic fibrosis transmembrane conductance regulator. Infect. Immun. *71*, 6049–6050.

Tukel, C., Raffatellu, M., Humphries, A.D., Wilson, R.P., Andrews-Polymenis, H.L., Gull, T., Figueiredo, J.F., Wong, M.H., Michelsen, K.S., Akcelik, M., *et al.* (2005). CsgA is a pathogen-associated molecular pattern of *Salmonella enterica* serotype Typhimurium that is recognized by Toll-like receptor 2. Mol. Microbiol. *58*, 289–304.

Tukel, C., Wilson, R.P., Nishimori, J.H., Pezeshki, M., Chromy, B.A., and Baumler, A.J. (2009). Responses to amyloids of microbial and host origin are mediated through toll-like receptor 2. Cell Host Microbe *6*, 45–53.

Turner, L.R., Lara, J.C., Nunn, D.N., and Lory, S. (1993). Mutations in the consensus ATP-binding sites of XcpR and PilB eliminate extracellular protein secretion and pilus biogenesis in Pseudomonas aeruginosa. J. Bacteriol. *175*, 4962–4969.

Vidal, O., Longin, R., Prigent-Combaret, C., Dorel, C., Hooreman, M., and Lejeune, P. (1998). Isolation of an Escherichia coli K-12 mutant strain able to form biofilms on inert surfaces: involvement of a new ompR allele that increases curli expression. J. Bacteriol. *180*, 2442–2449.

Waksman, G., and Hultgren, S.J. (2009). Structural biology of the chaperone-usher pathway of pilus biogenesis. Nat. Rev. Microbiol. *7*, 765–774.

Weening, E.H., Barker, J.D., Laarakker, M.C., Humphries, A.D., Tsolis, R.M., and Baumler, A.J. (2005). The *Salmonella enterica* serotype Typhimurium lpf, bcf, stb, stc, std, and sth fimbrial operons are required for intestinal persistence in mice. Infect. Immun. *73*, 3358–3366.

Wilson, R.L., Elthon, J., Clegg, S., and Jones, B.D. (2000). *Salmonella enterica* serovars *gallinarum* and pullorum expressing *Salmonella enterica* serovar Typhimurium type 1 fimbriae exhibit increased invasiveness for mammalian cells. Infect. Immun. *68*, 4782–4785.

Wolfgang, M., Lauer, P., Park, H.S., Brossay, L., Hebert, J., and Koomey, M. (1998). PilT mutations lead to simultaneous defects in competence for natural transformation and twitching motility in piliated *Neisseria gonorrhoeae*. Mol. Microbiol. *29*, 321–330.

Wolfgang, M., van Putten, J.P., Hayes, S.F., Dorward, D., and Koomey, M. (2000). Components and dynamics of fiber formation define a ubiquitous biogenesis pathway for bacterial pili. EMBO J. *19*, 6408–6418.

Wu, S.S., Wu, J., and Kaiser, D. (1997). The Myxococcus xanthus pilT locus is required for social gliding motility although pili are still produced. Mol. Microbiol. *23*, 109–121.

Yoshida, T., Kim, S.R., and Komano, T. (1999). Twelve pil genes are required for biogenesis of the R64 thin pilus. J. Bacteriol. *181*, 2038–2043.

Zhang, X.L., Tsui, I.S., Yip, C.M., Fung, A.W., Wong, D.K., Dai, X., Yang, Y., Hackett, J., and Morris, C. (2000). *Salmonella enterica* serovar Typhi uses type IVB pili to enter human intestinal epithelial cells. Infect. Immun. *68*, 3067–3073.

Zogaj, X., Bokranz, W., Nimtz, M., and Romling, U. (2003). Production of cellulose and curli fimbriae by members of the family Enterobacteriaceae isolated from the human gastrointestinal tract. Infect. Immun. *71*, 4151–4158.

New Insights into the Role and Formation of Flagella in *Salmonella*

9

Rasika M. Harshey*

Abstract

The flagellum of *Salmonella enterica* serovar Typhimurium is the best studied of all flagellar systems. The major function of the flagellum is to enable swimming and chemotaxis in liquid media, and swarming on surfaces. New structural information, along with biochemical, physicochemical and genetic analyses has greatly accelerated our understanding of the self-assembly of this highly sophisticated nano-machine. The study of swarming motility is a relatively new field, but has begun to reveal new roles for the flagellum, new functions for motility genes and new regulatory circuits that control the decision between motility and sessility. Morphological and functional similarities between flagella and needle complexes, discovery of partial flagellar structures that likely function in export rather than motility, and a rapidly accumulating genome database are gradually illuminating the evolutionary origins of the flagellum.

Introduction

Moving bacteria are fascinating to watch. Their seemingly effortless motion, which allows them to swim up chemical gradients or swarm over surfaces, belies the complexity of the flagellar organelles and mechanisms that propel them. The structure and assembly of flagella have been best studied in *Salmonella*, while motor function and chemotaxis have been most deeply studied in *E. coli*. However, the information is largely interchangeable between the two organisms, and will be treated as such in this review. Excellent recent reviews on various aspects of flagella and motility are available (Apel and Surette, 2008; Berg, 2003; Chevance and Hughes, 2008; Chilcott and Hughes, 2000; Ferris and Minamino, 2006; Harshey, 2003; Hazelbauer *et al.*, 2008; Kojima and Blair, 2004; Macnab, 2003; Minamino *et al.*, 2008; Pallen and Matzke, 2006). Here I will review the current state of the field after summarizing the background information.

The flagellum

Salmonella has 6–8 flagella per cell, distributed around the cell surface (peritrichous arrangement) (Fig. 9.1A). Each flagellum is a complex structure made of ~ 25 different proteins, each protein present in multiple copies from a few to several thousand. The flagellum can be divided into three major parts: the basal body (embedded in the cytoplasmic membrane), the external hook, and the filament (Fig. 9.1B). The basal body has two parts – a moving rotor and a stationary stator, which together constitute the motor. The moving parts include structures in the cytoplasm (C-ring), which connect to the flagellar parts outside the cell body through a membrane ring (MS) and a rod that spans the periplasmic space, penetrating the peptidoglycan layer and extending to the outer membrane. The non-moving parts include the force generating units or stators embedded in the cytoplasmic membrane (Mot proteins). Two rings (L and P) that assemble around the distal rod are also thought to be non-moving; they form a pore through which the rod exits the outer membrane. The hook is a curved structure that serves as a flexible coupler between the basal body and the filament, enabling the synchronous

*Corresponding author

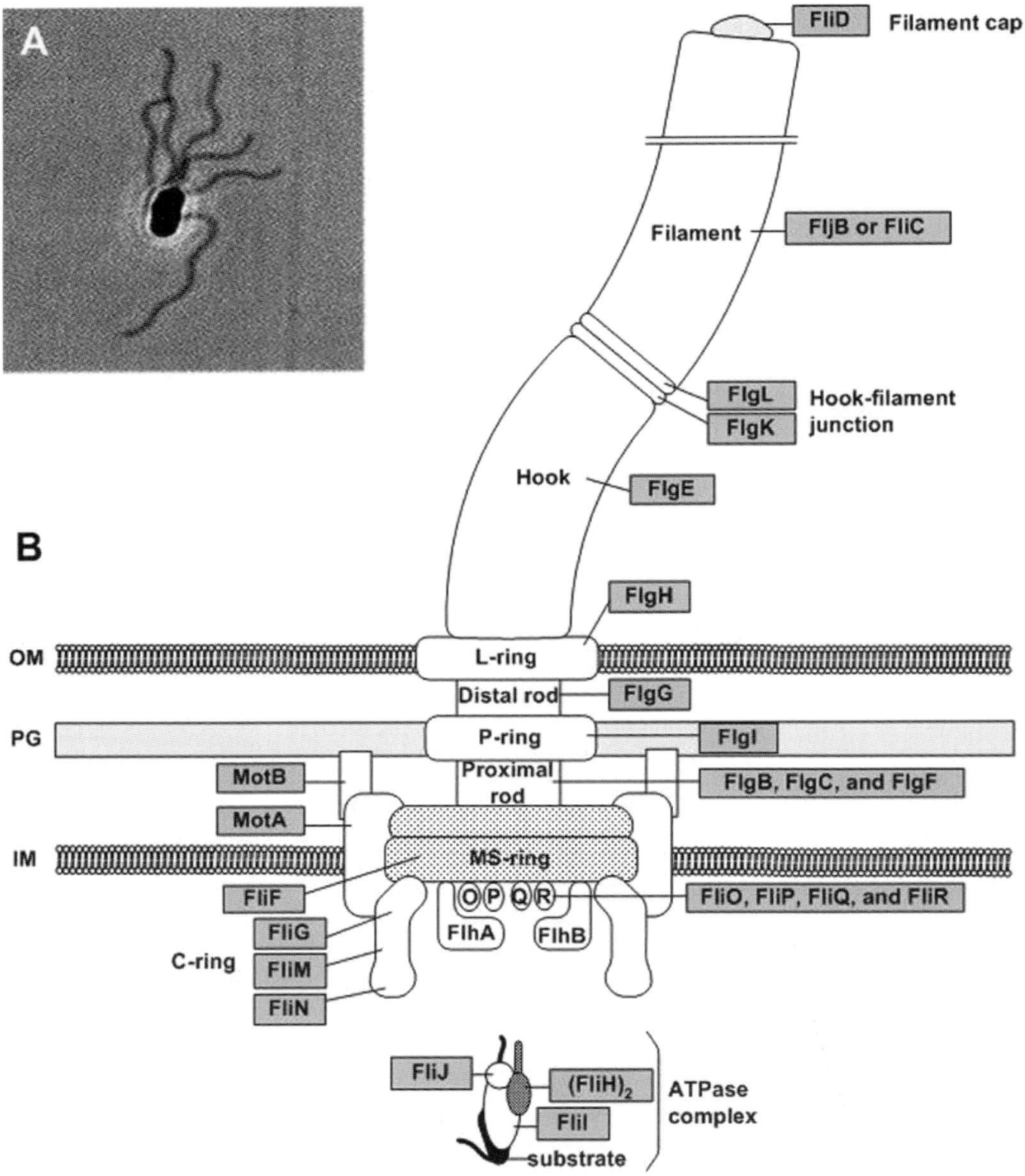

Figure 9.1 The flagellum of *Salmonella enterica*. (A) Peritrichous flagella on a single bacterium, visualized by silver staining. (B) Schematic of a flagellum, showing the membrane-embedded basal body and external parts. The inner diameter of the cytoplasmic C-ring is about 45 nm, and that of the MS ring is about 25 nm. The outer diameter of the filament is 20 nm, while the diameter of the inner channel is only 25 Å. The filament protein consists of either FliC or FljB, which are alternately transcribed. The rod cap (FlgJ) and the hook cap (FlgD) are not shown, because they are transiently associated with the flagellum during rod and hook polymerization, respectively. FliK (also not shown) is secreted during rod–hook polymerization as a molecular ruler that couples rod–hook length to the flagellar secretion specificity switch at FlhB (see Fig. 9.3B). The cytoplasmic ATPase complex delivers substrates to the Type III secretion system comprised of membrane proteins FlhA and B, and FliO, P, Q and R. OM, outer membrane; PG, peptidoglycan layer; IM, inner or cytoplasmic membrane.

rotation of the multiple flagella. While the hook is flexible, the filament is rigid in order to execute its propeller function. A smooth transition between the two mechanically distinct structures is enabled by two hook-associated proteins or HAPs, which form a short hook–filament junction that probably acts as a structural adaptor. The filament is several cell lengths long, composed of thousands of subunits of a single protein (either FliC or FljB) that is capped at its tip by a scaffold protein. The helical shape of the filament allows it to function like an Archimedes screw when rotated. The filament is firmly anchored to the basal body, resisting a pulling force of ~60 pN, which is estimated to be at least a 100 times more than the tensile stress experienced by the filament during swimming (Darnton and Berg, 2008).

How the motor works

The stator complex

The motor uses a transmembrane proton current (PMF or proton motive force) to drive rotary motion, at speeds reaching up to 18,000 rpm. The current is conducted through 10–12 stators assembled around the perimeter of the rotor. Each proton-conducting stator is composed of four MotA and two MotB protein molecules (Kojima and Blair, 2004; Reid *et al.*, 2006) (Fig. 9.2A). The stators are anchored to the peptidoglycan layer through MotB, and interact with the rotor through the cytoplasmic domain of MotA. The crystal structure of the C-terminal periplasmic fragment of MotB has been solved (Kojima *et al.*, 2009). MotA has four transmembrane (TM) segments and MotB has one. Of the 18 TM segments in a $MotA_4MotB_2$ complex, the proton channel is thought to consist of two of the four MotA TMs and the single MotB TM. This arrangement may allow two ion channels to form per stator complex. Mutational results imply that the proton pathway is contained within the stator and includes residue Asp-32 of MotB, close to the cytoplasmic side of the membrane. A structural model based on the cross-linking results shows clear channels to hold water, reaching from the periplasm to the Asp32 residues. Upon protonation, Asp32 might hydrogen bond to a carbonyl of a conserved residue in MotA, inducing a bend in this part of the TM in MotA. The mechanism of rotation is not understood in detail but is thought to involve conformational changes in the stator complexes driven by proton association/dissociation at the Asp-32 residue, which generate torque through interaction with FliG and drive rotation (Kim *et al.*, 2008). The rotor is only responsible for setting the direction of rotation. The switch in direction between clockwise (CW) and counterclockwise (CCW) rotation states of the flagellum is thought to involve changes in the topography and/or chemical properties at the rotor–stator interface, which are influenced by conformational changes taking place at the bottom of the C-ring.

The C-ring and switching

The rotor can switch between CW and CCW directions, as defined by an observer looking down the filament toward the cell body. Both rotation and switching are controlled by the FliG protein in the C ring (Fig. 9.2A,B). The other two proteins in the C ring – FliM and FliN – influence switching by binding the phosphorylated form of CheY (CheY~P), a response regulator of the chemotaxis system. CW and CCW states are essentially two thermal states of the switch separated by a barrier, the default state being CCW (Berg, 2003). The switch is highly cooperative (Bai *et al.*, 2010), responding to increasing CheY~P with a Hill coefficient greater than 10 (Cluzel *et al.*, 2000). CheY~P binding stabilizes the CW conformation, changing the orientation of FliG relative to MotA in such a way that the force exerted by MotA on FliG changes sign.

FliG, while interacting with cytoplasmic loops of MotA, is also directly mounted on the MS-ring, which is made of FliF. The stoichiometries of FliF, FliG, FliM, and FliN, are 25, 25, 34 and 110, respectively. FliN is a tetramer (Paul and Blair, 2006). Crystal structures have been solved for most of the FliG, FliN and FliM proteins (Blair, 2006). New electron microscopic images of the C-ring are at a resolution almost sufficient to allow crystal structures to be docked into the map (Thomas *et al.*, 2006). The new images show that, at least for a CW-locked motor, the symmetry of the MS ring is between 24- and 26-fold. While the most proximal part of the C ring has the same symmetry as the MS ring, that of the middle,

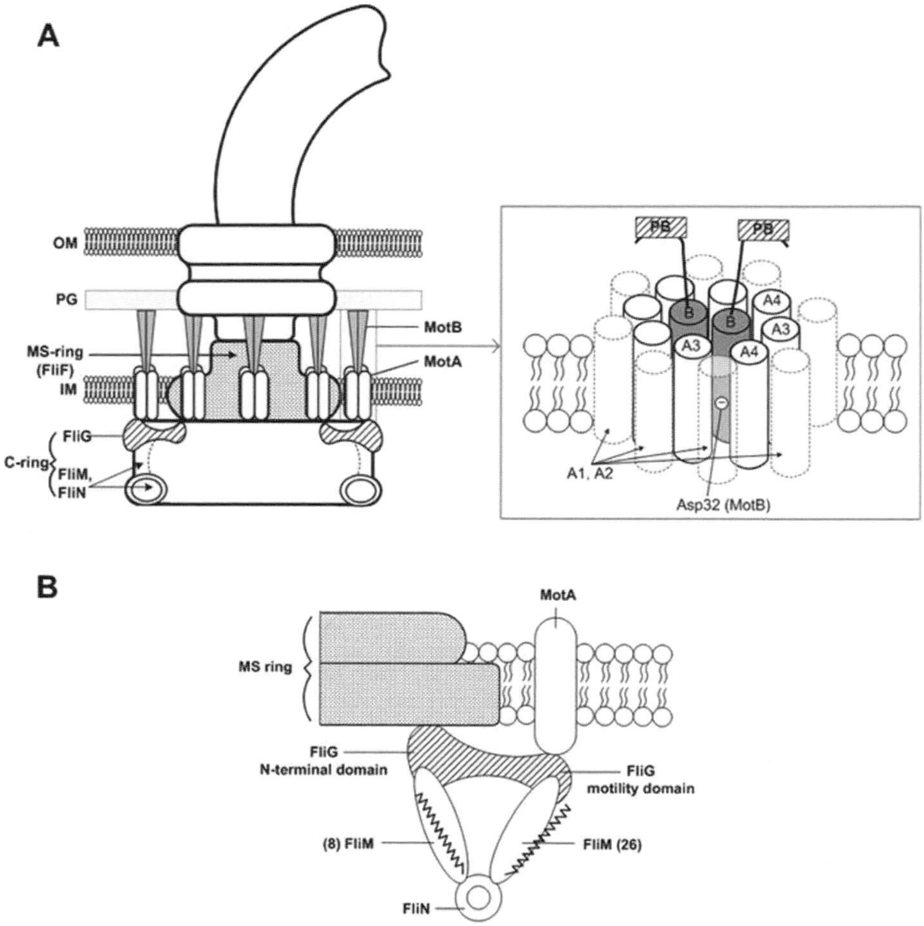

Figure 9.2 Proteins that function in rotation and switching. (A) MotA and MotB form 10–12 stator complexes per motor, each with a composition of $MotA_4MotB_2$. These are anchored to the cell wall by a peptidoglycan-binding motif (PB) in the periplasmic domain of MotB. MotA interacts with FliG, which is a constituent of the C-ring or 'switch complex' comprising FliG, FliM, and FliN. FliG is mounted on the MS-ring directly. Arrangement of membrane segments in the $MotA_4MotB_2$ complex shown on the right was deduced from cross-linking experiments. The TM segments contributing to two ion channels per stator complex are shown in solid lines. Adapted from Kim *et al.*, 2008. (B) FliG is cross-hatched and dog-bone shaped, with the N-terminal domain facing the C-ring and C-terminal domain contacting MotA. Two associations of FliM monomers with FliG are hypothesized (see text). The N-terminal sequence of FliM that binds CheY~P is shown as a squiggly line. The FliN tetramer is a doughnut. Adapted from Manson (2007).

where FliM is located, is between 32- and 36-fold. A resolution to this paradox of mismatched symmetry between FliG and FliM is the proposal of a 26 + 8 arrangement of FliM subunits (Brown *et al.*, 2007; Manson, 2007). For example, the MS ring of a particular basal body has 26-fold symmetry, meaning that 26 FliF subunits attach to 26 FliG subunits, then 26 FliM subunits can contact a hydrophobic region of FliG near the motility or MotA domain, while the remaining 6–10 FliM subunits could face inward to interact with the N-terminal region of FliG that attaches to FliF (Fig. 9.2B). Consistent with such an arrangement, Thomas *et al.* (Thomas *et al.*, 2006) have also found that, at the level of FliM, there was a lower electron density inside the high-density outer wall of the C ring. The role of the inward-facing FliM subunits might be to stabilize the connection between the MS and C rings.

Flagella assembly

The secretion apparatus

The flagellum is built from the inside out, and its components are translocated through a ~ 25 Å wide central channel to the distal, growing end of the flagellum. The flagellar filament grows to a length of up to 15 μm by polymerization of as many as 30,000 flagellin subunits. A protein export apparatus located on the cytoplasmic side of the basal body uses a secretion mechanism called type III secretion, which is part of a family of type III systems (T3SSs) whose key features are: (i), N-terminal peptide secretion signals that are often structurally disordered (Minamino *et al.*, 2008); (ii), substrate-associated secretion chaperones that stabilize or target their cognate secretion substrates; and (iii), ATP hydrolysis that is coupled to the delivery of the substrate for PMF-dependent secretion from the cell (Chevance and Hughes, 2008) (Fig. 9.3A). Virulence factors in Gram-negative plant and animal pathogens are also secreted by related T3SSs (Galan and Wolf-Watz, 2006). The flagellar export apparatus consists of seven integral membrane proteins (FlhA, FlhB, FliF, FliO, FliP, FliQ, FliR), three soluble components that form the ATPase complex (FliI, FliH, FliJ), and four cytoplasmic proteins (FlgN, FliA, FliS, FliT; not shown in Fig. 9.3A) that act as substrate-specific chaperones.

FliI, which is an ATPase, exists in the cytoplasm as a stable, heterotrimeric complex with FliH. FliH is a regulator, which controls the ATPase activity of FliI. FliJ is postulated to be a general chaperone, which prevents its substrates from premature aggregation in the cytoplasm. The central, catalytic domain of FliI shows extensive similarity to the catalytic α and β subunits of F_0F_1-ATP synthase (Imada *et al.*, 2007). FliI self-assembles into a homo-hexameric ring and FliH seems to play an important role in effective docking of FliI to FlhA and FlhB (Minamino *et al.*, 2008). FliJ also interacts with the cytoplasmic domains of FlhA and FlhB. When the catalytic site of FliI is mutated to have no ATPase activity, the binding of the $(FliH)_2$FliI complex to the export gate completely inhibits the export process, suggesting that the energy of ATP hydrolysis is used to facilitate the release of the complex from the export gate. FliI itself plays a critical role in substrate recognition based on its observed interaction with both FlgN, a flagellar cytoplasmic chaperone specific for the hook–filament junction proteins FlgK and FlgL (see Fig. 9.4), as well as the FlgN–FlgK complex.

New insights into the energy source for secretion

All T3S machines share homologues of the integral membrane components FliF, FlhA, FlhB, FliP, FliQ and FliR. They also share homologues of the cytoplasmic component FliN and the ATPase complex components FliH, FliI and FliJ. Secretion rates through these machines are high. Flagellin subunits can be secreted at the rate of ~10,000 amino acid residues per second. A similarly fast secretion rate is found during transfer of the virulence factor SipA into a host cell through the non-flagellar T3SS of the injectisome or needle complex of *Salmonella* Typhimurium (Schlumberger *et al.*, 2005). These high rates of secretion cannot be accounted for by activities measured for T3S-associated ATPases. New results suggest that secretion is driven solely by PMF in both systems. For example, in both *Yersinia* T3S and *Salmonella* flagellar export, treatment with a protonophore abolished secretion despite the presence of the ATPase complex (Minamino and Namba, 2008; Paul *et al.*, 2008; Wilharm *et al.*, 2004). It is thought that PMF may somehow bias one-dimensional thermal Brownian motion of unfolded polypeptide chains through the gate, acting as a thermal ratchet in this rapid translocation process.

Since proton influx is used to drive secretion, the primary role of the FliHIJ ATPase complex is likely to shuttle substrates to the integral-membrane secretion system and present them as unfolded substrates for efficient PMF-dependent secretion (Chevance and Hughes, 2008). Thus, the PMF is required for both flagellar secretion and rotation, whereas ATP hydrolysis might be

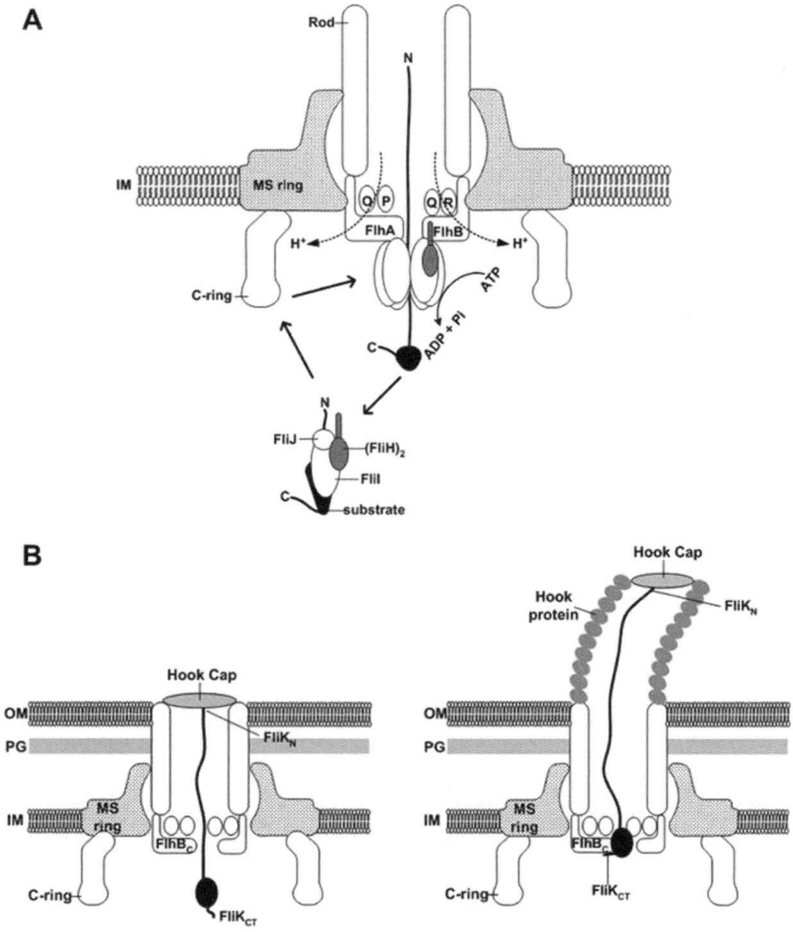

Figure 9.3 Flagellar Type III secretion and hook length control. (A) FlhA, FlhB, FliO, FliP, FliQ and FliR are integral membrane proteins postulated to be located within a central pore of the MS ring (see Fig. 9.1). FlhA and FlhB have large cytoplasmic domains, and are depicted as the mounting platform for the ATPase FliI and its regulator FliH. A general chaperone, FliJ, which prevents its substrates from premature aggregation in the cytoplasm, interacts with the FliH/I complex in the cytoplasm. FliI forms a heterotrimer with the FliH dimer in the cytoplasm, and after interaction of this complex with C-ring proteins (indicated by arrows), hexamerizes and docks to the FlhA–FlhB platform. ATP hydrolysis is coupled to disassembly of the FliH–FliI complex from both the export gate and the substrate protein. Successive unfolding and translocation of the polypeptide is driven by one-dimensional Brownian diffusion strongly biased by PMF (dotted arrows). (B) FliK is periodically exported during hook assembly. Interaction of $FliK_N$ with the assembled hook and hook cap is proposed to elicit a pause in FliK secretion. When the hook length reaches ~55 nm (or rod and hook together reach ~90 nm), $FliK_{CT}$ is near the C-terminus of FlhB protein when the pause occurs, causing auto-cleavage of FlhB and flipping the switch to a late secretion mode.

limited to substrate delivery, a division of labour that would increase the efficiency of assembly. Recent investigations have suggested roles for the ATPase and secretion chaperones in sorting secretion substrates, which would also increase efficiency in the assembly process (Evans *et al.*, 2006; Stafford *et al.*, 2007).

Role of integral membrane export components

FlhA and FlhB have substantial cytoplasmic domains, which project into the space within the C ring, and they play an important role in docking the FliI ATPase as described above. There is evidence that the TM region of FlhA interacts

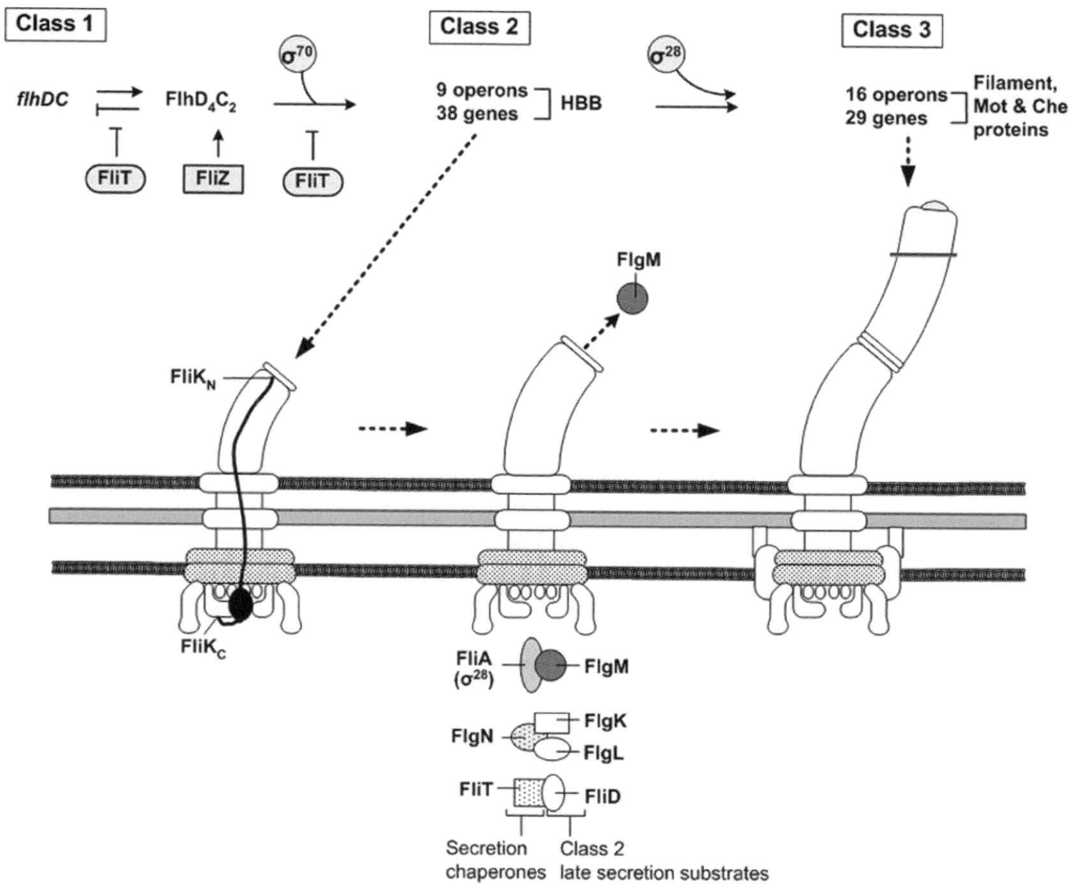

Figure 9.4 Regulation of flagellar assembly. Transcription is organized in three temporal classes. The class 1 flagellar master operon *flhDC* is under the control of many positive and negative global regulatory signals. FlhD and FlhC form a heteromultimeric complex $FlhD_4C_2$ that promotes σ^{70}-dependent transcription from class 2 flagellar promoters. $FlhD_4C_2$ auto-represses *flhDC* transcription. Class 2 promoters direct the transcription of genes required for hook–basal body (HBB) assembly. In response to completion of the HBB, the anti-σ^{28} factor FlgM is exported outside, allowing σ^{28} to transcribe the late class 3 promoters, which produce the filament, Mot proteins, and chemotaxis proteins. Several chaperones direct the export of late secretion substrates. These chaperones play a dual role in that depletion of exported substrates from the cytoplasm sets them free to regulate gene expression. FliZ positively regulates $FlhD_4C_2$ protein levels in a post-translational manner. Adapted from Chevance and Hughes (2008).

with the MS ring protein FliF. FlhB also acts as an export switch to control late flagellar protein export (see below). A FliR–FlhB fusion is still functional, indicating that FliR and FlhB are physically associated with each other in the membrane. The roles of the other membrane proteins FliOPQR are not yet understood.

Role of the C ring in export

The C ring proteins, which are responsible for torque generation and switching direction, are also essential for flagellar assembly. In the absence of either FliG, FliM or FliN, export of rod/hook-type proteins into the periplasm does not occur. Interestingly, the FliJ component of the FliHIJ ATPase complex interacts with FliM of the C ring and FlhA of the integral membrane T3SS, whereas FliH interacts with FliN of the C ring (Gonzalez-Pedrajo *et al.*, 2006; McMurry *et al.*, 2006; Paul *et al.*, 2006). FliJ–FliM interaction may facilitate the recruitment of export substrates and specific chaperone–substrate complex such as FlgN–FlgK, FlgN–FlgL, and FliT–FliD (see Fig. 9.4) to the $FliH_2FliI$ complexes that are prebound to the surface of the C-ring. Thus, the C ring seems to provide the binding sites for the $FliH_2FliI$–chaperone–substrate complexes for their concentration near the export gate so that they can efficiently dock to the FlhA–FlhB platform (Erhardt and Hughes, 2010; Konishi *et al.*, 2009).

Mechanism of flagellar protein export

Flagellar proteins are exported through a channel that is only 25 Å wide, so they must be largely unfolded for entry into and translocation through the channel. Protein synthesis inhibitors do not abolish export, indicating that export is not obligatorily linked to translation, and can occur post-translationally (Hirano *et al.*, 2003). This suggests that the amino acid sequences of these proteins suffice to mediate their recognition and export. The export signal lies within the N-terminal region of the substrates, but is not obvious at the primary sequence level. Although the mechanism of flagellar protein export is not yet completely understood, gradually accumulating experimental data allow the following scheme to be drawn (Fig. 9.3A). First, a $FliH_2FliI$–FliJ–substrate complex assembles in the cytoplasm. This complex binds to FliN in the C ring, effectively increasing its local concentration for enabling efficient docking to the FlhA–FlhB platform. FliJ facilitates docking by interacting with FliM and releasing the $FliH_2FliI$–substrate complex from the C ring. Specific interactions between the $FliH_xFliI_6$ complex and the FlhA–FlhB platform facilitate initial entry of the N-terminal segment of the substrate into the export gate. The energy of ATP hydrolysis seems to be used to promote the release of the $FliH_xFliI_6$ complex from the export gate for the next cycle of docking and entry. PMF drives translocation of the unfolded substrate through the narrow channel of the export gate by one-dimensional Brownian motion.

Rod and hook length control

Nine proteins are required for the structure and assembly of the rod and hook: proximal rod (FliE, FlgB, FlgC and FlgF), distal rod (FlgG), rod cap (FlgJ), hook (FlgE), hook cap (FlgD) and hook-length control (FliK). The length of the rod is 25 nm and the hook 55 nm. Two key players – FliK and FlhB – determine the rod-hook length and the switch to late secretion. A second regulatory protein, RflH/Flk also acts to inhibit the export of filament-type proteins before rod-hook completion (Aldridge *et al.*, 2006a). FliK was first implicated in length control when null mutations in *fliK* were found to produce uncontrolled hook elongation or polyhooks. These mutants failed to make the switch to late or filament secretion, and were non-motile. Motile revertants of the *fliK* mutants mapped to the *flhB* locus, but had lost normal hook-length control. These results implied that an interaction between FliK and FlhB was required for switching secretion specificity from rod-hook to filament, with FlhB controlling the secretion specificity and FliK controlling the length specificity.

FliK contains an unstable N-terminal region ($FliK_N$), a stably folded region near the C-terminus and a partially unfolded region at the C-terminus ($FliK_{CT}$) (Minamino *et al.*, 2004). FliK is a monomer in solution, and has an elongated structure (Minamino *et al.*, 2004). It is known to interact strongly with the hook-capping protein FlgD and to a lesser extent with the hook protein FlgE (Moriya *et al.*, 2006). Suppressor analysis suggests that the C-terminus of FliK interacts with FlhB to flip the secretion specificity switch. FlhB consists of a hydrophobic N-terminal domain and a large cytoplasmic C-terminal domain ($FlhB_C$). $FlhB_C$ undergoes autocleavage between the asparagine and proline residues of an asparagine–proline–threonine–histidine (NPTH) motif that is conserved among FlhB proteins and T3SS FlhB homologues of plant and animal pathogens. Amino acid substitutions within the NPTH motif prevent autocleavage and lock the export apparatus in the rod/hook-type specificity state, indicating that autocleavage is essential for the switch in substrate specificity (Fraser *et al.*, 2003). Recent reports on the crystal structures of the cytoplasmic domains of four FlhB T3SS homologues suggest that autocleavage does not affect protein folding but rather results in localized electrostatic and conformational changes that may influence interactions with other export apparatus components (Deane *et al.*, 2008).

In a ruler model for rod-hook length control, FliK has been proposed to be the ruler (Ferris and Minamino, 2006) (Fig. 9.3B). The model proposes that FliK is secreted during hook elongation, allowing a periodic or temporal assessment of hook length. Interaction of the FliK N-terminus with the assembled hook and hook cap is proposed to elicit a pause in FliK secretion. If the C-terminus of FliK is near the FlhB protein when the pause occurs, their interaction causes auto-cleavage of FlhB, flipping the switch to a late secretion mode. Presumably, this happens only when the hook has reached its complete length of 55 nm. The ruler model was originally proposed for the *Yersinia enterocolitica* T3 needle-length control system (Journet *et al.*, 2003). *Y. enterocolitica* has a FliK homologue called YscP. Insertions and deletions in YscP resulted in longer and shorter needle structures, respectively. A domain within the C-terminus of YscP identified as being responsible for the secretion specificity switch is also present in FliK (Agrain *et al.*, 2005). When a similar insertion/deletion study was performed on FliK of *Salmonella* Typhimurium, results similar to *Y. enterocolitica* were observed (Shibata *et al.*, 2007). There is now good evidence for FliK serving as a ruler in *Salmonella* Typhimurium (Erhardt *et al.*, 2010).

How is rod length controlled? The subunits on the flagellar filament are arranged such that they appear as 11 rows of protofilaments along the surface of a cylindrical lattice. The same 11-protofilament structure applies to the hook, which is continuous with the rod. Each layer of these two structures grows by the addition of 5.5 residues per turn, with 2 turns per helical layer. The proximal rod proteins FlgB, FlgC and FlgF polymerize to six subunits each (half a helical layer) followed by an estimated 26 subunits (two helical layers) of the distal rod protein FlgG. Like the hook protein FlgE and the filament protein FliC, which have the intrinsic ability to polymerize indefinitely in the absence of control mechanisms or when subunit availability is not limiting, FlgG has recently been shown to be capable of continuous polymerization (Chevance *et al.*, 2007). However, rod length control is part of the hook length control mechanism. Mutants in *flgG* resulted in rod structures that grew to about 80 nm, which is the length of a normal rod + hook. Interestingly, the FliK protein now regulated rod length in the 80 nm-rod strains. Loss of FliK in these strains resulted in polyrod structures of up to 1 μm in length. The *flgG* mutants that gave rise to 80 nm rods had two other interesting features: the rods were decorated with multiple P rings, and these structures failed to penetrate the outer membrane, so that their flagellar filaments now grew in the periplasm. Thus, outer-membrane penetration is apparently coupled to proper rod-growth termination. It is hypothesized that the normal rod length of 25 nm positions the tip of the rod at the outer membrane. A two-helical layer stack of FlgG subunits is the substrate for P ring formation, which is probably the mechanism for distal rod length control. Since P ring subunits are secreted through the general or Sec pathway, rod completion and P ring formation are not coupled. The P ring is the substrate for formation of the L ring, which results in an outer-membrane pore at the tip of the rod. This is thought to be the signal for the initiation of hook polymerization outside of the cell.

Late secretion switch: filament length control

The most important morphological checkpoint during the process of flagellar assembly is completion of the hook (Fig. 9.4). Until this point, filament formation and flagellar function are inhibited by the flagellum-specific negative transcriptional regulator FlgM (see below) (Chilcott and Hughes, 2000). Upon completion of the 55 nm long hook structure, FliK-dependent proteolytic cleavage of FlhB signals the switch to late or filament-type proteins, allowing FlgM to be secreted out of the cells through the hook–basal body structure, causing relief of this inhibition. Unlike the rod-type proteins, filament-type proteins (FlgK, FlgL, FliD, FliC, FlgM) require chaperones for efficient secretion. The switch to late secretion as well as the specific chaperones for these late proteins are thought to reduce substrate competition, allowing for the efficient assembly of the extracellular components of the flagellum (Aldridge and Hughes, 2001).

The capping protein FliD is essential for filament polymerization, as it selectively permits certain proteins such as FlgM to pass, while enabling the filament subunits to polymerize (see Fig. 9.1). The FliD cap is assembled from five molecules to form a flat-top pentameric structure with flexible leg domains (Minamino *et al.*, 2008; Yonekura *et al.*, 2000). Similar capping structures composed of FlgD and FlgJ are associated with the tip of the growing hook and rod, respectively, at earlier stages of assembly. Once the hook has reached its mature length, the hook cap is discarded and successively replaced first by the HAP (hook-associated-protein) FlgK, next by the HAP protein FlgL and finally by the capping protein FliD. The HAP proteins stay in place, while the FliD moves continuously to accommodate the growing filament.

In vitro, filaments polymerize to lengths much longer than the 10 μm length observed *in vivo*, limited only by subunit availability. This suggests that the final filament length *in vivo* is probably controlled by hindered diffusion. In support of this idea, the rate of *in vivo* polymerization slows exponentially with increasing length, whereas that of *in vitro* polymerization does not (Chevance and Hughes, 2008). Mutation of FlgM, which increases expression of filament subunit genes and hence subunit availability, does not significantly increase filament length *in vivo* (Wang *et al.*, 2005).

Controlling filament numbers

In *Salmonella*, cells average between six and eight flagella. The flagellar regulon includes more than 60 genes, which are transcribed hierarchically and temporally in three classes – early, middle and late (Fig. 9.4). Each temporal class is associated with a distinct set of promoters: Pclass1, Pclass2 and Pclass3 (Chilcott and Hughes, 2000). The single Pclass1 promoter encodes two proteins, FlhC and FlhD, which form the master regulatory complex $FlhD_4C_2$ (Wang *et al.*, 2006b). In concert with σ^{70}, $FlhD_4C_2$ activates the Pclass2 promoters, while also autoregulating its own synthesis by inhibiting the Pclass1 promoter. Pclass2 promoters control the middle genes, which encode for the structural subunits of the basal body, hook, and the two regulatory proteins, FlgM and FliA (σ^{28}). FlgM functions as an anti-σ^{28} that binds to and sequesters σ^{28}, preventing late gene expression prior to completion of the basal body and hook. After the hook is completed, and the late secretion switch is triggered, σ^{28} functions as a chaperone for FlgM secretion (Aldridge *et al.*, 2006b). Free of FlgM, σ^{28} now activates Pclass3 promoters, leading to late flagellar gene expression. This checkpoint in the flagellar assembly pathway ensures that transcription of the gene for the flagellar filament, whose translation is energetically the most costly step in the pathway, ensues only when there is a functional motor and an appropriately constructed hook structure onto which it can be assembled.

Besides the broad transcription classes in the flagellar regulon, a much more detailed temporal program has been elucidated using GFP (green fluorescent protein) as a reporter (Kalir *et al.*, 2001). Within the class 2 cluster, the promoters were found to be turned on sequentially, with significant delays, in the order corresponding to the spatial position of the gene products during flagellar assembly. A separation of class 3 genes into two kinetic groups was seen, with the filament structural operons activated first, and the chemotaxis operons going on only after a substantial delay. Thus, the hardware for the flagellar propeller is expressed before the chemotaxis navigation system. The genes for motor torque generation were in the late class 3 group. The mechanism underlying the temporal order of promoter activation within classes 2 and 3 is thought to be due to a hierarchical ranking of the promoters in affinity. As the concentration of the relevant transcription factor (FlhDC, FliA) gradually increases in the cell, it first binds and activates the operons with the highest affinity sites, and only later does it bind and activate operons with lower affinity sites. This finding explains why, even in the absence of FlgM, this back-up plan of gene expression maintains the temporal order of flagellar assembly. Quantitative *in vivo* measurements of the three promoter activities (class1–3) have been used to derive a blueprint of the flagella gene network (Kalir and Alon, 2004).

In addition to the transcription factors, other proteins play an important role in fine-tuning flagellar gene regulation (Fig. 9.4). These include two chaperones, FlgN and FliT, and a novel transcription factor FliZ. FlgN assists the secretion of the HAP proteins FlgK and FlgL (Bennett *et al.*, 2001). It also enhances FlgM translation from class 3 transcripts. Both FlgK and FlgL bind to and inhibit the regulatory activity of FlgN, suggesting that, like the σ^{28} -FlgM checkpoint, FlgN is also involved in a checkpoint coupled to completion of the hook (Aldridge *et al.*, 2003). FliT negatively regulates $FlhD_4C_2$ activity by binding to it (Yamamoto and Kutsukake, 2006); FliT is also the chaperone for FliD. Upon completion of the hook, FliD is secreted to the tip of the hook. Depletion of FliD from the cytoplasm permits FliT to bind to $FlhD_4C_2$, preventing it from activating Pclass2 promoters, whose transcription is no longer required. Similar to the chaperone functions of σ^{28} and FlgN, the activity of FliT is also likely coupled to hook completion. In contrast to FliT, FliZ positively regulates Pclass2 activity by stabilizing the $FlhD_4C_2$ complex (Saini *et al.*, 2008).

The emerging picture is that all of these regulators make flagellar gene expression at all levels of the hierarchy responsive to the rate of secretion through growing and completed flagellar structures, which in turn regulates both the timing and magnitude of gene expression (Brown *et al.*, 2008). A recent model predicts that the σ^{28}-FlgM checkpoint does not simply involve a binary decision regarding the expression of Pclass3-dependent genes, but rather involves a continuous or rheostatic mechanism that fine-tunes expression based on the amount of protein being secreted – the more abundant the flagella, the more protein is secreted. This new model for flagellar assembly (Brown *et al.*, 2008) proposes that the role of σ^{28} and FlgM in coupling late gene expression to the completion of the first few basal body-hook structures is relevant only to the assembly of the first few flagella. When cells have pre-existing flagella, the σ^{28}-FlgM checkpoint is no longer critical since FlgM is being actively secreted from cells at all times, and expression no longer proceeds in a strictly hierarchical manner, as is the case during the Fla^- to Fla^+ transition. Instead, it is proposed that all three promoter classes are being continuously either up- or downregulated in response to late protein secretion, which serves as a proxy for the integrated assembly state. Regulation of flagella assembly is clearly much more dynamic and complex than originally thought.

Why multiple flagella?

Multiple flagella apparently do not help a cell swim faster, since a bundle of several flagella was observed to produce just a little more torque than that produced by a single flagellum (Darnton *et al.*, 2007). Why then do cells have multiple flagella? One speculation is that, since there is a lag of several generations between turning on flagellar synthesis and completing the first new flagellum, multiple flagella provide a reservoir of extra flagella, allowing cells to maintain motility while dividing quickly. Another possibility is that having multiple, distributed flagella allows cells to change directions more efficiently when they tumble, facilitating more effective searches of the environment. Multiple flagella appear to play a more critical role during swarming (see below), since over-production of flagella by induction of *flhDC* improves swarming dramatically while affecting swimming only marginally (our unpublished data). During swarming, neighbouring cells may share flagella to coordinate group motility and overcome surface friction.

Chemotaxis

This is a vast subject and will be described only briefly, mainly because the following segment on swarming is impacted by this pathway. For recent reviews on chemotaxis see Baker *et al.*, 2006; Hazelbauer *et al.*, 2008; Sourjik, 2004 and/or Vladimirov and Sourjik, 2009. The swimming movement of *Salmonella* and *E. coli* has been described as a 'random walk', in which periods of smooth swimming (or runs) are interrupted by short re-orientations (or tumbles), corresponding to CCW and CW rotation of the flagellar motors, respectively. The chemotaxis system enables the bacteria to bias their random walk towards higher concentrations of attractants and avoid higher concentrations of repellents by sensing temporal changes in chemoeffector concentrations. The direction of swimming is still chosen randomly, but movement in a favourable direction suppresses tumbles, resulting in longer runs. These bacteria can sense a variety of amino acids, sugars and dipeptides, as well as pH, temperature and redox state. Some receptors, such as those for serine (Tsr) and aspartate (Tar), are highly abundant and number several thousand molecules per cell. They are classified as 'major' receptors. Minor receptors are much less abundant, with only a few hundred copies per cell; they recognize ribose and galactose (Trg), dipeptides (Tap), and redox potential (Aer). Two new chemoreceptors have been recently discovered in *Salmonella* (Frye *et al.*, 2006; Wang *et al.*, 2006a). Each receptor is a homodimer, which assembles into a trimer-of-dimers and then into larger arrays clustered at the cell poles. The discovery of receptor clustering was unexpected (Maddock and Shapiro, 1993), and has galvanized the chemotaxis field. While the details are still emerging, several levels of organization within the receptors contribute to their high performance as a signalling network (Hazelbauer *et al.*, 2008). The receptors transmit signals from their periplasmic ligand-binding domain via a piston-like movement through the transmembrane domain to their long cytoplasmic signalling domain (Fig. 9.5). The histidine kinase CheA is located at the tip of the cytoplasmic domain, linked to the receptors through CheW. As in many other signalling systems, signalling in chemotaxis relies on protein phosphorylation. The activity of CheA is modulated by both binding of chemoeffectors to the receptors and by the level of receptor methylation. Attractants inhibit the kinase while repellents activate it. The phosphate signal is passed from CheA~P to two response regulators – to CheY for motor control, and to CheB for sensory adaptation. In motor control, CheY~P binds to the flagellar motor, enhancing the probability of CW rotation, which causes random changes in direction (see 'Switching' above). CheZ is a phosphatase that hydrolyzes CheY~P, ensuring termination of the transduced signal so that CheY~P levels are in accordance with receptor-modulated CheA activity.

Chemoreceptors are methyl-accepting chemotaxis proteins (MCPs), which undergo reversible methylation at four to six glutamyl residues in

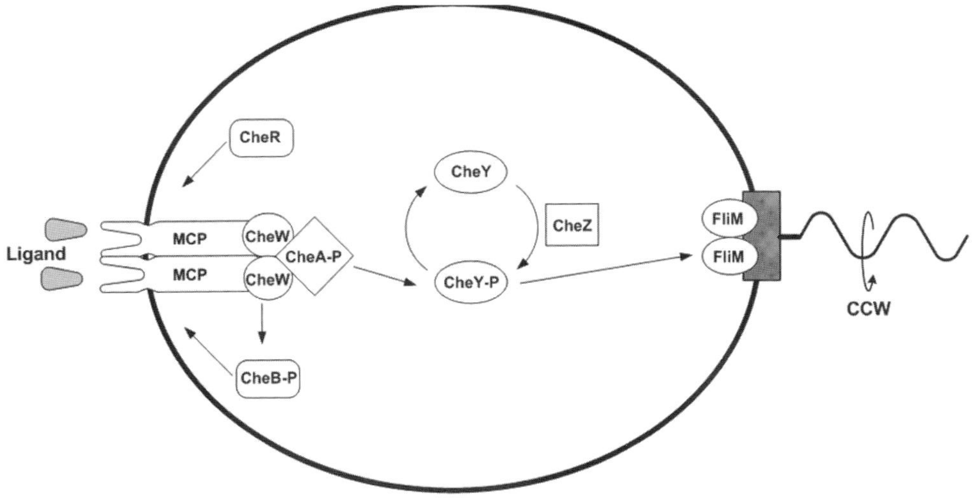

Figure 9.5 Chemotaxis pathway. Attractant and repellent ligands are detected by dimeric transmembrane receptors or MCPs, which are connected via an adaptor protein CheW to a histidine kinase CheA. A large array (not shown) of MCP–CheW-CheA complexes is located at the cell pole. Autophosphorylation activity of CheA is inhibited by attractant binding and enhanced by repellent binding to receptors. The phosphoryl group from CheA is rapidly transferred to the response regulator CheY. CheY~P diffuses to the flagellar motors and changes the direction of motor rotation from CCW to CW. CheZ phosphatase ensures a rapid turnover of CheY~P, to quickly re-adjust bacterial behaviour. Adaptation is mediated by methyltransferase CheR and methylesterase CheB, which add or remove methyl groups at four specific glutamyl residues on each receptor monomer. Receptor modification increases CheA activity and decreases sensitivity to attractants by counteracting the conformational and functional effects of ligand occupancy, re-establishing the signalling state of the chemoreceptor before ligand occupancy.

their cytoplasmic domain. These reactions are carried out by the adaptation enzymes CheR (methyltransferase) and CheB (methylesterase). The activity of CheB is modulated by phosphorylation, while that of CheR by the signalling state of the chemoreceptors. When attractant concentration is high, the receptors are in their 'kinase-off' signalling conformation, where they have a high capacity for methylation and low capacity for demethylation (since CheB~P is not being produced). Conversely, in the 'kinase-on' conformation, they have a low attractant affinity, low methylation capacity and high demethylation capacity. Therefore, these two adaptation enzymes continuously update the methylation record of the receptors.

While the signal generated by ligand occupancy is transduced to the motor within 200 ms, the adaptation response takes place over a much longer period. The slow rates of modification provide the bacteria with a temporal memory device when deciding whether to continue running in a favourable direction. This is because the extent of receptor methylation provides a record of the chemoeffector concentration encountered by the bacterium in its immediate past, which can be compared with the current concentration as measured by the extent of receptor ligand occupancy. Abundant quantitative data on the chemotaxis pathway have led to several mathematical models of chemotaxis, making this pathway the bellwether of signal transduction (Bray *et al.*, 2007).

Swarming

Swarming is defined as flagella-dependent group motility on the surface of solid media (Harshey, 2003) (Fig. 9.6A,B). The percentage of agar is critical for enabling swarming. Some bacteria like *Vibrio parahaemolyticus* and *Proteus mirabilis* can swarm readily on higher percentage agar (1.5–3% or 'hard' agar), while others like *Salmonella, E. coli, Serratia, Pseudomonas* and *Bacillus* swarm only on lower percentage agar (0.5–0.8% or 'low' agar). Swarming on hard-agar is enabled by differentiation of the bacteria into specialized swarm cells that are elongated and have increased flagella. Low-agar swarmers generally do not display a similar differentiated morphology (Tolker-Nielsen *et al.*, 2000; Wang *et al.*, 2004). I will call the former bacteria 'complex swarmers' and the latter 'simple swarmers'. In complex or high agar swarmers, swarm cell differentiation can be generated by increasing the viscosity of liquid media, suggesting that cells experience a higher viscosity on the surface, but there is no accurate measure of this parameter on agar surfaces (Belas *et al.*, 1986; Belas and Suvanasuthi, 2005; Kawagishi *et al.*, 1996; McCarter *et al.*, 1988). Polysaccharides secreted by the swarm cells are implicated in extracting water from the agar, probably contributing to surface viscosity and to keeping the colony hydrated (Fraser and Hughes, 1999; Rauprich *et al.*, 1996). Simple swarmers are critically dependent on secreted surfactants and wet surface conditions for movement (Caiazza *et al.*, 2005; Kearns and Losick, 2003; Matsuyama *et al.*, 1992; Toguchi *et al.*, 2000; Wang *et al.*, 2005). The chemotaxis pathway is required for maintaining a moist surface environment in organisms like *Salmonella* and *E. coli* (Burkart *et al.*, 1998), the ability to switch motor direction being important for generating wetness (Mariconda *et al.*, 2006). It is assumed that secreted osmotic agents are responsible for generating wet conditions, but how motor switching influences this process is not understood (see below).

Challenges of surface movement

Moving on a surface presents bacteria with several challenges, the most important of which is availability of water, since flagella push against water to effect thrust. If the flagella detect insufficient water, they prudently turn down their own expression (Wang *et al.*, 2005). Downregulation of flagellar gene expression is due to inhibition of secretion of FlgM, suggesting that the system may be monitoring hindered diffusion rates, which are influenced by water availability. If water is available, bacteria must overcome surface friction. While many simple swarmers secrete powerful surfactants that lower surface tension and allow rapid colony expansion, *Salmonella* and *E. coli* do not appear to elaborate such surfactants. Lipopolysaccharides have been proposed to play an equivalent role in these bacteria (Toguchi *et al.*, 2000), which are more fastidious than other simple swarmers in their rich-nutrition requirement

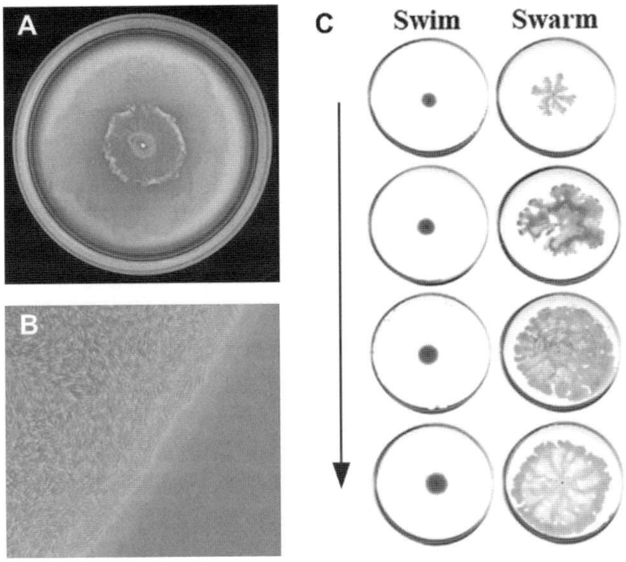

Figure 9.6 Swarming and chemotaxis. (A) A typical swarming experiment, where wild-type *Salmonella* bacteria have been inoculated on the centre of a 0.6% agar plate and incubated at 37°C for 8 hr. (B) Phase-contrast image showing the advancing edge of the swarming colony. The outermost region has an ~4-cell wide monolayer; behind this edge, cells are multi-layered. (C) Rescue of swarming in a non-chemotactic strain devoid of all chemoreceptors by providing increasing amounts of CheY~P from an inducible plasmid. Arrow indicates increasing concentration of the inducer. Adapted from Mariconda *et al*. (2006).

for swarming, and need to be supplemented with glucose to observe swarming reliably. The glucose requirement could be related to synthesis of compounds with surfactant properties, similar to nutritional requirements which affect production of the surfactant rhamnolipid in *P. aeruginosa* (Shrout *et al.*, 2006). Swarming behaviour was found to be prevalent among a large number of *Salmonella enterica* isolates, leading to speculation that the *in vitro* growth conditions that permit swarming may be in some aspects physiologically relevant to the nutrient-rich environment of the gastrointestinal tract, which the majority of *Salmonellae* favour regardless of the specific host (Kim and Surette, 2005).

The challenges of moving against surface friction and in a crowded environment also suggest that the bacteria must encode motility functions solely dedicated to surface movement. Indeed one such function appears to be carried out by FliL, not required for swimming but essential for swarming in *Salmonella* and *E. coli* (Attmannspacher *et al.*, 2008). In the absence of FliL, rotating flagella break within the rod. We have hypothesized that swarmers have an elevated proton motive force and hence a higher motor torque and that FliL protects the rod against the increased torsional stress. This hypothesis was arrived at by eliminating several competing hypotheses, and is based on the finding that swarmers have an elevated expression of enzymes involved in the *de novo* biosynthetic pathways of central metabolism and energy production (Kim and Surette, 2004). We expect more such surface-specific motility functions to be discovered, and that they will report not only on bacterial physiology on the surface but also on surface-related challenges to movement.

Surface growth promotes altered cell physiology but not swarm cell differentiation

Proteomic experiments have found that swarm cells of *Salmonella* are metabolically different compared to either 'vegetative' swimmer cells or surface grown cells propagated on non-swarming hard-agar in the same nutrient environment (Kim and Surette, 2004). Microarray experiments, however, found that gene expression patterns were similar between surface grown cells, irrespective of their swarming status, but that one-third of the genome was differentially regulated in swarmers versus broth grown cells; thus, changes in gene expression were surface-specific, not swarm cell-specific (Wang *et al.*, 2004). The discrepancy between the proteomic and microarray data remains to be resolved. The microarray data did not find flagellar gene expression to be upregulated in *Salmonella* swarmers, although a 2-fold increase in filament subunit gene expression was observed. To test if there were unfinished basal bodies in swimmers to which filaments get added, perhaps making swarmers more flagellated, we marked basal bodies with FliM-GFP and labelled filaments with Cy3 fluorescent dye. When we matched each internal fluorescent FliM dot with the external filament, we found that both swimmers and swarmers had 8–10 fluorescent FliM dots, and all dots were coincident with an external filament, i.e. there were no empty basal bodies in the swimmers, and swarmers were no more flagellated than swimmers (our unpublished data).

Swarming and chemotaxis: the flagellum as a drought sensor

Chemotaxis is not essential for swarming (Burkart *et al.*, 1998). Indeed, swarm cells suspended in a drop of liquid swim smoothly for long periods of time, as if they have suppressed chemotaxis (our unpublished data; Harshey, 2003). Yet mutants in the chemotaxis pathway do not swarm. There is a two-part explanation for this phenotype, which is still missing an important piece of the puzzle. The first part is that the chemotaxis mutants are defective in switching between the CW and CCW states of the motor. Suppressors that restore swarming in the *che* mutants mapped to FliM and had destabilized the switch i.e. had an increased frequency of switching between the CW and CCW states (Mariconda *et al.*, 2006). These suppressors were still non-chemotactic, yet they could swarm. The importance of switching was also demonstrated in experiments where providing increasing CheY from an inducible plasmid to a CCW-biased mutant resulted in restoration of swarming as its motor began to experience increasing reversals (Fig. 9.6C). Thus, the chemotaxis system, which controls CheY~P production, modulates switch behaviour during swarming as well, but the switching controls some function other than chemotaxis. The second part of the explanation for the non-swarming phenotype of *che* mutants is that their colonies have a flat and dry look in contrast to the glistening and hydrated morphology of a wild-type colony, they are inhibited in the export of the class 3 negative regulator FlgM, and have decreased flagellation (Wang *et al.*, 2005). When lawns of *che* mutants were sprayed with a fine mist of water, FlgM export resumed, as did normal flagellation and swarming, suggesting that it was primarily the lack of hydration that was responsible for the poor-flagellation phenotype. These studies uncovered an important role for FlgM as a reporter for external water conditions, and may explain why 80% of FlgM made by the cell is exported outside. Interestingly, while deletion of *flgM* restored class 3 gene expression, it did not restore normal flagellation or swarming. Thus, in spite of normal expression of class 3 genes in the *flgM* mutant, the transcripts were not being translated. While this could be explained as being due to the system monitoring hindered diffusion or secretion rates through the flagellum, the sensing mechanism involves other layers of control, since elimination of the Mot proteins appears to overcome inhibition of flagellar gene expression under these conditions (our unpublished data). These data suggest at least two mechanisms that operate on a surface to shut down the most energetically expensive step when cells encounter drought: FlgM-dependent inhibition of class 3 transcription by inhibition of FlgM export, and FlgM-independent inhibition of translation-secretion of flagellin by an as yet unknown mechanism. Thus, the flagellum is a drought sensor.

The two-part explanation for the non-swarming defect of *che* mutants tells us that motor switching is responsible for hydrating the colony, which in turn is important for normal flagellar secretion, but it does not tell us how switching controls hydration. We had proposed that switching plays a purely mechanical role in liberating flagella stuck on a relatively dry surface (Wang *et al.*, 2005). Once liberated, they stir the external milieu, perhaps stripping LPS from neighbouring cells. LPS could then serve as both surfactant and wetting agent. Evidence for this model is lacking. Another idea is that osmotic agents such as salts and sugars get out through the flagellum, and that switching is somehow important for this.

Where do bacteria swarm?

Complex swarmers like *Vibrio* and *Rhodospirillum* (Jiang *et al.*, 1997; McCarter *et al.*, 1988) probably live where surface movement is an essential aspect of their life cycle, for example on shells or skin of marine animals, or on salt sediments. The flagellum is implicated in sensing 'viscosity' or 'a hard surface' and turns on new flagella synthesis in these bacteria. Simple swarmers probably do not encounter such hard surfaces. For them, a hydrated surface is critical and a normal set of flagella appear to suffice. What kind of surfaces do these bacteria swarm on, out in the wild? Given how fastidious most of the simple swarmers are in the laboratory, and their need for rich media to swarm, one would guess that they might swarm on decaying animal carcasses or ripe fruits that provide a gelatinous, agar-like substrate for movement.

It would be interesting to see if common mechanisms underlie the deduced drought- and viscosity-sensing roles for the flagellum in simple and complex swarmers, respectively.

Cell density and group behaviour

A hallmark of the initiation of swarming movement is its dependence on cell density (Rauprich *et al.*, 1996). This dependence could be related to quorum-sensing regulation, but there is no evidence as yet implicating known quorum-sensing pathways in *Salmonella* swarming. In *Serratia*, surfactant production is dependent on quorum sensing (Eberl *et al.*, 1996). We have proposed that the lag associated with build-up of cell density is the time required to secrete sufficient osmotic agents to allow adequate hydration (Toguchi *et al.*, 2000). Another important feature of movement is that cells stay in a group. Isolated swarmers get stranded and do not move. Perhaps this too could be related to the amount of water surrounding a single cell versus a group of cells. Alternatively, this might be a reflection of shared flagellar bundles between neighbouring cells that generate greater thrust required for movement on a surface. Dramatic scanning electron micrographs of *P. mirabilis* showing highly organized flagella that appear to be interwoven in phase to form helical connections between adjacent swarmer cells suggest that cells in a crowd may indeed share flagellar bundles (Jones *et al.*, 2004). However, this needs to be demonstrated by imaging live moving cells.

Swarming and antibiotic resistance

Simple swarmers exhibit elevated resistance to multiple antibiotics (Kim *et al.*, 2003; Kim and Surette, 2003; Lai *et al.*, 2009; Overhage *et al.*, 2008). This resistance was reported to be linked specifically to swarming, and was apparently not exhibited either by swimming bacteria or by bacteria growing on a hard agar surface non-conducive to swarming. It was also shown to be an adaptive change, and not a selection for antibiotic-resistant mutants, since these cells were killed with lethal doses of antibiotic when inoculated in fresh liquid media (Kim *et al.*, 2003; Overhage *et al.*, 2008), reminiscent of the adaptive resistance seen in bacterial biofilms (Stewart, 2002).

We have found that this adaptive resistance is governed by the dense, multilayered architecture of the swarming colony, as well as its speed of migration (Butler *et al.*, 2010). The resistance is <u>not</u> a property of swarming-specific physiology as concluded earlier (Kim *et al.*, 2003; Kim and Surette, 2003), but rather afforded by high cell densities. Known quorum sensing pathways do not play a role in adaptive resistance. While all cells – swarmers, non-swarmers, and broth-grown swimmers – can tolerate higher antibiotic concentrations at high cell densities, the ability to move gives swarmers an added advantage in overriding the antibiotic. Individuals within the dense moving group, probably those directly exposed to the antibiotic, undergo cell death, protecting cells in the interior. We suggest that a similar mechanism underlies the adaptive resistance of bacterial biofilms.

Cyclic-di-GMP and motility

C-di-GMP is a newly recognized second messenger (Jenal and Malone, 2006; Romling *et al.*, 2005). First identified as a positive regulator of cellulose synthase in *Gluconacetobacter xylinus* (Ross *et al.*, 1991), c-di-GMP enhances biosynthesis of capsular and fimbrial components required for biofilm formation, while inhibiting flagella and pili that allow movement (Simm *et al.*, 2004; Wolfe and Visick, 2008). Steady state levels of c-di-GMP are thought to be maintained by diguanylate cyclases (DGCs) which synthesize c-di-GMP, and phosphodiesterases (PDEs) which degrade it. DGC activity depends on a conserved GGDEF domain, while PDE activity requires an EAL or HD-GYP domain (Ryan *et al.*, 2006). Such domains are associated with many diverse input and output domains, suggesting that they receive a variety of signals and respond through a variety of different mechanisms. Many bacteria possess multiple proteins with these domains (Galperin *et al.*, 2001). *Salmonella enterica* has five GGDEF proteins, eight EAL proteins, and seven proteins with both GGDEF and EAL domains (Simm *et al.*, 2007; Solano *et al.*, 2009). Proteins with both domains usually exhibit one or the other activity. Despite the existence of multiple such proteins, the observance of specific phenotypes associated with mutation of individual proteins suggests that c-di-GMP levels may be modulated in a localized manner. A c-di-GMP receptor or PilZ domain protein mediates many of the responses, although not all organisms have such proteins (Amikam and Galperin, 2006). New c-di-GMP receptor proteins probably await identification.

C-di-GMP inhibits flagellar motility in a variety of ways, either at the level of gene expression, flagellar assembly or function (Wolfe and Visick, 2008). In *Salmonella*, YhjH is an EAL domain protein and YcgR a PilZ domain protein. The PDE and c-di-GMP -receptor functions of these proteins have been confirmed (Ryjenkov *et al.*, 2006; Schmidt *et al.*, 2005). Both proteins are members of the late or class 3 flagellar regulon (Frye *et al.*, 2006; Ko and Park, 2000; Wang *et al.*, 2006a). Mutation of *yhjH* reduces motility, while that of *ycgR* does not. However, *ycgR* suppresses the motility defect of *yhjH*, suggesting that in the absence of YhjH, YcgR interacts with the resultant increased levels of c-di-GMP. No specific DGC has been identified to be associated with the motility phenotype (Simm *et al.*, 2007). A *yhjH* mutant in *E. coli* was shown to have a strong counter clockwise (CCW) bias, suggesting a role for the c-di-GMP–YcgR complex in motor switching (Girgis *et al.*, 2007). The activities of these proteins are also implicated in controlling motor rotation in *E. coli* (Ko and Park, 2000). Ko and Park found that absence of H-NS results in paralysed flagella, a defect that could be suppressed either by a mutation in *ycgR* or by providing *yhjH* in multiple copies. It has been proposed that the c–di-GMP-YcgR complex might inhibit motor rotation by interfering with the proper association of the proton-conducting Mot proteins with rotor protein FliG (Wolfe and Visick, 2008). YcgR is specific to c-di-GMP-dependent regulation of flagellum-based motility, and does not appear to influence other c-di-GMP-associated phenotypes such as rdar morphotype (red dry and rough), abiotic surface adherence and biofilm formation (Ryjenkov *et al.*, 2006).

In *P. aeruginosa*, c-di-GMP accumulation enhances biofilm formation by increasing extracellular polysaccharide (EPS) synthesis and simultaneously inhibiting swarming; the latter involves interference with chemoreceptor methylation to decrease flagellar reversals (Caiazza *et al.*, 2007; Kuchma *et al.*, 2007; Merritt *et al.*, 2007). It has been suggested that decreasing motor switching rates provides the cells with a longer window of opportunity to interact with the surface. Alternatively or in addition, a decreased switching would decrease surface hydration and hence inhibit swarming, as observed for such mutants in *S. enterica* (Mariconda *et al.*, 2006; Wang *et al.*, 2005).

We have found that the absence of YhjH inhibits chemotaxis, and that this inhibition is more pronounced when YcgR is overexpressed from a plasmid (Paul *et al.*, 2010). Swarming is completely inhibited under the latter conditions. In liquid media the bacteria were predominantly smooth swimming. Tethering experiments showed a pronounced shift to CCW rotation with a significant slowing of motor rotation. Inhibition of chemotaxis was therefore at the level of either production or activity of the chemotaxis response regulator CheY~P, which acts at the switch to change the default CCW bias to CW. Expression of a GFP-tagged c-di-GMP-binding protein YcgR in a *yhjH* mutant background resulted in punctate fluorescent dots along the cell membrane. Presence of the puncta was dependent on the presence of flagellar basal bodies, but not on the presence of chemotaxis signalling components. We have demonstrated that YcgR interacts with the flagellar switch-complex proteins FliG and FliM, most strongly in the presence of c-di-GMP. This interaction reduces the efficiency of torque generation and induces CCW motor bias. Inhibition of chemotaxis may represent a novel strategy to prepare for a sedentary existence by disfavouring migration away from a substrate on which a biofilm is to be formed. We are currently exploring whether this system is controlling the smooth swimming behaviour of swarmer cells.

Motility and virulence

Flagella contribute to the virulence of pathogenic bacteria through chemotaxis, and through adhesion to and invasion of host surfaces. Adherence of *Salmonella* flagella to surfaces contributes to food contamination in some serovars (Berger *et al.*, 2009), and its strong antigenic properties elicit an inflammatory response during host infection (Ramos *et al.*, 2004).

In *Salmonella*, expression of flagella is controlled by phase variation, a site-specific DNA recombination mechanism by which the organism alternately expresses two different types of flagellin subunit proteins, FljB and FliC (Fig. 9.1B). The biological significance of flagellar phase variation is not known, and there are conflicting reports on its role in virulence. Using mutants that are locked in one or the other phase, organisms that have switched to the $FliC^+$ phase were shown to have a selective advantage in the mouse model, although no such advantage was seen in invasion of epithelial cells or the induction of enteropathogenesis (Ikeda *et al.*, 2001). In another study, flagella were not necessary for the virulence of *S. enterica* in the mouse model (Lockman and Curtiss, 1990). Colonization of rabbit appendix was observed to depend on access to the intestinal mucosa and on motility (Marchetti *et al.*, 2004). Expression of flagellar genes and loci encoding virulence factors are part of regulons that are dependent on FlhDC and on two-component sensor kinase and response regulators such as PhoP–PhoQ or BarA–SirA (Chilcott and Hughes, 2000; Goodier and Ahmer, 2001). How these regulators co-ordinate their activities to control pathogenesis during the course of infection is not clear. Motility is thought to be turned down after entry into epithelial cells, before the bacteria disseminate into deeper tissues (Hughes and Galan, 2002), yet flagella have been reported to facilitate escape from macrophages (Sano *et al.*, 2007).

The pioneer studies conducted with flagellin that had been isolated from *Salmonella enterica* serovar Adelaide showed that flagellin is a potent stimulator of antibody responses, a hallmark of Th2 T helper responses (Parish, 1996, see also Chapter 12 of this book). Recent studies indicate that the primary signalling events, which initiate Th2-type responses, are mediated by Toll-like receptor TLR5 (Chappell *et al.*, 2009; Feuillet *et al.*, 2006). Flagellin consists of a conserved domain that is widespread in bacterial species and is dedicated to filament polymerization. Mammalian hosts detect this conserved domain on flagellin monomers through TLR5, which triggers proinflammatory and adaptive immune responses. TLR5 is expressed by a variety of cells including monocytes, dendritic cells (DCs), epithelial cells, and mast cells. The identification of flagellin as a major proinflammatory determinant of enteropathogenic *Salmonella* provides new opportunities to control the human mucosal innate-adaptive immune systems.

Flagellar origins

The flagellum has served as poster-child for arguments about 'intelligent design'. Proponents of intelligent design invoke 'Irreducible Complexity' as an argument against evolution. According to this argument, biological systems with multiple required components could not have evolved gradually, because intermediates lacking some components would be non-functional. However, a consideration of the diversity of flagellar structures, homologies of component parts to other functional modules, and existence of partial structures that can perform functions other than motility (such as secretion) shows that designing an evolutionary model to account for the origin of the ancestral flagellum requires no great conceptual leap.

Any attempt to explain the origin of bacterial flagella has to account for both their complexity and diversity. For example, flagellar filaments show considerable variability in their higher-order structures (Galkin *et al.*, 2008), have highly variable surface-exposed domains (Beatson *et al.*, 2006), can be right- or left-handed, flexible or rigid, straight or curly (Shibata *et al.*, 2005), and can have post-translational modifications such as glycosylation or methylation (Logan, 2006). Some systems deploy a single flagellin, whereas others, like *S. enterica,* exploit two different flagellins, and in some cases, up to six different flagellins are found in a single flagellum (Ely *et al.*, 2000). The flagella of Gram-positive bacteria lack P- and L-rings (Kubori *et al.*, 1997), and in spirochaetes the flagella rotate inside the periplasm (Li *et al.*, 2000). Some flagella rotate using a proton gradient, others use a sodium-ion gradient (McCarter, 2004). In *Sinorhizobium meliloti* and *Rhodobacter sphaeroides*, the flagellum rotates unidirectionally, with stops and starts instead of CW and CCW events to re-orient cells (Armitage and Macnab, 1987; Attmannspacher *et al.*, 2005).

A recent publication on flagellar evolution by Liu and Ochman (Liu and Ochman, 2007) concluded that all flagellar genes originated from a single gene that underwent successive duplications and subsequent diversification. That several flagellar proteins (e.g. FliG, FliM, FliI, FliC) are known to have completely different 3D structures (Brown *et al.*, 2002; Brown *et al.*, 2005; Imada *et al.*, 2007), renders the notion of common ancestry of <u>all</u> flagellar genes untenable (Doolittle and Zhaxybayeva, 2007; Snyder *et al.*, 2009). However, there <u>is</u> a conserved core set of ~20 proteins that is shared by all bacterial flagella (Pallen and Matzke, 2006). If one accepts that all current flagellar systems diverged from their last common ancestor, then this core set (comprising the rod-hook-filament) appears to have evolved by gene duplication-and-diversification, starting from two proteins that were capable of polymerization – a proto-flagellin and a proto-rod/hook protein. Outside this core set, several flagellar subunits share common ancestry with components from non-flagellar systems. For example, FliK resembles YscP from the Ysc–Yop T3SS, and the flagellar σ factor, FliA, shows both sequence and structural homology to several other σ factors. FlgA, which plays a role in assembly of the P ring, is homologous to CpaB involved in type IV pilus assembly (Iyer and Aravind, 2004). Similarly, FlgJ contains a C-terminal amidase 4 domain shared with other bacterial proteins, which mediate hydrolysis of peptidoglycan (Nambu *et al.*, 1999).

The three devices which are central to functioning of the flagellum in *S. enterica* – the rotor-stator, the T3SS, and the chemotaxis system – all contain parts that occur elsewhere in the biological world. MotA/MotB are homologous to components of the Ton and Tol systems (ExbB/ExbD and TolQ/TolR). These systems use ion-motive force to drive active transport of substrates across the outer membrane (Zhai *et al.*, 2003). FliM and FliN share a C-terminal SpoA domain with each other and with components of non-flagellar T3SSs (the YscQ-like proteins) (Pallen *et al.*, 2005b). The ATPase FliI shows significant homology to the catalytic subunits from F-type and V-type ATPases (Vogler *et al.*, 1991). At least nine flagellar proteins share common ancestry with core components of non-flagellar T3SSs. The main components of the chemotaxis apparatus are shared between bacterial and archaeal flagellar systems, even though the rest of their flagellar machinery is unrelated (Faguy and Jarrell, 1999; Szurmant and Ordal, 2004). Some bacteria, such as *Aquifex aeolicus*, assemble flagella but lack the chemotaxis machinery (Faguy and Jarrell, 1999). Many chemotaxis proteins contain kinase domains and response regulator domains, which are found in other bacterial regulatory proteins.

Genome sequencing projects have begun to reveal that the flagellar filament is homologous to the non-flagellar T3SS filament EspA, which functions in adhesion and delivery of virulence effectors (Pallen *et al.*, 2005a). Vestigial flagellar genes or regulons have been discovered in several bacterial species. At least one degenerate flagellar system (from *Brucella melitensis*) still has a cryptic role in infection, even though it is no longer capable of mediating flagellar motility (Fretin *et al.*, 2005). The non-motile aphid endosymbiont *Buchnera aphidicola* lacks genes for filament assembly, but expresses hundreds of hook–basal-body complexes on its cell surface (Maezawa *et al.*, 2006). The abundant presence of these partial structures suggests that they must have some role other than motility, probably the export of proteins essential to the symbiotic lifestyle of this bacterium. It is not difficult to imagine several functional shifts in the evolution of the flagellum, from an early organelle of adhesion or targeted secretion to today's curved structure capable of generating a propulsive force. By even the most conservative estimate, there must be thousands of different bacterial flagellar systems, perhaps even millions (Venter *et al.*, 2004). The 'irreducible complexity' of flagella will continue to be steadily reduced as other examples of potential evolutionary intermediaries will be found as we sequence an increasing proportion of the biosphere.

Acknowledgements

I thank my graduate student Jae-Min Lee for help with the illustrations. Work on *Salmonella* motility in my laboratory is supported by the National Institutes of Health (GM 57400).

References

Agrain, C., Callebaut, I., Journet, L., Sorg, I., Paroz, C., Mota, L.J., and Cornelis, G.R. (2005). Characterization of a Type III secretion substrate specificity switch (T3S4) domain in YscP from Yersinia enterocolitica. Mol. Microbiol. *56*, 54–67.

Aldridge, P., and Hughes, K.T. (2001). How and when are substrates selected for type III secretion? Trends Microbiol. *9*, 209–214.

Aldridge, P., Karlinsey, J., and Hughes, K.T. (2003). The type III secretion chaperone FlgN regulates flagellar assembly via a negative feedback loop containing its chaperone substrates FlgK and FlgL. Mol. Microbiol. *49*, 1333–1345.

Aldridge, P., Karlinsey, J.E., Becker, E., Chevance, F.F., and Hughes, K.T. (2006a). Flk prevents premature secretion of the anti-sigma factor FlgM into the periplasm. Mol. Microbiol. *60*, 630–643.

Aldridge, P.D., Karlinsey, J.E., Aldridge, C., Birchall, C., Thompson, D., Yagasaki, J., and Hughes, K.T. (2006b). The flagellar-specific transcription factor, sigma28, is the Type III secretion chaperone for the flagellar-specific anti-sigma28 factor FlgM. Genes Dev *20*, 2315–2326.

Amikam, D., and Galperin, M.Y. (2006). PilZ domain is part of the bacterial c-di-GMP binding protein. Bioinformatics *22*, 3–6.

Apel, D., and Surette, M.G. (2008). Bringing order to a complex molecular machine: the assembly of the bacterial flagella. Biochim. Biophys. Acta. *1778*, 1851–1858.

Armitage, J.P., and Macnab, R.M. (1987). Unidirectional, intermittent rotation of the flagellum of Rhodobacter sphaeroides. J. Bacteriol. *169*, 514–518.

Attmannspacher, U., Scharf, B., and Schmitt, R. (2005). Control of speed modulation (chemokinesis) in the unidirectional rotary motor of Sinorhizobium meliloti. Mol. Microbiol. *56*, 708–718.

Attmannspacher, U., Scharf, B.E., and Harshey, R.M. (2008). FliL is essential for swarming: motor rotation in absence of FliL fractures the flagellar rod in swarmer cells of *Salmonella enterica*. Mol. Microbiol. *68*, 328–341.

Bai, F., Branch, R.W., Nicolau, D.V., Jr., Pilizota, T., Steel, B.C., Maini, P.K., and Berry, R.M. (2010). Conformational spread as a mechanism for cooperativity in the bacterial flagellar switch. Science *327*, 685–689.

Baker, M.D., Wolanin, P.M., and Stock, J.B. (2006). Signal transduction in bacterial chemotaxis. Bioessays *28*, 9–22.

Beatson, S.A., Minamino, T., and Pallen, M.J. (2006). Variation in bacterial flagellins: from sequence to structure. Trends Microbiol. *14*, 151–155.

Belas, R., Simon, M., and Silverman, M. (1986). Regulation of lateral flagella gene transcription in *Vibrio parahaemolyticus*. J. Bacteriol. *167*, 210–218.

Belas, R., and Suvanasuthi, R. (2005). The ability of Proteus mirabilis to sense surfaces and regulate virulence gene expression involves FliL, a flagellar basal body protein. J. Bacteriol. *187*, 6789–6803.

Bennett, J.C., Thomas, J., Fraser, G.M., and Hughes, C. (2001). Substrate complexes and domain organization of the *Salmonella* flagellar export chaperones FlgN and FliT. Mol. Microbiol. *39*, 781–791.

Berg, H.C. (2003). The rotary motor of bacterial flagella. Annu Rev. Biochem. *72*, 19–54.

Berger, C.N., Shaw, R.K., Brown, D.J., Mather, H., Clare, S., Dougan, G., Pallen, M.J., and Frankel, G. (2009). Interaction of *Salmonella enterica* with basil and other salad leaves. ISME J. *3*, 261–265.

Blair, D.F. (2006). Fine structure of a fine machine. J. Bacteriol. *188*, 7033–7035.

Bray, D., Levin, M.D., and Lipkow, K. (2007). The chemotactic behavior of computer-based surrogate bacteria. Curr. Biol. *17*, 12–19.

Brown, J.D., Saini, S., Aldridge, C., Herbert, J., Rao, C.V., and Aldridge, P.D. (2008). The rate of protein secretion dictates the temporal dynamics of flagellar gene expression. Mol. Microbiol. *70*, 924–937.

Brown, P.N., Hill, C.P., and Blair, D.F. (2002). Crystal structure of the middle and C-terminal domains of the flagellar rotor protein FliG. EMBO J. *21*, 3225–3234.

Brown, P.N., Mathews, M.A., Joss, L.A., Hill, C.P., and Blair, D.F. (2005). Crystal structure of the flagellar rotor protein FliN from *Thermotoga maritima*. J. Bacteriol. *187*, 2890–2902.

Brown, P.N., Terrazas, M., Paul, K., and Blair, D.F. (2007). Mutational analysis of the flagellar protein FliG: sites of interaction with FliM and implications for organization of the switch complex. J. Bacteriol. *189*, 305–312.

Burkart, M., Toguchi, A., and Harshey, R.M. (1998). The chemotaxis system, but not chemotaxis, is essential for swarming motility in *Escherichia coli*. Proc. Natl. Acad. Sci. U.S.A. *95*, 2568–2573.

Butler, M.T., Wang, Q., and Harshey, R.M. (2010). Cell density and mobility protect swarming bacteria against antibiotics. Proc. Natl. Acad. Sci. U.S.A. *107*, 3776–3781.

Caiazza, N.C., Merritt, J.H., Brothers, K.M., and O'Toole, G.A. (2007). Inverse regulation of biofilm formation and swarming motility by Pseudomonas aeruginosa PA14. J. Bacteriol. *189*, 3603–3612.

Caiazza, N.C., Shanks, R.M., and O'Toole, G.A. (2005). Rhamnolipids modulate swarming motility patterns of Pseudomonas aeruginosa. J. Bacteriol. *187*, 7351–7361.

Chappell, L., Kaiser, P., Barrow, P., Jones, M.A., Johnston, C., and Wigley, P. (2009). The immunobiology of avian systemic salmonellosis. Vet. Immunol. Immunopathol. *128*, 53–59.

Chevance, F.F., and Hughes, K.T. (2008). Coordinating assembly of a bacterial macromolecular machine. Nat. Rev. Microbiol. *6*, 455–465.

Chevance, F.F., Takahashi, N., Karlinsey, J.E., Gnerer, J., Hirano, T., Samudrala, R., Aizawa, S., and Hughes, K.T. (2007). The mechanism of outer membrane penetration by the eubacterial flagellum and implications for spirochete evolution. Genes Dev *21*, 2326–2335.

Chilcott, G.S., and Hughes, K.T. (2000). Coupling of flagellar gene expression to flagellar assembly in *Salmonella enterica* serovar Typhimurium and Escherichia coli. Microbiol. Mol. Biol. Rev. *64*, 694–708.

Cluzel, P., Surette, M., and Leibler, S. (2000). An ultrasensitive bacterial motor revealed by monitoring signaling proteins in single cells. Science *287*, 1652–1655.

Darnton, N.C., and Berg, H.C. (2008). Bacterial flagella are firmly anchored. J. Bacteriol. *190*, 8223–8224.

Darnton, N.C., Turner, L., Rojevsky, S., and Berg, H.C. (2007). On torque and tumbling in swimming Escherichia coli. J. Bacteriol. *189*, 1756–1764.

Deane, J.E., Graham, S.C., Mitchell, E.P., Flot, D., Johnson, S., and Lea, S.M. (2008). Crystal structure of Spa40, the specificity switch for the Shigella flexneri type III secretion system. Mol. Microbiol. *69*, 267–276.

Doolittle, W.F., and Zhaxybayeva, O. (2007). Evolution: reducible complexity – the case for bacterial flagella. Curr. Biol. *17*, R510–512.

Eberl, L., Winson, M.K., Sternberg, C., Stewart, G.S., Christiansen, G., Chhabra, S.R., Bycroft, B., Williams, P., Molin, S., and Givskov, M. (1996). Involvement of N-acyl-L-hormoserine lactone autoinducers in controlling the multicellular behaviour of *Serratia liquefaciens*. Mol. Microbiol. *20*, 127–136.

Ely, B., Ely, T.W., Crymes, W.B., Jr., and Minnich, S.A. (2000). A family of six flagellin genes contributes to the *Caulobacter crescentus* flagellar filament. J. Bacteriol. *182*, 5001–5004.

Erhardt, M., Hirano, T., Su, Y., Paul, K., Wee, D.H., Mizuno, S., Aizawa, S.I., and Hughes, K.T. (2010). The role of the FliK molecular ruler in hook-length control in *Salmonella enterica*. Mol. Microbiol. 75, 1272–1284.

Erhardt, M., and Hughes, K.T. (2010). C-ring requirement in flagellar type III secretion is bypassed by FlhDC upregulation. Mol. Microbiol. *75*, 376–393.

Evans, L.D., Stafford, G.P., Ahmed, S., Fraser, G.M., and Hughes, C. (2006). An escort mechanism for cycling of export chaperones during flagellum assembly. Proc. Natl. Acad. Sci. U.S.A. *103*, 17474–17479.

Faguy, D.M., and Jarrell, K.F. (1999). A twisted tale: the origin and evolution of motility and chemotaxis in prokaryotes. Microbiology *145*, 279–281.

Ferris, H.U., and Minamino, T. (2006). Flipping the switch: bringing order to flagellar assembly. Trends Microbiol. *14*, 519–526.

Feuillet, V., Medjane, S., Mondor, I., Demaria, O., Pagni, P.P., Galan, J.E., Flavell, R.A., and Alexopoulou, L. (2006). Involvement of Toll-like receptor 5 in the recognition of flagellated bacteria. Proc. Natl. Acad. Sci. U.S.A. *103*, 12487–12492.

Fraser, G.M., Hirano, T., Ferris, H.U., Devgan, L.L., Kihara, M., and Macnab, R.M. (2003). Substrate specificity of type III flagellar protein export in *Salmonella* is controlled by subdomain interactions in FlhB. Mol. Microbiol. *48*, 1043–1057.

Fraser, G.M., and Hughes, C. (1999). Swarming motility. Curr. Opin. Microbiol. *2*, 630–635.

Fretin, D., Fauconnier, A., Kohler, S., Halling, S., Leonard, S., Nijskens, C., Ferooz, J., Lestrate, P., Delrue, R.M., Danese, I., *et al.* (2005). The sheathed flagellum of Brucella melitensis is involved in persistence in a murine model of infection. Cell Microbiol. *7*, 687–698.

Frye, J., Karlinsey, J.E., Felise, H.R., Marzolf, B., Dowidar, N., McClelland, M., and Hughes, K.T. (2006). Identification of new flagellar genes of *Salmonella enterica* serovar Typhimurium. J. Bacteriol. *188*, 2233–2243.

Galan, J.E., and Wolf-Watz, H. (2006). Protein delivery into eukaryotic cells by type III secretion machines. Nature *444*, 567–573.

Galkin, V.E., Yu, X., Bielnicki, J., Heuser, J., Ewing, C.P., Guerry, P., and Egelman, E.H. (2008). Divergence of quaternary structures among bacterial flagellar filaments. Science *320*, 382–385.

Galperin, M.Y., Nikolskaya, A.N., and Koonin, E.V. (2001). Novel domains of the prokaryotic two-component signal transduction systems. FEMS Microbiol. Lett. *203*, 11–21.

Girgis, H.S., Liu, Y., Ryu, W.S., and Tavazoie, S. (2007). A comprehensive genetic characterization of bacterial motility. PLoS Genet 3, 1644–1660.

Gonzalez-Pedrajo, B., Minamino, T., Kihara, M., and Namba, K. (2006). Interactions between C ring proteins and export apparatus components: a possible mechanism for facilitating type III protein export. Mol. Microbiol. *60*, 984–998.

Goodier, R.I., and Ahmer, B.M. (2001). SirA orthologs affect both motility and virulence. J. Bacteriol. *183*, 2249–2258.

Harshey, R.M. (2003). Bacterial motility on a surface: many ways to a common goal. Annu. Rev. Microbiol. *57*, 249–273.

Hazelbauer, G.L., Falke, J.J., and Parkinson, J.S. (2008). Bacterial chemoreceptors: high-performance signaling in networked arrays. Trends Biochem. Sci 33, 9–19.

Hirano, T., Minamino, T., Namba, K., and Macnab, R.M. (2003). Substrate specificity classes and the recognition signal for *Salmonella* type III flagellar export. J. Bacteriol. *185*, 2485–2492.

Hughes, E.A., and Galan, J.E. (2002). Immune response to *Salmonella*: location, location, location? Immunity *16*, 325–328.

Ikeda, J.S., Schmitt, C.K., Darnell, S.C., Watson, P.R., Bispham, J., Wallis, T.S., Weinstein, D.L., Metcalf, E.S., Adams, P., O'Connor, C.D., and O'Brien, A.D. (2001). Flagellar phase variation of *Salmonella enterica* serovar Typhimurium contributes to virulence in the murine typhoid infection model but does not influence *Salmonella*-induced enteropathogenesis. Infect. Immun. *69*, 3021–3030.

Imada, K., Minamino, T., Tahara, A., and Namba, K. (2007). Structural similarity between the flagellar type III ATPase FliI and F1-ATPase subunits. Proc. Natl. Acad. Sci. U.S.A. *104*, 485–490.

Iyer, L.M., and Aravind, L. (2004). The emergence of catalytic and structural diversity within the beta-clip fold. Proteins *55*, 977–991.

Jenal, U., and Malone, J. (2006). Mechanisms of cyclic-di-GMP signaling in bacteria. Annu Rev. Genet *40*, 385–407.

Jiang, Z.Y., Gest, H., and Bauer, C.E. (1997). Chemosensory and photosensory perception in purple photosynthetic bacteria utilize common signal transduction components. J. Bacteriol. *179*, 5720–5727.

Jones, B.V., Young, R., Mahenthiralingam, E., and Stickler, D.J. (2004). Ultrastructure of Proteus mirabilis swarmer cell rafts and role of swarming in catheter-associated urinary tract infection. Infect. Immun. *72*, 3941–3950.

Journet, L., Agrain, C., Broz, P., and Cornelis, G.R. (2003). The needle length of bacterial injectisomes is determined by a molecular ruler. Science *302*, 1757–1760.

Kalir, S., and Alon, U. (2004). Using a quantitative blueprint to reprogram the dynamics of the flagella gene network. Cell *117*, 713–720.

Kalir, S., McClure, J., Pabbaraju, K., Southward, C., Ronen, M., Leibler, S., Surette, M.G., and Alon, U. (2001). Ordering genes in a flagella pathway by analysis of expression kinetics from living bacteria. Science *292*, 2080–2083.

Kawagishi, I., Imagawa, M., Imae, Y., McCarter, L., and Homma, M. (1996). The sodium-driven polar flagellar motor of marine *Vibrio* as the mechanosensor that regulates lateral flagellar expression. Mol. Microbiol. *20*, 693–699.

Kearns, D.B., and Losick, R. (2003). Swarming motility in undomesticated *Bacillus subtilis*. Mol. Microbiol. *49*, 581–590.

Kim, E.A., Price-Carter, M., Carlquist, W.C., and Blair, D.F. (2008). Membrane segment organization in the stator complex of the flagellar motor: implications for proton flow and proton-induced conformational change. Biochemistry *47*, 11332–11339.

Kim, W., Killam, T., Sood, V., and Surette, M.G. (2003). Swarm-cell differentiation in *Salmonella enterica* serovar Typhimurium results in elevated resistance to multiple antibiotics. J. Bacteriol. *185*, 3111–3117.

Kim, W., and Surette, M.G. (2003). Swarming populations of *Salmonella* represent a unique physiological state coupled to multiple mechanisms of antibiotic resistance. Biol. Proced. Online *5*, 189–196.

Kim, W., and Surette, M.G. (2004). Metabolic differentiation in actively swarming *Salmonella*. Mol. Microbiol. *54*, 702–714.

Kim, W., and Surette, M.G. (2005). Prevalence of surface swarming behavior in *Salmonella*. J. Bacteriol. *187*, 6580–6583.

Ko, M., and Park, C. (2000). Two novel flagellar components and H-NS are involved in the motor function of Escherichia coli. J. Mol. Biol. *303*, 371–382.

Kojima, S., and Blair, D.F. (2004). The bacterial flagellar motor: structure and function of a complex molecular machine. Int. Rev. Cytol. *233*, 93–134.

Kojima, S., Imada, K., Sakuma, M., Sudo, Y., Kojima, C., Minamino, T., Homma, M., and Namba, K. (2009). Stator assembly and activation mechanism of the flagellar motor by the periplasmic region of MotB. Mol. Microbiol. *73*, 710–718.

Konishi, M., Kanbe, M., McMurry, J.L., and Aizawa, S. (2009). Flagellar formation in C-ring-defective mutants by overproduction of FliI, the ATPase specific for flagellar type III secretion. J. Bacteriol. *191*, 6186–6191.

Kuchma, S.L., Brothers, K.M., Merritt, J.H., Liberati, N.T., Ausubel, F.M., and O'Toole, G.A. (2007). BifA, a cyclic-Di-GMP phosphodiesterase, inversely regulates biofilm formation and swarming motility by Pseudomonas aeruginosa PA14. J. Bacteriol. *189*, 8165–8178.

Lai, S., Tremblay, J., and Deziel, E. (2009). Swarming motility: a multicellular behaviour conferring antimicrobial resistance. Environ. Microbiol. *11*, 126–136.

Li, C., Motaleb, A., Sal, M., Goldstein, S.F., and Charon, N.W. (2000). Spirochete periplasmic flagella and motility. J. Mol. Microbiol. Biotechnol. *2*, 345–354.

Liu, R., and Ochman, H. (2007). Stepwise formation of the bacterial flagellar system. Proc. Natl. Acad. Sci. U.S.A. *104*, 7116–7121.

Lockman, H.A., and Curtiss, R., 3rd (1990). *Salmonella* Typhimurium mutants lacking flagella or motility remain virulent in BALB/c mice. Infect. Immun. *58*, 137–143.

Logan, S.M. (2006). Flagellar glycosylation – a new component of the motility repertoire? Microbiology *152*, 1249–1262.

Macnab, R.M. (2003). How bacteria assemble flagella. Annu. Rev. Microbiol. *57*, 77–100.

Maddock, J.R., and Shapiro, L. (1993). Polar location of the chemoreceptor complex in the *Escherichia coli* cell. Science *259*, 1717–1723.

Maezawa, K., Shigenobu, S., Taniguchi, H., Kubo, T., Aizawa, S., and Morioka, M. (2006). Hundreds of flagellar basal bodies cover the cell surface of the endosymbiotic bacterium Buchnera aphidicola sp. strain APS. J. Bacteriol. *188*, 6539–6543.

Manson, M.D. (2007). How 34 pegs fit into 26 + 8 holes in the flagellar motor. J. Bacteriol. *189*, 291–293.

Marchetti, M., Sirard, J.C., Sansonetti, P., Pringault, E., and Kerneis, S. (2004). Interaction of pathogenic bacteria with rabbit appendix M cells: bacterial motility is a key feature *in vivo*. Microbes Infect *6*, 521–528.

Mariconda, S., Wang, Q., and Harshey, R.M. (2006). A mechanical role for the chemotaxis system in swarming motility. Mol. Microbiol. *60*, 1590–1602.

Matsuyama, T., Kaneda, K., Nakagawa, Y., Isa, K., Hara-Hotta, H., and Yano, I. (1992). A novel extracellular cyclic lipopeptide which promotes flagellum-dependent and -independent spreading growth of *Serratia marcescens*. J. Bacteriol. *174*, 1769–1776.

McCarter, L., Hilmen, M., and Silverman, M. (1988). Flagellar dynamometer controls swarmer cell differentiation of *V. parahaemolyticus*. Cell *54*, 345–351.

McCarter, L.L. (2004). Dual flagellar systems enable motility under different circumstances. J. Mol. Microbiol. Biotechnol. *7*, 18–29.

McMurry, J.L., Murphy, J.W., and Gonzalez-Pedrajo, B. (2006). The FliN–FliH interaction mediates localization of flagellar export ATPase FliI to the C ring complex. Biochemistry *45*, 11790–11798.

Merritt, J.H., Brothers, K.M., Kuchma, S.L., and O'Toole, G.A. (2007). SadC reciprocally influences biofilm formation and swarming motility via modulation of exopolysaccharide production and flagellar function. J. Bacteriol. *189*, 8154–8164.

Minamino, T., Imada, K., and Namba, K. (2008). Mechanisms of type III protein export for bacterial flagellar assembly. Mol. Biosyst. *4*, 1105–1115.

Minamino, T., and Namba, K. (2008). Distinct roles of the FliI ATPase and proton motive force in bacterial flagellar protein export. Nature *451*, 485–488.

Minamino, T., Saijo-Hamano, Y., Furukawa, Y., Gonzalez-Pedrajo, B., Macnab, R.M., and Namba, K. (2004). Domain organization and function of *Salmonella* FliK, a flagellar hook-length control protein. J. Mol. Biol. *341*, 491–502.

Moriya, N., Minamino, T., Hughes, K.T., Macnab, R.M., and Namba, K. (2006). The type III flagellar export specificity switch is dependent on FliK ruler and a molecular clock. J. Mol. Biol. 359, 466–477.

Nambu, T., Minamino, T., Macnab, R.M., and Kutsukake, K. (1999). Peptidoglycan-hydrolyzing activity of the FlgJ protein, essential for flagellar rod formation in *Salmonella* Typhimurium. J. Bacteriol. *181*, 1555–1561.

Overhage, J., Bains, M., Brazas, M.D., and Hancock, R.E. (2008). Swarming of Pseudomonas aeruginosa is a complex adaptation leading to increased production of virulence factors and antibiotic resistance. J. Bacteriol. *190*, 2671–2679.

Pallen, M.J., Beatson, S.A., and Bailey, C.M. (2005a). Bioinformatics, genomics and evolution of non-flagellar type-III secretion systems: a Darwinian perspective. FEMS Microbiol. Rev. *29*, 201–229.

Pallen, M.J., and Matzke, N.J. (2006). From The Origin of Species to the origin of bacterial flagella. Nat. Rev. Microbiol. *4*, 784–790.

Pallen, M.J., Penn, C.W., and Chaudhuri, R.R. (2005b). Bacterial flagellar diversity in the post-genomic era. Trends Microbiol. *13*, 143–149.

Parish, C.R. (1996). Immune deviation: a historical perspective. Immunol. Cell Biol. *74*, 449–456.

Paul, K., and Blair, D.F. (2006). Organization of FliN subunits in the flagellar motor of Escherichia coli. J. Bacteriol. *188*, 2502–2511.

Paul, K., Erhardt, M., Hirano, T., Blair, D.F., and Hughes, K.T. (2008). Energy source of flagellar type III secretion. Nature *451*, 489–492.

Paul, K., Harmon, J.G., and Blair, D.F. (2006). Mutational analysis of the flagellar rotor protein FliN: identification of surfaces important for flagellar assembly and switching. J. Bacteriol. *188*, 5240–5248.

Paul, K., Nieto, V., Carlquist, W.C., Blair, D.F., and Harshey, R.M. (2010). The c-di-GMP binding protein YcgR controls flagellar motor direction and speed to affect chemotaxis by a 'backstop brake' mechanism. Mol. Cell *38*, 128–139.

Ramos, H.C., Rumbo, M., and Sirard, J.C. (2004). Bacterial flagellins: mediators of pathogenicity and host immune responses in mucosa. Trends Microbiol. *12*, 509–517.

Rauprich, O., Matsushita, M., Weijer, C.J., Siegert, F., Esipov, S.E., and Shapiro, J.A. (1996). Periodic phenomena in *Proteus mirabilis* swarm colony development. J. Bacteriol. *178*, 6525–6538.

Reid, S.W., Leake, M.C., Chandler, J.H., Lo, C.J., Armitage, J.P., and Berry, R.M. (2006). The maximum number of torque-generating units in the flagellar motor of Escherichia coli is at least 11. Proc. Natl. Acad. Sci. U.S.A. *103*, 8066–8071.

Romling, U., Gomelsky, M., and Galperin, M.Y. (2005). C-di-GMP: the dawning of a novel bacterial signalling system. Mol. Microbiol. *57*, 629–639.

Ross, P., Mayer, R., and Benziman, M. (1991). Cellulose biosynthesis and function in bacteria. Microbiol. Rev. *55*, 35–58.

Ryan, R.P., Fouhy, Y., Lucey, J.F., and Dow, J.M. (2006). Cyclic di-GMP signaling in bacteria: recent advances and new puzzles. J. Bacteriol. *188*, 8327–8334.

Ryjenkov, D.A., Simm, R., Romling, U., and Gomelsky, M. (2006). The PilZ domain is a receptor for the second messenger c-di-GMP: the PilZ domain protein YcgR controls motility in enterobacteria. J. Biol. Chem. *281*, 30310–30314.

Saini, S., Brown, J.D., Aldridge, P.D., and Rao, C.V. (2008). FliZ Is a posttranslational activator of FlhD4C2-dependent flagellar gene expression. J. Bacteriol. *190*, 4979–4988.

Sano, G., Takada, Y., Goto, S., Maruyama, K., Shindo, Y., Oka, K., Matsui, H., and Matsuo, K. (2007). Flagella facilitate escape of *Salmonella* from oncotic macrophages. J. Bacteriol. *189*, 8224–8232.

Schlumberger, M.C., Muller, A.J., Ehrbar, K., Winnen, B., Duss, I., Stecher, B., and Hardt, W.D. (2005). Real-time imaging of type III secretion: *Salmonella* SipA injection into host cells. Proc. Natl. Acad. Sci. U.S.A. *102*, 12548–12553.

Schmidt, A.J., Ryjenkov, D.A., and Gomelsky, M. (2005). The ubiquitous protein domain EAL is a cyclic diguanylate-specific phosphodiesterase: enzymatically active and inactive EAL domains. J. Bacteriol. *187*, 4774–4781.

Shibata, S., Alam, M., and Aizawa, S. (2005). Flagellar filaments of the deep-sea bacteria *Idiomarina loihiensis* belong to a family different from those of *Salmonella* Typhimurium. J. Mol. Biol. *352*, 510–516.

Shibata, S., Takahashi, N., Chevance, F.F., Karlinsey, J.E., Hughes, K.T., and Aizawa, S. (2007). FliK regulates flagellar hook length as an internal ruler. Mol. Microbiol. *64*, 1404–1415.

Shrout, J.D., Chopp, D.L., Just, C.L., Hentzer, M., Givskov, M., and Parsek, M.R. (2006). The impact of quorum sensing and swarming motility on Pseudomonas aeruginosa biofilm formation is nutritionally conditional. Mol. Microbiol. *62*, 1264–1277.

Simm, R., Lusch, A., Kader, A., Andersson, M., and Romling, U. (2007). Role of EAL-containing proteins in multicellular behavior of *Salmonella enterica* serovar Typhimurium. J. Bacteriol. *189*, 3613–3623.

Simm, R., Morr, M., Kader, A., Nimtz, M., and Romling, U. (2004). GGDEF and EAL domains inversely regulate cyclic di-GMP levels and transition from sessility to motility. Mol. Microbiol. *53*, 1123–1134.

Snyder, L.A., Loman, N.J., Futterer, K., and Pallen, M.J. (2009). Bacterial flagellar diversity and evolution: seek simplicity and distrust it? Trends Microbiol. *17*, 1–5.

Solano, C., Garcia, B., Latasa, C., Toledo-Arana, A., Zorraquino, V., Valle, J., Casals, J., Pedroso, E., and Lasa, I. (2009). Genetic reductionist approach for dissecting individual roles of GGDEF proteins within the c-di-GMP signaling network in *Salmonella*. Proc. Natl. Acad. Sci. U.S.A. *106*, 7997–8002.

Sourjik, V. (2004). Receptor clustering and signal processing in *E. coli* chemotaxis. Trends Microbiol. *12*, 569–576.

Stafford, G.P., Evans, L.D., Krumscheid, R., Dhillon, P., Fraser, G.M., and Hughes, C. (2007). Sorting of early and late flagellar subunits after docking at the membrane ATPase of the type III export pathway. J. Mol. Biol. *374*, 877–882.

Stewart, P.S. (2002). Mechanisms of antibiotic resistance in bacterial biofilms. Int. J. Med. Microbiol. *292*, 107–113.

Szurmant, H., and Ordal, G.W. (2004). Diversity in chemotaxis mechanisms among the bacteria and archaea. Microbiol. Mol. Biol. Rev. *68*, 301–319.

Thomas, D.R., Francis, N.R., Xu, C., and DeRosier, D.J. (2006). The three-dimensional structure of the flagellar rotor from a clockwise-locked mutant of *Salmonella enterica* serovar Typhimurium. J. Bacteriol. *188*, 7039–7048.

Toguchi, A., Siano, M., Burkart, M., and Harshey, R.M. (2000). Genetics of swarming motility in *Salmonella enterica* serovar Typhimurium: critical role for lipopolysaccharide. J. Bacteriol. *182*, 6308–6321.

Tolker-Nielsen, T., Christensen, A.B., Holmstrom, K., Eberl, L., Rasmussen, T.B., Sternberg, C., Heydorn, A., Molin, S., and Givskov, M. (2000). Assessment of flhDC mRNA levels in Serratia liquefaciens swarm cells. J. Bacteriol. *182*, 2680–2686.

Venter, J.C., Remington, K., Heidelberg, J.F., Halpern, A.L., Rusch, D., Eisen, J.A., Wu, D., Paulsen, I., Nelson, K.E., Nelson, W., *et al.* (2004). Environmental genome shotgun sequencing of the Sargasso Sea. Science *304*, 66–74.

Vladimirov, N., and Sourjik, V. (2009). Chemotaxis: how bacteria use memory. Biol. Chem *390*, 1097–1104.

Vogler, A.P., Homma, M., Irikura, V.M., and Macnab, R.M. (1991). *Salmonella* Typhimurium mutants defective in flagellar filament regrowth and sequence similarity of FliI to F0F1, vacuolar, and archaebacterial ATPase subunits. J. Bacteriol. *173*, 3564–3572.

Wang, Q., Frye, J.G., McClelland, M., and Harshey, R.M. (2004). Gene expression patterns during swarming in *Salmonella* Typhimurium: genes specific to surface growth and putative new motility and pathogenicity genes. Mol. Microbiol. *52*, 169–187.

Wang, Q., Mariconda, S., Suzuki, A., McClelland, M., and Harshey, R.M. (2006a). Uncovering a large set of genes that affect surface motility in *Salmonella enterica* serovar Typhimurium. J. Bacteriol. *188*, 7981–7984.

Wang, Q., Suzuki, A., Mariconda, S., Porwollik, S., and Harshey, R.M. (2005). Sensing wetness: a new role for the bacterial flagellum. EMBO J. *24*, 2034–2042.

Wang, S., Fleming, R.T., Westbrook, E.M., Matsumura, P., and McKay, D.B. (2006b). Structure of the Escherichia coli FlhDC complex, a prokaryotic heteromeric regulator of transcription. J. Mol. Biol. *355*, 798–808.

Wilharm, G., Lehmann, V., Krauss, K., Lehnert, B., Richter, S., Ruckdeschel, K., Heesemann, J., and Trulzsch, K. (2004). Yersinia enterocolitica type III secretion depends on the proton motive force but not on the flagellar motor components MotA and MotB. Infect. Immun. *72*, 4004–4009.

Wolfe, A.J., and Visick, K.L. (2008). Get the message out: cyclic-Di-GMP regulates multiple levels of flagellum-based motility. J. Bacteriol. *190*, 463–475.

Yamamoto, S., and Kutsukake, K. (2006). FliT acts as an anti-FlhD2C2 factor in the transcriptional control of the flagellar regulon in *Salmonella enterica* serovar Typhimurium. J. Bacteriol. *188*, 6703–6708.

Yonekura, K., Maki, S., Morgan, D.G., DeRosier, D.J., Vonderviszt, F., Imada, K., and Namba, K. (2000). The bacterial flagellar cap as the rotary promoter of flagellin self-assembly. Science *290*, 2148–2152.

Zhai, Y.F., Heijne, W., and Saier, M.H., Jr. (2003). Molecular modeling of the bacterial outer membrane receptor energizer, ExbBD/TonB, based on homology with the flagellar motor, MotAB. Biochim. Biophys. Acta. *1614*, 201–210.

Salmonella-secreted Virulence Factors 10

Fred Heffron*, George Niemann, Hyunjin Yoon, Afshan Kidwai, Roslyn N.E. Brown, Jason D. McDermott, Richard Smith and Joshua N. Adkins

Abstract

Research in the past 20 years has shown that *Salmonella* precisely manipulates their host by hierarchical secretion of virulence factors (effectors). More than 40 secreted virulence factors have been identified in *Salmonella*, but the function and mammalian targets of only a few are known. Effectors are directed to specific sub-cellular compartments and mammalian targets, and they mediate a diverse array of activities. Thus, the first half of this review focuses upon our understanding of effector mechanisms and their roles during infection. However, the known effector repertoire is incomplete and the second half of this review places an emphasis on discovery. Computer analysis identified common secretion motifs and predicted that as many as 300 additional proteins may be secreted by *Salmonella*. In fact, mass spectrometry analysis identified a more complete secretome and found many novel, uncharacterized effector proteins. Several effectors identified in this study were small proteins of only 30–100 amino acids in length, suggesting that they are not enzymes but agonists or antagonists of specific host factors. One surprise from the mass spectrometry analysis was the identification of proteins that are secreted to mammalian cells via outer membrane vesicles. Complete characterization of the bewildering array of secreted proteins will take many years.

Introduction

Non-typhoidal *Salmonella* serovars such as Typhimurium, as discussed here, do not generally cause sepsis but are a leading cause of gastroenteritis in the US and a major cause of childhood mortality in developing countries. Typhoid fever, caused by typhoidal *Salmonella* serovars such as Typhi or Paratyphi A, is still common outside the developed nations. This may be surprising given the availability of effective vaccines, but they are too costly or logistically difficult to disseminate in many countries (Galen *et al.*, 2009). In this short review we discuss secreted virulence factors that directly affect *Salmonella* interaction with its host.

Salmonella secretes protein to subvert host defences but also to reduce virulence thereby permitting the bacteria to persist longer and to more successfully disperse. The type III secretion system (TTSS) is the best known and studied of the mechanisms that enable secretion from the bacterial cytoplasm to the host cell cytoplasm. Other secretion systems include outer membrane vesicles, which are present in all Gram-negative bacteria examined to date, two-partner secretion, and type VI secretion, and these will also be addressed. Excellent reviews of *Salmonella* secreted effectors have focused on themes such as actin rearrangements, vesicular trafficking, ubiquitination, and the activities of the virulence factors themselves (Ramsden *et al.*, 2007b; Steele-Mortimer, 2008; McGhie *et al.*, 2009; Santos *et al.*, 2009). This short review is based on *Salmonella* Typhimurium infection of mice because it is a model of typhoid like disease in humans. We have organized effectors in terms of events that happen during the infection cycle and how secreted effectors may be involved as shown in Fig. 10.1. We also highlight recent results of a proteomic study that identified the spectrum of proteins

*Corresponding author

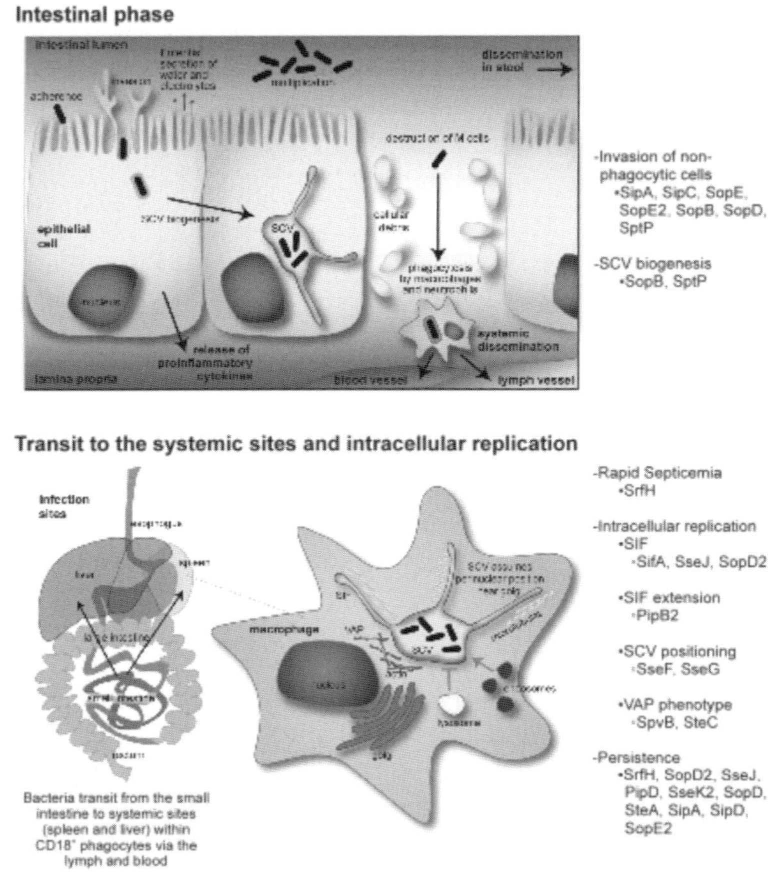

Figure 10.1 Steps to systemic infection and associated type III secreted effectors. The top panel shows the initial interaction between *Salmonella*, the intestinal epithelium, and the M cells. *Salmonella* Typhimurium may become systemic in the mouse via two routes of infection – either within CD18 positive cells that pass into the bloodstream from the lumen of the intestine or via destruction of M cells and passage through the lymphatic system. Few of the *Salmonella* genes that manipulate host cell trafficking have been identified; *srfH/sseI* is the only example to date. The bottom panel shows the events observed within macrophages. Once within macrophages, and perhaps neutrophils, *Salmonella* creates a unique vacuolar space (SCV) that avoids anti-microbial components and permits replication. At least 10 type III effectors associate with the SCV: SptP, PipB, PipB2, SifB, SifA, SseJ, SseF, SseG, SopD, and SopD2. Five effectors inhibit innate immune signalling – AvrA, SptP, SseL, SspH1, and SrfH. Effector functions are poorly understood, and it is likely that many possess redundant activities.

secreted into acidic minimal medium from various *Salmonella* genetic mutants. In the course of this work we used a machine learning approach to select the most likely secreted effectors from the several hundred proteins we detected. An adaptive immune response is required for clearance of a systemic *Salmonella* Infection. However, the response is usually delayed or weak indicating that *Salmonella* actively subverts host immunity both in humans and mice.

Animal models of infection

Mice are the most widely used *in vivo* model system for *Salmonella* infection. C57/BL6 and BALB/c are common strains of mice used to study *Salmonella* infection, but they both carry a mutation in the natural resistance-associated macrophage protein 1 (Nramp1/Slc11a1), making them suitable only as an acute infection model. The *Nramp1* gene encodes a metal ion transporter, is expressed uniquely in cells of myeloid origin, and inhibits growth of some intracellular pathogens apparently by reducing both iron and manganese concentrations within the phagosome (Liu *et al.*, 1995; Vidal *et al.*, 1995). Mouse strains that are Nramp$^+$ such as 129/SvJ are better suited for persistence studies. These mice will not resolve a persistent infection for weeks following oral infection and the bacteria are resident in the mesenteric lymph nodes, spleen and liver (Monack *et al.*, 2004). In either the mouse model or in human infection the initial contact between *Salmonella* and the intestinal wall results in an inflammatory response with release of pro-inflammatory cytokines that attract neutrophils and macrophages to the site of infection. *Salmonella* specifically infects these professional phagocytic cells and lives within them. Some of the infected CD18 positive cells carry *Salmonella* from the lumen of the intestine to systemic sites of infection via the lymph and blood (Gerichter, 1960; Vazquez-Torres *et al.*, 1999; Worley *et al.*, 2006). Normally mice do not show a very strong inflammatory response in the intestine following oral infection, unlike infection of *Salmonella* Typhimurium in humans. However, the inflammatory response to *Salmonella* is exacerbated if the mice are first treated with streptomycin, which thus provides a mouse model for gastroenteritis and inflammatory bowel disease (Stecher *et al.*, 2004; Hapfelmeier and Hardt, 2005). The antibiotic removes a part of the intestinal flora that normally dampens the inflammatory response, although the precise change in the microbiome is not yet known (ibid and B. Finlay, personal communication). An adaptive immune response is required for clearance of a systemic *Salmonella* infection in both humans and mice. However, the response is usually delayed or weak, indicating that *Salmonella* actively subverts host immunity.

Type III secretion

Interactions between the bacterial pathogens and the host are dependant on the presence of bacterial surface proteins and secreted virulence factors. Many of the known secreted effectors are injected into the eukaryotic cell via a type III secretion system. This is a complex structure evolutionarily related to bacterial flagella that uses the proton motive force (Paul *et al.*, 2008) to inject specific proteins into the eukaryotic cell cytoplasm via a protein tube, i.e. the needle complex. The two type III secretion systems encoded by *Salmonella* are about 40 kilobases in length and were acquired independently by horizontal transmission (see Fig. 10.2). They are located at 63 minutes (SPI-1) and at 31 minutes (SPI-2) on the chromosome. Both encode the structural components of their respective apparatus, and some effectors. However, the effectors are not well conserved and often have little to no homology with proteins of known function. The lack of conservation of even a secretion signal and the fact that many effectors are not co-located on the chromosome with a secretion apparatus has made it difficult to identify them. Table 10.1 provides a summary of all known *Salmonella* effectors and their primary characteristics and function. The following paragraphs elaborate on these effectors in greater detail.

Effectors during progression of a systemic *Salmonella* infection

The intestinal phase of infection and SPI-1 secretion

The roles of SPI-1 and SPI-2 TTSS systems are distinct. SPI-1 secreted effectors are essential for invasion of epithelial cells and promote intestinal inflammation and gastroenteritis. The SPI-1 TTSS apparatus is also required to kill the M cells that sample antigens and occlude access to gut-associated lymphoid tissue (Jones *et al.*, 1994). While this is the major route by which *Salmonella* Typhimurium becomes systemic in mice, the bacteria may also be carried to the bloodstream within CD18 positive cells. *Salmonella* causes inflammation by exposing cells to bacterial products that bind to toll-like receptors (TLRs), nod-like receptors (NLR), cytoplasmic and potentially other receptors as well (Monack, 2008). The interaction starts an inflammatory cascade leading to secretion of pro-inflammatory cytokines and chemokines as discussed in greater detail below and in Chapter 12 of this book. Chemokines recruit professional phagocytes to the lumen of the intestine. In humans this results in inflammatory diarrhoea, but mice do not normally present this

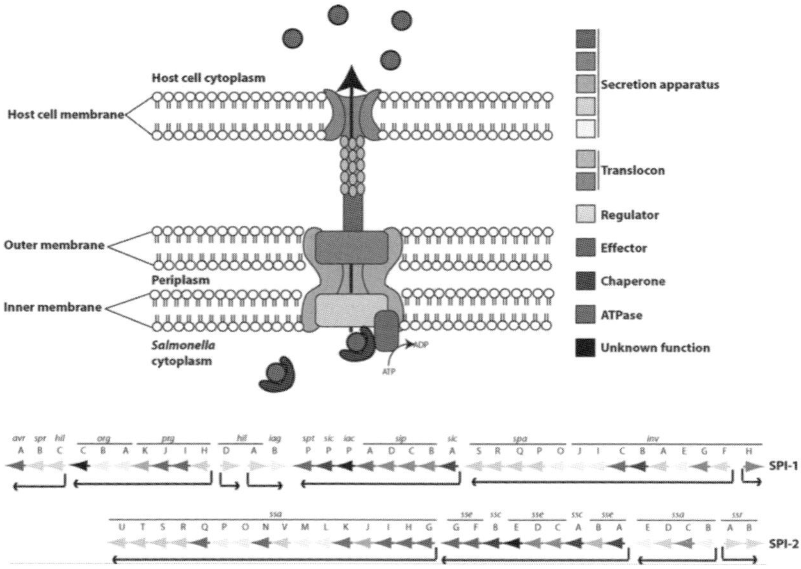

Figure 10.2 Comparison of the two type III secretion systems in *Salmonella*. Reprinted, with permission, from Kaoru Geddes (PhD thesis, Oregon Health and Science University, 2008). A colour version of this figure is located in the plate section at the back of this book.

phenotype. The precise reason is unknown, but TLRs are expressed at lower levels or are even absent on mouse enterocytes lining the small intestine and may account for the disparate phenotypes (Bambou *et al.*, 2004). The SPI-1 effectors SipA, SipC, SopB, SopD, SopE, SopE2, and SptP are important at the early intestinal stage of infection and are discussed in more detail below.

SPI-1 effectors SopE, SopE2, SopB and SptP

Microscopic observation of *Salmonella* infecting epithelial cells first revealed an amazing series of events (Yokoyama *et al.*, 1987). *Salmonella* comes into contact with a host cell, which evaginates by sending out membrane ruffles, which captures a bacterium, and the bacterium is internalised. This is descriptive of what is observed following cell contact, but membrane ruffles are not required for uptake (Hanisch *et al.*, 2009). Four secreted effectors mediate cell invasion: SopE, SopE2, SopB, and SptP.

SopE and SopE2 both function as guanine exchange factors (GEF)s for Rac, Cdc42, RhoA and RhoG leading to localized membrane ruffling and macropinocytosis. SopE2 more strongly activates Cdc42, although SopE activation of Rac alone is sufficient for internalization (reviewed in Zhou and Galan, 2001; Cain *et al.*, 2008). Surprisingly, neither SopE nor SopE2 share primary sequence homology to a mammalian counterpart, but their three dimensional structure resembles the cognate GEF (Buchwald *et al.*, 2002; Huang *et al.*, 2009). SopE and SopE2 also activate nuclear factor kappa B (NF-κB) via mitogen-activated protein kinases (MAPK) JNK, p38, and extracellular signal-regulated kinase (ERK) (Molero *et al.*, 2009). SopE also activates caspase I via cellular Rho GTPases Rac-1 and Cdc42, and is thus one of the inflammatory stimulators (Muller *et al.*, 2009).

SopB is a phosphoinositide phosphatase of PI(4,5)P2, the products of which activate SH3-containing guanine nucleotide exchange factor SGEF (Hernandez *et al.*, 2004). This in turn activates the small GTP-binding protein RhoG, which remodels actin via Arp2/3 (Patel and Galan, 2006). Hydrolysis of PI(4,5)P2 coincides with recruitment of the host soluble NSF attachment receptor (SNARE) protein VAMP8, which promotes membrane fission and macropinosome formation to engulf bacteria (Dai *et al.*, 2007).

Both SPI-1 and SPI-2 TTSSs secrete SptP, but SPI-2 does so approximately five fold less efficiently (Geddes *et al.*, 2005). SptP is a guanine activating protein (GAP) that deactivates Rho, Rac, and Cdc42 to reverse the effects of SopE, SopE2, and SopB. In addition, there are other mechanisms that limit the effects of SopE and SopE2. SopE has a very short half life in the cell; thus, while injected at the same rate as SptP, the short half life of SopE allows reversal of Cdc42 activation (Kubori and Galan, 2003). There is a temporal order of secretion in which SopE, the GEF, is injected first and SptP, the GAP, second (Winnen *et al.*, 2008). These examples demonstrate exquisite regulation, a property that probably applies to many secreted effector proteins. SptP has a second function encoded in its C-terminal portion. It associates with the *Salmonella*-containing vacuole (SCV) and encodes a C-terminal tyrosine phosphate phosphatase activity that dephosphorylates the AAA+ ATPase valosin-containing protein (VCP/p97) (Humphreys *et al.*, 2009). The protein is important for membrane fusion and presumably helps to block fusion with specific vesicles.

SPI-1 secreted effectors SipA, SipC and SopD

Recently, amazing pictures of *Salmonella* injecting type III effectors have shown that SipA is present at about 10^3 molecules per bacterium prior to cell invasion and that it is injected until the cytoplasmic store is depleted (Schlumbeger *et al.*, 2005). The micrographs showed localization of the effector within COS7 cells at the point of bacterial contact. It is not yet known how many other effector proteins are stored in the bacterium prior to cell contact allowing an immediate response. SipA injection into enterocytes activates protein kinase C alpha (PKCα) signaling pathways that lead to the secretion of IL-8 via Jun and p38 MAPK. IL-8 is a CXC chemokine and a potent chemoattractant for both neutrophils and macrophages (Silva *et al.*, 2004). Specifically, SipA induces CXC chemokine expression resulting in phosphorylation of IL-8-transcription regulatory proteins c-Jun and p38 (Figueiredo *et al.*, 2009). The pro-inflammatory stimulation is thought to recruit professional phagocytes which serve as a replicative niche for *Salmonella* (Dunlap *et al.*, 1992; Geddes *et al.*, 2007). SipA also increases cell invasion frequency by modulating the formation of F-actin (Galkin *et al.*, 2002; Lilic *et al.*, 2003). SipA prevents actin depolymerization by displacing and preventing F-actin interaction with cellular actin depolymerizing protein ADF and cofilin. Additionally, SipA potentiates SipC actin polymerization and both of these effectors limit host membrane ruffling to the point of bacterial contact (Hayward and Koronakis, 2002; Le Clainche and Drubin, 2004; McGhie *et al.*, 2004; Patel and Galan, 2005; Perrett and Jepson, 2009). SipC interacts with actin via two separate domains. The N-terminus bundles actin whereas the C-terminus nucleates actin assembly leading to filament growth from barbed ends (Hayward and Koronakis, 1999). However, SipC's primary function is to form part of the pore in the eukaryotic membrane (Chang *et al.*, 2005; Chang *et al.*, 2007). SipC and SipA exemplify the complexity of type III effectors, an emerging theme in *Salmonella* pathogenesis.

The function of SopD is less clear. SopD was first identified in *Salmonella* Dublin as a SPI-1 secreted effector that, along with SopB, induced inflammation in the ileal mucosa of calves (Jones *et al.*, 1998). SopD increases invasion efficiency and is recruited to the site of cell invasion by a process that requires the inositol phosphatase activity of SopB (Bakowski *et al.*, 2007).

In humans, inflammatory diarrhea is accompanied by an increased white blood cell count in the stool. Diarrhea improves the conditions for *Salmonella* replication in the lower bowel by eliminating some of the normal flora and by providing an increase in nutrients. Mice do not normally present gastroenteritis, but the influx of phagocytic cells provides an intracellular niche for a systemic infection via the blood and lymph (Vazquez-Torres *et al.*, 1999; Stecher *et al.*, 2007). *Salmonella* Typhi is also reported to enter the bloodstream quickly after oral infection in humans, presumably via carriage in monocytes (Gerichter, 1960).

Following oral infection, IL-1ß, IL-18, IL-17 and other factors are secreted as chemoattractants for professional phagocytes and to increase the microbicidal activity of neighbouring cells (see also Chapter 12 of this book). A multiply deleted strain missing all SPI-1 effectors was constructed and the effectors were then replaced one by one to show the strong effect for SipA. SipA induces

Table 10.1 *Salmonella* secreted effectors

Effector	Gene location	Homologues	Activity	Host cell target(s)
***Salmonella* effectors that require the SPI-1 TTSS for their translocation**				
SipA (SspA) (STM2882)	SPI-1	*Shigella* IpaA	Actin binding/ stabilizing	Actin,T-plastin
SipB (SspB) (STM2885)	SPI-1	*Shigella* IpaB	SPI-1 TTSS translocon component	Cholesterol
SipC (SspC) (STM2884)	SPI-1	*Shigella* IpaC	SPI-1 TTSS translocon component, actin nucleation and bundling	Actin
SipD (SspD) (STM2883)	SPI-1	*Shigella* IpaD	SPI-1 TTSS needle tip	
SopA (STM2066)	Outside SPI-1	Putative EHEC effector (GenBank NP_309587.1)	E3 ubiquitin ligase	HsRMA-1
SopB (SigD) (STM1091)	SPI-5	*Shigella* spp. IpgD, cellular 4-phosphatases, synaptojanin	Inositol polyphosphate phosphatase	Cdc42, key substrate probably phosphatidyl inositol-4–5 bisphosphate [PI(4,5)P2]

Infection phenotypes	Stage of infection	References
Contributes to membrane ruffles and actin rearrangements associated with invasion. Not required for entry, but increases uptake efficiency. Acts as a catalyst for actin polymerization. C-terminus interacts with F-actin and stabilizes filaments. Prevents actin depolymerization by displacing and preventing F–actin interaction with cellular actin depolymerizing proteins ADF and cofilin. Internal F1 and F2 domains drive SifA focus formation and SipA–SipA interactions, respectively. Potentiates SipC actin polymerization, triggers PMN transmigration, maintains perinuclear SCV positioning by cooperating with SPI-2 effector SifA, disrupts tight junctions, induces CXC chemokine expression	Invasion and phagocytosis; replication and SCV positioning; pro/anti-inflammatory responses	Zhou *et al.* (1999a), Zhou *et al.* (1999b), Lee *et al.* (2000), McGhie *et al.* (2001), Galkin *et al.* (2002), Hayward and Koronakis (2002), Zhang *et al.* (2002), Lilic *et al.* (2003), Cain *et al.* (2004), McGhie *et al.* (2004), Silva *et al.* (2004), Hapfelmeier and Hardt (2005), Raffatellu *et al.* (2005), Boyle *et al.* (2006), Patel and Galan (2006), Bakowski *et al.* (2007), Brawn *et al.* (2007), Schlumberger *et al.* (2007), Wall *et al.* (2007), Cain *et al.* (2008), Popp *et al.* (2008), Winnen *et al.* (2008), Figueiredo *et al.* (2009), Perrett and Jepson (2009)
SPI-1 effector delivery, apoptosis of phagocytes	Invasion and phagocytosis	Hersh *et al.* (1999), Hernandez *et al.* (2003), Hayward *et al.* (2005), Fink and Cookson (2007), Layton and Galyov (2007), Mueller *et al.* (2008)
SPI-1 effector delivery, N-terminus bundles actin filaments. C-terminus nucleates actin assembly leading to filament growth from barbed ends, induces membrane ruffling	Invasion and phagocytosis	Hayward and Koronakis (1999), McGhie *et al.* (2001), Hayward and Koronakis (2002), Cain *et al.* (2004), Cain *et al.* (2008), Mueller *et al.* (2008)
Regulates SPI-1 translocon assembly and effector secretion	Invasion and phagocytosis	Mueller *et al.* (2008)
Target of ubiquitin ligase domain unknown. However, recognized, ubiquitinated, and degraded through HsRMA-1 mediated pathway. Degradation permits *Salmonella* escape from vacuole into cytoplasm of epithelial cells. Induces PMN transmigration, localizes to mitochondria	SVC maturation and trafficking	Zhang *et al.* (2002), Layton *et al.* (2005), Raffatellu *et al.* (2005), Zhang *et al.* (2005), Layton and Galyov (2007), Diao *et al.* (2008)
Invasion: Promotes membrane fission and macropinosome formation, recruits SNARE protein VAMP8 for phagocytosis of bacteria, activates Akt at the plasma membrane which may promote epithelial cell survival. During entry Cdc42 interaction thought to localize effector at plasma membrane. SCV: maintains perinuclear SCV positioning, recruits Rab5 to SCV which in turn recruits Vps34 (a PI3-kinase that phosphorylates PI to PI3P). Monoubiquitination at 9 different lysine residues triggers cellular relocalization from plasma membrane to SCV. E3 ligase that ubiquitinates SopB unknown. Inflammation: triggers nitric oxide production in macrophages, promotes fluid secretion in calf ileal loops, disrupts tight junctions	Invasion and phagocytosis; SVC maturation and trafficking; replication and SCV positioning; pro/anti-inflammatory responses	Norris *et al.* (1998), Steele-Mortimer *et al.* (2000), Marcus *et al.* (2001), Zhou *et al.* (2001), Terebiznik *et al.* (2002), Zhang *et al.* (2002), Cain *et al.* (2004), Hernandez *et al.* (2004), Patel and Galan (2005), Raffatellu *et al.* (2005), Boyle *et al.* (2006), Dukes *et al.* (2006), Patel and Galan (2006), Bakowski *et al.* (2007), Dai *et al.* (2007), Kuijl *et al.* (2007), Layton and Galyov (2007), Mason *et al.* (2007), Bakowski *et al.* (2008), Bujny *et al.* (2008), Cain *et al.* (2008), Mallo *et al.* (2008), Wasylnka *et al.* (2008), Patel *et al.* (2009)

Table 10.1 continued

Effector	Gene location	Homologues	Activity	Host cell target(s)
SopD (STM2945)	Outside SPI-1/SPI-2	*Salmonella* SopD2		
SopE[a]	Phage SopEϕ	*Salmonella* SopE2	Guanine exchange factor (GEF) mimic	Rac-1, Cdc42
SopE2 (STM1855)	In vicinity of phage remnants	*Salmonella* SopE	GEF mimic	Cdc42
***Salmonella* effectors that are secreted by both SPI-1 and SPI-2 TTSS**				
AvrA (STM2865)	SPI-1	*Yersinia* YopJ/P	Cysteine protease, deubiquitinase, acetyltransferase	MKK4/7 IκBα, β-catenin
GtgE (STM1055)	Phage Gifsy-2			
PagJ (STM2585)	Phage Gifsy-1			
SlrP (STM0800)	Outside SPI-1/SPI-2	N-terminus: *Salmonella* SifA, SifB, SrfH, SseJ, SspH1/H2. Leucine rich repeat: *Salmonella* SspH1/2, GogB. C-terminus: *Salmonella* SspH1/2, *Yersinia* YopM, *Shigella* IpaH7.8/9.8	E3 ubiquitin ligase	Thioredoxin
SpvC (PSLT038)	pSLT (*Salmonella* Typhimurium)	*Shigella* OspF, *Pseudomonas syringae* HopAI1	Phosphothreonine lyase	
SpvD (PSLT037)	pSLT (*Salmonella* Typhimurium)			

Infection phenotypes	Stage of infection	References
Promotes membrane fission and macropinosome formation, works cooperatively with SopB and promotes invasion of T84 cells, required for optimal replication in mouse macrophages, induces fluid secretion in infected ileum	Invasion and phagocytosis; pro/anti-inflammatory responses	Zhang *et al.* (2002), Jiang *et al.* (2004), Raffatellu *et al.* (2005), Bakowski *et al.* (2007), Layton and Galyov (2007)
Induces membrane ruffling and proinflammatory responses, promotes fusion of SCV with early endosomes, disrupts tight junctions. After translocation ubiquitinated and degraded by a proteosome mediated pathway.	Invasion and phagocytosis; SCV maturation and trafficking; pro/anti-inflammatory responses	Hardt and Galan (1997), Mukherjee *et al.* (2001), Zhou *et al.* (2001), Kubori and Galan (2003), Kubori and Galan (2003), Cain *et al.* (2004), Patel and Galan (2005), Boyle *et al.* (2006), Patel and Galan (2006), Cain *et al.* (2008), Winnen *et al.* (2008)
Induces membrane ruffling and proinflammatory responses, increases macrophage iNos expression, disrupts tight junctions	Invasion and phagocytosis; Pro/anti-inflammatory responses	Bakshi *et al.* (2000), Stender *et al.* (2000), Zhang *et al.* (2002), Cain *et al.* (2004), Patel and Galan (2005), Raffatellu *et al.* (2005), Boyle *et al.* (2006), Patel and Galan (2006), Layton and Galyov (2007)
Inhibits inflammation, represses apoptosis and epithelial innate immunity, stabilizes tight junctions. Deubiquitinates and stabilizes IκBα and β-catenin in epithelial cells. Consequently β-catenin activated while NF-κB activation and apoptosis inhibited	Pro/anti-inflammatory responses	Ye *et al.* (2007), Jones *et al.* (2008), Liao *et al.* (2008)
		Ho *et al.* (2002), this study
		Figueroa-Bossi *et al.* (2001), this study
Confers host specificity, decreases thioredoxin activity, promotes cell death, interferes with antigen presentation in dendritic cells	Pro/anti-inflammatory responses	Miao *et al.* (1999), Tsolis *et al.* (1999), Miao *et al.* (2003), Rohde *et al.* (2007), Halici *et al.* (2008), Bernal-Bayard and Ramos-Morales (2009)
Required for full virulence in mice		Li *et al.* (2007), Chen *et al.* (2008), Mazurkiewicz *et al.* (2008)
		This study

Table 10.1 continued

Effector	Gene location	Homologues	Activity	Host cell target(s)
SptP (STM2878)	SPI-1	N-terminus: *Yersinia* YopE, *Pseudomonas aeruginosa* ExoS. C-terminus: cellular tyrosine phosphatases, *Yersinia* YopH	GTPase activating protein (GAP) mimic, tyrosine phosphatase	Cdc42, Rac-1, VCP, vimentin
SseK1 (STM4157)	Outside SPI-2	*Salmonella* SseK2/3, *Citrobacter rodentium* NleB, EHEC Z4328		
SspH1[b]	Phage Gifsy-3	N-terminus: *Salmonella* SifA, SifB, SrfH, SseJ, SlrP, SspH2. Leucine rich repeat: *Salmonella* SspH2, SlrP, GogB. C-terminus: *Salmonella* SspH2, SlrP, *Yersinia* YopM, *Shigella* IpaH7.8/9.8	E3 ubiquitin ligase	PKN1
SteA (STM1583)	Outside SPI-2			
SteB (STM1629)	Outside SPI-2		Putative dipicolinate reductase	
***Salmonella* effectors that require the SPI-2 TTSS for their translocation**				
CigR (STM3762)	SPI-3			
GogB (STM2584)	Phage Gifsy-1	N-terminus: *Yersinia* YopM, *Shigella* IpaH7.8/9.8, *Salmonella* SspH1/2, SlrP. C-terminus: *Yersinia* YP2634/ Y1471, rabbit EPEC OrfL, EHEC 0157:H7 Z1829		
GtgA (STM1026)	Phage Gifsy-2	GogA (STM2614), PipA (STM1087)		
PipB (STM1088)	SPI-5	*Salmonella* PipB2		

Infection phenotypes	Stage of infection	References
Antagonist of SopE/E2, returns host cytoskeleton to resting state following bacterial entry. N-terminal GAP activity deactivates Rho GTPases Rac and Cdc42. C-terminal tyrosine phosphatase activity targets the AAA+ ATPase VCP (valosin-containing protein). Dephosphorylation of VCP leads to increased bacterial replication. Downregulates proinflammatory responses. Degraded by proteosome mediated pathway. Ubiquitination likely, but not demonstrated	Replication and SCV positioning; pro/anti-inflammatory responses	Fu and Galan (1999), Murli *et al.* (2001), Kubori and Galan (2003), Kubori and Galan (2003), Lin *et al.* (2003), Cain *et al.* (2004), Patel and Galan (2005), Cain *et al.* (2008), Winnen *et al.* (2008), Humphreys *et al.* (2009)
		Kujat Choy *et al.* (2004)
LRR domain interacts with PKN1, down-regulates proinflammatory responses	Pro/anti-inflammatory responses	Miao *et al.* (2003), Miao *et al.* (2003), Haraga and Miller (2006), Rohde *et al.* (2007)
Required for efficient mouse spleen colonization		Geddes *et al.* (2005)
		Geddes *et al.* (2005)
Hypervirulence phenotype in 129/SVJ mice		Blanc-Potard *et al.* (1999), this study
		Coombes *et al.* (2005), Ramsden *et al.* (2007a)
		Smith and Ahmer (2003), this study
Targeted with PipB2 to detergent resistant lipid rafts on the SCV and SIF		Knodler *et al.* (2003), Knodler and Steele-Mortimer (2005), Steele-Mortimer (2008)

Table 10.1 continued

Effector	Gene location	Homologues	Activity	Host cell target(s)
PipB2 (STM2780)	Outside SPI-2	*Salmonella* PipB		Kinesin-1
SifA (STM1224)	Outside SPI-2	N-terminus: *Salmonella* SrfH, SifB, SseJ, SspH1/2, SlrP. C-terminus: EPEC EspM2/3, *Shigella* IpgB1	GEF mimic, Rab GTPase antagonist	SKIP, Rab9
SifB (STM1602)	Outside SPI-2	N-terminus: *Salmonella* SifA, SrfH, SseJ, SspH1/2, SlrP		
SopD2 (STM0972)	Outside SPI-2	*Salmonella* SopD		
SpiC (SsaB) (STM1393)	SPI-2			Hook3, TassC
SpvB (PSLT039)	pSLT (*S.* Typhimurium)	N-terminus: *Photorhabdus luminescens* TcaC, *Bacillus cereus* VipB2, *Clostridium botulinum* C2 toxin component 1/C3 toxin, *Clostridium perfringens* IotA	ADP ribosyl transferase, inhibits actin polymerization *in vitro*, depolymerizes F-actin upon transfection	Actin
SseB (STM1398)	SPI-2		SPI-2 TTSS needle tip	
SseC (STM1400)	SPI-2		SPI-2 TTTS translocon component	
SseD (STM1401)	SPI-2		SPI-2 TTTS translocon component	

Infection phenotypes	Stage of infection	References
Recruits kinesin-1 to SCV, promotes SIF extension by affecting late endosome/lysosone distribution, targeted with PipB to detergent resistant lipid rafts on the SCV and SIF, contributes to cell to cell spread, interferes with antigen presentation in dendritic cells	Replication and SCV positioning; pro/anti-inflammatory responses	Knodler *et al.* (2003), Knodler and Steele-Mortimer (2005), Henry *et al.* (2006), Halici *et al.* (2008), Szeto *et al.* (2009)
Required for full virulence in mice, N-terminus interacts with SKIP, C-terminus has WxxxE motif and structure characteristic of a GEF mimic, inhibits Rab9 interaction with SKIP, required for SCV membrane integrity and SIF formation, co-expression with SseJ promotes SIF formation in cells, maintains perinuclear SCV positioning, redirects exocytic vesicles to SCV, interferes with antigen presentation in dendritic cells	Replication and SCV positioning; pro/anti-inflammatory responses	Stein *et al.*, (1996), Brumell *et al.* (2001a), Brumell *et al.* (2001b), Ruiz-Albert *et al.* (2002), Miao *et al.* (2003), Harrison *et al.* (2004), Boucrot *et al.* (2005), Knodler and Steele-Mortimer (2005), Alto *et al.* (2006), Henry *et al.* (2006), Kuhle *et al.* (2006), Brawn *et al.* (2007), Handa *et al.* (2007), Ramsden *et al.* (2007a), Arbeloa *et al.* (2008), Halici *et al.* (2008), Jackson *et al.* (2008), Ohlson *et al.* (2008), Steele-Mortimer (2008)
Targeted to SCV and SIF		Freeman *et al.* (2003), Miao *et al.* (2003)
Contributes to SIF formation with SifA and SseF/G, required for efficient bacterial replication in macrophages and mice, interferes with antigen presentation in dendritic cells	Replication and SCV positioning; pro/anti-inflammatory responses	Jiang *et al.* (2004), Halici *et al.* (2008)
Interferes with vesicular trafficking via interactions with Hook3 and TassC, role in VAP and SIF formation, controls order of protein export through SPI-2 TTSS, prevents dendritic cells from activating T cells, regulates expression of flagella which affects pro-inflammatory MAPK signalling	SCV maturation and trafficking; pro/anti-inflammatory responses	Uchiya *et al.* (1999), Lee *et al.* (2002), Shotland *et al.* (2003), Tobar *et al.* (2006), Steele-Mortimer (2008), Uchiya and Nikai (2008), Uchiya *et al.* (2009)
Inhibition of VAP and SIF formation, apoptosis of infected cells via caspase-3, required for full virulence in mice	SCV maturation and trafficking; replication and SCV positioning	Lesnick *et al.* (2001), Miao *et al.* (2003), Birmingham *et al.* (2005), Ramsden *et al.* (2007a), Browne *et al.* (2008)
		Nikolaus *et al.* (2001), Mueller *et al.* (2008)
		Nikolaus *et al.* (2001)
		Nikolaus *et al.* (2001)

Table 10.1 continued

Effector	Gene location	Homologues	Activity	Host cell target(s)
SseF (STM1404)	SPI-2			
SseG (STM1405)	SPI-2			
SrfH (SseI/GtgB) (STM1051)	Phage Gifsy-2	N-terminus: *Salmonella* SifA/B, SlrP, SseJ, SspH1/2		Filamin, TRIP6, IQGAP1
SseJ (STM1613)	Outside SPI-2	N-terminus: *Salmonella* SifA/B, SlrP, SrfH, SspH1/2	Deacylase, phospholipase A, glycerophospholipid–cholesterol acyltransferase	Cholesterol
SseK2 (STM2137)	Outside SPI-2	*Salmonella* SseK1/3, *Citrobacter rodentium* NleB, EHEC Z4328		
SseK3[c]	ST64B coliform bacteriophage	*Salmonella* SseK1/2, *Citrobacter rodentium* NleB, EHEC Z4328		
SseL (STM2287)	Outside SPI-2		Cysteine protease with deubiquitinase activity	IκBα
SspH2 (STM2241)	In vicinity of phage remnants	N-terminus: *Salmonella* SspH1, SlrP, SifA/B, SrfH, SseJ. LRR: *Salmonella* GogB. C-terminus: *Yersinia* YopM, *Shigella* IpaH7.8/9.8	Inhibits actin polymerization *in vitro*	Filamin, profilin
SteC (STM1698)	Outside SPI-2	Human kinase Raf-1	Serine/Threonine kinase	
SteD (STM2139)	Outside SPI-2			

Infection phenotypes	Stage of infection	References
Contributes to SIF formation with SifA, SopD2 and SseG, immunoprecipitates with SseG, recruits dynein to SCV, maintains perinuclear SCV positioning, required for formation of microtubule bundles around SCV, redirects exocytic transport vesicles to SCV	Replication and SCV positioning	Kuhle and Hensel (2002), Kuhle *et al.* (2004), Abrahams *et al.* (2006), Deiwick *et al.* (2006), Kuhle *et al.* (2006), Ramsden *et al.* (2007a)
Contributes to SIF formation with SifA, SopD2 and SseF, immunoprecipitates with SseF, recruits dynein to SCV, maintains perinuclear SCV positioning, required for formation of microtubule bundles around SCV, redirects exocytic transport vesicles to SCV	Replication and SCV positioning	Kuhle and Hensel (2002), Salcedo and Holden (2003), Kuhle *et al.* (2004), Deiwick *et al.* (2006), Kuhle *et al.* (2006), Ramsden *et al.* (2007a)
Co-localizes with polymerizing actin cytoskeleton, promotes phagocyte motility and rapid septicemia after oral infection		Miao *et al.* (2003), Worley *et al.* (2006), McLaughlin *et al.* (2009)
Required for full virulence in mice, over-expression negatively regulates SIF formation in infected cells, destabilizes SCV in Δ*sifA* mutants, co-expression with SifA promotes SIF formation	SCV maturation and trafficking; replication and SCV positioning	Ruiz-Albert *et al.* (2002), Freeman *et al.* (2003), Miao *et al.* (2003), Birmingham *et al.* (2005), Ohlson *et al.* (2005), Ramsden *et al.* (2007a), Lossi *et al.* (2008), Nawabi *et al.* (2008), Ohlson *et al.* (2008)
		Kujat Choy *et al.*, (2004)
		Kujat Choy *et al.* (2004)
Impairs IκBα ubiquitination but not phosphorylation thereby blocking NF-κB activation, inhibits macrophage apoptosis, downregulates inflammatory responses	Pro/anti-inflammatory responses	Rytkonen *et al.* (2007), Le Negrate *et al.* (2008)
Remodels SCV-associated F actin; interferes with antigen presentation in dendritic cells, N-terminus localizes protein to apical membrane of epithelial cells, internal LRR domain interacts with C-terminal E3 ubiquitin ligase domain and inhibits ligase activity. Localization to membrane thought to break this association and restrict ubiquitin ligase to particular cellular regions	SCV maturation and trafficking; pro/anti-inflammatory responses	Miao *et al.*, (2003), Rohde *et al.* (2007), Halici *et al.* (2008), Quezada *et al.* (2009)
Kinase activity required for actin remodelling, associated with VAP phenotype	SCV maturation and trafficking	Geddes *et al.* (2005), Poh *et al.* (2008)
Hypervirulence phenotype in 129/SVJ mice		Unpublished

Table 10.1 continued

Effector	Gene location	Homologues	Activity	Host cell target(s)
***Salmonella* effectors secreted independent of TTSS**				
PagK1 (STM1867)				
PagK2 (STM2585A)				
PagC (STM1246)				
PagD (STM1244)				
SrfN (STM0082)				
SssA (STM0359)				
SssB (STM1478)		YdgH domain conserved in Enterobacteriaceae		
ZirS (STM1668)				

[a]Found in *Salmonella* Typhimurium SL1344 but not 14028 or LT2.

[b]Found in *Salmonella* Typhimurium 14028 but not SL1344 or LT2.

[c]Found in *Salmonella* Typhimurium 14028 and SL1344 but not LT2.

PMN, polymorphonuclear; PSLT, virulence plasmid from *S. enterica* serovar Typhimurium LT2; SCV, *Salmonella* containing vacuole; SIF, *Salmonella* induced filaments; STM, *S. enterica* serovar Typhimurium LT2; VAP, SCV-associated actin polymerization.

polymorphonuclear (PMN) migration in polarized cells through PKC-α-mediated secretion of a pathogen-elicited epithelial chemoattractant, presumably the CXC chemokines (see above).

Non-type III secretion in the intestine

The proteins ZirT/ZirS (STM1668/1669) are conserved throughout the *Salmonella* genus and form a two-partner secretion system in which ZirT provides a pore through which ZirS is secreted (Gal-Mor *et al.*, 2008). This system is expressed during intestinal colonization and in fecal pellets, where it reduces *Salmonella* virulence. ZirS is related to a soluble lytic murein transglycosylase that is involved in cell remodeling, although how this might impact *Salmonella* virulence is unclear. *Salmonella* also encodes a type VI secretion system. The type VI secretion system was horizontally acquired and is located on an island at 3' on the chromosome (Bingle *et al.*, 2008). One of the key components is a homologue of *icmF* (intracellular multiplication) from *Legionella* (Purcell and Shuman, 1998; Segal *et al.*, 1998; Parsons and Heffron, 2005). IcmF is required for *Salmonella* virulence and was originally reported as a component of a type IV secretion system, but with the exception of DotU, the other type IV components are missing (Filloux *et al.*, 2008). IcmF and presumably the rest of the type VI secretion system is expressed extracellularly and, like ZirT/S, appears to function as an anti-virulence factor (Parsons and Heffron, 2005). Anti-virulence factors are thought to increase intra- and inter-host dissemination by preventing acute infections that are quickly cleared from the host.

Infection phenotypes	Stage of infection	References
Vesicle effector		Unpublished
Vesicle effector		Unpublished
Vesicle effector		Unpublished
Vesicle effector		Unpublished
Vesicle effector		Unpublished
Unknown secretion mechanism		Unpublished
Unknown secretion mechanism, hypervirulence phenotype in 129/SVJ mice		Unpublished
Two-partner secretion with ZirT (STM1669)		Gal-Mor *et al.* (2008)

The intracellular phase of infection including dissemination

Intracellular *Salmonella* resides and replicates within a vesicle known as the *Salmonella* containing vacuole (SCV), and it modifies the cellular architecture to permit replication by altering vesicular trafficking as described in detail below. Following internalization of a bacterium, effectors are secreted via SPI-2 into epithelial cells or professional phagoctyes such as macrophages. SPI-2 TTS promotes *Salmonella* survival and replication within phagocytes. Since these cells constitute the main reservoirs at systemic sites, a SPI-2 mutant fails to persist after infection (Lineham *et al.*, 2005). Worley *et al.* reported that some effectors (SrfH or SseI) are secreted exclusively to macrophages and not epithelial cells (Worley *et al.*, 2000). However, direct visualization of effector secretion within the spleens of infected mice did not show cell specific secretion, although only 6 effectors were examined, excluding SrfH (Geddes *et al.*, 2007). Secretion of SPI-1 effectors is not necessary for internalization by macrophages nor for survival within these cells in an *in vitro* model. However *in vivo*, the SPI-1 secretion apparatus and the SPI-1 central regulators HilA, HilD, InvF, and HilC are all required for persistence (Lawley *et al.*, 2006; Yoon *et al.*, 2009). Whether this discrepancy is a consequence of growth in cell types other than professional phagocytes is not yet clear.

Once within the cell *Salmonella* must alter the cellular architecture to permit *Salmonella* replication and it does this by completely altering vesicular traffic (Drecktrah *et al.*, 2007; reviewed in Bakowski *et al.*, 2008; see also Chapter 11 of this book). Vesicular transport initiates at the Golgi where an amino acid sequence determines a protein's glycosylation pattern, which in turn directs transport within the cell. At least one effector, SteA, is targeted to the Golgi, which is not surprising given the central role the Golgi plays in assembling and releasing vesicles. Vesicular compartments bud from the *trans*-Golgi and are transported along the microtubules. Vesicles are distinguished by both v-and t-SNAREs, which fuse membranes, and by specific monomeric G proteins called Rabs, which are related to the cytoplasmic portion of G-coupled receptors. Rabs are the primary distinguishing feature of different vesicular compartments and attach to vesicles via prenylation of a C-terminal cysteine. For fusion, the target vesicle must contain the appropriate t-SNARE, Rab, and Rab binding proteins. There are more than 60 Rabs encoded within the mammalian genome and at least that number of v-, t-SNARE, and Rab binding proteins. When Rabs are active they associate with specific vesicles and permit fusion although the precise steps are still being investigated. *Salmonella* has evolved strategies to avoid anti-microbial components by blocking fusion with vesicles that contain NADPH oxidase and inducible nitric oxide synthase while allowing fusion with other vesicles that presumably contain nutrients (reviewed in Fang and Vazgquez-Torres, 2002) Rab manipulation likely facilitates these phenotypes. Fig. 10.3 reprinted from Smith *et al.* (Smith *et al.*, 2007), illustrates the complexity of Rab recruitment to a vesicle containing either wild type *Salmonella* or *Salmonella* within an atypical phagosome. It is not yet clear how the various *Salmonella* effectors alter the Rab profile of the SCV, but the process appears to be complex. Following phagocytosis of *Salmonella,* early endosomal markers such as EEA1, Rab5, transferrin receptor and Rab11 are all recruited to the SCV (Steele-Mortimer *et al.*, 1999). Further maturation results in the recruitment of vesicular ATPase, Rab7, Rab11a, Rab11b, and lysosomal associated membrane protein LAMP1. Vesicular ATPase acidifies the SCV, and Rab7 recruits the dynein-dynactin motor complex through Rab7 interacting lysosomal protein (RILP). The function of the other markers is unknown, but they are thought to help stabilize the vacuole (Garcia-del Portillo and Finlay, 1995; Meresse *et al.*, 1999; Hashim *et al.*, 2000). Additionally, the mannose-6-phosphate receptor, a late endosomal marker, is excluded from the SCV (Garcia-del Portillo and Finlay, 1995). The final stages of SCV maturation are characterized by vacuole-associated actin polymerization (VAP) and *Salmonella* induced filaments (SIF), membrane tubules that radiate throughout the cell. The function of the filamentous vesicles that are the product of this maturation could be manifold such as allowing cell-to-cell spread, nutrient acquisition, as well as avoidance of anti-microbial mechanisms. For more details on SIF function, see Chapter 11 of this book.

While the SPI-1 secretion system is not required for uptake or replication in professional phagocytes some of its effectors synergize with SPI-2 effectors in the creation of the SCV. The SPI-1 effector SopB affects internalization as well as SCV biogenesis. After translocation, the effector initially localizes to the host plasma membrane and increases the efficiency of uptake as previously described. SopB is then mono ubiquitinated at nine different lysine residues. This relocalizes SopB to the nascent SCV where it recruits Vps34 and Rab5 (Knodler *et al.*, 2009). Both SopB and Vps34 are phoshpoinositide phosphatases and are thought to enrich the SCV with PI3P.

SPI-2 secreted effectors SifA and SseJ

Ten effectors are recruited to the cytoplasmic face of the SCV – SptP, PipB, PipB2, SopD, SopD2, SseF, SseG, SseJ, SifB, and SifA. Of these only *sifB* and *pipB* mutants do not have virulence phenotypes. The virulence phenotype of the remaining eight vary considerably. The effect of the SPI-2 effectors, particularly SifA, on the intracellular life style of *Salmonella* is described in more detail in Chapter 11 of this book. Here, we will discuss the basic characteristics and functions of these effectors.

SifA was the first SPI-2 effector discovered via penicillin selection for non-dividing intracellular bacteria. It plays a central role in intracellular replication both in epithelial cells and macrophages (Stein *et al.*, 1996). SifA is necessary for the formation of *Salmonella* induced filamentous (SIF) vesicles after which it is named (Stein *et al.*, 1996; Brumell *et al.*, 2001a). The gene is located in the polyamine transport (*pot*) operon, but it is regulated independently. SifA contains a

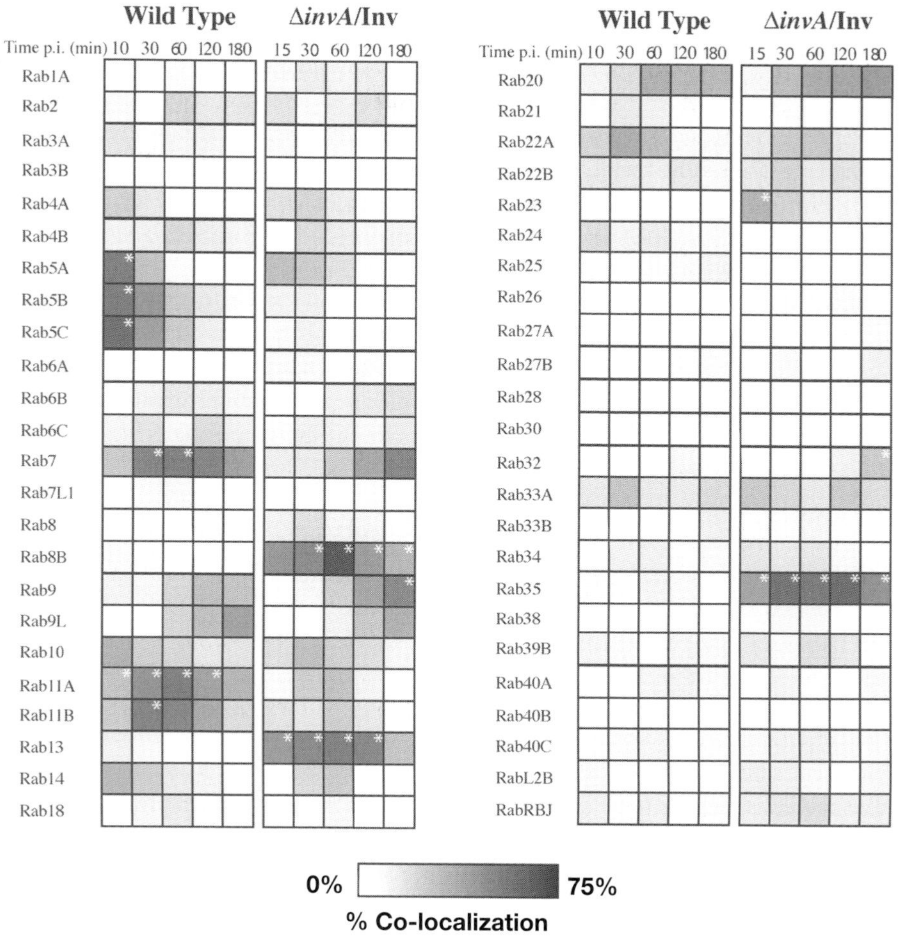

Figure 10.3 Rab GTPase localization on wild-type and Δ*invA*/Inv *Salmonella* Typhimurium phagosomes. InvA is a component of the SPI-1 TTSS. An *invA* mutant is deficient for secretion and is non-invasive. Expression of the invasin (*inv*) locus from *Yersinia pseudotuberculosis* allowed the bacteria to enter cells via an alternate mechanism, but they were targeted to late endosomes and unable to replicate. Wild-type and Δ*invA*/Inv *Salmonella* Typhimurium phagosomes were enumerated for their association with 48 Rab GTPases at the indicated times. The scale represents 0 to 75% of vacuoles co-localizing with a Rab GTPase as indicated by degree of shading.*$P < 0.05$, comparing differences between wild-type and Δ*invA*/Inv vacuoles. Reprinted from Smith *et al.* (2007).

carboxy-terminal CaaX motif, which is prenylated by geranylgeranyl transferase-1 and thus is attached to the SCV membrane in the same way as Rabs (Reinicke *et al.*, 2005). It also contains a WxxxE motif that mimics a RhoA GAP and consequently localizes RhoA to the SCV (Alto *et al.*, 2006; Alto and Dixon, 2008; Ohlson *et al.*, 2008). SifA kinesin interacting protein (SKIP) attaches the SCV to microtubules. SifA also interacts with Rab7. The Rab7 interaction uncouples binding with Rab interacting lysosomal protein (RILP), preventing dynein recruitment and movement of the SCV.

The effector SseJ is expressed at high levels following *Salmonella* internalization and is also found on the cytoplasmic face of the SCV. Purified SseJ has cholesterol deacylase and acyltransferase activity *in vitro*. SseJ catalytic-triad mutants that reduce deacylase activity are attenuated for virulence in mice, indicating that SseJ enzymatic activity contributes to intracellular replication in host tissues (Freeman *et al.*, 2003). One function of SseJ must be to alter the composition of the SCV membrane. SseJ interacts with RhoA, SifA, and SKIP (Ohlson *et al.*, 2005; Lossi *et al.*, 2008; Nawabi *et al.*, 2008; Ohlson *et al.*, 2008). The function of SseJ appears to be to counteract some of the effects of SifA, because a strain containing a mutation in *sifA* escapes the vacuole but in the double mutant with *sseJ Salmonella* remains within the vacuole (Ruiz-Albert *et al.*, 2002). Furthermore, co-expression of *sseJ* and *sifA* in a mammalian cell results in filamentous vesicles similar to those observed following *Salmonella* infection, but expression individually does not (ibid).

SPI-2 effectors SseF, SseG, PipB, PipB2 and SopD2

SseF and *sseG* encode effectors of about 30 and 25 kDa, respectively. SseF and SseG interact and, along with SifA, direct the SCV to the perinuclear/Golgi region. SseF contains three transmembrane domains suggesting that it may be inserted directly into membranes. Both are required for virulence but no activity is ascribed to them (Abrahams *et al.*, 2006; Deiwick *et al.*, 2006; Ramsden *et al.*, 2007a; Ramsden *et al.*, 2007b).

PipB and PipB2 are related effectors that associate with detergent resistant lipid rafts on the SCV and at similar vesicular membranes found elsewhere in the cell (Knodler *et al.*, 2003). PipB is not required for virulence in an acute infection model, but PipB2 is required. PipB2 extends SIF vesicles during intracellular replication and relocates late endosomes/lysosomes from a perinuclear location to the cell periphery (Knodler and Steele-Mortimer, 2005). SIF extension may occur by direct interaction between PipB2 and kinesin (Henry *et al.*, 2006). Interestingly, ectopic expression of PipB2 overcomes dominant active mutations in Rab7 and Rab34 that position vesicles near the microtubule organizing center suggesting that PipB2 acts downstream of these two Rabs (Knodler and Steele-Mortimer, 2005). SopD2 shares 47% identity with the SPI-1 effector SopD. SopD2 is required for virulence as a mutant is impaired in SIF formation (Jiang *et al.*, 2004), but the precise function of SopD2 is unknown. In summary, 10 effectors associate with the SCV, but little is understood about their mechanisms and how they alter the intracellular environment to allow intracellular parasitism. Better understanding of their function will provide new insights into how vesicular traffic is redirected in the cell.

SPI-2 effectors SpvB and SteC

Salmonella begins replication about 2–6 hours after cell infection depending to some extent on the cell type infected. Actin accumulates around the vesicle by a process requiring SpvB and SteC although a mutation in SteC, related to the mammalian Raf kinase, is fully virulent (Geddes *et al.*, 2005; Poh *et al.*, 2008). SpvB ADP-ribosylates actin and depolymerizes actin filaments when expressed in cultured epithelial cells (Gotoh *et al.*, 2003). It is possible that SpvB and SteC serve redundant functions in infected cells and will not show a phenotype unless both are deleted.

Signal transduction pathways central to multiple intracellular pathogens

The phosphoinositide activity of the SPI-1 effector SopB is known to activate AKT1 (Marcus et al., 2001; Molero et al., 2009), a serine-threonine kinase that promotes bacterial invasion, inhibits apoptosis, and aids in the maturation of the SCV. The first phenotype is attributable to PAK4. AKT1 controls actin dynamics by activating PAK4 which links Rho GTPases Cdc42 and Rac1 to the actin cytoskeleton (Kuijl et al., 2007). Since Cdc42 and Rac1 are also targeted by SopE and SopE2, *Salmonella* appears to manipulate Rho GTPases from multiple directions to promote internalization. Activated AKT1 inhibits apoptosis via several mechanisms, but the particular pathway is unknown. For example, AKT1 stimulates IKK to phosphorylate and activate the p65 subunit of NF-κB which controls the expression of several anti-apoptotic proteins such as FLIP, XIAP, survivin, and IAP-2. AKT1 also inhibits cell apoptosis via interaction with mitochondria-related Bcl-2 family proteins (He et al., 2007). The effects on SCV maturation occur through AKT1 activation of AS160, the Rab14-GAP, which blocks Rab14 mediated phagosome-lysosome fusion. In human macrophages AKT1 inhibitors blocked *Salmonella* replication but did not affect bacterial viability, demonstrating that AKT1 activation may be required for intracellular replication. Moreover, chemical and siRNA inhibitors of AKT1 blocked the intracellular replication of multiple intracellular pathogens. These results suggest that AKT1 signaling networks play a central role, but infection of a mouse model that expressed a dominant negative version of AKT1 was not described (Kuijl et al., 2007).

Alteration in vesicular traffic and human disease

The alteration in vesicular traffic mediated by *Salmonella* can be compared to some of the human genetic disorders associated with defects in vesicular transport (Schweitzer *et al.*, 2009). Several neurodegenerative diseases including Huntington's and Niemann-Pick Type C show gross abnormalities in trafficking including an increase in expression of autophagy, although the function of the mutant gene is not known. Oculocutaneous albinism can be caused by a mutation in Rab38, which is expressed in the skin and alveolar type II cells, resulting in a defect in exocytosis of surfactant and melanin. Gaucher's disease is the most common lysosomal storage disease and an autosomal recessive deficiency of the lysosomal enzyme, acid beta-glucosidase (glucocerebrosidase). Glucocerebrosidase hydrolyzes glucosylceramide into glucose and ceramide. Ceramide is a precursor of sphingolipids, second messenger, and regulator of vesicular traffic (van Blitterswijk *et al.*, 2003). Surprisingly, the filamentous lysosomes observed in patients with Gaucher's disease are remarkably similar to those observed for *Salmonella* infection (Reuser and Drost, 2006).

Dissemination from the intestine and systemic infection

Following the initial interaction between *Salmonella* and a host cell, the cell can be considered to be at an unstable equilibrium in which it can survive and mount an active inflammatory response to infection or it can undergo cell death (Micheau and Tschopp, 2003; Bruey *et al.*, 2007). Cell death, with the exception of apoptosis, results in massive release of cytokines and provides an alarm signal to the surrounding tissue. There are many factors that contribute to an outcome. Following release of inflammatory cytokines, chemotactic macrophages may be killed by direct interaction between SipB and caspase 1 (Chen *et al.*, 1996; Monack *et al.*, 1996, 2000). Additionally, *Salmonella* kills macrophages depending on its growth phase (i.e. late log which induces SPI-1, or stationary phase which induces SPI-2) as a result of expression of different effectors after cell infection (van der Velden *et al.*, 2000). In the ileum, SPI-2 is induced prior to intestinal penetration, suggesting that SPI-2 can be expressed under multiple conditions and that the interaction between *Salmonella* and cells in the intestine could also depend on the SPI-2 effectors (Brown *et al.*, 2005). *Salmonella* causes cell death by pyroptosis, a more inflammatory variant of apoptosis, in which DNA cleavage takes place and cell lysis results from the formation of membrane pores (Fink and Cookson, 2007; Fink *et al.*, 2008). *Salmonella* also activates the inflammasome via flagellin release to the cytoplasm, which stimulates platelet activating factor (PAF), a Nod like protein that activates caspase 1 (Mariathasan *et al.*, 2004; Franchi *et al.*, 2006; Miao *et al.*, 2006; Sun *et al.*, 2007). Despite all of these pro-inflammatory pathways, there is little doubt that critical, but undiscovered, determinants of *Salmonella* pathogenesis block premature cell death.

Salmonella effectors that direct the course of infection

Salmonella reaches the bloodstream within minutes after oral infection, although the number of bacteria is quite low – about 1 in a hundred thousand Typhimurium bacilli fed to a mouse (Vazquez-Torres *et al.*, 1999; Worley *et al.*, 2006), and for Typhi in humans (Gerichter, 1960). There are two pathways for dissemination from the intestinal lumen to the blood – a direct way in which the bacteria are carried in phagocytic cells that takes minutes to a few hours, and an indirect way in which the bacteria pass up the lymphatic system which takes at least 24 hours (Vazquez-Torres *et al.*, 1999). SrfH (SseI) promotes the rapid trip to the bloodstream via interaction with either TRIP6, IQGAP1, or both adaptors, to alter infected cell motility (Worley *et al.*, 2006; McLaughlin *et al.*, 2009). Both of these proteins modulate cellular motility and are important in metastasis of cancer (Yi and Beckerle, 1998; Li *et al.*, 2000). Identification of IQGAP1 as the mammalian target of SrfH was based on affinity purification using the SrfH protein from *Salmonella* Typhimurium SL1344, whereas the strain used to identify TRIP6 was *Salmonella* Typhimurium

14028. There is one amino acid difference between the two alleles and the difference may determine the specificity of interaction (M. Worley, personal communication). More experiments will be necessary before any conclusions can be drawn. It is likely that this process is more complicated than initially appreciated. A *srfH* mutant can still disseminate to the blood stream rapidly but at a five fold lower frequency. However, when the SPI-2 apparatus is mutated, dissemination does not occur suggesting that additional redundant effectors may be involved. Infected cells are not only programmed not to move but also to release cytokines that attract additional phagocytic cells. Therefore, *Salmonella* must not only alter infected cell motility but must also reduce the inflammatory response. The fact that a pathogen directs its course of infection by secretion of virulence factors is a new way of looking at infection. SrfH is the first effector described that modulates infected cell motility, thereby directing the course of infection. The major route by which *Salmonella* gains entrance to the blood stream is through the M cells, Peyer's patches, and mesenteric lymph nodes and lymphatics. No doubt several effectors will be required for this infection route but they have not been identified. Stochastic restriction occurs at several steps during infection, limiting bacteria at subsequent steps to those that first pass a barrier (bottle neck). Consequently, these genes cannot be identified from a mixed infection.

Growth in the spleen

Starting at 4 hours after cell infection, the SCV progressively fills with bacteria until they appear as large cytoplasmic holes filled with bacteria. Infection of neighbouring, non-infected cells is not well understood. Detailed studies have examined the systemic spread of virulent *Salmonella* bacteria distinguished by specific sequence tags. It was observed that an individual bacterium developed into a small, localized colony of infected cells suggesting that adjacent cells take up bacterial progeny once the original infected cell died. The bacteria then spread to areas distant from the initial colony suggesting that there may be an additional mechanism of dissemination, perhaps by carriage within phagocytic cells (Grant *et al.*, 2008). Not all interactions between *Salmonella* and phagocytic cells are successful as about 5–10% of infecting bacteria are non-replicating, found in tight vesicles, and cannot form colonies if re-isolated (Buchmeier and Heffron, 1991; D. Holden, pers. Comm.).

Avoiding innate and adaptive immune surveillance

How *Salmonella* evades the innate and adaptive immune response is only now being studied. This new direction is a consequence of a change in the animal model from mice that are very sensitive to *Salmonella* infection such as C57/BL6 and BALB/c to ones that are resistant but nevertheless develop a persistent infection. In 129/SvJ mice *Salmonella* Typhimurium infection is not cleared for weeks, long after an adaptive immune response should arise (Monack *et al.*, 2004).

Several studies have shown that the T cell response to *Salmonella* Typhimurium in mice is delayed and weak (Hess *et al.*, 1996; Mittrucker and Kaufmann, 2000; Mittrucker *et al.*, 2002; Ravindran and McSorley, 2005). An effective adaptive immune response against *Salmonella* requires many steps, including processing of *Salmonella* proteins and presentation by dendritic cells to naïve T cells in secondary lymphoid organs. Dendritic cells (DC) are the initial antigen presenting cells of the immune system and possess the unique ability to stimulate naïve T-cells directly. It has recently been shown that *Salmonella* kills a subset of DCs (CD8alpha+), perhaps creating a more favourable environment for *Salmonella* to persist, but how it kills these cells and the specific CD8alpha+ functions are not yet known (Wick, 2007). Three signals are necessary for optimal stimulation of T cells – engagement of the T cell receptor, a co-receptor (CD28 on the T cell, B7 on the antigen presenting cell or APC) and cytokines that determine the outcome of T cell differentiation. The cytokines are produced as a consequence of pathogen-activated molecular patterns (PAMPs) interacting with cell receptors – toll like receptors (TLR), NLRs and possible other receptors – and are crucial to initiate an appropriate adaptive immune response. The cytokines act in both an autocrine and paracrine fashion, ensuring activation of dendritic cells to produce the appropriate cytokines, even if they are not infected. Dendritic cells exposed to pathogens express a variety of cytokines that tailor the immune response to the infecting pathogen, and there are multiple dendritic cell subsets which further refine the response. To clear a *Salmonella* infection requires CD4, CD8 and B cells, providing many targets for *Salmonella* to disrupt the immune response (Mittrucker and Kaufmann, 2000).

Anti-inflammatory effectors that target NF-κB activation: SptP, AvrA, SseL, SspH1 and SrfH

In contrast to SPI-1 effectors, SPI-2 effectors are generally anti-inflammatory and target the innate and adaptive immune systems to promote immune evasion. To date five effectors have been identified that directly block signalling that leads to NF-κB activation – the SPI-1/2 (1/2 indicates that the effector can be secreted by either type III system) effectors SptP, SspH1 and AvrA, and the SPI-2 effectors SseL and SrfH (SseI).

SptP is a GAP mimic described in detail above and functions in part by blocking the MAPK pathways. AvrA is one of the few highly conserved effectors, first identified as an anti-virulence factor in the plant pathogen *Pseudomonas syringiae*. It is an anti-virulence factor in *P. syringiae* because the plant directly recognizes this effector by a specific receptor similar to the Nod like receptors of mammalian cells. Recognition stimulates the plant to drop the infected leaf (Napoli and Staskawicz, 1987). *Salmonella* AvrA has acetyltransferase activity toward specific mitogen-activated protein kinase kinases (MAPKKs) and potently inhibits c-Jun N-terminal kinase (JNK) and NF-κB signalling pathways (Jones *et al.*, 2008). SspH1 is a ubiquitin ligase of PKN1 (Rohde *et al.*, 2007) a protein kinase involved in the NF-κB pathway that is activated upon cell infection. However, SspH1 activity is not thought to cause PKN1 degradation because the interaction between SspH1 and PKN1 in mammalian cells leads to activation of the kinase and subsequent inhibition of NF-κB signalling (Haraga and Miller, 2006)

The effector SrfH (SseI) also inhibits the inflammatory response by inhibiting the signal transduction pathways that lead from Nod1, Nod2, and perhaps other pathogen recognition receptors to NF-κB (D. Philpott, personal communication). Expression of SPI-2 effector SseL in mamalian cells suppresses IκB activation and degradation by de-ubiquitination, blocking its degradation by the proteosome, thus inhibiting NF-κB activation (Neish et al,. 2000; Le Negrate *et al.*, 2008). In fact, mice infected with SseL-deficient bacteria mount stronger inflammatory responses than those infected with the parent strain. Following macrophage infection *in vitro*, the secretion of SseL delays cell death (van der Velden *et al.*, 2000; Rytkonen *et al.*, 2007). The identification of *Salmonella* genes that alter the immune response may have important consequences for vaccine design. In the long term it may be possible to tailor the immune response based on the role of specific proteins. Since a strain deleted for sseL gave a stronger immune response to *Salmonella* infection, it is a logical candidate for an attenuated vaccine strain.

Effectors block antigen processing and presentation

The classical MHC class I pathway of antigen processing and presentation is present in nearly all cell types. Foreign protein in the cell cytoplasm is degraded to peptides in the proteosome, transported to the endoplasmic reticulum (ER) via the transporter associated with antigen processing (TAP), loaded onto MHC-I, and then transported to the cell surface. *Salmonella* inhibits MHC-1 presentation in infected intestinal epithelial cells but not macrophages (Harding and Pfeifer, 1994), but the precise mechanism is not known. In contrast, the MHC class II pathway is present in only three cell types: macrophages, B cells, and dendritic cells. Peptides presented by MHC-II molecules are derived from extracellular proteins. As a result, peptide loading occurs by a different mechanism. Extracellular proteins are endocytosed, trafficked to lysozomes, and processed to peptides by cathepsin. MHC-II is loaded with the invariant chain in the ER, traffics to vesicles containing antigen, antigen replaces the MHC-II invariant chain, and the complex traffics to the cellular membrane. A screen for effectors that inhibited MHC-1 and MHC-2 dendritic cell antigen presentation identified SifA, SspH2, SlrP, PipB2, and SopD2, whereas SseF and SseG contributed to a lesser extent to this phenotype (Cheminay *et al.*, 2005; Halici *et al.*, 2008). Porgador and his collaborators observed that mutations in a conserved unknown operon, the *yej* operon, increased MHC-I antigen presentation (Qimron *et al.*, 2004). The expression was strongly dependent on the two-component regulator PhoP/Q and, given the fact that PhoP/Q regulates many of the proteins secreted via outer membrane vesicles (see below), it is possible that one of these conserved proteins is secreted via this mechanism. Ubiquitination of the MHC-II-peptide with concomitant reduction in surface expression has been observed following *Salmonella* infection, and the ubiquitination is dependent on an unidentified SPI-2 secreted effector(s) (Lapaque *et al.*, 2009). Ubiquitination led to removal of mature αβ dimers of peptide-carrying MHC-II from the cell surface, and affected all class II isotypes similarly. We compared MHC-I expression on the surface of infected and uninfected J774 macrophages and found no difference (D. Ellefson, D. Parker, and F. Heffron, unpublished observations).

T cell activation and replication are inhibited by *Salmonella*

Salmonella infection leads to a T cell response that is delayed or weak (Hess *et al.*, 1996; Mittrucker and Kaufmann, 2000; Mittrucker *et al.*, 2002; Ravindran and McSorley, 2005). A Th1 response characterized by IFN-γ and IL-12 expression increases the microbicidal activity of professional phagocytes and is necessary for *Salmonella* clearance. Injection of IFN-γ restricts the growth of *Salmonella* in mice (Muotiala and Makela, 1990) and, in fact, patients with a genetic deficiency in either IFN-γ or IL-12 are predisposed to *Salmonella* infection (de Jong *et al.*, 1998; Picard *et al.*, 2002; Fieschi *et al.*, 2005). Strong protective immunity to *Salmonella* follows infection with an attenuated strain such as an *aroA*$^{-}$ derivative, although it takes about one month to develop (Maskell and Hormaeche, 1985). *Salmonella* protective antigens have not been defined. However, the strongest CTL response to *Salmonella* following infection is against heat shock proteins, while the strongest antibody response is against flagella, cell surface proteins such as porins, and proteins that are expressed at the highest levels (Leclerc *et al.*, 1994; Kaufmann and Hess, 1999; Arockiasamy and Krishnaswamy, 2000; McSorley and Jenkins, 2000; McSorley *et al.*, 2000; Singh *et al.*, 2003). Presumably, cell surface proteins cross-link multiple antibody receptors on the B cell and are processed and displayed to T cells, resulting in a strong antibody response. The strong CD8 response to heat shock proteins may reflect their high level of expression following phagocytosis by the antigen presenting cell (Buchmeier and Heffron, 1990).

Two different approaches have been used to study how *Salmonella* infection may impact T cell activation and replication – *in vitro* and in the mouse. *In vitro* studies show that T cells fail to replicate upon contact with infected APCs even if the MHC-I or MHC-II has been exogenously loaded with its cognate peptide (van der Velden *et al.*, 2005). Furthermore, exposure of T cells to *Salmonella* results in down regulation of the T cell receptor ß-chain (van der Velden *et al.*, 2008). This step requires direct contact between the bacterium and the T cell, but does not depend on SPI-1 or SPI-2 secretion. One possible explanation is the secretion of *Salmonella* outer membrane vesicles containing virulence factors, which may be strongly induced upon contact with mammalian cells. The alternative *in vivo* approach produced results showing that other factors also come into play. Live or heat killed *Salmonella* injected into mice resulted in immediate activation of *Salmonella* specific T cells, as judged by expression of the activation marker CD69. However, the activated T cells were lost from the mouse after a few days, suggesting that activated T cells were depleted as a consequence of *Salmonella* infection (Moon and McSorley, 2009; Srinivasan *et al.*, 2009). Therefore, these cells were unable to enter the memory pool or become effector cells. The depletion appears to be independent of T cell antigen specificity, but dependent on factors secreted by SPI-2.

Identification of new effectors

Mass spectrometry

We have used mass spectrometry to identify many of the proteins that *Salmonella* secretes into media using growth conditions that partially mimic the intracellular environment (see Fig. 10.4). We identified 78% of the known type III repertoire. To validate new effectors, we made fusions between candidate novel effectors and the calmodulin-dependent CyaA' fragment of *Bordetella pertussis* (Sory *et al.*, 1995). Following infection of cells we measured cAMP levels to determine if the effector fusions reached the eukaryotic cell cytoplasm. For the identification of novel secreted proteins we also relied on a machine-learning algorithm (SIEVE) (Samudrala *et al.*, 2009). SIEVE scores the likelihood with which a given protein is secreted based on a number of separate factors derived from the protein sequence (see below for a more complete description). Coupling SIEVE with the MS identified proteins proved to be an excellent strategy for selecting candidates for secretion analysis. In fact, all but one of the 204 secreted proteins identified with high confidence by MS with a SIEVE score ≥1.5 were secreted into mammalian cells using the cAMP assay. The only exception was PipA. PipA contributes to enteropathogenesis and is located in a pathogenicity island; perhaps the conditions we employed were not optimal for its detection (Wood *et al.*, 1998). Using this approach we identified six new type III secreted effectors listed in Table 10.1. In addition, we found effectors secreted by an unknown mechanism, and several others secreted in outer membrane vesicles, given their presence in vesicular bodies that contain OmpA. The newly identified type III effectors include: STM3762 (*cigR*; SPI-2), STM2585 (*steE*; SPI-1/2), PSLT037 (*spvD*; SPI-1/2), STM1026 (*gtgA*; SPI-2), STM1055 (*steE*) and STM2139 (*steD*; SPI-2). *GtgE* has been previously shown to be required for virulence in mice (Ho *et al.*, 2002). Non-polar deletions of each of these have been constructed and assayed for virulence by persistence in 129/SvJ mice. Deletion of *spvD*, *steE*, *gtgE*, and *steD* significantly attenuated colonization of mouse spleens by day 7 post infection. Our data agree with the previously reported virulence phenotype for GtgE, but not for SpvD. SpvD was shown to be dispensable for virulence in Balb/C mice (Matsui *et al.*, 2001; Ho *et al.*, 2002). The disparity may be attributable to different mouse strains. No new effectors were found to be secreted by only SPI-1, presumably because the samples were not prepared under SPI-1 inducing conditions. However, we identified some known SPI-1 effectors including SipA and SopB and the SPI-1 translocon proteins SipB, SipC and SipD. This may reflect inefficient secretion via SPI-2 or low level expression of SPI-1 in acidic minimal media, the condition we employed.

However, the majority of proteins identified by MS were either outer membrane proteins or highly expressed cytoplasmic proteins such as ribosomal and heat shock proteins, suggesting some leakage of intracellular contents. Surprisingly, analysis of more than 108 samples resulted in detection of the same subset of about 30 abundant *Salmonella* proteins within the secretome, while other proteins of even higher abundance were never detected, including the cytoplasmic protein DnaK. The level of expression of *Salmonella* proteins is based on spectral counts in previous MS experiments (Adkins *et al.*, 2006).

It has become clear that *Salmonella* uses mechanisms other than type III secretion to subvert host defenses, including outer membrane vesicles (Wai *et al.*, 2003; Spano and Galan, 2008; Deatherage *et al.*, 2009). Proteins that are secreted independently from SPI-2 identified in our MS study include PagC, PagD, PagK1, PagK2, SrfN, STM0359 (SssA) and STM1478 (YdgH; SssB), although there are likely to be more. We observed that PagC, PagD, PagK1, PagK2, and SrfN are present in a vesicle that can be purified via differential centrifugation and that also contains OmpA, a marker commonly used to identify outer membrane vesicles (OMVs) (Henry *et al.*, 2004; Kim *et al.*, 2008; Deatherage *et al.*, 2009). These secreted proteins may not only interfere with the infected cell but also with surrounding cells (Spano and Galan, 2008). Immunohistochemistry has shown that Pag proteins secreted in OMVs are present in the cytoplasm of the mammalian cell in a punctate pattern but separate from the internalized bacteria.

With regards to the specific subset of abundant ribosomal and heat shock proteins observed in the supernatant, our earlier study did suggest that these abundant proteins may be processed for secondary functions (Manes *et al.*, 2007). Many of these are partially conserved in mammals. Expression of a conserved peptide on an antigen

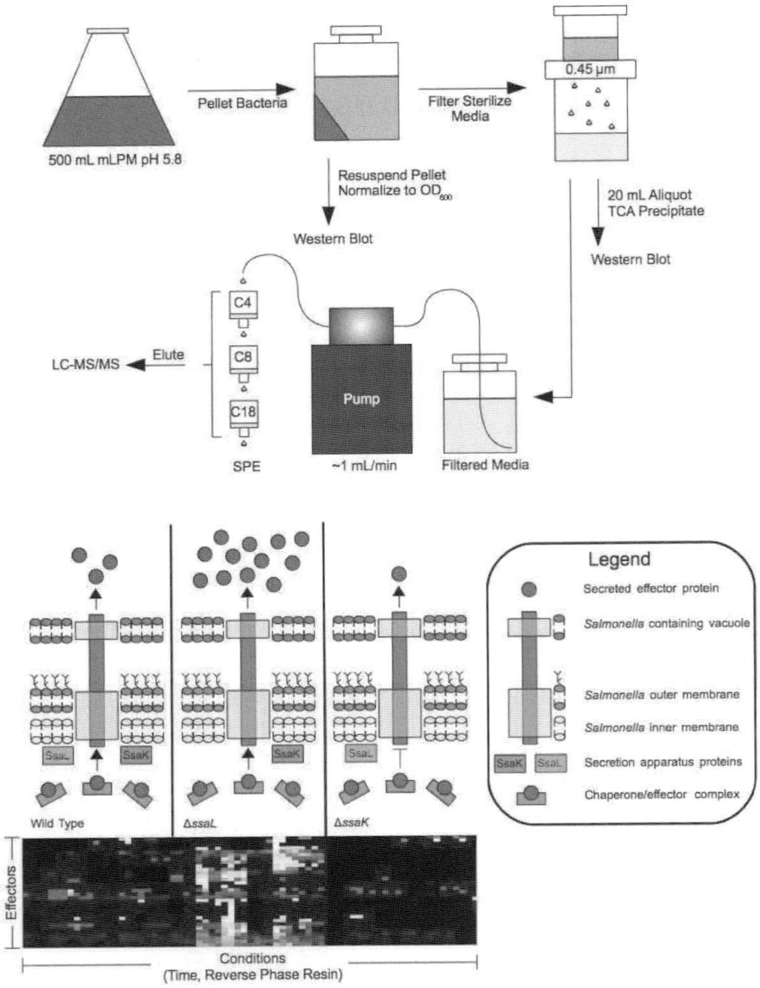

Figure 10.4 Identification of the *Salmonella* secretome by mass spectrometry. *Salmonella* was grown in minimal acidic media – conditions that partially mimic the intracellular environment. After 8 hours of growth the bacteria were removed by centrifugation and filtered to remove residual bacteria. Supernatant was tested to verify that DnaK, an abundant cytoplasmic protein, was not present. The cleared supernatant was passed through three reverse phase columns in series to collect proteins and peptides from the supernatant. The least hydrophobic column, C4, is preferred for longer peptides and proteins because they can be eluted more easily. Two mutations were used in the study – a mutation of SPI-2 encoded *ssaL* (which still permits secretion into media) and a mutation in *ssaK,* an essential structural component of the secretion apparatus. SsaL is a component of a barrier structure that controls protein secretion, and its deletion results in a dramatic increase of secretion under the conditions shown. At the bottom of the figure is a heat map for known type III effectors in which the amount of protein recovered is indicated by the intensity. As shown, most known effectors were secreted in the *ssaL* mutant but not in the *ssaK* mutant, validating the approach.

presenting cell may break tolerance resulting in autoimmunity, particularly in patients that are predisposed because of their haplotype (HLA B27). We speculate that the secreted proteins we observe are in fact secreted at high levels by phagocytosed bacteria within APCs and break tolerance resulting in reactive arthritis (Lacoste *et al.*, 2007; Girschick *et al.*, 2008; Townes *et al.*, 2008). The complete elucidation of the proteins secreted within outer membrane vesicles by *Salmonella* and other pathogens may explain the strong association between Gram-negative bacterial infections and reactive arthritis.

Computational Identification of Type III Secreted Proteins in *Salmonella*

The discovery of novel secreted effectors has been hindered by the lack of sequence similarity between them. There is one family of related effectors in *Salmonella* (Miao *et al.*, 1999) but most have little or no sequence similarity with other effectors in any pathogen. Computational mining of genomic sequence information has been used to examine sequence features of effectors in several pathogens to identify features that direct type III secretion (Guttman *et al.*, 2002; Petnicki-Ocwieja *et al.*, 2002; Greenberg and Vinatzer, 2003). These approaches have been marginally successful in *Salmonella*. To address the lack of broadly applicable computational methods to identify secreted substrates from protein sequences we developed a data integration method using machine learning techniques that can accurately identify secreted substrates of type III systems in a range of bacteria (Samudrala *et al.*, 2009). Machine learning describes a set of computational approaches in which a method is 'trained' on a set of known examples to learn the relationships between the input features from each example that allow the best discrimination between the positive and negative example sets. Briefly, our study found that type III secreted substrates could be identified with high specificity and sensitivity using the SVM-based Identification and Evaluation of Virulence Effectors (SIEVE) algorithm, that integrates several genomic-level features of proteins. Initially, a very large number of criteria was used for each known effector, but this long list was subsequently reduced to the minimal number that provided correct prediction with reasonable sensitivity and specificity. We found that a combination of evolutionary conservation and distribution, amino acid composition and amino acid sequence in the N-terminal 30 residues of the proteins allowed confident classification of known type III substrates (effectors). Furthermore, the method was found to have >90% sensitivity and specificity when trained on a set of substrates from one genus of bacteria (e.g. the plant pathogen *Pseudomonas syringae*) and then applied to another genus that is not highly related (*Salmonella* Typhimurium). Two other studies reiterated our findings and developed similar approaches, although they did not focus on *Salmonella* (Arnold *et al.*, 2009; Lower and Schneider, 2009). Using SIEVE, we were able to determine a highly degenerate consensus signal for type III secretion that is located in the N termini of secreted substrates. This putative secretion signal sequence is detectable by SIEVE among many known effectors in *Salmonella* Typhimurium, as well as in effectors from other pathogens. However, SIEVE detects patterns that are apparently conserved within a variety of secreted proteins aside from those secreted via the type III system. For example, PagD has a very high SIEVE score and is secreted in outer membrane vesicles.

As mentioned above, validation of our predictions resulted in greater than 90% accuracy for the SIEVE method and aided in identification of six novel type III secreted effectors in *Salmonella* Typhimurium. From our analysis we estimate that there may be as many as 300 substrates targeted to the type III secretion systems in *Salmonella* Typhimurium. This number seems surprisingly large, but is supported by the discovery of a large complement of effectors secreted via the type IV system in *Legionella pneumophila* (Burstein *et al.*, 2009). We have developed a web server (http://www.sysbep.org/sieve/) that allows the analysis of entire uploaded genomes by SIEVE (McDermott *et al.*, manuscript in preparation). Even if there are only 100 effectors secreted by *Salmonella*, they will eventually need to be studied in detail. This is a daunting task that will provide a challenge for the current and next generation of scientists.

Acknowledgements

Support for this work was provided by the National Institute of Allergy and Infectious Diseases NIH/DHHS through interagency agreement Y1-A1-8401-01 and by grant NIH/NIAID A1022933-22A1 to F.H. Significant portions of this work were performed in the Environmental Molecular Science Laboratory, a United States Department of Energy (DOE) office of Biological and Environmental Research (BER) national scientific user facility at Pacific Northwest National

Laboratory (PNNL) in Richland, WA. PNNL is operated for the DOE by Battelle Memorial Institute under contract DE-AC05-76RLO-1830. This work used instrumentation and capabilities developed under support from the National Center for Research Resources (Grant RR 018522 to RDS) and the DOE's Office of Biological and Environmental Research.

References

Abrahams, G.L., Muller, P., and Hensel, M. (2006). Functional dissection of SseF, a type III effector protein involved in positioning the *Salmonella*-containing vacuole. Traffic *7*, 950–965.

Adkins, J.N., Mottaz, H.M., Norbeck, A.D., Gustin, J.K., Rue, J., Clauss, T.R., Purvine, S.O., Rodland, K.D., Heffron, F., and Smith, R.D. (2006). Analysis of the *Salmonella* Typhimurium proteome through environmental response toward infectious conditions. Mol. Cell Proteomics 5, 1450–1461.

Alto, N.M., and Dixon, J.E. (2008). Analysis of Rho-GTPase mimicry by a family of bacterial type III effector proteins. Methods Enzymol. *439*, 131–143.

Alto, N.M., Shao, F., Lazar, C.S., Brost, R.L., Chua, G., Mattoo, S., McMahon, S.A., Ghosh, P., Hughes, T.R., Boone, C., and Dixon, J.E. (2006). Identification of a bacterial type III effector family with G protein mimicry functions. Cell *124*, 133–145.

Arbeloa, A., Bulgin, R.R., MacKenzie, G., Shaw, R.K., Pallen, M.J., Crepin, V.F., Berger, C.N., and Frankel, G. (2008). Subversion of actin dynamics by EspM effectors of attaching and effacing bacterial pathogens. Cell Microbiol. *10*, 1429–1441.

Arnold, R., Brandmaier, S., Kleine, F., Tischler, P., Heinz, E., Behrens, S., Niinikoski, A., Mewes, H.W., Horn, M., and Rattei, T. (2009). Sequence-based prediction of type III secreted proteins. PLoS Pathog. *5*, e1000376.

Arockiasamy, A., and Krishnaswamy, S. (2000). Homology model of surface antigen OmpC from *Salmonella* Typhi and its functional implications. J. Biomol. Struct. Dyn. *18*, 261–271.

Bakowski, M.A., Braun, V., and Brumell, J.H. (2008). *Salmonella*-containing vacuoles: directing traffic and nesting to grow. Traffic *9*, 2022–2031.

Bakowski, M.A., Cirulis, J.T., Brown, N.F., Finlay, B.B., and Brumell, J.H. (2007). SopD acts cooperatively with SopB during *Salmonella enterica* serovar Typhimurium invasion. Cell Microbiol. *9*, 2839–2855.

Bakshi, C.S., Singh, V.P., Wood, M.W., Jones, P.W., Wallis, T.S., and Galyov, E.E. (2000). Identification of SopE2, a *Salmonella* secreted protein which is highly homologous to SopE and involved in bacterial invasion of epithelial cells. J. Bacteriol. *182*, 2341–2344.

Bambou, J.C., Giraud, A., Menard, S., Begue, B., Rakotobe, S., Heyman, M., Taddei, F., Cerf-Bensussan, N., and Gaboriau-Routhiau, V. (2004). *In vitro* and ex vivo activation of the TLR5 signaling pathway in intestinal epithelial cells by a commensal Escherichia coli strain. J. Biol. Chem. *279*, 42984–42992.

Bernal-Bayard, J., and Ramos-Morales, F. (2009). *Salmonella* type III secretion effector SlrP is an E3 ubiquitin ligase for mammalian thioredoxin. J. Biol. Chem. *284*, 27587–27595.

Bingle, L.E., Bailey, C.M., and Pallen, M.J. (2008). Type VI secretion: a beginner's guide. Curr. Opin Microbiol. *11*, 3–8.

Birmingham, C.L., Jiang, X., Ohlson, M.B., Miller, S.I., and Brumell, J.H. (2005). *Salmonella*-induced filament formation is a dynamic phenotype induced by rapidly replicating *Salmonella enterica* serovar Typhimurium in epithelial cells. Infect. Immun. *73*, 1204–1208.

Blanc-Potard, A.B., Solomon, F., Kayser, J., and Groisman, E.A. (1999). The SPI-3 pathogenicity island of *Salmonella enterica*. J. Bacteriol. *181*, 998–1004.

Boucrot, E., Henry, T., Borg, J.P., Gorvel, J.P., and Meresse, S. (2005). The intracellular fate of *Salmonella* depends on the recruitment of kinesin. Science *308*, 1174–1178.

Boyle, E.C., Brown, N.F., and Finlay, B.B. (2006). *Salmonella enterica* serovar Typhimurium effectors SopB, SopE, SopE2 and SipA disrupt tight junction structure and function. Cell Microbiol. *8*, 1946–1957.

Brawn, L.C., Hayward, R.D., and Koronakis, V. (2007). *Salmonella* SPI1 effector SipA persists after entry and cooperates with a SPI2 effector to regulate phagosome maturation and intracellular replication. Cell Host Microbe *1*, 63–75.

Brown, N.F., Szeto, J., Jiang, X., Coombes, B.K., Finlay, B.B., and Brumell, J.H. (2006). Mutational analysis of *Salmonella* translocated effector members SifA and SopD2 reveals domains implicated in translocation, subcellular localization and function. Microbiology *152*, 2323–2343.

Brown, N.F., Vallance, B.A., Coombes, B.K., Valdez, Y., Coburn, B.A., and Finlay, B.B. (2005). *Salmonella* Pathogenicity Island 2 Is Expressed Prior to Penetrating the Intestine. PLoS Pathog *1*, e32.

Browne, S.H., Hasegawa, P., Okamoto, S., Fierer, J., and Guiney, D.G. (2008). Identification of *Salmonella* SPI-2 secretion system components required for SpvB-mediated cytotoxicity in macrophages and virulence in mice. FEMS Immunol. Med. Microbiol. *52*, 194–201.

Bruey, J.M., Bruey-Sedano, N., Luciano, F., Zhai, D., Balpai, R., Xu, C., Kress, C.L., Bailly-Maitre, B., Li, X., Osterman, A., Matsuzawa, S., Terskikh, A.V., Faustin, B., and Reed, J.C. (2007). Bcl-2 and Bcl-XL regulate proinflammatory caspase-1 activation by interaction with NALP1. Cell *129*, 45–56.

Brumell, J.H., Kujat-Choy, S., Brown, N.F., Vallance, B.A., Knodler, L.A., and Finlay, B.B. (2003). SopD2 is a novel type III secreted effector of *Salmonella* Typhimurium that targets late endocytic compartments upon delivery into host cells. Traffic *4*, 36–48.

Brumell, J.H., Rosenberger, C.M., Gotto, G.T., Marcus, S.L., and Finlay, B.B. (2001a). SifA permits survival and replication of *Salmonella* Typhimurium in murine macrophages. Cell Microbiol. 3, 75–84.

Brumell, J.H., Tang, P., Mills, S.D., and Finlay, B.B. (2001b). Characterization of *Salmonella*-induced filaments(Sifs) reveals a delayed interaction between *Salmonella*-containing vacuoles and late endocytic compartments. Traffic *2*, 643–653.

Bruno, V.M., Hannemann, S., Lara-Tejero, M., Flavell, R.A., Kleinstein, S.H., and Galan, J.E. (2009). *Salmonella* Typhimurium type III secretion effectors stimulate innate immune responses in cultured epithelial cells. PLoS Pathog *5*, e1000538.

Buchmeier, N.A., and Heffron, F. (1990). Induction of *Salmonella* stress proteins upon infection of macrophages. Science *248*, 730–732.

Buchmeier, N.A., and Heffron, F. (1991). Inhibition of macrophage phagosome–lysosome fusion by *Salmonella* Typhimurium. Infect. Immun. *59*, 2232–2238.

Buchwald, G., Friebel, A., Galan, J.E., Hardt, W.D., Wittinghofer, A., and Scheffzek, K. (2002). Structural basis for the reversible activation of a Rho protein by the bacterial toxin SopE. EMBO J. *21*, 3286–3295.

Bujny, M.V., Ewels, P.A., Humphrey, S., Attar, N., Jepson, M.A., and Cullen, P.J. (2008). Sorting nexin-1 defines an early phase of *Salmonella*-containing vacuole-remodeling during *Salmonella* infection. J. Cell Sci. *121*, 2027–2036.

Burstein, D., Zusman, T., Degtyar, E., Viner, R., Segal, G., and Pupko, T. (2009). Genome-scale identification of Legionella pneumophila effectors using a machine learning approach. PLoS Pathog *5*, e1000508.

Cain, R.J., Hayward, R.D., and Koronakis, V. (2004). The target cell plasma membrane is a critical interface for *Salmonella* cell entry effector-host interplay. Mol. Microbiol. *54*, 887–904.

Cain, R.J., Hayward, R.D., and Koronakis, V. (2008). Deciphering interplay between *Salmonella* invasion effectors. PLoS Pathog *4*, e1000037.

Chang, J., Chen, J., and Zhou, D. (2005). Delineation and characterization of the actin nucleation and effector translocation activities of *Salmonella* SipC. Mol. Microbiol. *55*, 1379–1389.

Chang, J., Myeni, S.K., Lin, T.L., Wu, C.C., Staiger, C.J., and Zhou, D. (2007). SipC multimerization promotes actin nucleation and contributes to *Salmonella*-induced inflammation. Mol. Microbiol. *66*, 1548–1556.

Cheminay, C., Mohlenbrink, A., and Hensel, M. (2005). Intracellular *Salmonella* inhibit antigen presentation by dendritic cells. J. Immunol. *174*, 2892–2899.

Chen, L., Wang, H., Zhang, J., Gu, L., Huang, N., Zhou, J.M., and Chai, J. (2008). Structural basis for the catalytic mechanism of phosphothreonine lyase. Nat. Struct. Mol. Biol. *15*, 101–102.

Chen, L.M., Kaniga, K., and Galan, J.E. (1996). *Salmonella* spp. are cytotoxic for cultured macrophages. Mol. Microbiol. *21*, 1101–1115.

Coombes, B.K., Wickham, M.E., Brown, N.F., Lemire, S., Bossi, L., Hsiao, W.W., Brinkman, F.S., and Finlay, B.B. (2005). Genetic and molecular analysis of GogB, a phage-encoded type III-secreted substrate in *Salmonella enterica* serovar Typhimurium with autonomous expression from its associated phage. J. Mol. Biol. *348*, 817–830.

Dai, S., Zhang, Y., Weimbs, T., Yaffe, M.B., and Zhou, D. (2007). Bacteria-generated PtdIns(3)P recruits VAMP8 to facilitate phagocytosis. Traffic *8*, 1365–1374.

de Jong, R., Altare, F., Haagen, I.A., Elferink, D.G., Boer, T., van Breda Vriesman, P.J., Kabel, P.J., Draaisma, J.M., van Dissel, J.T., Kroon, F.P., Casanova, J.L., and Ottenhoff, T.H. (1998). Severe mycobacterial and *Salmonella* infections in interleukin-12 receptor-deficient patients. Science *280*, 1435–1438.

Deatherage, B.L., Lara, J.C., Bergsbaken, T., Rassoulian Barrett, S.L., Lara, S., and Cookson, B.T. (2009). Biogenesis of bacterial membrane vesicles. Mol. Microbiol. 72, 1395–1407.

Deiwick, J., Salcedo, S.P., Boucrot, E., Gilliland, S.M., Henry, T., Petermann, N., Waterman, S.R., Gorvel, J.P., Holden, D.W., and Meresse, S. (2006). The translocated *Salmonella* effector proteins SseF and SseG interact and are required to establish an intracellular replication niche. Infect. Immun. *74*, 6965–6972.

Diao, J., Zhang, Y., Huibregtse, J.M., Zhou, D., and Chen, J. (2008). Crystal structure of SopA, a *Salmonella* effector protein mimicking a eukaryotic ubiquitin ligase. Nat. Struct Mol. Biol. *15*, 65–70.

Drecktrah, D., Knodler, L.A., Howe, D., and Steele-Mortimer, O. (2007). *Salmonella* trafficking is defined by continuous dynamic interactions with the endolysosomal system. Traffic *8*, 212–225.

Dukes, J.D., Lee, H., Hagen, R., Reaves, B.J., Layton, A.N., Galyov, E.E., and Whitley, P. (2006). The secreted *Salmonella dublin* phosphoinositide phosphatase, SopB, localizes to PtdIns(3)P-containing endosomes and perturbs normal endosome to lysosome trafficking. Biochem. J. *395*, 239–247.

Dunlap, N.E., Benjamin, W.H., Jr., Berry, A.K., Eldridge, J.H., and Briles, D.E. (1992). A 'safe-site' for *Salmonella* Typhimurium is within splenic polymorphonuclear cells. Microb Pathog *13*, 181–190.

Fang, F., and Vazquez-Torres, A. (2002). *Salmonella* selectively stops traffic. Trends Microbiol. *10*, 391–392.

Fieschi, C., Allez, M., and Casanova, J.L. (2005). High risk of infectious disease caused by *Salmonellae* and mycobacteria infections in patients with Crohn disease treated with anti-interleukin-12 antibody. Clin. Infect. Dis. *40*, 1381.

Figueiredo, J.F., Lawhon, S.D., Gokulan, K., Khare, S., Raffatellu, M., Tsolis, R.M., Baumler, A.J., McCormick, B.A., and Adams, L.G. (2009). *Salmonella enterica* Typhimurium SipA induces CXC-chemokine expression through p38MAPK and JUN pathways. Microbes Infect *11*, 302–310.

Figueroa-Bossi, N., Uzzau, S., Maloriol, D., and Bossi, L. (2001). Variable assortment of prophages provides a transferable repertoire of pathogenic determinants in *Salmonella*. Mol. Microbiol. *39*, 260–271.

Filloux, A., Hachani, A., and Bleves, S. (2008). The bacterial type VI secretion machine: yet another player for protein transport across membranes. Microbiology *154*, 1570–1583.

Fink, S.L., Bergsbaken, T., and Cookson, B.T. (2008). Anthrax lethal toxin and *Salmonella* elicit the common cell death pathway of caspase-1-dependent pyroptosis via distinct mechanisms. Proc. Natl. Acad. Sci. U.S.A. *105*, 4312–4317.

Fink, S.L., and Cookson, B.T. (2007). Pyroptosis and host cell death responses during *Salmonella* infection. Cell Microbiol. *9*, 2562–2570.

Franchi, L., Amer, A., Body-Malapel, M., Kanneganti, T.D., Ozoren, N., Jagirdar, R., Inohara, N., Vandenabeele, P., Bertin, J., Coyle, A., *et al.* (2006). Cytosolic flagellin requires Ipaf for activation of caspase-1 and interleukin 1beta in *Salmonella*-infected macrophages. Nat. Immunol. 7, 576–58Freeman, J.A., Ohl, M.E., and Miller, S.I. (2003). The *Salmonella enterica* serovar Typhimurium translocated effectors SseJ and SifB are targeted to the *Salmonella*-containing vacuole. Infect. Immun. *71*, 418–427.

Fu, Y., and Galan, J.E. (1999). A *Salmonella* protein antagonizes Rac-1 and Cdc42 to mediate host-cell recovery after bacterial invasion. Nature *401*, 293–297.

Gal-Mor, O., Gibson, D.L., Baluta, D., Vallance, B.A., and Finlay, B.B. (2008). A novel secretion pathway of *Salmonella enterica* acts as an antivirulence modulator during salmonellosis. PLoS Pathog *4*, e1000036.

Galen, J.E., Pasetti, M.F., Tennant, S., Ruiz-Olvera, P., Sztein, M.B., and Levine, M.M. (2009). *Salmonella enterica* serovar Typhi live vector vaccines finally come of age. Immunol. Cell Biol. *87*, 400–412.

Galkin, V.E., Orlova, A., VanLoock, M.S., Zhou, D., Galan, J.E., and Egelman, E.H. (2002). The bacterial protein SipA polymerizes G-actin and mimics muscle nebulin. Nat. Struct. Biol. *9*, 518–521.

Garcia-del Portillo, F., and Finlay, B.B. (1995). Targeting of *Salmonella* Typhimurium to vesicles containing lysosomal membrane glycoproteins bypasses compartments with mannose 6-phosphate receptors. J. Cell Biol. *129*, 81–97.

Geddes, K., Worley, M., Niemann, G., and Heffron, F. (2005). Identification of new secreted effectors in *Salmonella enterica* serovar Typhimurium. Infect. Immun. *73*, 6260–6271.

Geddes, K., Cruz, F., and Heffron, F. (2007). Analysis of Cells Targeted by *Salmonella* Type III Secretion *In Vivo*. PLoS Pathog *3*, e196.

Gerichter, C.B. (1960). The dissemination of *Salmonella* Typhi, S. Paratyphi A and S. Paratyphi B through the organs of the white mouse by oral infection. J. Hyg *58*, 307–319.

Girschick, H.J., Guilherme, L., Inman, R.D., Latsch, K., Rihl, M., Sherer, Y., Shoenfeld, Y., Zeidler, H., Arienti, S., and Doria, A. (2008). Bacterial triggers and autoimmune rheumatic diseases. Clin. Exp. Rheumatol. *26*, S12–7.

Gotoh, H., Okada, N., Kim, Y.G., Shiraishi, K., Hirami, N., Haneda, T., Kurita, A., Kikuchi, Y., and Danbara, H. (2003). Extracellular secretion of the virulence plasmid-encoded ADP-ribosyltransferase SpvB in *Salmonella*. Microb Pathog *34*, 227–238.

Grant, A.J., Restif, O., McKinley, T.J., Sheppard, M., Maskell, D.J., and Mastroeni, P. (2008). Modelling within-host spatiotemporal dynamics of invasive bacterial disease. PLoS Biol. *6*, e74.

Greenberg, J.T., and Vinatzer, B.A. (2003). Identifying type III effectors of plant pathogens and analyzing their interaction with plant cells. Curr. Opin Microbiol. *6*, 20–28.

Guttman, D.S., Vinatzer, B.A., Sarkar, S.F., Ranall, M.V., Kettler, G., and Greenberg, J.T. (2002). A functional screen for the type III (Hrp) secretome of the plant pathogen Pseudomonas syringae. Science *295*, 1722–1726.

Halici, S., Zenk, S.F., Jantsch, J., and Hensel, M. (2008). Functional analysis of the *Salmonella* pathogenicity island 2-mediated inhibition of antigen presentation in dendritic cells. Infect. Immun. *76*, 4924–4933.

Handa, Y., Suzuki, M., Ohya, K., Iwai, H., Ishijima, N., Koleske, A.J., Fukui, Y., and Sasakawa, C. (2007). Shigella IpgB1 promotes bacterial entry through the ELMO-Dock180 machinery. Nat. Cell Biol. *9*, 121–128.

Hanisch, J., Ehinger, J., Ladwein, M., Rohde, M., Derivery, E., Bosse, T., Steffen, A., Bumann, D., Misselwitz, B., Hardt, W.D., Gautreau, A., Stradal, T.E., and Rottner, K. (2009). Molecular dissection of *Salmonella*-induced membrane ruffling versus invasion. Cell Microbiol. *12*, 84–98.

Hapfelmeier, S., and Hardt, W.D. (2005). A mouse model for S. Typhimurium-induced enterocolitis. Trends Microbiol. *13*, 497–503.

Haraga, A., and Miller, S.I. (2006). A *Salmonella* type III secretion effector interacts with the mammalian serine/threonine protein kinase PKN1. Cell Microbiol. *8*, 837–846.

Harding, C.V., and Pfeifer, J.D. (1994). Antigen expressed by *Salmonella* Typhimurium is processed for class I major histocompatibility complex presentation by macrophages but not infected epithelial cells. Immunology *83*, 670–674.

Hardt, W.D., and Galan, J.E. (1997). A secreted *Salmonella* protein with homology to an avirulence determinant of plant pathogenic bacteria. Proc. Natl. Acad. Sci. U.S.A. *94*, 9887–9892.

Harrison, R.E., Brumell, J.H., Khandani, A., Bucci, C., Scott, C.C., Jiang, X., Finlay, B.B., and Grinstein, S. (2004). *Salmonella* impairs RILP recruitment to Rab7 during maturation of invasion vacuoles. Mol. Biol. Cell *15*, 3146–3154.

Hashim, S., Mukherjee, K., Raje, M., Basu, S.K., and Mukhopadhyay, A. (2000). Live *Salmonella* modulate expression of Rab proteins to persist in a specialized compartment and escape transport to lysosomes. J. Biol. Chem. *275*, 16281–16288.

Hayward, R.D., Cain, R.J., McGhie, E.J., Phillips, N., Garner, M.J., and Koronakis, V. (2005). Cholesterol binding by the bacterial type III translocon is essential for virulence effector delivery into mammalian cells. Mol. Microbiol. *56*, 590–603.

Hayward, R.D., and Koronakis, V. (1999). Direct nucleation and bundling of actin by the SipC protein of invasive *Salmonella*. EMBO J. *18*, 4926–4934.

Hayward, R.D., and Koronakis, V. (2002). Direct modulation of the host cell cytoskeleton by *Salmonella* actin-binding proteins. Trends Cell Biol. *12*, 15–20.

He, L., Kim, B.Y., Kim, K.A., Kwon, O., Kim, S.O., Bae, E.Y., Lee, M.S., Kim, M.S., Jung, M., Moon, A., *et al.* (2007). NF-kappaB inhibition enhances caspase-3 degradation of Akt1 and apoptosis in response to cauptothecin. Cell Signal *19*, 1713–1721.

Henry, T., Couillault, C., Rockenfeller, P., Boucrot, E., Dumont, A., Schroeder, N., Hermant, A., Knodler, L.A., Lecine, P., Steele-Mortimer, O., Borg, J.P., Gorvel, J.P., and Meresse, S. (2006). The *Salmonella* effector protein PipB2 is a linker for kinesin-1. Proc. Natl. Acad. Sci. U.S.A. *103*, 13497–13502.

Henry, T., Pommier, S., Journet, L., Bernadac, A., Gorvel, J.P., and Lloubes, R. (2004). Improved methods for producing outer membrane vesicles in Gram-negative bacteria. Res Microbiol. *155*, 437–446.

Hernandez, L.D., Hueffer, K., Wenk, M.R., and Galan, J.E. (2004). *Salmonella* modulates vesicular traffic by altering phosphoinositide metabolism. Science *304*, 1805–1807.

Hernandez, L.D., Pypaert, M., Flavell, R.A., and Galan, J.E. (2003). A *Salmonella* protein causes macrophage cell death by inducing autophagy. J. Cell Biol. *163*, 1123–1131.

Hersh, D., Monack, D.M., Smith, M.R., Ghori, N., Falkow, S., and Zychlinsky, A. (1999). The *Salmonella* invasin SipB induces macrophage apoptosis by binding to caspase-1. Proc. Natl. Acad. Sci. U.S.A. *96*, 2396–2401.

Hess, J., Ladel, C., Miko, D., and Kaufmann, S.H. (1996). *Salmonella* Typhimurium aroA- infection in gene-targeted immunodeficient mice: major role of CD4+ TCR-alpha beta cells and IFN-gamma in bacterial clearance independent of intracellular location. J. Immunol. *156*, 3321–3326.

Ho, T.D., Figueroa-Bossi, N., Wang, M., Uzzau, S., Bossi, L., and Slauch, J.M. (2002). Identification of GtgE, a novel virulence factor encoded on the Gifsy-2 bacteriophage of *Salmonella enterica* serovar Typhimurium. J. Bacteriol. *184*, 5234–5239.

Huang, Z., Sutton, S.E., Wallenfang, A.J., Orchard, R.C., Wu, X., Feng, Y., Chai, J., and Alto, N.M. (2009). Structural insights into host GTPase isoform selection by a family of bacterial GEF mimics. Nat. Struct Mol. Biol. *16*, 853–860.

Humphreys, D., Hume, P.J., and Koronakis, V. (2009). The *Salmonella* effector SptP dephosphorylates host AAA+ ATPase VCP to promote development of its intracellular replicative niche. Cell Host Microbe *5*, 225–233.

Jackson, L.K., Nawabi, P., Hentea, C., Roark, E.A., and Haldar, K. (2008). The *Salmonella* virulence protein SifA is a G protein antagonist. Proc. Natl. Acad. Sci. U.S.A. *105*, 14141–14146.

Jiang, X., Rossanese, O.W., Brown, N.F., Kujat-Choy, S., Galan, J.E., Finlay, B.B., and Brumell, J.H. (2004). The related effector proteins SopD and SopD2 from *Salmonella enterica* serovar Typhimurium contribute to virulence during systemic infection of mice. Mol. Microbiol. *54*, 1186–1198.

Jones, B.D., Ghori, N., and Falkow, S. (1994). *Salmonella* Typhimurium initiates murine infection by penetrating and destroying the specialized epithelial M cells of the Peyer's patches. J. Exp. Med. *180*, 15–23.

Jones, M.A., Wood, M.W., Mullan, P.B., Watson, P.R., Wallis, T.S., and Galyov, E.E. (1998). Secreted effector proteins of *Salmonella* dublin act in concert to induce enteritis. Infect. Immun. *66*, 5799–5804.

Jones, R.M., Wu, H., Wentworth, C., Luo, L., Collier-Hyams, L., and Neish, A.S. (2008). *Salmonella* AvrA Coordinates Suppression of Host Immune and Apoptotic Defences via JNK Pathway Blockade. Cell Host Microbe *3*, 233–244.

Kaufmann, S.H., and Hess, J. (1999). Impact of intracellular location of and antigen display by intracellular bacteria: implications for vaccine development [In Process Citation]. Immunol. Lett. *65*, 81–84.

Kim, J.Y., Doody, A.M., Chen, D.J., Cremona, G.H., Shuler, M.L., Putnam, D., and DeLisa, M.P. (2008). Engineered bacterial outer membrane vesicles with enhanced functionality. J. Mol. Biol. *380*, 51–66.

Knodler, L.A., and Steele-Mortimer, O. (2005). The *Salmonella* effector PipB2 affects late endosome/lysosome distribution to mediate Sif extension. Mol. Biol. Cell *16*, 4108–4123.

Knodler, L.A., Vallance, B.A., Hensel, M., Jackel, D., Finlay, B.B., and Steele-Mortimer, O. (2003). *Salmonella* type III effectors PipB and PipB2 are targeted to detergent-resistant microdomains on internal host cell membranes. Mol. Microbiol. *49*, 685–704.

Knodler, L.A., Winfree, S., Drecktrah, D., Ireland, R., and Steele-Mortimer, O. (2009). Ubiquitination of the bacterial inositol phosphatase, SopB, regulates its biological activity at the plasma membrane. Cell Microbiol. *11*, 1652–1670.

Kotani, K., Ogawa, W., Hino, Y., Kitamura, T., Ueno, H., Sano, W., Sutherland, C., Granner, D.K., and Kasuga, M. (1999). Dominant negative forms of Akt (protein kinase B) and atypical protein kinase Clambda do not prevent insulin inhibition of phosphoenolpyruvate carboxykinase gene transcription. J. Biol. Chem. *274*, 21305–21312.

Kubori, T., and Galan, J.E. (2003). Temporal regulation of *Salmonella* virulence effector function by proteasome-dependent protein degradation. Cell *115*, 333–342.

Kuhle, V., Abrahams, G.L., and Hensel, M. (2006). Intracellular *Salmonella enterica* redirect exocytic transport processes in a *Salmonella* pathogenicity island 2-dependent manner. Traffic *7*, 716–730.

Kuhle, V., and Hensel, M. (2002). SseF and SseG are translocated effectors of the type III secretion system of *Salmonella* pathogenicity island 2 that modulate aggregation of endosomal compartments. Cell Microbiol. *4*, 813–824.

Kuhle, V., Jackel, D., and Hensel, M. (2004). Effector proteins encoded by *Salmonella* pathogenicity island 2 interfere with the microtubule cytoskeleton after translocation into host cells. Traffic *5*, 356–370.

Kuijl, C., Savage, N.D., Marsman, M., Tuin, A.W., Janssen, L., Egan, D.A., Ketema, M., van den Nieuwendijk, R., van den Eeden, S.J., Geluk, A., Poot, A., van der Marel, G., Beijersbergen, R.L., Overkleeft, H., Ottenhoff, T.H., and Neefjes, J. (2007). Intracellular bacterial growth is controlled by a kinase network around PKB/AKT1. Nature *450*, 725–730.

Kujat Choy, S.L., Boyle, E.C., Gal-Mor, O., Goode, D.L., Valdez, Y., Vallance, B.A., and Finlay, B.B. (2004). SseK1 and SseK2 are novel translocated proteins of *Salmonella enterica* serovar Typhimurium. Infect. Immun. 72, 5115–5125.

Lacoste, M.G., Cargnelutti, D.E., Tamashiro, H., and Di Genaro, M.S. (2007). Study of predominant bacterial antigens triggering antibody response in *Salmonella* reactive arthritis: apropos of a case. Jpn. J. Infect. Dis. *60*, 220–224.

Lapaque, N., Hutchinson, J.L., Jones, D.C., Meresse, S., Holden, D.W., Trowsdale, J., and Kelly, A.P. (2009). *Salmonella* regulates polyubiquitination and surface expression of MHC class II antigens. Proc. Natl. Acad. Sci. U.S.A. *106*, 14052–14057.

Lawley, T.D., Chan, K., Thompson, L.J., Kim, C.C., Govoni, G.R., and Monack, D.M. (2006). Genome-wide screen for *Salmonella* genes required for long-term systemic infection of the mouse. PLoS Pathog *2*, e11.

Layton, A.N., Brown, P.J., and Galyov, E.E. (2005). The *Salmonella* translocated effector SopA is targeted to the mitochondria of infected cells. J. Bacteriol. *187*, 3565–3571.

Layton, A.N., and Galyov, E.E. (2007). *Salmonella*-induced enteritis: molecular pathogenesis and therapeutic implications. Expert Rev. Mol. Med. *9*, 1–17.

Le Clainche, C., and Drubin, D.G. (2004). Actin lessons from pathogens. Mol. Cell *13*, 453–454.

Le Negrate, G., Faustin, B., Welsh, K., Loeffler, M., Krajewska, M., Hasegawa, P., Mukherjee, S., Orth, K., Krajewski, S., Godzik, A., Guiney, D.G., and Reed, J.C. (2008). *Salmonella* secreted factor L deubiquitinase of *Salmonella* Typhimurium inhibits NF-kappaB, suppresses IkappaBalpha ubiquitination and modulates innate immune responses. J. Immunol. *180*, 5045–5056.

Leclerc, C., Lo-Man, R., Fayolle, C., Charbit, A., Clement, J.M., Martineau, P., O'Callaghan, D., and Hofnung, M. (1994). Molecular and cellular parameters controlling the immunogenicity of foreign B- or T-cell epitopes expressed by recombinant vectors. Dev Biol. Stand *82*, 193–199.

Lee, A.H., Zareei, M.P., and Daefler, S. (2002). Identification of a NIPSNAP homologue as host cell target for *Salmonella* virulence protein SpiC. Cell Microbiol. *4*, 739–750.

Lee, C.A., Silva, M., Siber, A.M., Kelly, A.J., Galyov, E., and McCormick, B.A. (2000). A secreted *Salmonella* protein induces a proinflammatory response in epithelial cells, which promotes neutrophil migration. Proc. Natl. Acad. Sci. U.S.A. *97*, 12283–12288.

Lesnick, M.L., Reiner, N.E., Fierer, J., and Guiney, D.G. (2001). The *Salmonella* spvB virulence gene encodes an enzyme that ADP-ribosylates actin and destabilizes the cytoskeleton of eukaryotic cells. Mol. Microbiol. *39*, 1464–1470.

Li, E., Stupack, D.G., Brown, S.L., Klemke, R., Schlaepfer, D.D., and Nemerow, G.R. (2000). Association of p130CAS with phosphatidylinositol-3-OH kinase mediates adenovirus cell entry. J. Biol. Chem. *275*, 14729–14735.

Li, H., Xu, H., Zhou, Y., Zhang, J., Long, C., Li, S., Chen, S., Zhou, J.M., and Shao, F. (2007). The phosphothreonine lyase activity of a bacterial type III effector family. Science *315*, 1000–1003.

Liao, A.P., Petrof, E.O., Kuppireddi, S., Zhao, Y., Xia, Y., Claud, E.C., and Sun, J. (2008). *Salmonella* type III effector AvrA stabilizes cell tight junctions to inhibit inflammation in intestinal epithelial cells. PLoS One *3*, e2369.

Lilic, M., Galkin, V.E., Orlova, A., VanLoock, M.S., Egelman, E.H., and Stebbins, C.E. (2003). *Salmonella* SipA polymerizes actin by stapling filaments with non-globular protein arms. Science *301*, 1918–1921.

Lin, S.L., Le, T.X., and Cowen, D.S. (2003). SptP, a *Salmonella* Typhimurium type III-secreted protein, inhibits the mitogen-activated protein kinase pathway by inhibiting Raf activation. Cell Microbiol. *5*, 267–275.

Linehan, S.A., Rytkonen, A., Yu, X.J., Liu, M., and Holden, D.W. (2005). SlyA regulates function of *Salmonella* pathogenicity island 2 (SPI-2) and expression of SPI-2- associated genes. Infect. Immun. *73*, 4354–4362.

Liu, J., Fujiwara, T.M., Buu, N.T., Sanchez, F.O., Cellier, M., Paradis, A.J., Frappier, D., Skamene, E., Gros, P., Morgan, K., *et al.* (1995). Identification of polymorphisms and sequence variants in the human homologue of the mouse natural resistance-associated macrophage protein gene. Am. J. Hum. Genet. *56*, 845–853.

Lossi, N.S., Rolhion, N., Magee, A.I., Boyle, C., and Holden, D.W. (2008). The *Salmonella* SPI-2 effector SseJ exhibits eukaryotic activator-dependent phospholipase A and glycerophospholipid: cholesterol acyltransferase activity. Microbiology *154*, 2680–2688.

Lower, M., and Schneider, G. (2009). Prediction of type III secretion signals in genomes of gram-negative bacteria. PLoS One *4*, e5917.

Mallo, G.V., Espina, M., Smith, A.C., Terebiznik, M.R., Aleman, A., Finlay, B.B., Rameh, L.E., Grinstein, S., and Brumell, J.H. (2008). SopB promotes phosphatidylinositol 3-phosphate formation on *Salmonella* vacuoles by recruiting Rab5 and Vps34. J. Cell Biol. *182*, 741–752.

Manes, N.P., Gustin, J.K., Rue, J., Mottaz, H.M., Purvine, S.O., Norbeck, A.D., Monroe, M.E., Zimmer, J.S., Metz, T.O., Adkins, J.N., Smith, R.D., and Heffron, F. (2007). Targeted protein degradation by *Salmonella* under phagosome-mimicking culture conditions investigated using comparative peptidomics. Mol. Cell Proteomics *6*, 717–727.

Marcus, S.L., Wenk, M.R., Steele-Mortimer, O., and Finlay, B.B. (2001). A synaptojanin-homologous region of *Salmonella* Typhimurium SigD is essential for inositol phosphatase activity and Akt activation. FEBS Lett *494*, 201–207.

Mariathasan, S., Newton, K., Monack, D.M., Vucic, D., French, D.M., Lee, W.P., Roose-Girma, M., Erickson, S., and Dixit, V.M. (2004). Differential activation of the inflammasome by caspase-1 adaptors ASC and Ipaf. Nature *430*, 213–218.

Maskell, D.J., and Hormaeche, C.E. (1985). Relapse following cessation of antibiotic therapy for mouse typhoid in resistant and susceptible mice infected with *Salmonellae* of differing virulence. J. Infect. Dis. *152*, 1044–1049.

Mason, D., Mallo, G.V., Terebiznik, M.R., Payrastre, B., Finlay, B.B., Brumell, J.H., Rameh, L., and Grinstein, S. (2007). Alteration of epithelial structure and function associated with PtdIns(4,5)P2 degradation by a bacterial phosphatase. J. Gen. Physiol. *129*, 267–283.

Matsui, H., Bacot, C.M., Garlington, W.A., Doyle, T.J., Roberts, S., and Gulig, P.A. (2001). Virulence plasmid-borne *spvB* and *spvC* genes can replace the 90-kilobase plasmid in conferring virulence to *Salmonella enterica* serovar Typhimurium in subcutaneousley inoculated mice. J. Bacteriol. *183*, 4652–4658.

Mazurkiewicz, P., Thomas, J., Thompson, J.A., Liu, M., Arbibe, L., Sansonetti, P., and Holden, D.W. (2008). SpvC is a *Salmonella* effector with phosphothreonine lyase activity on host mitogen-activated protein kinases. Mol. Microbiol. *67*, 1371–1383.

McCormick, B.A. (2007). Bacterial-induced hepoxilin A3 secretion as a pro-inflammatory mediator. FEBS J. *274*, 3513–3518.

McGhie, E.J., Brawn, L.C., Hume, P.J., Humphreys, D., and Koronakis, V. (2009). *Salmonella* takes control: effector-driven manipulation of the host. Curr. Opin Microbiol. *12*, 117–124.

McGhie, E.J., Hayward, R.D., and Koronakis, V. (2001). Cooperation between actin-binding proteins of invasive *Salmonella*: SipA potentiates SipC nucleation and bundling of actin. EMBO J. *20*, 2131–2139.

McGhie, E.J., Hayward, R.D., and Koronakis, V. (2004). Control of actin turnover by a *Salmonella* invasion protein. Mol. Cell *13*, 497–510.

McLaughlin, L.M., Govoni, G.R., Gerke, C., Gopinath, S., Peng, K., Laidlaw, G., Chien, Y.H., Jeong, H.W., Li, Z., Brown, M.D., Sacks, D.B., and Monack, D. (2009). The *Salmonella* SPI2 effector SseI mediates long-term systemic infection by modulating host cell migration. PLoS Pathog *5*, e1000671.

McSorley, S.J., Cookson, B.T., and Jenkins, M.K. (2000). Characterization of CD4+ T cell responses during natural infection with *Salmonella* Typhimurium. J. Immunol. *164*, 986–993.

McSorley, S.J., and Jenkins, M.K. (2000). Antibody is required for protection against virulent but not attenuated *Salmonella enterica* serovar Typhimurium. Infect. Immun. *68*, 3344–3348.

Meresse, S., Steele-Mortimer, O., Finlay, B.B., and Gorvel, J.P. (1999). The rab7 GTPase controls the maturation of *Salmonella* Typhimurium-containing vacuoles in HeLa cells. EMBO J. *18*, 4394–4403.

Miao, E.A., Alpuche-Aranda, C.M., Dors, M., Clark, A.E., Bader, M.W., Miller, S.I., and Aderem, A. (2006). Cytoplasmic flagellin activates caspase-1 and secretion of interleukin 1beta via Ipaf. Nat. Immunol. *7*, 569–575.

Miao, E.A., Brittnacher, M., Haraga, A., Jeng, R.L., Welch, M.D., and Miller, S.I. (2003). *Salmonella* effectors translocated across the vacuolar membrane interact with the actin cytoskeleton. Mol. Microbiol. *48*, 401–415.

Miao, E.A., Scherer, C.A., Tsolis, R.M., Kingsley, R.A., Adams, L.G., Baumler, A.J., and Miller, S.I. (1999). *Salmonella* Typhimurium leucine-rich repeat proteins are targeted to the SPI1 and SPI2 type III secretion systems. Mol. Microbiol. *34*, 850–864.

Micheau, O., and Tschopp, J. (2003). Induction of TNF receptor I-mediated apoptosis via two sequential signaling complexes. Cell *114*, 181–190.

Mittrucker, H.W., and Kaufmann, S.H. (2000). Immune response to infection with *Salmonella* Typhimurium in mice. J. Leukoc. Biol. *67*, 457–463.

Mittrucker, H.W., Kohler, A., and Kaufmann, S.H. (2002). Characterization of the murine T-lymphocyte response to *Salmonella enterica* serovar Typhimurium infection. Infect. Immun. *70*, 199–203.

Molero, C., Rodriguez-Escudero, I., Aleman, A., Rotger, R., Molina, M., and Cid, V.J. (2009). Addressing the effects of *Salmonella* internalization in host cell signaling on a reverse-phase protein array. Proteomics *9*, 3652–3665.

Monack, D.M., Bouley, D.M., and Falkow, S. (2004). *Salmonella* Typhimurium persists within macrophages in the mesenteric lymph nodes of chronically infected Nramp1+/+ mice and can be reactivated by IFNgamma neutralization. J. Exp. Med. *199*, 231–241.

Monack, D.M., Hersh, D., Ghori, N., Bouley, D., Zychlinsky, A., and Falkow, S. (2000). *Salmonella* exploits caspase-1 to colonize Peyer's patches in a murine typhoid model. J. Exp. Med. *192*, 249–258.

Monack, D.M., Raupach, B., Hromockyj, A.E., and Falkow, S. (1996). *Salmonella* Typhimurium invasion induces apoptosis in infected macrophages. Proc. Natl. Acad. Sci. U.S.A. *93*, 9833–9838.

Monack, D.M. (2008). The inflammasome: a key player in the inflammation triggered in response to bacterial pathogens. J. Pediatr. Gastroenterol. Nutr. *46*(Suppl 1), E14.

Moon, J.J., and McSorley, S.J. (2009). Tracking the dynamics of *Salmonella* specific T cell responses. Curr. Top. Microbiol. Immunol. *334*, 179–198.

Mrsny, R.J., Gewirtz, A.T., Siccardi, D., Savidge, T., Hurley, B.P., Madara, J.L., and McCormick, B.A. (2004). Identification of hepoxilin A3 in inflammatory events: a required role in neutrophil migration across intestinal epithelia. Proc. Natl. Acad. Sci. U.S.A. *101*, 7421–7426.

Mueller, C.A., Broz, P., and Cornelis, G.R. (2008). The type III secretion system tip complex and translocon. Mol. Microbiol. *68*, 1085–1095.

Mukherjee, K., Parashuraman, S., Raje, M., and Mukhopadhyay, A. (2001). SopE acts as an Rab5-specific nucleotide exchange factor and recruits non-prenylated Rab5 on *Salmonella*-containing phagosomes to promote fusion with early endosomes. J. Biol. Chem. *276*, 23607–23615.

Muller, A.J., Hoffmann, C., Galle, M., Van Den Broeke, A., Heikenwalder, M., Falter, L., Misselwitz, B., Kremer, M., Beyaert, R., and Hardt, W.D. (2009). The S. Typhimurium effector SopE induces caspase-1 activation in stromal cells to initiate gut inflammation. Cell Host Microbe *6*, 125–136.

Mumy, K.L., Bien, J.D., Pazos, M.A., Gronert, K., Hurley, B.P., and McCormick, B.A. (2008). Distinct isoforms of phospholipase A2 mediate the ability of *Salmonella enterica* serotype Typhimurium and Shigella flexneri to induce the transepithelial migration of neutrophils. Infect. Immun. *76*, 3614–3627.

Muotiala, A., and Makela, P.H. (1990). The role of IFN-gamma in murine *Salmonella* Typhimurium infection. Microb Pathog *8*, 135–141.

Murli, S., Watson, R.O., and Galan, J.E. (2001). Role of tyrosine kinases and the tyrosine phosphatase SptP in the interaction of *Salmonella* with host cells. Cell Microbiol. *3*, 795–810.

Napoli, C., and Staskawicz, B. (1987). Molecular characterization and nucleic acid sequence of an avirulence gene from race 6 of Pseudomonas syringae pv. glycinea. J. Bacteriol. *169*, 572–578.

Nawabi, P., Catron, D.M., and Haldar, K. (2008). Esterification of cholesterol by a type III secretion effector during intracellular *Salmonella* infection. Mol. Microbiol. *68*, 173–185.

Neish, A.S., Gewirtz, A.T., Zeng, H., Young, A.N., Hobert, M.E., Karmali, V., Rao, A.S., and Madara, J.L. (2000). Prokaryotic regulation of epithelial responses by inhibition of IkappaB-alpha ubiquitination. Science *289*, 1560–1563.

Nikolaus, T., Deiwick, J., Rappl, C., Freeman, J.A., Schroder, W., Miller, S.I., and Hensel, M. (2001). SseBCD proteins are secreted by the type III secretion system of *Salmonella* pathogenicity island 2 and function as a translocon. J. Bacteriol. *183*, 6036–6045.

Norris, F.A., Wilson, M.P., Wallis, T.S., Galyov, E.E., and Majerus, P.W. (1998). SopB, a protein required for virulence of *Salmonella* dublin, is an inositol phosphate phosphatase [see comments]. Proc. Natl. Acad. Sci. U.S.A. *95*, 14057–14059.

Ohlson, M.B., Fluhr, K., Birmingham, C.L., Brumell, J.H., and Miller, S.I. (2005). SseJ deacylase activity by *Salmonella enterica* serovar Typhimurium promotes virulence in mice. Infect. Immun. *73*, 6249–6259.

Ohlson, M.B., Huang, Z., Alto, N.M., Blanc, M.P., Dixon, J.E., Chai, J., and Miller, S.I. (2008). Structure and function of *Salmonella* SifA indicate that its interactions with SKIP, SseJ, and RhoA family GTPases induce endosomal tubulation. Cell Host Microbe *4*, 434–446.

Parsons, D.A., and Heffron, F. (2005). sciS, an icmF homolog in *Salmonella enterica* serovar Typhimurium, limits intracellular replication and decreases virulence. Infect. Immun. *73*, 4338–4345.

Patel, J.C., and Galan, J.E. (2005). Manipulation of the host actin cytoskeleton by *Salmonella*–all in the name of entry. Curr. Opin Microbiol. *8*, 10–15.

Patel, J.C., and Galan, J.E. (2006). Differential activation and function of Rho GTPases during *Salmonella*-host cell interactions. J. Cell Biol. *175*, 453–463.

Patel, J.C., Hueffer, K., Lam, T.T., and Galan, J.E. (2009). Diversification of a *Salmonella* virulence protein function by ubiquitin-dependent differential localization. Cell *137*, 283–294.

Paul, K., Erhardt, M., Hirano, T., Blair, D.F., and Hughes, K.T. (2008). Energy source of flagellar type III secretion. Nature *451*, 489–492.

Perrett, C.A., and Jepson, M.A. (2009). Regulation of *Salmonella*-induced membrane ruffling by SipA differs in strains lacking other effectors. Cell Microbiol. *11*, 475–487.

Petnicki-Ocwieja, T., Schneider, D.J., Tam, V.C., Chancey, S.T., Shan, L., Jamir, Y., Schechter, L.M., Janes, M.D., Buell, C.R., Tang, X., Collmer, A., and Alfano, J.R. (2002). Genomewide identification of proteins secreted by the Hrp type III protein secretion system of Pseudomonas syringae pv. tomato DC3000. Proc. Natl. Acad. Sci. U.S.A. *99*, 7652–7657.

Picard, C., Fieschi, C., Altare, F., Al-Jumaah, S., Al-Hajjar, S., Feinberg, J., Dupuis, S., Soudais, C., Al-Mohsen, I.Z., Genin, E., *et al.* (2002). Inherited interleukin-12 deficiency: IL12B genotype and clinical phenotype of 13 patients from six kindreds. Am. J. Hum. Genet. *70*, 336–348.

Poh, J., Odendall, C., Spanos, A., Boyle, C., Liu, M., Freemont, P., and Holden, D.W. (2008). SteC is a *Salmonella* kinase required for SPI-2-dependent F-actin remodelling. Cell Microbiol. *10*, 20–30.

Popp, D., Yamamoto, A., Iwasa, M., Nitanai, Y., and Maeda, Y. (2008). Single molecule polymerization, annealing and bundling dynamics of SipA induced actin filaments. Cell Motil. Cytoskeleton *65*, 165–177.

Purcell, M., and Shuman, H.A. (1998). The Legionella pneumophila icmGCDJBF genes are required for killing of human macrophages. Infect. Immun. *66*, 2245–2255.

Qimron, U., Madar, N., Mittrucker, H.W., Zilka, A., Yosef, I., Bloushtain, N., Kaufmann, S.H., Rosenshine, I., Apte, R.N., and Porgador, A. (2004). Identification of *Salmonella* Typhimurium genes responsible for interference with peptide presentation on MHC class I molecules: Deltayej *Salmonella* mutants induce superior CD8+ T-cell responses. Cell Microbiol. *6*, 1057–1070.

Quezada, C.M., Hicks, S.W., Galan, J.E., and Stebbins, C.E. (2009). A family of *Salmonella* virulence factors functions as a distinct class of autoregulated E3 ubiquitin ligases. Proc. Natl. Acad. Sci. U.S.A. *106*, 4864–4869.

Raffatellu, M., Wilson, R.P., Chessa, D., Andrews-Polymenis, H., Tran, Q.T., Lawhon, S., Khare, S., Adams, L.G., and Baumler, A.J. (2005). SipA, SopA, SopB, SopD, and SopE2 contribute to *Salmonella enterica* serotype Typhimurium invasion of epithelial cells. Infect. Immun. *73*, 146–154.

Ramsden, A.E., Holden, D.W., and Mota, L.J. (2007a). Membrane dynamics and spatial distribution of *Salmonella*-containing vacuoles. Trends Microbiol. *15*, 516–524.

Ramsden, A.E., Mota, L.J., Munter, S., Shorte, S.L., and Holden, D.W. (2007b). The SPI-2 type III secretion system restricts motility of *Salmonella*-containing vacuoles. Cell Microbiol. *9*, 2517–2529.

Ravindran, R., and McSorley, S.J. (2005). Tracking the dynamics of T-cell activation in response to *Salmonella* infection. Immunology *114*, 450–458.

Reinicke, A.T., Hutchinson, J.L., Magee, A.I., Mastroeni, P., Trowsdale, J., and Kelly, A.P. (2005). A *Salmonella* Typhimurium effector protein SifA is modified by host cell prenylation and S-acylation machinery. J. Biol. Chem. *280*, 14620–14627.

Reuser, A.J., and Drost, M.R. (2006). Lysosomal dysfunction, cellular pathology and clinical symptoms: basic principles. Acta. Paediatr. Suppl. *95*, 77–82.

Rohde, J.R., Breitkreutz, A., Chenal, A., Sansonetti, P.J., and Parsot, C. (2007). Type III secretion effectors of the IpaH family are E3 ubiquitin ligases. Cell Host Microbe *1*, 77–83.

Ruiz-Albert, J., Yu, X.J., Beuzon, C.R., Blakey, A.N., Galyov, E.E., and Holden, D.W. (2002). Complementary activities of SseJ and SifA regulate dynamics of the *Salmonella* Typhimurium vacuolar membrane. Mol. Microbiol. *44*, 645–661.

Rytkonen, A., Poh, J., Garmendia, J., Boyle, C., Thompson, A., Liu, M., Freemont, P., Hinton, J.C., and Holden, D.W. (2007). SseL, a *Salmonella* deubiquitinase required for macrophage killing and virulence. Proc. Natl. Acad. Sci. U.S.A. *104*, 3502–3507.

Salcedo, S.P., and Holden, D.W. (2003). SseG, a virulence protein that targets *Salmonella* to the Golgi network. EMBO J. *22*, 5003–5014.

Samudrala, R., Heffron, F., and McDermott, J.E. (2009). Accurate prediction of secreted substrates and identification of a conserved putative secretion signal for type III secretion systems. PLoS Pathog 5, e1000375.

Santos, R.L., Raffatellu, M., Bevins, C.L., Adams, L.G., Tukel, C., Tsolis, R.M., and Baumler, A.J. (2009). Life in the inflamed intestine, *Salmonella* style. Trends Microbiol. *17*, 498–506.

Schlumberger, M.C., Kappeli, R., Wetter, M., Muller, A.J., Misselwitz, B., Dilling, S., Kremer, M., and Hardt, W.D. (2007). Two newly identified SipA domains(F1, F2) steer effector protein localization and contribute to *Salmonella* host cell manipulation. Mol. Microbiol. *65*, 741–760.

Schlumberger, M.C., Muller, A.J., Ehrbar, K., Winnen, B., Duss, I., Stecher, B., and Hardt, W.D. (2005). Real-time imaging of type III secretion: *Salmonella* SipA injection into host cells. Proc. Natl. Acad. Sci. U.S.A. *102*, 12548–12553.

Schweitzer, J.K., Krivda, J.P., and D'Souza-Schorey, C. (2009). Neurodegeneration in Niemann-Pick Type C disease and Huntington's disease: impact of defects in membrane trafficking. Curr. Drug Targets *10*, 653–665.

Segal, G., Purcell, M., and Shuman, H.A. (1998). Host cell killing and bacterial conjugation require overlapping sets of genes within a 22-kb region of the Legionella pneumophila genome. Proc. Natl. Acad. Sci. U.S.A. *95*, 1669–1674.

Shotland, Y., Kramer, H., and Groisman, E.A. (2003). The *Salmonella* SpiC protein targets the mammalian Hook3 protein function to alter cellular trafficking. Mol. Microbiol. *49*, 1565–1576.

Silva, M., Song, C., Nadeau, W.J., Matthews, J.B., and McCormick, B.A. (2004). *Salmonella* Typhimurium SipA-induced neutrophil transepithelial migration: involvement of a PKC-alpha-dependent signal transduction pathway. Am. J. Physiol. Gastrointest. Liver Physiol. *286*, G1024–G1031.

Singh, S.P., Williams, Y.U., Miller, S., and Nikaido, H. (2003). The C-terminal domain of *Salmonella enterica* serovar Typhimurium OmpA is an immunodominant antigen in mice but appears to be only partially exposed on the bacterial cell surface. Infect. Immun. *71*, 3937–3946.

Smith, A.C., Heo, W.D., Braun, V., Jiang, X., Macrae, C., Casanova, J.E., Scidmore, M.A., Grinstein, S., Meyer, T., and Brumell, J.H. (2007). A network of Rab GTPases controls phagosome maturation and is modulated by *Salmonella enterica* serovar Typhimurium. J. Cell Biol. *176*, 263–268.

Smith, J.N., and Ahmer, B.M. (2003). Detection of other microbial species by *Salmonella*: expression of the SdiA regulon. J. Bacteriol. *185*, 1357–1366.

Sory, M.P., Boland, A., Lambermont, I., and Cornelis, G.R. (1995). Identification of the YopE and YopH domains required for secretion and internalization into the cytosol of macrophages, using the cyaA gene fusion approach. Proceedings of the National Academy of Sciences of the United States of America *92*, 11998–12002.

Spano, S., and Galan, J.E. (2008). A novel pathway for exotoxin delivery by an intracellular pathogen. Curr. Opin Microbiol. *11*, 15–20.

Srinivasan, A., Nanton, M., Griffin, A., and McSorley, S.J. (2009). Culling of activated CD4 T cells during typhoid is driven by *Salmonella* virulence genes. J. Immunol. *182*, 7838–7845.

Stecher, B., Hapfelmeier, S., Muller, C., Kremer, M., Stallmach, T., and Hardt, W.D. (2004). Flagella and chemotaxis are required for efficient induction of *Salmonella enterica* serovar Typhimurium colitis in streptomycin-pretreated mice. Infect. Immun. *72*, 4138–4150.

Stecher, B., Robbiani, R., Walker, A.W., Westendorf, A.M., Barthel, M., Kremer, M., Chaffron, S., Macpherson, A.J., Buer, J., Parkhill, J., Dougan, G., von Mering, C., and Hardt, W.D. (2007). *Salmonella enterica* serovar Typhimurium exploits inflammation to compete with the intestinal microbiota. PLoS Biol. 5, 2177–2189.

Steele-Mortimer, O. (2008). The *Salmonella*-containing vacuole: moving with the times. Curr. Opin. Microbiol. *11*, 38–45.

Steele-Mortimer, O., Knodler, L.A., Marcus, S.L., Scheid, M.P., Goh, B., Pfeifer, C.G., Duronio, V., and Finlay, B.B. (2000). Activation of Akt/protein kinase B in epithelial cells by the *Salmonella* Typhimurium effector sigD. J. Biol. Chem. *275*, 37718–37724.

Steele-Mortimer, O., Meresse, S., Gorvel, J.P., Toh, B.H., and Finlay, B.B. (1999). Biogenesis of *Salmonella* Typhimurium-containing vacuoles in epithelial cells involves interactions with the early endocytic pathway. Cell Microbiol. *1*, 33–49.

Stein, M.A., Leung, K.Y., Zwick, M., Garcia-del Portillo, F., and Finlay, B.B. (1996). Identification of a *Salmonella* virulence gene required for formation of filamentous structures containing lysosomal membrane glycoproteins within epithelial cells. Mol. Microbiol. *20*, 151–164.

Stender, S., Friebel, A., Linder, S., Rohde, M., Mirold, S., and Hardt, W.D. (2000). Identification of SopE2 from *Salmonella* Typhimurium, a conserved guanine nucleotide exchange factor for Cdc42 of the host cell. Mol. Microbiol. *36*, 1206–1221.

Sun, Y.H., Rolan, H.G., and Tsolis, R.M. (2007). Injection of flagellin into the host cell cytosol by *Salmonella enterica* serotype Typhimurium. J. Biol. Chem. *282*, 33897–33901.

Szeto, J., Namolovan, A., Osborne, S.E., Coombes, B.K., and Brumell, J.H. (2009). *Salmonella*-containing vacuoles display centrifugal movement associated with cell-to-cell transfer in epithelial cells. Infect. Immun. *77*, 996–1007.

Terebiznik, M.R., Vieira, O.V., Marcus, S.L., Slade, A., Yip, C.M., Trimble, W.S., Meyer, T., Finlay, B.B., and Grinstein, S. (2002). Elimination of host cell PtdIns(4,5)P(2) by bacterial SigD promotes membrane fission during invasion by *Salmonella*. Nat. Cell Biol. *4*, 766–773.

Tobar, J.A., Carreno, L.J., Bueno, S.M., Gonzalez, P.A., Mora, J.E., Quezada, S.A., and Kalergis, A.M. (2006). Virulent *Salmonella enterica* serovar Typhimurium evades adaptive immunity by preventing dendritic cells from activating T cells. Infect. Immun. *74*, 6438–6448.

Townes, J.M., Deodhar, A.A., Laine, E.S., Smith, K., Krug, H.E., Barkhuizen, A., Thompson, M.E., Cieslak, P.R., and Sobel, J. (2008). Reactive arthritis following culture-confirmed infections with bacterial enteric pathogens in Minnesota and Oregon: a population-based study. Ann Rheum Dis. *67*, 1689–1696.

Tsolis, R.M., Townsend, S.M., Miao, E.A., Miller, S.I., Ficht, T.A., Adams, L.G., and Baumler, A.J. (1999). Identification of a putative *Salmonella enterica* serotype Typhimurium host range factor with homology to IpaH and YopM by signature-tagged mutagenesis. Infect. Immun. *67*, 6385–6393.

Uchiya, K., Barbieri, M.A., Funato, K., Shah, A.H., Stahl, P.D., and Groisman, E.A. (1999). A *Salmonella* virulence protein that inhibits cellular trafficking. EMBO J. *18*, 3924–3933.

Uchiya, K., and Nikai, T. (2008). *Salmonella* virulence factor SpiC is involved in expression of flagellin protein and mediates activation of the signal transduction pathways in macrophages. Microbiology *154*, 3491–3502.

Uchiya, K.I., Sugita, A., and Nikai, T. (2009). Involvement of SPI-2-encoded SpiC in flagellum synthesis in *Salmonella enterica* serovar Typhimurium. BMC Microbiol. *9*, 179.

van Blitterswijk, W.J., van der Luit, A.H., Veldman, R.J., Verheij, M., and Borst, J. (2003). Ceramide: second messenger or modulator of membrane structure and dynamics? Biochem. J. *369*, 199–211.

van der Velden, A.W., Copass, M.K., and Starnbach, M.N. (2005). *Salmonella* inhibit T cell proliferation by a direct, contact-dependent immunosuppressive effect. Proc. Natl. Acad. Sci. U.S.A. *102*, 17769–17774.

van der Velden, A.W., Lindgren, S.W., Worley, M.J., and Heffron, F. (2000). *Salmonella* pathogenicity island 1-independent induction of apoptosis in infected macrophages by *Salmonella enterica* serotype Typhimurium. Infect. Immun. *68*, 5702–5709.

van der Velden, A.W., Dougherty, J.T., and Starnbach, M.N. (2008). Down-modulation of TCR expression by *Salmonella enterica* serovar Typhimurium. J. Immunol. *180*, 5569–5574.

Vazquez-Torres, A., Jones-Carson, J., Baumler, A.J., Falkow, S., Valdivia, R., Brown, W., Le, M., Berggren, R., Parks, W.T., and Fang, F.C. (1999). Extraintestinal dissemination of *Salmonella* by CD18-expressing phagocytes. Nature *401*, 804–808.

Vidal, S., Tremblay, M.L., Govoni, G., Gauthier, S., Sebastiani, G., Malo, D., Skamene, E., Olivier, M., Jothy, S., and Gros, P. (1995). The Ity/Lsh/Bcg locus: natural resistance to infection with intracellular parasites is abrogated by disruption of the Nramp1 gene. J. Exp. Med. *182*, 655–666.

Wai, S.N., Lindmark, B., Soderblom, T., Takade, A., Westermark, M., Oscarsson, J., Jass, J., Richter-Dahlfors, A., Mizunoe, Y., and Uhlin, B.E. (2003). Vesicle-mediated export and assembly of pore-forming oligomers of the enterobacterial ClyA cytotoxin. Cell *115*, 25–35.

Wall, D.M., Nadeau, W.J., Pazos, M.A., Shi, H.N., Galyov, E.E., and McCormick, B.A. (2007). Identification of the *Salmonella enterica* serotype Typhimurium SipA domain responsible for inducing neutrophil recruitment across the intestinal epithelium. Cell Microbiol. *9*, 2299–2313.

Wasylnka, J.A., Bakowski, M.A., Szeto, J., Ohlson, M.B., Trimble, W.S., Miller, S.I., and Brumell, J.H. (2008). Role for myosin II in regulating positioning of *Salmonella*-containing vacuoles and intracellular replication. Infect. Immun. *76*, 2722–2735.

Wick, M.J. (2007). Monocyte and dendritic cell recruitment and activation during oral *Salmonella* infection. Immunol. Lett *112*, 68–74.

Winnen, B., Schlumberger, M.C., Sturm, A., Schupbach, K., Siebenmann, S., Jenny, P., and Hardt, W.D. (2008). Hierarchical effector protein transport by the *Salmonella* Typhimurium SPI-1 type III secretion system. PLoS One 3, e2178.

Wood, M.W., Jones, M.A., Watson, P.R., Hedges, S., Wallis, T.S., and Galyov, E.E. (1998). Identification of a pathogenicity island required for *Salmonella* enteropathogenicity. Mol. Microbiol. *29*, 883–891.

Worley, M.J., Ching, K.H., and Heffron, F. (2000). *Salmonella* SsrB activates a global regulon of horizontally acquired genes. Mol. Microbiol. *36*, 749–761.

Worley, M.J., Nieman, G.S., Geddes, K., and Heffron, F. (2006). *Salmonella* Typhimurium disseminates within its host by manipulating the motility of infected cells. Proc. Natl. Acad. Sci. U.S.A. *103*, 17915–17920.

Ye, Z., Petrof, E.O., Boone, D., Claud, E.C., and Sun, J. (2007). *Salmonella* effector AvrA regulation of colonic epithelial cell inflammation by deubiquitination. Am. J. Pathol. *171*, 882–892.

Yi, J., and Beckerle, M.C. (1998). The human TRIP6 gene encodes a LIM domain protein and maps to chromosome 7q22, a region associated with tumorigenesis. Genomics *49*, 314–316.

Yokoyama, H., Ikedo, M., Kohbata, S., Ezaki, T., and Yabuuchi, E. (1987). An ultrastructural study of HeLa cell invasion with *Salmonella* Typhi GIFU 10007. Microbiol. Immunol. *31*, 1–11.

Yoon, H., McDermott, J.E., Porwollik, S., McClelland, M., and Heffron, F. (2009). Coordinated regulation of virulence during systemic infection of *Salmonella enterica* serovar Typhimurium. PLoS Pathog *5*, e1000306.

Zhang, S., Santos, R.L., Tsolis, R.M., Stender, S., Hardt, W.D., Baumler, A.J., and Adams, L.G. (2002). The *Salmonella enterica* serotype Typhimurium effector proteins SipA, SopA, SopB, SopD, and SopE2 act in concert to induce diarrhea in calves. Infect. Immun. *70*, 3843–3855.

Zhang, Y., Higashide, W., Dai, S., Sherman, D.M., and Zhou, D. (2005). Recognition and ubiquitination of *Salmonella* type III effector SopA by a ubiquitin E3 ligase, HsRMA1. J. Biol. Chem. *280*, 38682–38688.

Zhou, D., Chen, L.M., Hernandez, L., Shears, S.B., and Galan, J.E. (2001). A *Salmonella* inositol polyphosphatase acts in conjunction with other bacterial effectors to promote host cell actin cytoskeleton rearrangements and bacterial internalization. Mol. Microbiol. *39*, 248–259.

Zhou, D., and Galan, J. (2001). *Salmonella* entry into host cells: the work in concert of type III secreted effector proteins. Microbes Infect 3, 1293–1298.

Zhou, D., Mooseker, M.S., and Galan, J.E. (1999a). An invasion-associated *Salmonella* protein modulates the actin-bundling activity of plastin. Proc. Natl. Acad. Sci. U.S.A. *96*, 10176–10181.

Zhou, D., Mooseker, M.S., and Galan, J.E. (1999b). Role of the S. Typhimurium actin-binding protein SipA in bacterial internalization. Science *283*, 2092–2095.

The Intracellular Lifestyle of *Salmonella enterica* and Novel Approaches to Understand the Adaptation to Life within the *Salmonella*-containing Vacuole

11

Roopa Rajashekar and Michael Hensel*

Abstract

Salmonella enterica is a facultative intracellular pathogen that resides in a unique membrane-bound compartment, referred to as *Salmonella*-containing vacuole or SCV. Within the SCV, *Salmonella* is able to survive the antimicrobial activities of phagocytic cells and can rapidly multiply in a variety of host cells. Intracellular life of *Salmonella* is dependent on a large number of virulence traits, but the function of the type III secretion system (T3SS) encoded by *Salmonella* Pathogenicity Island 2 (SPI2) is of central importance. Although more than 20 effector proteins have been identified as translocated by the SPI2-T3SS, the molecular function and contribution to intracellular live is only known for a few of these proteins. Intracellular *Salmonella* modify basic functions of the host cell such as the structure of the microtubule cytoskeleton and induce a massive reorganization of vesicular transport and the endosomal system. A unique phenomenon is the SPI2-dependent induction of extensive tubular membrane aggregations of endosomal or Golgi-derived vesicles. The SCV itself has features of a novel organelle and the fate of this compartment is controlled by the pathogen. Previous observations indicated that the SCV is arrested in the state of late endosomal compartment, but recent studies using advanced ultrastructural analyses and live cell studies indicate a complex and highly dynamic interaction of the intracellular *Salmonella* and their host cells.

Introduction

Salmonella enterica belongs to the group of bacterial pathogens that have evolved a facultative intracellular lifestyle. This remarkable ability allows *Salmonella* to grow in the environment, extracellularly inside a host organism, and inside eukaryotic cells where certain cell types permit intracellular replication of the pathogen. *Salmonella* has the ability to invade non-phagocytic eukaryotic cells by injection of effector proteins through a type III secretion system (T3SS) that is encoded by *Salmonella* Pathogenicity Island 1 (SPI1) (Schlumberger and Hardt, 2006). A more conventional route of uptake is phagocytosis, which does not require an active contribution of the bacteria. After internalization, *Salmonella* reside in a membrane-bound compartment, or parasitophorous vacuole, which is maintained throughout intracellular life. A large number of bacterial factors are required for successful intracellular pathogenesis, and recent observations showed that the intracellular pathogen can actively manipulate the host cell in order to maintain the parasitophorous vacuole, to avoid antimicrobial activities of the host cell and to acquire nutrients for intracellular replication. In this chapter, we will describe the intracellular lifestyle of *Salmonella*, the manipulations that alter normal functions of the host cell and the molecular function of virulence factors of *Salmonella*. We will also point out the current frontiers in our understanding of the intracellular lifestyle of *Salmonella*

*Corresponding author

and discuss how modern approaches such as live cell imaging and topographic electron microscopy may contribute to extend these frontiers.

Divergent lifestyles of intracellular pathogens

A facultative intracellular lifestyle is a common observation among pathogenic bacteria. It is widely considered that intracellular life on one hand protects bacteria from a variety of humoral immune responses of the host organisms and, on the other hand, allows the utilization of host cell molecules that may otherwise be limited in the host organisms. However, the benefits of intracellular life appear less prominent if one considers the activation of the mechanisms of host immune defence against intracellular organisms and the restricted access to nutrients for bacteria that reside within a membrane-bound compartment. *Salmonella* has evolved efficient strategies to cope with these challenges.

Bacterial pathogens have evolved two main concepts of intracellular life. Members of one subgroup have the ability to lyse the parasitophorous vacuole and escape into the host cell cytoplasm. These bacteria usually also show actin-based intracellular motility which allows bacteria to infect neighbouring cells in a tissue without the need to re-enter the extracellular space. *Listeria monocytogenes* and *Shigella* spp. are the best-studied pathogens in this group, and recent additions include *Rickettsia typhi*, *Burkholderia pseudomallei* and *Mycobacterium marinum* (reviewed in Gouin *et al.*, 2005). An obvious benefit of the intra-cytoplasmic lifestyle of bacteria is the direct access to metabolites and various nutrients of the host cell. One important disadvantage is the recognition by the immune system due to the presentation of foreign cytosolic proteins by major histocompatibility complex class I (MHC-I) and the cytotoxic killing of infected cells. Recent studies showed that intracellular motility not only enables bacteria to spread intercellularly, but also provides an important contribution to prevention of tagging by the ubiquitin system and proteasomal degradation (Birmingham and Brumell, 2006; Perrin *et al.*, 2004).

The other, larger, subgroup of intracellular bacteria has evolved various strategies to persist and replicate within membrane-bound compartments or 'pathogen-inhabited compartments' (PIC), and *Salmonella* belongs to this group. Bacteria that inhabit PICs are often are also invasive. However, an intracellular lifestyle within a PIC can also occur after passive uptake of bacteria by phagocytosis. One example is *Mycobacterium tuberculosis* that has no strategy for invasion, but is phagocytosed by cells of the host immune system. It is able to prevent killing by host cells due to the unique structure of the cell envelope, and transforms its parasitophorous vacuole into a replication-permissive PIC. The number of intracellular strategies probably equals the number of pathogens inhabiting PICs. Recent molecular and cellular analyses show that bacteria in PICs use highly evolved and rather divergent strategies to modify their host cells and to create their intracellular niches. At first glance, one can characterize the PIC of a given pathogen in order to understand the specific mechanisms. For example, *Legionella pneumophila* is a natural intracellular pathogen in amoebae, but can also infect the human lung if transmitted by contaminated aerosols from air condition or hot water devices (reviewed in Vogel and Isberg, 1999). This pathogen generates a unique PIC that is characterized by the presence of markers characteristic for the endoplasmic reticulum of the host cell (Kagan and Roy, 2002). Other bacteria such as *M. tuberculosis* appear to modify the normal maturation of a phagosome in an early stage, resulting in a PIC that is characterized by the absence of the canonical markers of late endosomes or lysosomes. Further individual strategies can be observed for the formation of the PIC of pathogens such as *Chlamydia* spp., *Coxiella* spp., *Bartonella* spp., *Francisella* spp., and many others, and it is recommended to consult excellent recent reviews for more details (Schaible and Haas, 2009). Although the formation of a PIC is a well-known virulence trait of *S. enterica*, the biogenesis of this compartment and the mechanisms are still topic of many current investigations.

Salmonella as a facultative intracellular pathogen

The ability of *S. enterica* to infect host cells and to survive and proliferate within a PIC has been observed soon after the discovery of this species as an infectious agent. Early observations revealed the role of intracellular life for the ability of *Salmonella* to cause systemic infections (Fields *et al.*, 1986). *Salmonella* is able to infect various eukaryotic host cells. The bacteria are invasive and can deploy the SPI1-encoded T3SS to initiate complex modifications of the host, resulting in a process termed macropinocytosis that ultimately leads to the uptake. *Salmonella* can proliferate in epithelial cells and non-activated macrophages. The survival in activated macrophages and persistence in dendritic cells have been observed, while the situation in fibroblasts appears to more diverse – here, a restriction of replication was seen (Garcia-del Portillo *et al.*, 2008).

Entry into host cells can either occur via bacteria-mediated invasion, or by phagocytosis. The bacteria are thought to primarily replicate in macrophages, as they are found in the lymphatic tissues and organs during systemic infection. *Salmonella* strains that are defective for macrophage replication are avirulent in mouse models of infection, which underscores the importance of bacterial survival and replication in macrophages for disease outcome (Fierer and Guiney, 2001). The ways of entry into the cell probably affects the initial phase of maturation of the compartment containing the bacterium. However, *Salmonella* internalized by either invasion or phagocytosis both induce the formation of a specialized PIC that is referred to as '*Salmonella*-containing vacuole' or SCV. The SCV has certain characteristics of a late endosomal compartment, such as the presence of lysosomal glycoproteins and the acidic luminal pH.

SCV – the intracellular habitat of *Salmonella*

Similar to the intracellular pathogens introduced earlier on, *Salmonella* is able to actively modify the biogenesis of a novel membrane-bound compartment in which the bacteria can survive and efficiently replicate. Although this compartment has some features in common with late endosomes, other properties are unique and may be the result of a fine-tuned interference with normal host cell functions. In the following, we will describe the features of the SCV. The model for the biogenesis of the SCV is shown in Fig. 11.1.

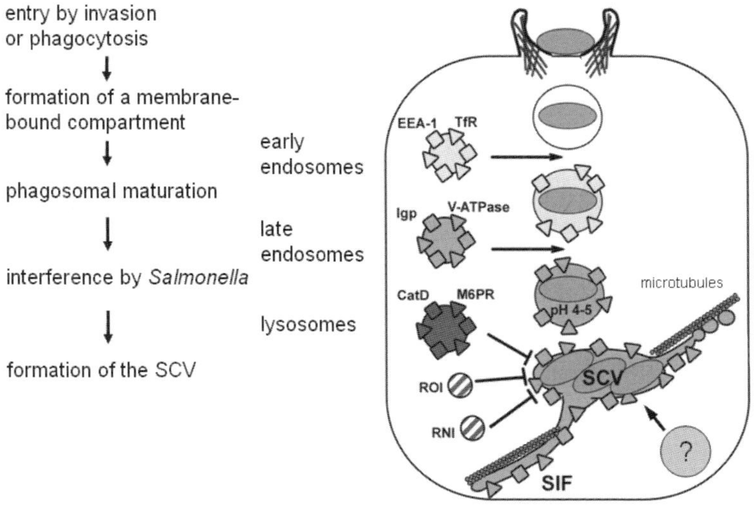

Figure 11.1 Model for the biogenesis of the *Salmonella*-containing vacuole (SCV). The sequential maturation of the SCV is shown. Some canonical markers of the maturation are indicated, i.e. early endosomal antigen 1 (EEA-1) and transferin receptor (TfR) for early endosomes, V-ATPase and lysosomal glycoproteins (lgp) for late endosomes and cathepsin D and mannose-6-phosphate receptor (M6PR) for lysosomes. In immune cells, the bacteria are also confronted with the antimicrobial activity of reactive oxygen intermediates (ROI) and reactive nitrogen intermediates (RNI). The subsequent formation of *Salmonella*-induced filaments (SIFs) is discussed in Fig. 11.3.

Salmonella is capable of infecting and replicating in many different cell types. Unlike non-phagocytic cells, *Salmonella* can enter macrophages by several endocytic processes, including SPI1-induced macropinocytosis. Following SPI1-induced macropinocytosis, a few SPI1-T3SS effectors, such as SipA, SopB, SopD and SopE2, persist within the cell and have recently been implicated in contributing to the intracellular stages of the infection process (Brawn *et al.*, 2007; Hernandez *et al.*, 2004; Patel *et al.*, 2009). Once intracellular, *Salmonella* remains inside the SCV, and persistence or proliferation may occur. If *Salmonella* has specific mechanisms to egress infected cells or, more specifically, for egress from the SCV, has only been partially studied (Szeto *et al.*, 2009). Within the SCV, the bacteria can persist intracellularly for hours to days, making it a unique phagosome with respect to the normal progression of phagolysosomal maturation and recycling. Although there has been some controversy within the field, several reports have shown that *Salmonella* can survive within macrophages in which the lysosomal compartments have fused with the SCV (Ishibashi and Arai, 1995; Oh and Straubinger, 1996). Consequently, the avoidance of phagolysosomal fusion is unlikely to be a major pathogenic strategy of *Salmonella*. Studies in various cell types have also demonstrated that the vacuole acidifies, however, depending on the mechanism of host cell entry, vacuolar acidification may be delayed in both macrophages and epithelial cells (Drecktrah *et al.*, 2006; Rathman *et al.*, 1996). The SCV has been shown to interact transiently with the early endocytic pathway and quickly recruits and looses early endocytic markers, such as EEA1 (early endosomal antigen 1), TfR (transferrin receptor) and the early endocytic trafficking GTPases Rab5 and Rab11. Several late endosomal markers are commonly associated with the SCV at later time points, including the GTPase Rab7, LAMP1 (lysosomal-associated membrane protein 1), LAMP2, LAMP3 and the vacuolar ATPase (Brumell and Grinstein, 2004; Steele-Mortimer, 2008). Other markers such as M6PR (mannose 6-phosphate receptor), LBPA (lysobisphosphatidic acid) and the lysosomal hydrolase cathepsin D appear transiently associated with the SCV. Furthermore, cholesterol has been reported to accumulate in the membrane of the SCV (Catron *et al.*, 2002). The presence or absence of different markers on the persistent SCV may simply indicate that they are variably detected rather than reflect whether or not the SCV has matured through a normal endocytic pathway. It has been suggested that the SCV can acidify and fuse with lysosomes. The ability of *Salmonella* to survive exposure to lysosomal contents is mediated by its resistance to antimicrobial peptides, nitric oxide and oxidative killing – important features for its survival within macrophages and for virulence (Chakravortty and Hensel, 2003; Vazquez-Torres and Fang, 2001). This is supported by the observation that *Salmonella* spp. mutants that are sensitive to these compounds are attenuated in virulence in the murine model, whereas knock-out mice deficient for the production of these compounds have increased susceptibility to *Salmonella* spp. (Chakravortty and Hensel, 2003; Vazquez-Torres and Fang, 2001). *Salmonella* sense the acidic environment of the SCV, resulting in the induction of various regulatory systems that promote intracellular survival, for example, by surface remodelling of the protein, carbohydrate and phospholipid components of the bacterial envelope (Bijlsma and Groisman, 2003). Such regulatory systems include OmpR/EnvZ, PhoP/PhoQ, RpoS/RpoE, PmrA/PmrB, Cya/Crp and cyclic diGMP, all of which confer resistance to antimicrobial peptides and oxidative stress (Groisman and Mouslim, 2006). The phagosomal environment is acidic, with a pH range of <5 to 5.5, has a concentration of magnesium and calcium in the 1 mM range and contains antimicrobial peptides, oxygen and nitrogen radicals that can damage the bacterial cell (Prost *et al.*, 2007). Various studies indicate that both pH and antimicrobial peptides are important signatures of the phagosomal environment, and such conditions activate many of the regulators that are implicated in *Salmonella* virulence (reviewed in Haraga *et al.*, 2008). It is likely that several sensory systems respond to the phagosome environment and cooperate to orchestrate the complex cascade of events that is necessary to alter the bacterial surface and promote intracellular survival (Groisman and Mouslim, 2006). The best characterized of these is the PhoQ sensor, which promotes resistance to antimicrobial peptides (Shi *et al.*, 2004). PhoQ contains a novel acidic domain that is bridged to the bacterial inner membrane by interactions with metal ions and binds to and initiates responses to antimicrobial peptides (Prost and Miller, 2008). It also responds to pH by structural changes that are determined by regions of the protein separate from the metal ion bridges involved in antimicrobial peptide sensing (Bader *et al.*, 2005; Cho *et al.*, 2006). After phagocytosis, *Salmonella* undergoes extensive bacterial surface remodelling, as has been shown for the lipid A component of lipopolysaccharide (LPS) during growth within macrophages (Prost and Miller, 2008). Bacterial molecules that the host can recognize as indicators of infection, such as the SPI1 T3SS and flagellin, are repressed and the LPS structure is altered (Trent *et al.*, 2001). Some of the specific surface modifications include a decrease in length of the O antigen, which is the repeating carbohydrate polymer of the LPS, alterations to the number of acyl chains in the structure of the lipid A component of LPS, and changes in the protein content of the outer membrane, the inner membrane and the peptidoglycan layer. Synthesis of enzymes that allow the bacteria to cope with oxidative and nitrogenous radicals also occurs. Microarray studies have shown that up to 919 *S. enterica* serovar Typhimurium genes are differentially regulated in response to the phagosomal environment, demonstrating that dramatic transcriptional and probably also post-transcriptional changes occur when *Salmonella* makes the transition from a nutrient-rich extracellular environment to the intracellular environment (Eriksson *et al.*, 2003).

Avoidance of host-derived antimicrobial radicals

During the intracellular life, *Salmonella* is exposed to a variety of antimicrobial defence mechanisms that would ultimately result in the killing and degradation of intracellular bacteria. The nature and efficiency of the antimicrobial mechanisms vary between the cell types infected by *Salmonella*. While activated macrophages can mount a large panel of highly reactive radicals and active degradative enzymatic activities, this capacity is less pronounced in resting macrophages, and epithelial cells are only able to activate a limited number of antimicrobial activities. In phagocytic cells, the main anti-microbial activity is attributed to the action of reactive oxygen intermediates (ROI) and reactive nitrogen intermediates (RNI) that are generated by the constitutively expressed NADPH oxidase (phox) and the inducible nitric oxide synthase (iNOS), respectively. Both ROI and RNI are short-lived, highly reactive molecules with the ability to damage various macromolecules of the internalized microbes. If both species of reactive intermediates are produced simultaneously, the reaction product peroxynitrate is formed that has an even higher capacity to kill microbes. The role of phox and iNOS in the control of *Salmonella* proliferation has been demonstrated on the cellular level as well as in infection studies in murine models. Being confronted with these potent defence mechanisms, intracellular *Salmonella* have to mount efficient counteractions. These include activation of repair mechanisms to compensate for the radical-induced damages, detoxification of radicals by expression of catalytic enzymes, but also avoidance of exposure to radicals by re-directing the radical-producing enzymes away from the SCV. A more detailed discussion of the immune response to intracellular *Salmonella* can be found in Chapter 12 of this book.

Virulence factors and their molecular mechanisms in controlling the intracellular fate of *Salmonella*

The ability to survive and proliferate within mammalian cells is dependent on the function of a large number of defensive and offensive virulence factors (reviewed in Groisman and Ochman, 1997). Early studies demonstrated that the ability to survive inside host cells is correlated to the ability of the pathogen to cause systemic infections (Fields *et al.*, 1986).

Role of the *Salmonella* Pathogenicity Island 2-encoded type III secretion system

The main factor for adaptation to intracellular life probably is *Salmonella* Pathogenicity Island 2 (SPI2). This genetic locus comprises 40 genes, of which 25 have been shown to contribute to the intracellular phenotypes. Genes within SPI2 encode for a type III secretion system (T3SS). T3SSs can be found in most pathogenic or symbiotic Gram-negative bacteria and can be considered as complex molecular machines for the injection of toxin-like proteins into eukaryotic host cells (reviewed in Galan and Wolf-Watz, 2006). T3SS and their effector proteins are involved in various host–microbe interactions ranging from invasion of non-phagocytic cells and the paralysis of phagocytes to the establishment of symbiotic relationships. Once inside the SCV, *Salmonella* sense the novel environment, and as a result the expression and assembly of the SPI2-encoded T3SS is induced (Cirillo *et al.*, 1998). Although the function of this T3SS in pathogenesis requires further molecular understanding, it has been shown to be essential for virulence in the mouse model of infection (Shea *et al.*, 1996). Mutants deficient in the SPI2-T3SS cannot replicate efficiently in tissue-culture cells and are highly attenuated in animal models of infection. It is likely that the role of this T3SS during disease is to promote intracellular replication within cells of the intestine during the acute phase of the infection, and in other organs during the persistent state. Current hypotheses for the function of the SPI2-T3SS propose the promotion of intracellular replication by altering host vesicular trafficking, so that useful metabolic molecules, such as amino acids and lipids, are routed to the SCV and the vesicular compartment membrane is expanded. The SPI2-encoded T3SS is active in intracellular *Salmonella* and the translocation of a large set of 20 and probably more effector proteins into host cells has been observed (recent overview in Haraga *et al.*, 2008). *Salmonella* mutant strains defective in the SPI2-T3SS and thus unable to translocate the cocktail of effector proteins are highly attenuated in systemic pathogenesis and, on the cellular level, show a highly reduced intracellular survival and proliferation. Yet, these mutant strains maintain an intracellular habitat that is similar to the SCV formed by wild-type *Salmonella*.

Effector proteins of the SPI2-T3SS and their contribution to intracellular life

To date, at least 20 *Salmonella* spp. effector proteins are known to be translocated by the SPI2-T3SS across the phagosomal membrane into the cytoplasm of the eukaryotic cell (Haraga *et al.*, 2008). However, their specific roles in promoting intracellular replication or modifying vesicular movement are not yet understood (for a list of SPI2-T3SS effectors, their functions and binding partners, see Haraga *et al.*, 2008 and Chapter 10 of this book). In addition, no individual translocated effector has definitively been shown to alter vesicular trafficking. Although it has been reported that SpiC alters endosome fusion *in vitro* (Uchiya *et al.*, 1999) and binds to proteins that are implicated in vesicular trafficking (Lee *et al.*, 2002; Shotland *et al.*, 2003), its role as a T3SS effector protein remains controversial (Freeman *et al.*, 2002; Yu *et al.*, 2002). As SpiC has not been universally observed to be translocated into eukaryotic cells, and is required for the translocation of several, if not all, SPI2-T3SS effectors as well as surface expression of translocon proteins, SpiC is more likely part of the SPI2 secretion apparatus rather than an effector (Yu *et al.*, 2002). The most important translocated effectors, by virtue of causing virulence defects when mutated, are SifA, SseF, SseG, SopD2, SseJ, and PipB2 (Garcia-del Portillo *et al.*, 1993; Henry *et al.*, 2006; Hensel *et al.*, 1998; Jiang *et al.*, 2004; Ohlson *et al.*, 2005). The observation that *S. enterica* serovar Typhimurium that lacks any single SPI2 T3SS effector protein cannot cause the same virulence attenuation in mice as a mutant strain that lacks the entire SPI2 T3SS suggests that many effectors function cooperatively to exert their effects on the host cell. Furthermore, the deletion or mutation of many effector genes has no virulence phenotype, which implies that their functions might be redundant. Early studies of SPI2 T3SS effectors, which primarily focused on determining their subcellular localization in mammalian cells, revealed that they might have specific targeting sequences that direct localization to endosomal compartments (Kuhle and Hensel, 2002), the Golgi apparatus (Salcedo and Holden, 2003), the actin cytoskeleton (Miao *et al.*, 2003) and the microtubule network (Kuhle *et al.*, 2004). This indicates that components of these host-cell structures might be the intracellular targets of the SPI2 T3SS.

Among the various effector proteins, SifA probably has the most prominent role (Beuzon *et al.*, 2000). Mutant strains defective in *sifA* fail to generate *Salmonella*-induced filaments (SIFs), and loss of the SCV and escape into the host cell cytoplasm has been observed. The contribution of SifA to *Salmonella* virulence and intracellular replication has been demonstrated by *in vitro* as well as by *in vivo* studies. A *sifA*⁻ strain is highly attenuated in a murine model and reduced in the replication in cultured cells. Furthermore, a *sifA*⁻ strain escapes into the cytoplasm of infected host cells (Beuzon *et al.*, 2000) where its fate is variable depending on the host cell type. The *sifA*⁻ strain cannot grow in macrophages as it is killed by the highly stringent environment of the macrophage cytoplasm (Beuzon *et al.*, 2002; Brumell *et al.*, 2001a). In contrast, the same mutant replicates very rapidly in the cytoplasm of epithelial cells, and high bacterial proliferation kills the host cell too rapidly to allow efficient net replication in an infected cell population. This observation is of interest in relation to previous studies by the group of W. Goebel that used microinjection of intracellular pathogens into the cytoplasm of various host cells (Goetz *et al.*, 2001). A defect in cytosolic replication was observed for microinjected *Salmonella* but not for *Listeria,* and these results were considered an indication of the specific adaptation to the nutritional conditions of intracellular pathogens with cytosolic or vacuolar lifestyle.

Regardless of the mechanism of SIF formation, the filaments may function to increase the size of the SCV to accommodate the growing numbers of *Salmonella* during intracellular replication (Beuzon *et al.*, 2000). SIF formation is dependent on SifA (Beuzon *et al.*, 2000; Brumell *et al.*, 2001b; Stein *et al.*, 1996), but also, to a lesser extent, on SseF, SseG (Guy *et al.*, 2000; Kuhle and Hensel, 2002), SopD2 (Jiang *et al.*, 2004) and PipB2 (Knodler and Steele-Mortimer, 2005). However, it may also be modulated by other effectors. SifA was recently shown to bind to the host cell protein SKIP (SifA and kinesin interacting protein) (Boucrot *et al.*, 2005). The same authors also found that, if SKIP was depleted by RNA interference in *S. enterica* serovar Typhimurium-infected cells, the cells failed to form SIFs, suggesting that SIF formation requires SKIP. In addition, SifA is a member of a family of T3SS-effector proteins that possesses the motif WxxxE [tryptophan (W)-variable (x)-xx- glutamate (E)], and for some of these effectors a function in mimicking small GTPases of the host cell was demonstrated (Alto *et al.*, 2006). As SifA, similar to GTPases, contains a carboxy-terminal Caax [cysteine (C)-aliphatic residue (a)-a-x] motif that is prenylated and *S*-acylated by host-cell enzymes (Boucrot *et al.*, 2003), it is possible that SifA uses a GTPase-type mechanism for membrane localization and, perhaps, SIF formation (Jackson *et al.*, 2008). Furthermore, the region of SKIP that interacts with SifA is a pleckstrin homology domain, which is commonly found in proteins that bind to signalling lipids and other regulatory molecules, including those that modulate Rho GTPases.

The roles of PipB2, SopD2, SseF and SseG in SIF formation are not entirely clear. *Salmonella* strains that lack any of these effectors do not induce SIF as efficiently as the wild-type strain (Guy *et al.*, 2000; Kuhle and Hensel, 2002). Instead, infection of cultured cells with these mutants tends to lead to the formation of 'pseudo-SIF', which extend from the SCV and co-localize with effectors, but do not contain LAMP1. This suggests that in the absence of these proteins, the ability to form SIF, which is defined as being entirely LAMP1-positive, is impaired. Furthermore, induction of SIF involves multiple steps that are orchestrated by different effectors. Transient expression of PipB2 in mammalian cells induces the movement of LAMP1-positive compartments to the cell periphery, which is probably the result of its interaction with the plus-end-directed microtubule motor kinesin (Knodler and Steele-Mortimer, 2005). This activity might contribute to the outward extension of the SCV, which would promote SIF formation. SopD2, which has homology to the SPI1-T3SS translocated effector SopD, has also been shown to cooperate with SifA to induce SIF, but by an as yet unidentified mechanism (Brumell *et al.*, 2003). SseF and SseG promote the aggregation of endosomal vesicles and recruit Golgi-derived exocytic vesicles to the SCV, which suggests that *Salmonella* is able to usurp both endocytic and exocytic cellular transport processes (Kuhle *et al.*, 2006; Salcedo and Holden, 2005). However, it is not known how these activities directly influence SIF formation. The deletion of SseJ and SpvB can cause an increase in SIF formation at later time points during the infection of cultured cells, which suggests that these proteins have SIF down-regulatory functions. SpvB is an actin-specific ADP-ribosyltransferase that promotes actin depolymerization (Tezcan-Merdol *et al.*, 2005). SseJ has homology to glycerophospholipid cholesterol acyl transferase enzymes, which can remove acyl chains from phospholipids and transfer them to cholesterol in a two-step deacylase-acyltransferase reaction (Nawabi *et al.*, 2008). In fact, purified SseJ has deacylase activity *in vitro* and has been shown to be required for *Salmonella* virulence in mice (Ohlson *et al.*, 2005). It has been proposed that SseJ and SifA have complementary roles in maintaining the integrity of the SCV membrane, because *sifA*-deficient *Salmonella* Typhimurium tends to lose the SCV membrane but does not do so if SseJ is also lacking (Ruiz-Albert *et al.*, 2002). This suggests that, in the absence of SifA, SseJ could cause damage to the SCV by its enzymatic activity. Nawabi and co-workers suggest that SseJ has phospholipase activity *in vivo* and that this function could be localized to the endosomal membrane, resulting in cleavage of phospholipids from the SCV membrane (Nawabi *et al.*, 2008). The cleavage of acyl chains from phospholipids could then promote curvature of the membrane or produce discrete lipid environments that serve as platforms for promoting vesicle fusion and binding scaffolding proteins, and such activity could modulate SIF formation. However, how these functions, and SIF formation in general, are important for *Salmonella*-induced disease is currently not understood. It is plausible that SIFs increase the size of the SCV in a specific and directional fashion that promotes bacterial replication and/or redirect nutrient-rich organelles to the SCV. In this regard, the SPI2 T3SS could help *Salmonella* to replicate inside the phagosome and gain important nutrients for growth, yet avoid some of the host-defence mechanisms and inflammatory responses that would result from their release into the cytoplasm.

Phenotypes of intracellular *Salmonella* and consequences of intracellular life

Redirecting traffic – induction of *Salmonella*-induced filaments and *Salmonella*-induced SCAMP3 tubules

The protection of intracellular *Salmonella* against antimicrobial effectors is only one of the many consequences of the interference with host cell transport processes. A peculiar phenotype is the ability of *Salmonella* to induce the aggregation of endosomal/lysosomal vesicles into extensive tubular structures referred to as '*Salmonella*-induced filaments' or SIFs (see above). SIFs were initially reported for *Salmonella*-infected epithelial cells but this phenotype also occurs in a variety of other host cell types, such as interferon γ-primed primary cells or cell culture macrophages and dendritic cells. The filaments originate from the SCV and extend throughout the cell, often forming extensive networks of variable appearance. SIF are LAMP1-positive and have a membrane composition similar to that of the SCV (Garcia-del Portillo *et al.*, 1993). SIF formation requires microtubules, but not the actin cytoskeleton, although these filaments can be decorated with actin (Brumell *et al.*, 2002; Kuhle *et al.*, 2004). Possible mechanisms that contribute to SIF formation include repetitive initiation of vesicular budding from the SCV, in which fission events are incomplete, or continuous fusion of endocytic vesicles with the SCV, which would result in endosomal tubulation or elongation of the SCV.

SIFs extend from the SCV and are characterized by the presence of markers that can also be found on the SCV, such as lysosomal glycoproteins (lgp) (Garcia-del Portillo *et al.*, 1993). SIF formation coincides with the onset of intracellular replication. Over time, a continuous centripetal extension of SIF occurs, but live cell analyses revealed that SIF are highly dynamic structures that can extend, retract, branch or fuse with other SIF (Drecktrah *et al.*, 2008; Rajashekar *et al.*, 2008). These dynamic properties were most pronounced in the early phase of infection, where growth and retraction had the velocity of microtubule-dependent motility of endosomes. In later stages of infection (>8 h after infection), SIF dynamics are reduced and an extensive, rather static network of SIF was observed. A screen for mutant strains defective in SIF induction identified SifA as the responsible virulence factor of *Salmonella* (Stein *et al.*, 1996), and subsequent work showed that SifA is an effector protein translocated by the SPI2-T3SS (Beuzon *et al.*, 2000; see above).

Studies on the fate of exocytic markers in *Salmonella*-infected cells indicated that the normal route of exocytic transport was also affected. A redirection of Golgi transport to the SCV instead of transport to the plasma membrane was observed (Kuhle *et al.*, 2006). More recently, the formation of a second tubular network was observed that is distinct from the SIF network. The second network is characterized by membrane proteins of the *trans*–Golgi network (TGN) such as SCAMP3, indicating the remodelling of exocytic transport (Mota *et al.*, 2009). Formation of this network is induced by wild-type *Salmonella*, and the network is termed '*Salmonella*-induced SCAMP-3 tubules' (SIST). SIST and SIF partially overlap and tubular membranes with endosomal and TGN markers were observed. In addition, tubular structures with endosomal markers lacking TGN markers, and vice versa, were found (Mota *et al.*, 2009). These researchers also found that the formation of both types of tubular networks depends on the function of the SPI2-T3SS since neither SIST nor SIF were induced in cells infected with a SPI2-T3SS-deficient strain. These observations demonstrate that as a consequence of the modification of host cell transport, aggregation of vesicles in both the endocytic and the exocytic pathway occurs. The particular appearance of tubular aggregates can be explained by the involvement of microtubules as transport routes for vesicular traffic. In the case of SIF induction, a specific involvement of microtubule motor proteins was observed, and a subset of the SPI2-T3SS effector proteins required for SIF induction interferes with motor proteins such as kinesin or dynein, or their adaptors or effectors (reviewed in Abrahams and Hensel, 2006; Bakowski *et al.*, 2008).

Nutritional adaptation to intracellular life

In contrast to bacteria that are highly adapted to life inside cells, most serotypes of *Salmonella enterica* are fully prototrophic bacteria that can use a variety of carbon compounds for biosynthesis of macromolecules and as sources for energy. Various researchers observed a rapid intracellular growth rate of *Salmonella* within host cells, and the generation time of intracellular *Salmonella* may be similar to that of bacteria growing in rich media without limitations for nutrients or trace elements. This observation indicates that intracellular bacteria have sufficient access to nutrients to allow rapid replication. The actual source of nutrients within eukaryotic host cells is still a matter of debate. Assuming that the SCV is a nutritionally poor environment, the bacteria may adjust their metabolism to the utilization of those compounds that are in sufficient amounts present in the SCV. For example, using a reporter strain of *S. enterica* serovar Typhimurium or microarray analyses, it has been observed that high affinity uptake systems of Mg^{2+} and iron are induced in intracellular bacteria (Eriksson *et al.*, 2003; Garcia-del Portillo *et al.*, 1992). Data for similar analyses of the carbon metabolisms are not available. We have investigated the role of various metabolic pathways in the intracellular replication of *S. enterica* serovar Typhimurium within macrophages (unpublished observations). These analyses indicated that none of the individual mutations or combinations of mutations caused a reduction in intracellular replication that was as strong as that of SPI2-T3SS deficient strains. An alternative model for the source of intracellular nutrition of *Salmonella* results from gross remodelling of the vesicular trafficking of the host cell. A series of studies described the rerouting of endocytic as well as exocytic vesicular transport in *Salmonella*-infected cells (reviewed in Abrahams and Hensel, 2006; Bakowski *et al.*, 2008; Steele-Mortimer, 2008). One prominent effect is the formation of SIF with characteristic late endosomal und lysosomal markers, and of SIST with exocytic markers (see above). Recent experiments with fluid tracers performed in live cell experiments indicate that endocytosed markers become closely associated with the SCV (Drecktrah *et al.*, 2007;

Rajashekar *et al.*, 2008). These observations suggest that intracellular *Salmonella* may have access to various vesicular contents that could provide a variety of nutrients, depending on the origin of the particular vesicle. At present, it is not clear if the vesicular content of endocytic or exocytic vesicles that fused into tubular aggregates is accessible by *Salmonella* residing within the SCV. Light microscopy does not allow sufficiently high resolution and higher resolution techniques, such as electron microscopy (EM), or biochemical approaches are required to resolve this question.

It will be an important challenge for future research to determine whether these nutrient are host-derived macromolecules or metabolites directly utilized by *Salmonella*, or if the bacteria directs host cell vesicular transport in order to 'hijack' vesicles containing internalized nutrients.

New models and open questions

Is the SCV a segregated compartment or does it interact with the vesicular systems of the host cell?

A number of recent approaches applied live cell imaging techniques to study the intracellular life of *Salmonella* with a high temporal and spatial resolution. The analysis of the SCV formation in living host cells showed that this compartment itself displays movement within the cell, and a role of the SPI2-T3SS and its effectors in controlling motility and positioning was found (reviewed in Steele-Mortimer, 2008). Drecktrah *et al.* were able to show that the SCV is in continuous contact with endocytosed material. The group used modern technologies of live cell imaging, where dynamic events can be captured in real-time, ruling out many artefacts of cell fixation, and highly sensitive confocal microscopy (spinning disc confocal with a sensitive CCD camera and new generation fluorescent tracers of cellular transport). They also observed that the acidification of phagolysosomes is delayed in *Salmonella*-infected epithelial cells, but the acidification of SCV containing wild-type *Salmonella* or bacteria deficient in SPI1 or SPI2 was similar (Drecktrah *et al.*, 2007). This could be an advantage for the bacteria as they can establish a foothold for replication by reducing the effectiveness of lysosomal hydrolases delivered to the SCV that require low pH for the maturation and activity. This suggests that the SCV does not initially form a specialized but isolated compartment, but rather starts out as an organelle that is in interaction with the endocytic and exocytic systems of the host. Virulence mechanisms of intracellular *Salmonella* could direct selective interaction with those vesicles that are beneficial for intracellular proliferation, while interaction with compartments that deliver antimicrobial activities could be efficiently avoided.

Is the intracellular compartment of *Salmonella* undergoing dynamic or gradual changes?

Work by two groups recently investigated the intracellular compartment of *Salmonella* and formation of SIFs in various living host cells (Drecktrah *et al.*, 2008; Rajashekar *et al.*, 2008). Both studies concluded that SIF formation is a highly dynamic process with rapid extension and retraction of the membrane tubules. The dynamic properties were most pronounced early after onset of SIF formation. Later, large SIF networks with reduced dynamics were found.

Rajashekar *et al.* (2008) were successful in establishing a live cell assay for studying dynamics of SIFs using spinning disc confocal microscopy and time lapse imaging of various cellular events after infection of HeLa cells at high temporal resolution. The work revealed a complete reorganization of the host endosomal system. Gross modification of the endosomes occurred, and by the end of a 10 h observation period, a major proportion of endosomal vesicles were incorporated into growing SIFs. This process was highly dynamic and detailed kinetics showed that SCV formation began as early as 20 min after infection by recruitment of late endosomes, and that SIF formation initiated as early as 3 h after cell invasion.

Regarding the process of tubular vesicle formation, the authors found that early SIFs (between 4 to 6 h post infection) are highly dynamic structures, and form a complex network and become stabilized and non-dynamic as infection progresses. The highly dynamic SIFs were growing, shrinking and branching. Several other phenotypes were observed such as endosome tubules with smaller diameter followed by tubules with larger diameter (Fig. 11.2), fragmentation of long SIF tubules and joining of tubular fragments of other existing SIFs. Live cell studies provided strong evidence for the role of microtubules as scaffold in SIF formation and dynamics. Depolymerization of microtubules using nocodazole was found to completely arrest SIF dynamics.

SIFs are therefore highly dynamic structures which are growing and retracting on a given cytoskeletal scaffold and require SPI2 effectors for formation. During the early phase of intracellular life of *Salmonella*, SIFs grow out from the SCV both by continual fusion of vesicles with the tip of the SIF and by pulling force generated by microtubule motors associated with the tips of the SIF. SIF growth can be directed toward minus ends or plus ends, depending on the proportion of dynein or kinesin motors, respectively, that are recruited. Every fusion step will introduce membrane material into the SIF, and partially and temporarily relax the internal stress. However, when the protruding end of a SIF is pulled by microtubule motor complexes too far without fusion, the elastic stress in the membrane increases and reaches a critical threshold. In addition, motor complexes might lose contact to the microtubule or the SIF membrane. Either event would result in the rapid collapse of the SIF, and relaxation of the membrane elastic stress. The higher velocity of SIF contraction compared to extension would be in line with such a model. Although SIF extension in both directions can be observed, extension toward the plus ends of microtubules dominates over the time–course of infection, indicating a preferential recruitment of kinesin or cargo transported by kinesin. The appearance of a stabilized SIF network in the later phase might indicate a consumption of vesicles available for fusion. A model for SIF dynamics is illustrated in Fig. 11.3.

What are the composition of the SCV and the ultrastructure of the membrane compartment?

Most of the current knowledge about the intracellular lifestyle of *Salmonella* and the biogenesis of the SCV is based on the interpretation of epifluorescence of fixed cells. Both techniques result in snapshots of usually rather dynamic and fast host-pathogen-interactions. While EM allows the analysis of intracellular events down to the molecular level, the resolution limits of light microscopy hamper the proper interpretation of the phenotypes induced by intracellular *Salmonella*. Some recent approaches deploy the high resolution of transmission electron microscopy (TEM) to investigate the ultrastructure of the intracellular compartments of *Salmonella*. Although only rarely found, Chakravortty *et al.* (2005) were able to detect needle-like structures on *Salmonella* residing within the SCV of infected macrophages. The observed structures might represent the SPI2-T3SS that is active under these conditions.

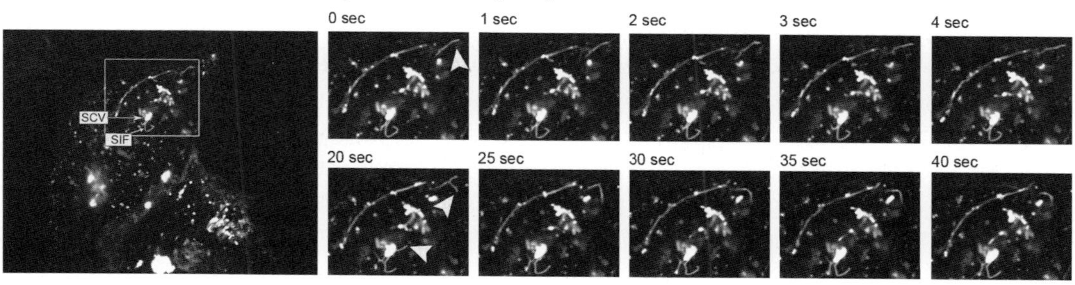

Figure 11.2 Examples for the dynamic properties of *Salmonella*-induced filaments (SIFs). An epithelial cell is shown that was transfected for expression of LAMP1-GFP and infected with *Salmonella enterica* serovar Typhimurium for 4 h. Live cell imaging was performed and the fluorescence signals of the GFP channel are displayed. The high magnification inserts show the morphological changes of selected SIFs over the indicated time frame.

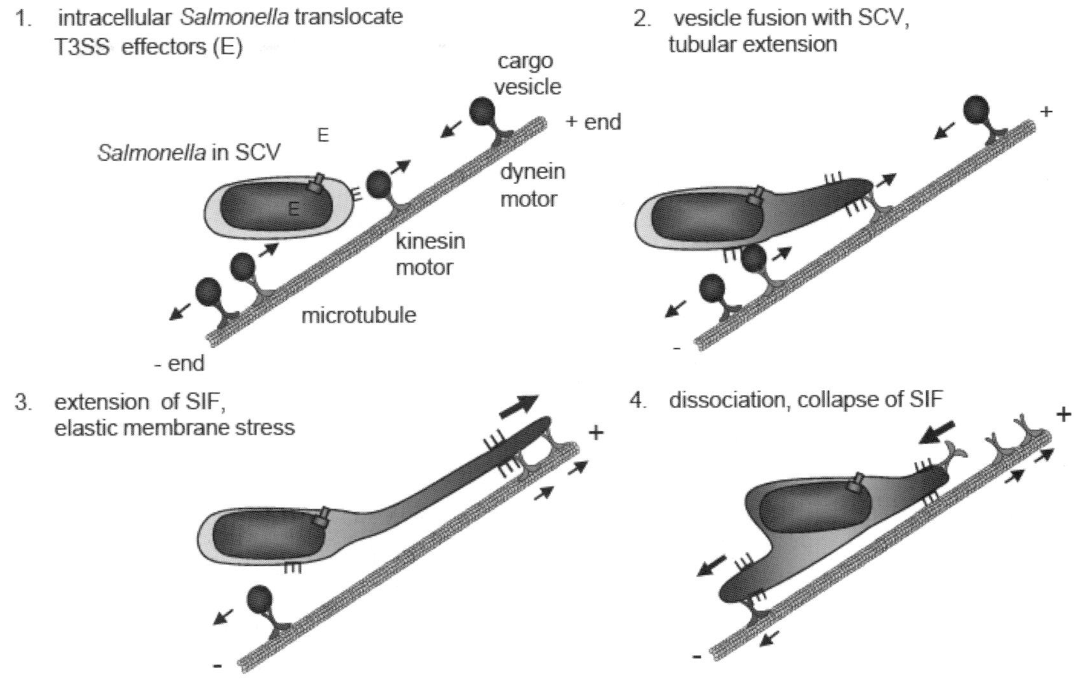

Figure 11.3 Model for the formation of dynamic tubular membrane structures by intracellular *Salmonella*. Modified from Rajashekar *et al*. (2008).

Further work on the ultrastructural analysis of the SCV resulted in a series of rather unexpected observations. A major proportion of SIFs formed in infected epithelial cells were not composed of a single membrane but rather indicated the presence of two membranes that appeared as a tube within a tube. Such tubular double membrane structures frequently had incorporated ribosomes and filamentous structures with the appearance of F-actin or microtubules. Other tubular structures showed the presence of multi-membrane vesicles. Similar to the observations made by live cell imaging approaches, EM analyses also revealed a high degree of variability in the appearance of SIFs as well as of the SCV (Fig. 11.4).

Conclusions and perspectives

In light of the new observations described above, current models for the biogenesis of SIFs have to be revised. The SCV is a highly dynamic compartment and by means of virulence factors such as the effectors of the SPI2-T3SS, intracellular *Salmonella* control the positioning of this compartment as well as fusion events with endosomal and exocytic vesicles. The pathogen also transforms the entire endomembrane system of the host cell. The recent observation of SIST as a second tubular membrane system raises the question if even more tubular membrane structures exist, and further membrane markers should be investigated. The dynamics and complexity of the resulting unique membrane structures will require further detailed studies on their biogenesis. In continuation of the previous studies that led to new insights into the intracellular lifestyle of *Salmonella*, both EM with high spatial resolution and live cell imaging with high temporal resolution will play important roles in this process.

Although ultrastructural analyses could lead to novel insights into the intracellular biology of *Salmonella*, there are currently two major limitations, i.e. the highly dynamic characteristics and the heterogeneity of the intracellular phenotypes. An important approach to overcome these limitations will be the correlative EM technique, which will address the ultrastructure of individual

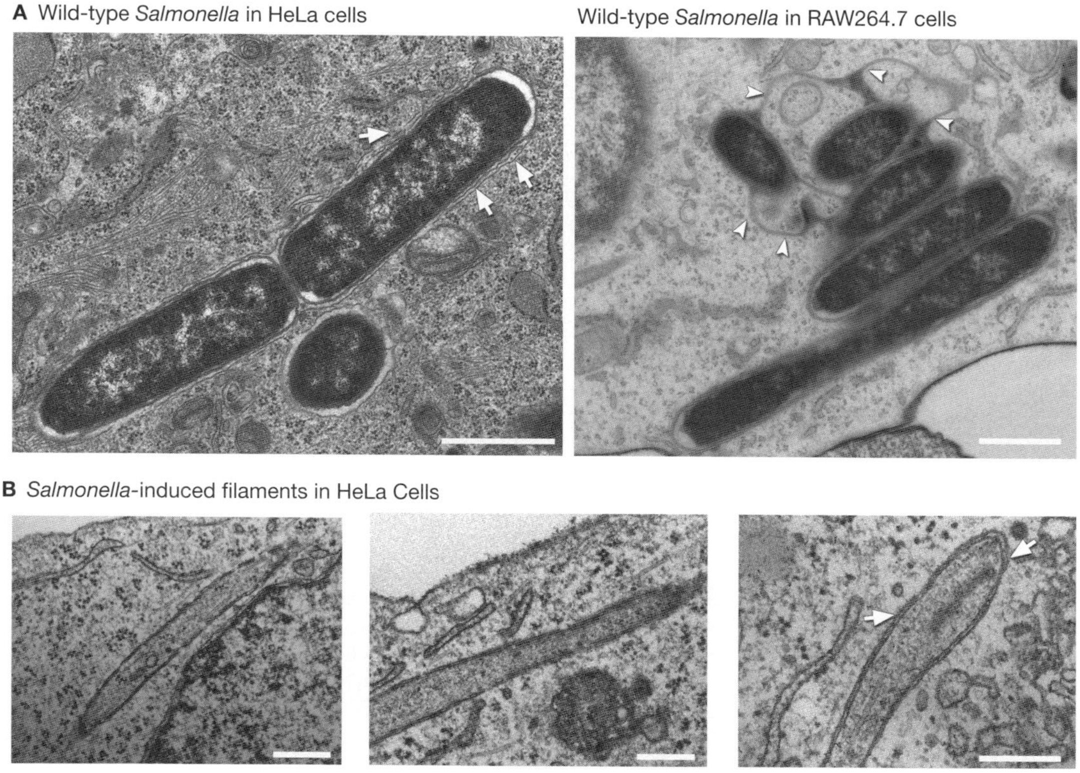

Figure 11.4 The ultrastructure of the *Salmonella*-containing vacuoles. (A) Representative examples for the membrane compartment of intracellular *Salmonella* in HeLa cells (left panel) or macrophages (right panel). Note the presence of multiple membranes in HeLa cells (left panel, indicated by arrows) and of large numbers of tubular membrane structures in RAW264.7 macrophages (right panel, arrowheads). Scale bars: 1 μm. (B) Ultrastructure of *Salmonella*-induced filaments (SIF) in infected epithelial cells. Note the presence of double membranes in SIFs (arrows). Scale bars: 200 nm. EM micrographs were kindly provided by Dr. David Liebl.

intracellular bacteria, SCVs or SIFs directly after observation in a live cell setup. Another limitation of ultrastructural investigations, i.e. the restriction to analysis of a rather small volume of the cell, might be circumvented by techniques such as EM tomography that might allow the three-dimensional reconstruction of novel membrane compartments with ultimate resolution. We envision that combinations of such state of the art approaches of imaging will unveil even more of the secrets of the intracellular life of *Salmonella*. Yet, the next challenges will be the molecular understanding of the contribution of the various virulence factors to the intracellular manipulations and to advance investigations from cultured cell models to infected tissues in living host organisms. We foresee many more unexpected insights into the pathobiology of *Salmonella*.

Acknowledgements

Work is our group was supported by grants of the Deutsche Forschungsgemeinschaft.

References

Abrahams, G.L., and Hensel, M. (2006). Manipulating cellular transport and immune responses: dynamic interactions between intracellular *Salmonella enterica* and its host cells. Cell Microbiol. *8*, 728–737.

Alto, N.M., Shao, F., Lazar, C.S., Brost, R.L., Chua, G., Mattoo, S., McMahon, S.A., Ghosh, P., Hughes, T.R., Boone, C., *et al.* (2006). Identification of a bacterial type III effector family with G protein mimicry functions. Cell *124*, 133–145.

Bader, M.W., Sanowar, S., Daley, M.E., Schneider, A.R., Cho, U., Xu, W., Klevit, R.E., Le Moual, H., and Miller,

S.I. (2005). Recognition of antimicrobial peptides by a bacterial sensor kinase. Cell *122*, 461–472.

Bakowski, M.A., Braun, V., and Brumell, J.H. (2008). *Salmonella*-containing vacuoles: directing traffic and nesting to grow. Traffic *9*, 2022–2031.

Beuzon, C.R., Meresse, S., Unsworth, K.E., Ruiz-Albert, J., Garvis, S., Waterman, S.R., Ryder, T.A., Boucrot, E., and Holden, D.W. (2000). *Salmonella* maintains the integrity of its intracellular vacuole through the action of SifA. EMBO J. *19*, 3235–3249.

Beuzon, C.R., Salcedo, S.P., and Holden, D.W. (2002). Growth and killing of a *Salmonella enterica* serovar Typhimurium *sifA* mutant strain in the cytosol of different host cell lines. Microbiology *148*, 2705–2715.

Bijlsma, J.J., and Groisman, E.A. (2003). Making informed decisions: regulatory interactions between two-component systems. Trends Microbiol. *11*, 359–366.

Birmingham, C.L., and Brumell, J.H. (2006). Autophagy recognizes intracellular *Salmonella enterica* serovar Typhimurium in damaged vacuoles. Autophagy *2*, 156–158.

Boucrot, E., Beuzon, C.R., Holden, D.W., Gorvel, J.P., and Meresse, S. (2003). *Salmonella* Typhimurium SifA effector protein requires its membrane-anchoring C-terminal hexapeptide for its biological function. J. Biol. Chem. *278*, 14196–14202.

Boucrot, E., Henry, T., Borg, J.P., Gorvel, J.P., and Meresse, S. (2005). The intracellular fate of *Salmonella* depends on the recruitment of kinesin. Science *308*, 1174–1178.

Brawn, L.C., Hayward, R.D., and Koronakis, V. (2007). *Salmonella* SPI1 effector SipA persists after entry and cooperates with a SPI2 effector to regulate phagosome maturation and intracellular replication. Cell Host Microbe *1*, 63–75.

Brumell, J.H., Goosney, D.L., and Finlay, B.B. (2002). SifA, a type III secreted effector of *Salmonella* Typhimurium, directs *Salmonella*-induced filament (Sif) formation along microtubules. Traffic 3, 407–415.

Brumell, J.H., and Grinstein, S. (2004). *Salmonella* redirects phagosomal maturation. Curr. Opin Microbiol. *7*, 78–84.

Brumell, J.H., Kujat-Choy, S., Brown, N.F., Vallance, B.A., Knodler, L.A., and Finlay, B.B. (2003). SopD2 is a novel type III secreted effector of *Salmonella* Typhimurium that targets late endocytic compartments upon delivery into host cells. Traffic *4*, 36–48.

Brumell, J.H., Rosenberger, C.M., Gotto, G.T., Marcus, S.L., and Finlay, B.B. (2001a). SifA permits survival and replication of *Salmonella* Typhimurium in murine macrophages. Cell Microbiol. *3*, 75–84.

Brumell, J.H., Tang, P., Mills, S.D., and Finlay, B.B. (2001b). Characterization of *Salmonella*-induced filaments (Sifs) reveals a delayed interaction between *Salmonella*-containing vacuoles and late endocytic compartments. Traffic *2*, 643–653.

Catron, D.M., Sylvester, M.D., Lange, Y., Kadekoppala, M., Jones, B.D., Monack, D.M., Falkow, S., and Haldar, K. (2002). The *Salmonella*-containing vacuole is a major site of intracellular cholesterol accumulation and recruits the GPI-anchored protein CD55. Cell Microbiol. *4*, 315–328.

Chakravortty, D., and Hensel, M. (2003). Inducible nitric oxide synthase and control of intracellular bacterial pathogens. Microbes Infect *5*, 621–627.

Chakravortty, D., Rohde, M., Jäger, L., Deiwick, J., and Hensel, M. (2005). Formation of a novel surface structure encoded by *Salmonella* Pathogenicity Island 2. EMBO J. *24*, 2043–2052.

Cho, U.S., Bader, M.W., Amaya, M.F., Daley, M.E., Klevit, R.E., Miller, S.I., and Xu, W. (2006). Metal bridges between the PhoQ sensor domain and the membrane regulate transmembrane signaling. J. Mol. Biol. *356*, 1193–1206.

Cirillo, D.M., Valdivia, R.H., Monack, D.M., and Falkow, S. (1998). Macrophage-dependent induction of the *Salmonella* pathogenicity island 2 type III secretion system and its role in intracellular survival. Mol. Microbiol. *30*, 175–188.

Drecktrah, D., Knodler, L.A., Howe, D., and Steele-Mortimer, O. (2007). *Salmonella* trafficking is defined by continuous dynamic interactions with the endolysosomal system. Traffic *8*, 212–225.

Drecktrah, D., Knodler, L.A., Ireland, R., and Steele-Mortimer, O. (2006). The mechanism of *Salmonella* entry determines the vacuolar environment and intracellular gene expression. Traffic *7*, 39–51.

Drecktrah, D., Levine-Wilkinson, S., Dam, T., Winfree, S., Knodler, L.A., Schroer, T.A., and Steele-Mortimer, O. (2008). Dynamic behavior of *Salmonella*-induced membrane tubules in epithelial cells. Traffic *9*, 2117–2129.

Eriksson, S., Lucchini, S., Thompson, A., Rhen, M., and Hinton, J.C. (2003). Unravelling the biology of macrophage infection by gene expression profiling of intracellular *Salmonella enterica*. Mol. Microbiol. *47*, 103–118.

Fields, P.I., Swanson, R.V., Haidaris, C.G., and Heffron, F. (1986). Mutants of *Salmonella* Typhimurium that cannot survive within the macrophage are avirulent. Proc. Natl. Acad. Sci. U.S.A. *83*, 5189–5193.

Fierer, J., and Guiney, D.G. (2001). Diverse virulence traits underlying different clinical outcomes of *Salmonella* infection. J. Clin. Invest. *107*, 775–780.

Freeman, J.A., Rappl, C., Kuhle, V., Hensel, M., and Miller, S.I. (2002). SpiC is required for translocation of *Salmonella* Pathogenicity Island 2 effectors and secretion of translocon proteins SseB and SseC. J. Bacteriol. *184*, 4971–4980.

Galan, J.E., and Wolf-Watz, H. (2006). Protein delivery into eukaryotic cells by type III secretion machines. Nature *444*, 567–573.

Garcia-del Portillo, F., Foster, J.W., Maguire, M.E., and Finlay, B.B. (1992). Characterization of the microenvironment of *Salmonella* Typhimurium-containing vacuoles within MDCK epithelial cells. Mol. Microbiol. *6*, 3289–3297.

Garcia-del Portillo, F., Nunez-Hernandez, C., Eisman, B., and Ramos-Vivas, J. (2008). Growth control in the *Salmonella*-containing vacuole. Curr. Opin Microbiol. *11*, 46–52.

Garcia-del Portillo, F., Zwick, M.B., Leung, K.Y., and Finlay, B.B. (1993). *Salmonella* induces the formation of filamentous structures containing lysosomal membrane glycoproteins in epithelial cells. Proc. Natl. Acad. Sci. U.S.A. *90*, 10544–10548.

Goetz, M., Bubert, A., Wang, G., Chico-Calero, I., Vazquez-Boland, J.A., Beck, M., Slaghuis, J., Szalay, A.A., and Goebel, W. (2001). Microinjection and growth of bacteria in the cytosol of mammalian host cells. Proc. Natl. Acad. Sci. U.S.A. *98*, 12221–12226.

Gouin, E., Welch, M.D., and Cossart, P. (2005). Actin-based motility of intracellular pathogens. Curr. Opin Microbiol. *8*, 35–45.

Groisman, E.A., and Mouslim, C. (2006). Sensing by bacterial regulatory systems in host and non-host environments. Nat. Rev. Microbiol. *4*, 705–709.

Groisman, E.A., and Ochman, H. (1997). How *Salmonella* became a pathogen. Trends Microbiol. *5*, 343–349.

Guy, R.L., Gonias, L.A., and Stein, M.A. (2000). Aggregation of host endosomes by *Salmonella* requires SPI2 translocation of SseFG and involves SpvR and the *fms-aroE* intragenic region. Mol. Microbiol. *37*, 1417–1435.

Haraga, A., Ohlson, M.B., and Miller, S.I. (2008). *Salmonellae* interplay with host cells. Nat. Rev. Microbiol. *6*, 53–66.

Henry, T., Couillault, C., Rockenfeller, P., Boucrot, E., Dumont, A., Schroeder, N., Hermant, A., Knodler, L.A., Lecine, P., Steele-Mortimer, O., *et al.* (2006). The *Salmonella* effector protein PipB2 is a linker for kinesin-1. Proc. Natl. Acad. Sci. U.S.A. *103*, 13497–13502.

Hensel, M., Shea, J.E., Waterman, S.R., Mundy, R., Nikolaus, T., Banks, G., Vazquez-Torres, A., Gleeson, C., Fang, F., and Holden, D.W. (1998). Genes encoding putative effector proteins of the type III secretion system of *Salmonella* Pathogenicity Island 2 are required for bacterial virulence and proliferation in macrophages. Mol. Microbiol. *30*, 163–174.

Hernandez, L.D., Hueffer, K., Wenk, M.R., and Galan, J.E. (2004). *Salmonella* modulates vesicular traffic by altering phosphoinositide metabolism. Science *304*, 1805–1807.

Ishibashi, Y., and Arai, T. (1995). *Salmonella* Typhi does not inhibit phagosome–lysosome fusion in human monocyte-derived macrophages. FEMS Immunol. Med. Microbiol. *12*, 55–61.

Jackson, L.K., Nawabi, P., Hentea, C., Roark, E.A., and Haldar, K. (2008). The *Salmonella* virulence protein SifA is a G protein antagonist. Proc. Natl. Acad. Sci. U.S.A. *105*, 14141–14146.

Jiang, X., Rossanese, O.W., Brown, N.F., Kujat-Choy, S., Galan, J.E., Finlay, B.B., and Brumell, J.H. (2004). The related effector proteins SopD and SopD2 from *Salmonella enterica* serovar Typhimurium contribute to virulence during systemic infection of mice. Mol. Microbiol. *54*, 1186–1198.

Kagan, J.C., and Roy, C.R. (2002). *Legionella* phagosomes intercept vesicular traffic from endoplasmic reticulum exit sites. Nat. Cell Biol. *4*, 945–954.

Knodler, L.A., and Steele-Mortimer, O. (2005). The *Salmonella* effector PipB2 affects late endosome/lysosome distribution to mediate Sif extension. Mol. Biol. Cell *16*, 4108–4123.

Kuhle, V., Abrahams, G.L., and Hensel, M. (2006). Intracellular *Salmonella enterica* redirect exocytic transport processes in a *Salmonella* pathogenicity island 2-dependent manner. Traffic *7*, 716–730.

Kuhle, V., and Hensel, M. (2002). SseF and SseG are translocated effectors of the type III secretion system of *Salmonella* pathogenicity island 2 that modulate aggregation of endosomal compartments. Cell Microbiol. *4*, 813–824.

Kuhle, V., Jäckel, D., and Hensel, M. (2004). Effector proteins encoded by *Salmonella* pathogenicity island 2 interfere with the microtubule cytoskeleton after translocation into host cells. Traffic *5*, 356–370.

Lee, A.H., Zareei, M.P., and Daefler, S. (2002). Identification of a NIPSNAP homologue as host cell target for *Salmonella* virulence protein SpiC. Cell Microbiol. *4*, 739–750.

Miao, E.A., Brittnacher, M., Haraga, A., Jeng, R.L., Welch, M.D., and Miller, S.I. (2003). *Salmonella* effectors translocated across the vacuolar membrane interact with the actin cytoskeleton. Mol. Microbiol. *48*, 401–415.

Mota, L.J., Ramsden, A.E., Liu, M., Castle, J.D., and Holden, D.W. (2009). SCAMP3 is a component of the *Salmonella*-induced tubular network and reveals an interaction between bacterial effectors and post-Golgi trafficking. Cell Microbiol. *11*, 1236–1253.

Nawabi, P., Catron, D.M., and Haldar, K. (2008). Esterification of cholesterol by a type III secretion effector during intracellular *Salmonella* infection. Mol. Microbiol. *68*, 173–185.

Oh, Y.K., and Straubinger, R.M. (1996). Intracellular fate of *Mycobacterium avium*: use of dual-label spectrofluorometry to investigate the influence of bacterial viability and opsonization on phagosomal pH and phagosome- lysosome interaction. Infect. Immun. *64*, 319–325.

Ohlson, M.B., Fluhr, K., Birmingham, C.L., Brumell, J.H., and Miller, S.I. (2005). SseJ deacylase activity by *Salmonella enterica* serovar Typhimurium promotes virulence in mice. Infect. Immun. *73*, 6249–6259.

Patel, J.C., Hueffer, K., Lam, T.T., and Galan, J.E. (2009). Diversification of a *Salmonella* virulence protein function by ubiquitin-dependent differential localization. Cell *137*, 283–294.

Perrin, A.J., Jiang, X., Birmingham, C.L., So, N.S., and Brumell, J.H. (2004). Recognition of bacteria in the cytosol of mammalian cells by the ubiquitin system. Curr. Biol. *14*, 806–811.

Prost, L.R., and Miller, S.I. (2008). The *Salmonellae* PhoQ sensor: mechanisms of detection of phagosome signals. Cell Microbiol. *10*, 576–582.

Prost, L.R., Sanowar, S., and Miller, S.I. (2007). *Salmonella* sensing of anti-microbial mechanisms to promote survival within macrophages. Immunol. Rev. *219*, 55–65.

Rajashekar, R., Liebl, D., Seitz, A., and Hensel, M. (2008). Dynamic remodeling of the endosomal system during formation of *Salmonella*-induced filaments by intracellular *Salmonella enterica*. Traffic *9*, 2100–2116.

Rathman, M., Sjaastad, M.D., and Falkow, S. (1996). Acidification of phagosomes containing *Salmonella* Typhimurium in murine macrophages. Infect. Immun. *64*, 2765–2773.

Ruiz-Albert, J., Yu, X.J., Beuzon, C.R., Blakey, A.N., Galyov, E.E., and Holden, D.W. (2002). Complementary activities of SseJ and SifA regulate dynamics of the *Salmonella* Typhimurium vacuolar membrane. Mol. Microbiol. *44*, 645–661.

Salcedo, S.P., and Holden, D.W. (2003). SseG, a virulence protein that targets *Salmonella* to the Golgi network. EMBO J. *22*, 5003–5014.

Salcedo, S.P., and Holden, D.W. (2005). Bacterial interactions with the eukaryotic secretory pathway. Curr. Opin Microbiol. *8*, 92–98.

Schaible, U.E., and Haas, A., eds (2009). Intracellular Niches for Microbes: A Pathogens Guide Through the Host Cell (Weinheim, Wiley-VCH).

Schlumberger, M.C., and Hardt, W.D. (2006). *Salmonella* type III secretion effectors: pulling the host cell's strings. Curr. Opin Microbiol. *9*, 46–54.

Shea, J.E., Hensel, M., Gleeson, C., and Holden, D.W. (1996). Identification of a virulence locus encoding a second type III secretion system in *Salmonella* Typhimurium. Proc. Natl. Acad. Sci. U.S.A. *93*, 2593–2597.

Shi, Y., Cromie, M.J., Hsu, F.F., Turk, J., and Groisman, E.A. (2004). PhoP-regulated *Salmonella* resistance to the antimicrobial peptides magainin 2 and polymyxin B. Mol. Microbiol. *53*, 229–241.

Shotland, Y., Kramer, H., and Groisman, E.A. (2003). The *Salmonella* SpiC protein targets the mammalian Hook3 protein function to alter cellular trafficking. Mol. Microbiol. *49*, 1565–1576.

Steele-Mortimer, O. (2008). The *Salmonella*-containing vacuole: moving with the times. Curr. Opin Microbiol. *11*, 38–45.

Stein, M.A., Leung, K.Y., Zwick, M., Garcia del Portillo, F., and Finlay, B.B. (1996). Identification of a *Salmonella* virulence gene required for formation of filamentous structures containing lysosomal membrane glycoproteins within epithelial cells. Mol. Microbiol. *20*, 151–164.

Szeto, J., Namolovan, A., Osborne, S.E., Coombes, B.K., and Brumell, J.H. (2009). *Salmonella*-containing vacuoles display centrifugal movement associated with cell-to-cell transfer in epithelial cells. Infect. Immun. *77*, 996–1007.

Tezcan-Merdol, D., Engstrand, L., and Rhen, M. (2005). *Salmonella enterica* SpvB-mediated ADP-ribosylation as an activator for host cell actin degradation. Int. J. Med. Microbiol. *295*, 201–212.

Trent, M.S., Pabich, W., Raetz, C.R., and Miller, S.I. (2001). A PhoP/PhoQ-induced Lipase (PagL) that catalyzes 3-O-deacylation of lipid A precursors in membranes of *Salmonella* Typhimurium. J. Biol. Chem. *276*, 9083–9092.

Uchiya, K., Barbieri, M.A., Funato, K., Shah, A.H., Stahl, P.D., and Groisman, E.A. (1999). A *Salmonella* virulence protein that inhibits cellular trafficking. EMBO J. *18*, 3924–3933.

Vazquez-Torres, A., and Fang, F.C. (2001). *Salmonella* evasion of the NADPH phagocyte oxidase. Microbes Infect *3*, 1313–1320.

Vogel, J.P., and Isberg, R.R. (1999). Cell biology of *Legionella pneumophila*. Curr. Opin Microbiol. *2*, 30–34.

Yu, X.J., Ruiz-Albert, J., Unsworth, K.E., Garvis, S., Liu, M., and Holden, D.W. (2002). SpiC is required for secretion of *Salmonella* Pathogenicity Island 2 type III secretion system proteins. Cell Microbiol. *4*, 531–540.

Anti-*Salmonella* Immunity: Highlighting New Research in Vaccines, Mucosal Immunology and Systemic Disease

12

Jennifer L. Bishop*, Ellen T. Arena, Kenneth W. Harder and B. Brett Finlay

Abstract

Enteric fever and non-typhoidal salmonelloses (NTS) are caused by a wide variety of *Salmonella enterica* serovars and are a serious health threat throughout the world. Immunity to systemic typhoid and NTS requires intricate crosstalk between both innate and adaptive immune cells spanning multiple organ systems. The development of a number of new mouse and *in vitro* culture models suitable for studying gastroenteritis has highlighted the complexity of mucosal responses and shown how a diverse subset of cells interacts within the intestinal architecture to elicit anti-*Salmonella* immunity. These cells include specific dendritic cell subsets, natural killer cells, TH17 skewed T helper cells and the repertoire of cytokines they produce, including IL-17, IL-23, IL-22 and IL-15. Furthermore, the importance of commensal microflora has been stressed in various *Salmonella* models, and new research has shown the various effects of prebiotics, probiotics and antibiotics on *Salmonella* pathogenesis. Systemic immune responses are also more explicitly understood, as the location and phenotype of cells harbouring intracellular bacteria become more defined. This chapter will review these recent advances and how they are being translated into new therapies and vaccine studies in the human population. For the reader's convenience, a list of frequently used abbreviations throughout this review is presented at the end of this chapter.

History of salmonellosis

Enteric fevers are caused by the human-specific pathogens *Salmonella enterica* serovars Typhi and Paratyphi. The name 'enteric fever' reflects the route of acquisition, as infection of the gastrointestinal (GI) tract through exposure to contaminated food and water eventually causes systemic dissemination of bacteria leading to disease. Enteric fever, also commonly known as typhoid fever, or simply typhoid, is characterized by a sustained high fever and flu-like symptoms. Until the 20th century, typhoid fever was a common household illness found around the globe and was often indistinguishable from other sicknesses with similar symptoms, such as malaria, dengue fever, and typhus. However, with the advancement of microbiological techniques in the late 1800s, the *Salmonella* 'bacillus' was isolated, and widespread efforts were made to identify, prevent and treat typhoid fever (Fig. 12.1, modified from Bishop, 2008).

Epidemiology of salmonellosis

Typhoid and paratyphoid fevers

Today, enteric fevers are most common in the developing world; the highest incidence of typhoid is found in South-East and Central Asia where it is endemic, and enteric fevers are a frequent disease in Africa, the Middle East, South America, and even parts of Southern and Eastern Europe. Children between the ages of 2 and 19 are the most widely affected population. While typhoid seems to be less severe in children than adults, the risk of death in youth is high (Crump *et al.*, 2008). Typhoid epidemics are less common, however, they can be severe. For example, in Tajikistan in 1997, more than 1200 cases of typhoid were reported

*Corresponding author

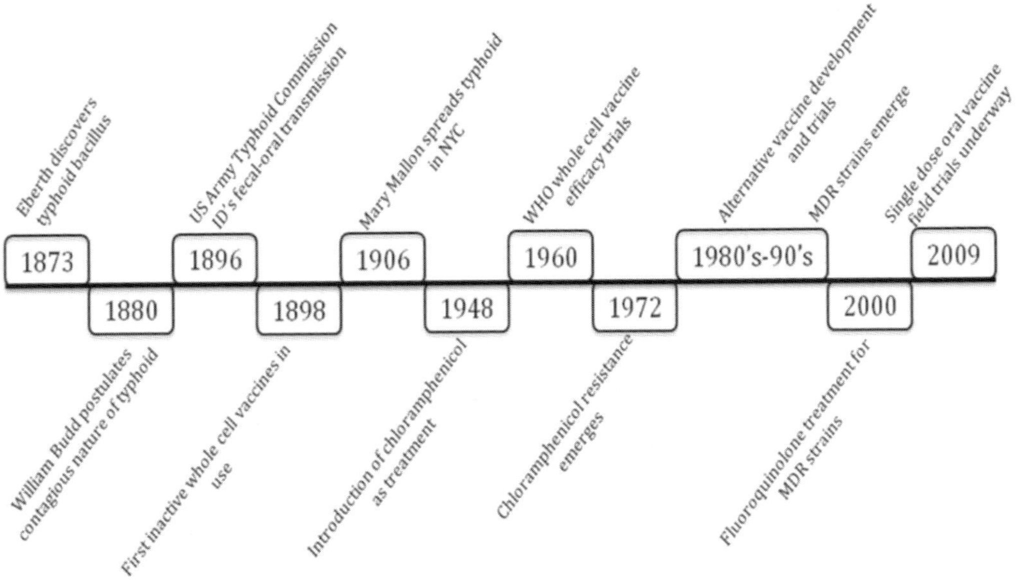

Figure 12.1 Salmonellosis: history, treatment and prevention.

weekly in the city of Dushanbe in the wake of a water sanitation plant malfunction (Mermin *et al.*, 1999). For the most part, *Salmonella* Typhi is the cause of most (around 80%) endemic and epidemic cases of typhoid fever globally (Connor and Schwartz, 2005). However, the emergence of Paratyphi strains, especially Paratyphi A, as major contributor to disease, is becoming apparent. In particular, in countries with the highest incidence of enteric fevers, including India, Pakistan, China, and the Philippines, recent reports have estimated Paratyphi A as the cause of between 35–65% of all enteric fever cases. Interestingly, in contrast to typhoid, paratyphoid is a disease that affects teenagers and young adults more so than young children in endemic areas (Fangtham and Wilde, 2008).

As of the year 2000, there were an estimated 22 million cases globally of typhoid fever each year, and roughly 10% of these cases were fatal. Paratyphoid cases were estimated to be 5.5 million (Crump *et al.*, 2004). Unfortunately however, these statistics are badly flawed and there are large discrepancies and gaps in reports of incidence or complications and fatality rates of typhoid especially from areas outside Asia. Mortality is especially hard to estimate, since generating data from hospitalization records include fatality rates only from the most severe illnesses or those who choose, or have access to, medical treatment. Similarly, data from patients involved in study cases are skewed since these individuals usually benefit from optimal clinical management, increasing their chances of survival. Combined with the fact that few reports to date exist on resultant complications from typhoid, case fatality rates could be much higher than statistics show (Crump *et al.*, 2008). Despite this, generalization in epidemiological trends can be found. For example, from the start of the antibiotic era around 1950, case fatality rates of patients with typhoid globally have dropped from approximately 11 to 6% and incidences of typhoid infections in developed countries have shifted from the general population to travellers, especially in those travelling to India (Butler *et al.*, 1985; Connor and Schwartz, 2005). Interestingly, while *Salmonella* Typhi is still the major cause of disease in endemic areas, travellers are more susceptible to Paratyphi infections, possibly due to the availability of travel vaccines that protect only against *Salmonella* Typhi infection. Certainly this statistic reflects the disproportionate increase in Paratyphi infections seen in areas of high enteric fever incidence (Connor and Schwartz, 2005).

Non-typhoidal salmonelloses

Estimates for numbers of non-typhoidal salmonellosis (NTS) cases worldwide vastly exceed those for enteric fevers; there are 1.3 billion cases annually, 300,000 of which are fatal (Pang *et al.*, 1995). Like enteric fevers, NTS is spread via the faecal–oral route. However, in contrast to typhoid and paratyphoid, NTS are caused by infection with various *S. enterica* serovars and are characterized by gastroenteritis or bacteraemia. Also unlike Typhi and Paratyphi, non-typhoidal *Salmonellae* are not human-restricted; many serovars, including Typhimurium, Enteritidis, Newport and Heidelberg have reservoirs in animal populations used for agriculture (Hohmann, 2001; Rabsch *et al.*, 2001). As such, these pathogens are among the most important linked to food-poisoning cases in developed countries. In the United States alone, the Center for Disease Control estimates the annual number of NTS cases to be 1.4 million. Approximately 600 cases of these are fatal, making NTS the leading cause of food-related death (Rabsch *et al.*, 2001).

Infections with various non-typhoidal serovars can also result in bacteraemia. This occurs in approximately 5% of NTS cases and is most prevalent in underdeveloped countries (Hohmann, 2001). Alarmingly, the number of invasive NTS cases is increasing, especially in individuals with severe immunosuppressive conditions or malignancies (Gordon, 2008). This trend is particularly evident in sub-Saharan Africa where NTS are in the top three most common types of blood infections in both children and adults. In many areas and populations the incidence of bacteremic NTS infection is higher than that of typhoid (Morpeth *et al.*, 2009). In particular, NTS infection rates can reach up to 60% in HIV/AIDS patients there (Gordon, 2002; Graham, 2002). NTS is also an important nosocomial infection in Africa, with outbreaks in paediatric wards especially prevalent. It is thought that a variety of risk factors exist in the hospital setting, including overcrowding, health status of patients (malnutrition, HIV infection), exposure to antimicrobials and unsanitary food sources, which favour outbreak conditions (Morpeth *et al.*, 2009).

Sources of infection and risk factors

Both enteric fevers and NTS infections are spread through the faecal–oral route and are usually contracted by ingestion of *S. enterica* serovars present in contaminated food or water sources. Human to human transmission can occur. Interestingly, typhoid and paratyphoid fevers seem to be spread in different ways. While typhoid is often contracted in the home through contact with carriers or poor hygiene and food sharing, paratyphoid is more associated with consumption of contaminated food from public sources (street vendors or produce from flooded areas) (Vollaard *et al.*, 2004). NTS-associated serovars are typically introduced to meat and dairy products from host faecal matter during processing. Interestingly, there has been a shift in developed countries in the serovars and food sources most associated with NTS infections. Namely, the eradication of *Salmonella* Gallinarum from poultry populations in the 1970's enabled other serovars, like Enteritidis and Typhimurium, to become the leading causes of food poisoning associated with contaminated eggs (Rabsch *et al.*, 2001).

Despite these shifts and different routes of acquisition, it remains clear that risk factors for contracting both enteric fevers and NTS are much the same as those identified in the late 20th century. Poor hygiene and limited access to sanitized food and water are the main prerequisites for infection. Recent studies have shown that socioeconomic and environmental factors, such as poor housing, inadequate food, and rainfall amounts correlate with disease (Gasem *et al.*, 2001; Luby *et al.*, 1998). Many of these factors reflect the robust nature of *S. enterica* in the environment, which is able to survive even in ice, dust and raw sewage for prolonged periods of time, increasing the risk of transmission in areas where sanitation is an issue. In developed countries, pre-existing health conditions, such as asthma or inflammatory bowel diseases, are also risk factors for contracting NTS (Doorduyn *et al.*, 2008).

Human susceptibility to *Salmonella* infections

Enteric fevers and NTS are diseases that primarily affect immunocompromised hosts in endemic areas. These include young children, the elderly and HIV/AIDS patients, or patients with autoimmune disorders. However, there is also a genetic basis for the predisposition to typhoidal and non-typhoidal *Salmonella* infections, and many of these genetic factors increase susceptibility to not only *Salmonella,* but to other intracellular pathogens. Most notably, individuals with genetic deficiencies in type 1 cytokine pathway function (IFNγR/IL-12/IL-23) are highly susceptible to both *Salmonella* and *Mycobacteria* infections. This cytokine axis will be discussed in detail in a further section, but IL-12 and IFNγ are required for full activation of innate immune responses against pathogens and are required to initiate T cell mediated immunity. IL-23 is an important factor for the induction of TNF α, IL-17 and nitric oxide, which enable macrophages to efficiently kill intracellular bacteria. Interestingly, patients with IFNγ receptor abnormalities are more susceptible to *Mycobacterium,* while IL-12/IL-23 deficiencies are more associated with *Salmonella* infections (van de Vosse *et al.*, 2004). In addition, polymorphisms in IFNγ genes are associated with the development of reactive arthritis after NTS infections and this condition is seen in women more often than men (Doorduyn *et al.*, 2008). Outside of cytokine network genes, polymorphisms in protein degradation pathway genes, like PARK2/PACRG, increase susceptibility to both *Salmonella* Typhi and Paratyphi A infections (Ali *et al.*, 2006). Interestingly, polymorphisms in the gene NRAMP1 (Sca11a), which controls susceptibility to *Salmonella* in mice, are not associated with increased susceptibility to typhoid in humans (Dunstan *et al.*, 2001).

Clinical disease, diagnosis and treatment of salmonellosis

Clinical signs and symptoms

Infectious doses of *S. enterica* that can cause NTS or typhoid fever have a wide range; ingestion of anywhere from 10–100 bacteria has been associated with food poisoning outbreaks, and typically 10^3–10^9 *Salmonella* Typhi are sufficient to cause enteric fever; however, this number is most likely higher for Paratyphi A (Fangtham and Wilde, 2008; Hornick *et al.*, 1970a,b). The overall course of infection is as follows: bacteria are swallowed and many resist the harsh environment of the stomach to adhere to and invade the epithelium of the small intestine. In NTS cases, gastrointestinal symptoms, such as nausea, vomiting and diarrhea, are typically self-limiting and last for a maximum of 7 days (Dhanoa and Fatt, 2009). In contrast, typhoid, paratyphoid, and invasive non-typhoidal infections spread to systemic sites as bacteria disseminate within phagocytes of the immune system, such as macrophages and dendritic cells (DC), to lymphoid follicles, the mesenteric lymph nodes (MLN), and the liver. After an asymptomatic incubation period of 7–14 days, secondary bacteraemia and blood infections can occur in cases of typhoid and invasive NTS. Organisms then spread to other organs, such as the bone marrow and gallbladder. During this time, infected patients will develop an incremental fever that can reach 40°C for a sustained period of time (up to two weeks), as well as other flu-like symptoms including chills, malaise and cough (House *et al.*, 2001). Often, rose spots are found on the torso of infected patients; this symptom along with enlarged spleen (splenomegaly) and liver (hepatomegaly), bradychardia, and thrombocytopenia were recently found to be predictive, significant features of typhoid infection (Kuvandik *et al.*, 2009). Historically, paratyphoid was thought to be a more mild disease; however, new data has shown it is just as severe as typhoid and importantly, the clinical features of the two diseases are indistinguishable (Fangtham and Wilde, 2008).

If left undiagnosed or untreated, typhoid/paratyphoid can progress and serious complications result that can affect almost every organ system, including the central nervous, cardiovascular, pulmonary, hepatobilliary, genito-urinary and skeletal systems. Some of the most severe complications are gastrointestinal bleeding or perforation, shock, coma, meningitis and encephalitis (Huang and DuPont, 2005). As aforementioned, statistical data on complication incidence is patchy at best; however, it is estimated that in endemic areas between 10–15% of typhoid and paratyphoid infections result in some complication (Connor and Schwartz, 2005).

Recovery from typhoid can take up to four months and relapses are not uncommon. It has been estimated that up to 10% of patients suffer relapsed infections and between 1–4% of typhoid sufferers will become chronic carriers of the bacteria (Caygill *et al.*, 1994; Marmion *et al.*, 1953). Carriage is divided into 4 separate categories: acute asymptomatic carriers (in early stages of infection), transient or convalescent carriers (individuals shedding bacteria in stool for less than 1 year, possibly those receiving antimicrobial treatment or in natural convalescence), chronic carriers (individuals shedding bacteria in stool for more than 1 year) and permanent carriage (Huang and DuPont, 2005; Sirinavin and Garner, 2000). Carriage is a serious complication of typhoid infection, not only because of the increased risk for transmission, but also because it predisposes patients to bowel, gallbladder and pancreatic cancers (Caygill *et al.*, 1994).

Diagnosis

Diagnosis of typhoid, paratyphoid and NTS infections, especially in endemic areas, is extremely difficult for a number of reasons. From a clinical perspective, enteric fevers and especially non-typhoidal bacteraemia, are characterized by non-specific symptoms and therefore are not easily distinguished from other diseases common in endemic areas, such as dengue fever, malaria or leishmaniasis. This is further complicated by the fact that typhoid and paratyphoid present with the exact same symptoms and often patients may suffer from concomitant infections with various bacteria, making the identification of the cause of infection impossible, except after organism culture.

Despite the requirement for laboratory tests to confirm infection, there are major obstacles in this type of diagnosis. For example, the gold standard in diagnosis for typhoid, paratyphoid and NTS is still culture of bacteria from bone marrow, a procedure that is highly invasive and requires a high skill level of medical personnel (Parry, 1998). Blood culture has become an acceptable replacement, but this test is subject to patients having access to timely diagnostic laboratory tests, which are often not readily available in developing regions (Butler *et al.*, 1978). Importantly, blood culture and, to a lesser extent, bone marrow culture of *Salmonella* are broadly affected by the use of antimicrobials, often leading to false-negative results even in symptomatic patients (Nagarajan *et al.*, 2009b). In addition, serological tests may not prove that a patient is currently infected with *Salmonella* Typhi; the presence of anti-*Salmonella* Typhi antibodies from prior infections, vaccinations or cross-reactive epitopes can easily skew the tests towards a false-positive result (Parry *et al.*, 1999). Furthermore, no serological tests to date are able to distinguish Typhi from Paratyphi infections. Commercially available updates to the poorly specific Widal test (developed in 1896 and still used widely to diagnose enteric fever) detect only *Salmonella* Typhi specific IgM and IgG and an antigen-based test kit for Paratyphi A is only available in the food market and not for use on human samples (Olsen *et al.*, 2004). However, more modern, PCR-based diagnostics have been developed. Based on the isolation of bacterial DNA from organisms isolated from blood cultures, older versions of these tests identified the presence of *Salmonella* by amplification of genes common to many serovars, including flagellin or *Salmonella* pathogenicity island (SPI) genes. Recently however, more specific PCR tests were designed to identify genes, such as STY0312 located on SPI-6, that are only present in *Salmonella* Typhi or Paratyphi A (see also Chapter 1 of this book). These improvements in specificity, combined with the fact that bacterial DNA isolation can be done from whole blood preparations and the test can detect dead bacteria, are making PCR diagnostics an affordable and rapid alternative to bacterial culture (Nagarajan *et al.*, 2009b).

Treatment

If properly diagnosed, treatment with antibiotics is still standard procedure for cases of typhoid, paratyphoid and both invasive and gastrointestinal NTS, despite widespread drug resistance. At the beginning of the antibiotic era in the 1950's, enteric fevers were treated with non-absorbable antibiotics, such as neomycin. Shortly afterwards, absorbable drugs like chloramphenicol, ampicillin, amoxicillin, trimethoprim and tetracycline came into widespread use (Sirinavin and Garner, 2000). However, the efficacy of these drugs lasted only a short period of time and resistance to chloramphenicol was seen in a variety of serovars as early as 1972. Multi-drug resistance (MDR) was widespread in areas of high incidence, like China, by the late 1980s (Fig. 12.1). As a result, a new class of antibiotic, the fluoroquinolones, which have a high level of activity against intracellular organisms like *Salmonella*, came into use. Alarmingly, however, chromosomal fluoroquinolone resistance developed rapidly; by 1993, 76% of clinical *Salmonella* isolates from South-East Asia were resistant to the drugs (Parry and Threlfall, 2008).

In the wake of this exponential increase in drug resistance, the question remains as to why typhoid infections are still treated with antibiotics. There is controversy in the literature as to the benefit of antibiotic treatment regimes and whether they are beneficial to patients with typhoid or paratyphoid fever. In the 1990's, it was shown that only long antibiotic regimens of between 21 and 28 days of treatment were sufficient to prevent shedding of *Salmonella* in stools of convalescent patients. A report in 2007 even went so far as to say that the health of patients on antibiotic regimens was not increased and that antibiotic treatments probably exacerbated the regional burden of enteric fever by prolonging shedding of bacteria into the environment and increasing transmission (Sirinavin and Garner, 2000). In contrast, other studies have postulated that with careful assessment of resistance on a regional level, classic antibiotics and fluoroquinolones can still be effective against susceptible bacteria (Gulig *et al.*, 1993; Parry, 2004).

Vaccinology

With increasing multi-drug resistance, as well as limited access to clean water and sanitary agriculture or animal husbandry practices in developing nations, it is essential that more efforts go towards prevention rather than treatment of salmonelloses. Vaccinology, therefore, has been a prime target for controlling the spread of *Salmonella* in both humans and livestock. Whole cell inactivated vaccines against typhoid were in use in England and Germany dating as far back as 1896 and are still licensed in many countries worldwide. They were the first vaccines to undergo large scale field trials directed by the World Health Organization in the 1960's and 1970's and did show protection in up to 80% of individuals (Guzman *et al.*, 2006). However, their deleterious side-effects, mainly high reactivity, has prevented widespread use (Engels *et al.*, 1998). Since then, development of more efficacious methods has been an ongoing struggle because (i), we have only limited knowledge of immunity and memory to systemic typhoid; (ii), many vaccine candidate mutants are either under- or over-attenuated; (iii), there is little cross-protection for one vaccine against other serovars and strains causing enteric fevers and NTS; and (iv), economic pressures often prevent the implementation of vaccine programs (Boyle *et al.*, 2007). Despite these obstacles, two typhoid vaccines have been licensed, the Ty21A live-attenuated vaccine and the Vi antigen subunit vaccine. In addition, there are a number of new vaccine candidates in clinical trials (Garmory *et al.*, 2002).

Ty21a live-attenuated vaccine

The Ty21a live-attenuated vaccine, or Vivotif® (Berna Biotech AG, Switzerland), was created in the 1970's by nitrosoguanidine chemical mutagenesis of the wild-type *Salmonella* Typhi strain Ty2 (Germanier and Furer, 1983). The most important characteristics of the mutant vaccine strain are that it lacks the Vi (virulence) antigen, a polysaccharide that protects the bacteria from phagocytosis and complement-mediated cell lysis, and GalE, an enzyme essential for the incorporation of galactose into the core O-polysaccharide of LPS. In addition, Ty21a has mutations in other important virulence genes, such as the central stress-response regulator *rpoS* and in *ilvD*, rendering it auxotrophic for isoleucine and valine. As a result of all these mutations, the Ty21a strain made for an ideal vaccine candidate, being strongly attenuated *in vivo* for colonization, dissemination, and shedding (Kopecko *et al.*, 2009).

Since its development, Ty21a has been proven as an extremely safe and tolerable vaccine. It is currently licensed for use in 56 countries worldwide and over 200 million vaccines have been administered with little to no adverse side effects. In addition, the multiple mutations in Ty21a increased the genetic stability of the strain, without threat of reversion to virulence (Kopecko *et al.*, 2009). However, there are significant drawbacks to its use. For example, the Ty21a vaccine is most efficacious when a multiple dosing regimen is followed, making it ill suited to use in endemic countries where cost and access to medical care are limiting factors. In a retrospective analysis, the cumulative effectiveness of this vaccine was only 51% with three doses, and efficacy can increase with four doses (Fraser *et al.*, 2007). In addition, Ty21a has a strict cold chain requirement for distribution. Maybe most importantly, it is not effective in children under 2 years old and cannot be used in immunocompromised individuals, pregnant women or patients already on antimicrobial drugs (Guzman *et al.*, 2006).

Furthermore, there are conflicting reports as to whether Ty21a offers cross-protection against Paratyphi infections. In theory, because Paratyphi A and B share the same O-12 antigen with *Salmonella* Typhi and protection against *Salmonella* is mediated by O-antigen specific serum IgG and secreted IgA antibodies, Ty21a could be efficacious in preventing paratyphoid fever. Indeed, cross-reactive antibodies against Paratyphi A and B are found in individuals vaccinated with Ty21a. These seem to be protective in certain regions, since 49% of individuals vaccinated with Ty21a in a study in Chile were protected against Paratyphi B, and a study on Israeli travellers showed increased protection against Paratyphi A. However, data from Indonesia has shown no protection against paratyphoid in Ty21a vaccines (Fangtham and Wilde, 2008; Levine *et al.*, 2007). Taken together, these drawbacks to Ty21a suggest that the populations most widely affected by typhoid fever in endemic areas are poor vaccine candidates.

Vi antigen subunit vaccine

The Vi antigen vaccine (Typhim Vi, Aventis Pasteur) is a subunit vaccine created by purification of the capsular polysaccharide Vi antigen and, unlike Ty21a, is administered via intramuscular injection in a single dose. It provides protection against typhoid by eliciting anti-Vi antibodies. The Vi vaccine is extremely safe and relatively effective. In the 1980's the World Health Organization began vaccine trials in South Africa and Nepal; in these studies and many others conducted since, the overall efficacy rate of the Vi vaccine was around 55% (Fraser *et al.*, 2007; Guzman *et al.*, 2006). In contrast to Ty21a, this subunit vaccine is a more cost effective choice for mass immunizations in underdeveloped nations. It is cheaply and easily produced and does not require the cold storage that Ty21a does. Also, its efficacy of producing high anti-Vi antibody titres within 2 weeks of immunization after a single dose makes it an attractive vaccine candidate in areas where multiple dosing is difficult (Whitaker *et al.*, 2009). However, like Ty21a, there are major drawbacks to the use of the Vi subunit vaccine. Most importantly, the vaccine induces only T cell-independent immune responses and generates circulating antibodies against only the Vi antigen. Full protection against *Salmonella* Typhi infection requires not only antibody responses against multiple antigens including LPS, flagellin, as well as the Vi antigen, but also significant cell mediated immunity. Importantly, circulating antibodies against *Salmonella* antigens do not prevent re-infection. This highlights the importance of T cell mediated immunological memory in protecting against *Salmonella* infections; the lack of stimulation of cell-mediated responses by the Vi vaccine therefore is a major obstacle. Immunization programs have attempted to remedy this issue by having Vi vaccine boosters every 2–3 years in endemic areas. However, actual protection from booster immunizations does not increase (Guzman *et al.*, 2006; Whitaker *et al.*, 2009). More promising, however, is the use of conjugate vaccines that link the Vi antigen to another, T cell-dependent protein carrier molecule. For example, the Vi–rEPA conjugate vaccine, which links the Vi antigen to recombinant exotoxin A from *Pseudomonas aeruginosa*, has shown drastic improvements in immunogenicity. This vaccine was recently shown to have 92% efficacy in a trial in Vietnam; importantly, it protected young children, as well as adults, and showed a booster effect with multiple doses (Konadu *et al.*, 2000).

New vaccine candidates

The major drawbacks of the currently licensed vaccines against *Salmonella* are that they are expensive, poorly immunogenic in children, show only moderate efficacy, offer questionable protection against Paratyphi strains, and/or do not produce long-lived immunity. As such, the possibilities of new vaccine candidates are being explored all the time. Most of these vaccines are live-attenuated vaccines, with the focus on reducing oral immunization to a single dose. Ideally, this type of vaccine strain would be able to colonize the gastrointestinal tract long enough to induce a strong immune response, but be highly attenuated *in vivo* thereafter.

Many of the newer *Salmonella* Typhi vaccine strain candidates are auxotrophic for amino acid synthesis; they lack the ability to make aromatic compounds, and deletion of '*aro*' genes is not complemented by any other metabolism pathway in *Salmonella*. As a result, these strains can survive only long enough to elicit an immune response but are strongly attenuated *in vivo* for sustained colonization and systemic spread (Guzman *et al.*, 2006). Several auxotrophic Typhi mutants, developed at the Center for Vaccine Development (CVD) in Maryland, have had vaccine potential including CVD 908 (Ty2 *aroC/aroD* mutant), CVD 906 (ISP1820 *aroA/aroD* mutant) and PBCC211 (CDC10–80 *aroA/aroD* mutant). However, all of these have been deemed unsuitable for general use because they caused vaccinaemia or other adverse reactions. Newer derivatives have combined the auxotrophic mutations with other essential virulence factor mutations to produce safer, yet more effective vaccines. These include CVD 909, which in addition to mutations in *aroA/aroD*, has a mutation in *htrA*, a heat-shock protein essential for stress response and intracellular survival in macrophages; and M01ZH09 (previously called ZH9), which is an *aroC/ssaV* [*Salmonella* Pathogenicity Island (SPI)–2 effector] mutant. Both have shown improved efficacy and low reactivity in phase 1 or 2 clinical trials (Tacket and Levine, 2007). Importantly, these vaccines have made significant improvements in the generation of cell mediated and memory responses against *Salmonella* Typhi. For example, it was recently shown that CVD 909 elicits strong $CD4^+$ and $CD8^+$ central and effector memory T cell responses and that these cells can migrate to the GI tract and secondary lymphoid tissue to generate an immune response (Wahid *et al.*, 2008). In addition, the creation of a Vi antigen positive CVD 909 has overcome the problem that most new live-attenuated vaccines pose: the lack of generation of serum Vi antibody titres (Tacket *et al.*, 2004).

In addition to auxotroph-based mutations, *Salmonella* strains with mutations in two-component regulatory systems or surface molecule expression also are strongly attenuated *in vivo* and are live vaccine candidates. For example, Ty800, a Ty2 derivative vaccine strain with mutations in *phoP-phoQ*, which is an essential regulon for intracellular bacterial survival and evasion of degradation in the phagolysosome, has been shown to be effective in clinical trials (Hohmann *et al.*, 1996). More recently, the *pmr*G-HM-D *Salmonella* Typhimurium mutant was shown to be protective in a mouse model of typhoid fever and, importantly, may be safe for use during pregnancy (Negi *et al.*, 2007). This strain combines mutations that cover multiple *pmr* systems in *Salmonella*, which are known to regulate antimicrobial peptide resistance and LPS modifications, as well as other two-component regulons. Recently, Roy Curtiss and colleagues have been evaluating *in vivo* attenuation in a number of regulon mutants and have found that mutating genes in a variety of regulators, such as *fur* and *rpoS*, may also provide new vaccine candidates (Curtiss *et al.*, 2009).

LPS and outer membrane protein mutants are still an attractive possibility for vaccine strains as well. For example the *rfaH Salmonella* Typhi mutant has a significantly altered LPS structure that has been called 'gently rough', indicating the heterogeneity of length of LPS chains it expresses. *rfaH* mutants show strong attenuation *in vivo* and are protective in animal models of typhoid in a single dose (Nagy *et al.*, 2004). Importantly, because the surface of the bacteria is so greatly altered in this strain, a high level of immunogenicity against highly conserved epitopes shared among many *Salmonella* serovars has been shown, suggesting that vaccines like the *rfaH* strain could provide cross-protection between typhoid, paratyphoid, and NTS (Nagy *et al.*, 2008). More recently, the deletion of *tolA* in *Salmonella* Typhimurium has also been shown to be effective in preventing disease. *TolA* encodes an envelope protein that contributes to virulence, membrane and LPS integrity, as well as bile and serum resistance in *Salmonella* (Paterson *et al.*, 2009).

Finally, efforts are being made to incorporate more modern technologies, such as DNA-based vaccines, into *Salmonella* vaccinology. DNA vaccines work by delivering plasmid DNA encoding specific protein antigens of choice into host cells, which are used as vehicles to produce the immunizing protein antigen. While there are many advantages to using DNA vaccines on their own, it has been shown that they work best in conjunction with live-attenuated or subunit vaccines, making DNA technology ideally suited for incorporation into *Salmonella* vaccine development (Lu *et al.*, 2008). To make the ideal *Salmonella* DNA vaccine, the antigen that is responsible for eliciting the strong $CD8^+$ (cytotoxic T lymphocyte/CTL) immune responses required for protective immunity must be found. While it is known that some *Salmonella* proteins, such as GroEL, FliC, and SipC do generate antibody responses *in vivo*, these proteins only give limited protection as subunit vaccines in animal models. Recently, the SPI-1 *Salmonella* inositol phosphatase effector, SopB, was identified in a bioinformatic screen as a potential DNA vaccine candidate. Because SopB is secreted by both *Salmonella* Typhi and Typhimurium and showed strong potential for binding to major histocompatibility complex (MHC) I alleles, using it in combination with already available anti-*Salmonella* vaccines may increase cell-mediated immune responses and importantly, provide cross-protection between typhoidal and non-typhoidal *Salmonella* infections (Nagarajan *et al.*, 2009a).

Animal models of salmonellosis

Murine typhoid

Salmonella Typhi is a pathogen that can cause disease only in humans (host-restricted). Therefore, we must utilize other *S. enterica* species that have a broader host range, yet cause similar symptoms to the human disease, to study typhoid fever in an animal model. Why host restriction occurs is a matter of debate, but genetic factors in the host and pathogen, as well as environmental factors, most likely play key roles (Mastroeni and Maskell, 2006). *Salmonella* Typhimurium is one of the top choices for modelling typhoid fever in mice because it causes systemic disease that closely mimics the human condition. This model can be modified in a number of ways, mainly by the route of infection (oral, intraperitoneal, intravenous) or infection strain (attenuated mutants, wild-type laboratory strains, and clinical isolates), enabling researchers to study the importance of distinct organ systems in bacterial pathogenesis and anti-*Salmonella* immunity. In general, mice are either genetically resistant or susceptible to systemic *Salmonella* Typhimurium infection, and this depends primarily on the expression of the *Slc11a1* (*Nramp1*) gene, which encodes a divalent cation transporter that affects innate immune responses and survival of intracellular bacteria. While susceptible mice have a relatively low threshold for infectious doses of *Salmonella* Typhimurium by the oral route and succumb to infection between 7–10 days, resistant mice survive and either clear the infection or develop into chronic, asymptomatic carriers. As such, using resistant and susceptible mice has allowed researchers to develop both acute and persistent models of typhoid fever and to study the immune responses, cell types, and organ systems most important in both models (Blackwell *et al.*, 2001).

Gastroenteritis/colitis

The classic murine model of *Salmonella* Typhimurium infection mimics the systemic disease seen in human typhoid fever and is not a model for the gastroenteritis associated with NTS food poisoning infections. In the typhoid model, while mice exhibit classic inflammation at infection foci in systemic sites, *Salmonella* Typhimurium poorly colonizes the GI tract and does not cause enterocolitis or diarrhea (Santos *et al.*, 2001). However, there are two intestinal disease models based on *Salmonella* Typhimurium infection. The first is a calf infection model of *Salmonella* Typhimurium; the animals develop enterocolitis that closely resembles human disease. Of course, the drawbacks to using such a large animal model are obvious, and improvements were made in which sections of intestine in calves were ligated into loops, allowing single animals to be used for multiple infection studies (infection doses, mutants, etc.) (Santos *et al.*, 2001). More recently, however, a new *Salmonella* Typhimurium murine model has been developed that also mimics *Salmonella*-induced human enterocolitis. Based on studies that showed that germ-free mice are susceptible to colitis after infection with *Salmonella* Typhimurium, this new model capitalizes on perturbations of the intestinal microbiota with streptomycin pre-treatment 24 hours prior to oral infection with *Salmonella* Typhimurium to cause gastroenteritis (Stecher *et al.*, 2005). The mice develop severe intestinal pathology, characterized by oedema and infiltration of inflammatory cells into the cecum and colon, however, they do not develop diarrhea. Importantly, this model does not induce small intestine inflammation that is often associated with NTS in humans. However, this is a large step forward in the search for more relevant animal models used to study *Salmonella* pathogenesis (Barthel *et al.*, 2003).

Salmonella pathogenesis in murine typhoid: an overview

During invasion of the GI tract and spread to systemic sites, *Salmonella* uses a variety of virulence factors for pathogenesis. While virulence genes can be found anywhere in the genome, within *Salmonella,* they are mostly clustered on SPI's. *Salmonella* has 18 SPIs, the first two of which are the most fully characterized. Encoded on SPI-1 and 2 are type-three secretion systems (TTSS) that are essential for *Salmonella* pathogenesis. These protein complexes act as 'molecular syringes' that inject bacterial encoded proteins, termed effectors, directly into host cells. These effectors modulate host cell signalling that promote either bacterial invasion of host cells (SPI-1 effectors) or intracellular survival and systemic spread of bacteria (SPI-2 effectors) (Gal-Mor and Finlay, 2006; Hensel, 2004). For example, SPI-1 encoded effectors are necessary to induce cytoskeletal rearrangements required for entry into non-phagocytic cells (Schlumberger and Hardt, 2006) and are only required for colonization of the intestinal tract, but not systemic spread, in the mouse typhoid model (Galan and Curtiss, 1989). In contrast, SPI-2 effectors are required for the maintenance of, and replication within, the *Salmonella* Containing Vacuole (SCV), a specialized compartment *Salmonella* form when phagocytosed that prevents degradation by the normal endocytic pathway (Galan, 2001). In addition, effectors of both SPI-1 and SPI-2 encoded TTSS are essential for modulating host immune responses against *Salmonella* (Table 12.1). Chapter 10 of this book elaborates in more detail on the effector molecules of *Salmonella*.

Immunity during the course of infection

Immediately after entering the gastrointestinal (GI) tract, a complex and concerted immune response involving epithelial cells and both innate and adaptive immune cells is mounted against pathogenic *Salmonella*. In the rest of this chapter, we will follow the course of infection as *Salmonella* travel from the GI tract to systemic sites of infection, detailing the complex interplay between immune signals that are required to control infection. We will provide an overview of key components that regulate immunity in the GI tract, and we will highlight new advances that have been made in understanding how mucosal defences respond to *Salmonella* infection. Importantly, we will explore how commensal bacteria and epithelial cell responses are emerging as primary mediators that determine the severity of Salmonellosis. In addition, we will discuss how protective immunity is generated in systemic sites of infection and the implications systemic disease have on *Salmonella* persistence, shedding, and transmission to new hosts.

Mucosal immunity in the gastrointestinal tract

Aside from nutrient absorption, the other critical function of the GI tract is the generation of immunological tolerance and immunity to pathogens. At the most basic level, the GI tract serves as physical barrier against invading pathogens. The single layer columnar epithelium that separates the GI tract lumen from underlying tissue is enhanced by microvillus membrane extensions on the apical surface of the cells, forming the filamentous brush border. This epithelial layer is in

Table 12.1 Most prominent *Salmonella* secreted effectors and their function in immune cells. See Chapter 10 of this book for an extended list

Location	Effector	Target cell	Function	Reference
SPI-1	SipA	Epithelial	Invasion Inflammation	Zhang *et al.* (2003b)
SPI-1	SipB	Epithelial PMN MΦ	Invasion Recruitment Apoptosis, increased IL-1β	Monack *et al.* (2000)
Chromosome	SopA	Epithelial PMN	Invasion Recruitment	Raffatellu *et al.* (2005), Wood *et al.* (2000), Zhang *et al.* (2006)
SPI-5	SopB (SigD)	Epithelial PMN MΦ	Invasion Recruitment Prevents lysosomal fusion	Dukes *et al.* (2006), Raffatellu *et al.* (2005), Zhang *et al.* (2002), Zhou (2001)
Chromosome Phage	SopD/E/E2	Epithelial PMN	Invasion Recruitment, inflammation	Bakshi *et al.* (2000), Bruno *et al.* (2009), Muller *et al.* (2009), Raffatellu *et al.* (2005), Zhang *et al.* (2002)
SPI-1	SptP	Epithelial	Decreased IL-8	Haraga and Miller (2003)
Gifsy-3 Prophage	Ssph1	Epithelial	Decreased NF-κB signalling	Haraga and Miller (2003)
Chromosome SPI-2 SPI-2 Chromosome	SifA SseF SseG PipB2	MΦ Epithelial	SIF formation SCV integrity Intracellular survival	Beuzon *et al.* (2000), Knodler and Steele-Mortimer (2005), Kuhle and Hensel (2002), Stein *et al.* (1996)
SPI-2	SpiC	Phagocytes	Vesicular trafficking Intracellular survival	Uchiya *et al.* (1999)

constant contact not only with food antigens, but also with an enormous community of commensal bacteria and fungi (the microbiota, discussed in more detail below) that must be distinguished from pathogens. Glycoproteins secreted by epithelial cells limit commensal and pathogenic organisms from passing through the epithelial barrier. Epithelial cells also secrete antimicrobial peptides (β-defensins, cathelicidins) and trefoil peptides that can directly kill bacteria, and they are key in transporting secretory (s)IgA to the lumen that opsonizes potential invaders. This barrier function of the GI tract prevents antigens from accessing immune-reactive areas of the GI tract, which generates a state of 'antigenic ignorance' within the gut. This is one key component of mucosal tolerance. Active tolerance also occurs in the GI tract when specialized immune cell types generate both antigen-specific and non-specific inflammatory responses against luminal antigens (Mason *et al.*, 2008). *Salmonella* specific immune responses generated in the GI tract will be covered in a later section.

Like other immune organs, such as the spleen and lymph nodes, the intestines are designed with a unique architecture essential in generating immune responses. Within the GI tract, immunity is initiated in inductive sites; these are highly organized follicles of lymphoid cells, such as the Peyer's Patches (PP), isolated lymphoid follicles (ILF), and mesenteric lymph nodes (MLN), where antigens are presented by professional antigen presenting cells (APC) to induce responses from naïve T and B cells. PP and ILF are embedded in the follicle-associated epithelium (FAE) that separates the lumen from the rest of the GI tract tissue. FAE is unique from other GI tract epithelium in that it has a reduced brush border and digestive enzyme and mucin secretion. While PP are permanent structures located in the terminal ileum, ILF can be found throughout the length of the GI tract, and their formation is induced during inflammatory responses (Mason *et al.*, 2008). Central to the function of PP are 'M' or microfold cells, that are unique, actively phagocytic epithelial cells that have the ability to transport food and microorganism antigens to underlying DCs, T cells and B cells located in the sub-epithelial dome (SED). Specific DC subsets are known to be associated with PP, and they include myeloid ($CD11c^+/CD11b^+$) and lymphoid ($CD11c^+/CD8^+/CD4^+$) DC's; lymphoid and plasmacytoid ($CD11c^+/CD11b^-/B220^+$) DCs are also found in interfollicular regions of the FAE (Iwasaki and Kelsall, 2001). Interestingly, PP myeloid DCs produce low levels of IL-12 and high levels of IL-10, skewing them towards an anti-inflammatory phenotype, enabling Th2 T cell responses, while lymphoid and plasmacytoid DCs are pro-inflammatory (Iwasaki and Kelsall, 1999).

While inductive sites in the GI tract are tightly organized lymphoid aggregates, effector sites, like the lamina propria (LP), are more diffuse and are integrated into tissue architecture. These are areas where innate and adaptive immune cells like DCs, macrophages, neutrophils, natural killer (NK) cells and T helper cells (Th1, Th2 and Th17 subsets) induce strong pro-inflammatory cytokine responses in order to prevent infection into deeper tissue. However, in the absence of infection, cells in effector sites promote tolerance and reduce inflammation. For example, DCs in the LP are mostly myeloid DCs that are central to the induction of IgA class switching and are required for T regulatory cells to home to the GI tract (Uematsu and Akira, 2009). LP DCs can also be characterized based on the expression of the integrin CD103. $CD103^+$ DCs in the LP are anti-inflammatory and these cells are the primary DC population known to migrate to the MLN, where they drive the expansion of T regulatory cells (Coombes *et al.*, 2007). In contrast, $CD103^-$ DCs in the LP express high levels of SIRPα and are highly inflammatory, inducing Th17 cell responses that are the driving force behind colitis (Fortin *et al.*, 2009). In addition, the LP is home to specialized NK cell subsets, which are also skewed to a tolerogenic phenotype. For example, while 'classic' NK cells are known to produce high levels of IFNγ and secrete granzymes like perforin, LP NK cells secrete much lower levels of these and produce IL-22, a cytokine required for epithelial repair during inflammatory responses (Cella *et al.*, 2009; Colonna, 2009).

The primary function of specialized immune cell subsets present in the GI tract is to promote tolerance to the constant onslaught of antigens sensed by the GI tract epithelium. This tolerance is maintained by tight regulation of the balance of pro and anti-inflammatory signals produced by epithelial cells and innate and adaptive immune cells. As mentioned above, DCs in the LP are skewed to produce lower levels of Th1 inducing cytokines, like IL-12 and more IL-10. In addition, the interplay between Th17 cells, T regulatory cells, and the IL-17/23 cytokine axis is essential in maintaining low levels of inflammation in the GI tract (Ahern *et al.*, 2008; Awasthi and Kuchroo, 2009). Naïve T cells present in PP and ILF change into Th17 cells in response to TGF-β and IL-6, and they secrete a variety of cytokines, including IL-17, IL-6, IL-21, IL-22, and TNFα. Th17 cells contribute significantly to the inflammation associated with various types of colitis in both murine models and inflammatory bowel disease (IBD) (Awasthi and Kuchroo, 2009).

Other cytokines, especially IL-23, produced by activated macrophages and DCs, maintain Th17 cell populations and can also contribute to inflammation by a positive feedback mechanism, inducing greater pro-inflammatory cytokine production by innate cells (Ahern *et al.*, 2008). In contrast, IL-27, which can act on a variety of cells including macrophages, DCs, mast cells, neutrophils and $CD4^{+}$ T cells, simultaneously promotes Th1 responses, prevents the development of Th17 cells and is protective in the DSS model of murine colitis (Troy *et al.*, 2009).

The role of commensal microbiota in anti-*Salmonella* immune responses

The trillions of microorganisms that live symbiotically in the human gastrointestinal tract outnumber our own cells by a factor of 10, and the genetic information they contain (the microbiome) is estimated to be 1000 times greater than our own genetic code (O'Hara and Shanahan, 2006). As such, commensal microbiota that live in our GI tract are an integral part of our physiology. This has been exemplified in studies of germ-free mice, which show serious defects in not only intestinal development and digestive function, but also in immune responses (Round, 2009). This fits, as the GI tract is the richest place in the body in terms of concentrations of immune cells. In fact, the development and function of most of the GI tract-associated lymphoid tissue (GALT), such as the PP, MLN, and FAE, is dependent on the presence of a diverse population of commensal microflora (Round, 2009).

Importantly, key components to regulating mucosal inflammation are governed, in part, by commensal bacteria. For example, ATP and DNA from commensal bacteria induce the production of IL-17 in the colon and accordingly, germ-free mice have significantly fewer Th17 cells and increased levels of T regulatory (Treg) cells in the GI tract (Atarashi *et al.*, 2008). However, Treg function seems to be impaired. This is evidenced by low IL-10 production by germ-free Tregs, and the fact that they cannot prevent GI tract inflammation when transferred into mice with experimental colitis (Strauch *et al.*, 2005). Low levels of IL-10 in these mice could be explained by the lack of short-chain fatty acids produced by microflora that are also known to induce the production of IL-10 (Saemann *et al.*, 2000). Therefore, the microbiota not only drive the development of specific immune cell types, they also seem to play a role in governing the function of these cells during immune responses.

Most exemplary, however, is the association between dysbiosis (an alteration in natural microflora) and inflammatory bowel disease (IBD) or enteric infections. Patients with IBD suffer from reoccurring inflammatory 'flares' in the intestine and one crucial contributing factor to this disease is an inappropriate immune response to commensal bacteria. This is evidenced by circulating antibodies in IBD patients to microflora species, as well as the amelioration of symptoms with the use of antibiotics that reduce commensal numbers or change the composition of the microbiota (Round and Mazmanian, 2009). In the case of enteric pathogens, it is clear that the microbiota is a key defence mechanism, as germ-free mice are highly susceptible to colonization and infection with pathogens like *Shigella*, *Listeria* and *Salmonella* (Stecher and Hardt, 2008). This observation lead to the development of numerous mouse models of gastroenteritis where disruption of the microbiota is a crucial first step in aiding these pathogens to overcome the 'colonization resistance' provided by commensal microflora (Sekirov and Finlay, 2009). In addition, recent studies have shown the protective effects probiotic organisms, or their products, can have on the outcome of *Salmonella* infections. For example, feeding mice Bimuno, a galactooligosaccharide produced by the commensal bacterium *Bifidobacterium bifidum*, reduces invasion of epithelial cells by *Salmonella*, as well as decreases colonization of the GI tract and reduces intestinal inflammation in the *Salmonella* gastroenteritis model (Searle *et al.*, 2009). *Lactobacillus rhumnosus* GG also reduces epithelial cell stress during *Salmonella* infection (Burkholder, 2009). Probiotics can also be protective against systemic *Salmonella* infection; treatment of mice with *Lactobacillus fermentum* ME-3 used in combination with the antibiotic ofloxacin, reduces bacterial numbers in the typhoid model of *Salmonella* infection (Truusalu *et al.*, 2008).

Recent studies on anti-*Salmonella* immune responses in the GI tract have focused extensively on the relationship between perturbations in microbiota ecology and the outcome of infection using the *Salmonella* gastroenteritis model. In this model, a high dose of streptomycin is used to overcome colonization resistance provided by the microbiota by eliminating almost 98% of commensal species, opening an enormous ecological niche for infecting *Salmonella* to readily colonize the small intestine and colon (Garner *et al.*, 2009; Sekirov and Finlay, 2009). Interestingly, this total abolishment of commensal organisms is not a requirement for *Salmonella* to establish infection; however, dysbiosis to some extent is required and has long lasting effects on the outcome of infection. For example, using low doses of antibiotics (including both streptomycin and vancomycin) that do not decrease commensal bacteria numbers, but do cause significant shifts in populations, increases susceptibility of C57BL/6 mice to *Salmonella* gastroenteritis (Sekirov *et al.*, 2008). Furthermore, the changes in microbiota composition that result from both low and high doses of antibiotic treatment decrease colonization resistance for an extended period of time; *Salmonella* can more easily establish infection in mice treated with streptomycin whose microbiota was allowed to recover to pre-treatment levels compared to untreated mice (Croswell *et al.*, 2009). Transmission and persistence of *Salmonella* infection is also altered by antibiotic treatment; changing the microbiota either before or during *Salmonella* infection will induce a 'super-shedding' phenotype in 129svJ mice, increasing transmission between animals in a persistent model of infection (Lawley *et al.*, 2008).

Extensive inflammation in the cecum and large intestine is a hallmark feature of *Salmonella* gastroenteritis, and this phenotype goes hand in hand with decreased colonization resistance provided by the microbiota. However, it is difficult to assess whether decreased colonization resistance leads to inflammation or vice versa. Studies in FVB mice, which are readily colonized in the GI tract by *Salmonella* without need of antibiotic treatment, suggest that induction of inflammation by *Salmonella* may be a prerequisite to decreasing colonization resistance. In these mice, *Salmonella* infection alone (even at sub-lethal doses) causes a 95% reduction in commensal flora in the cecum and massive population shifts when bacterial numbers rebound. This is accompanied by severe pathology and diarrhea. However, similar changes in commensal flora is not observed after infecting mice with the *invA Salmonella* mutant, which does not cause inflammation in the GI tract, suggesting that the ability to cause an inflammatory response is required to establish infection with or without commensal involvement (Barman *et al.*, 2008). This is further supported by a recent study that showed how the *Salmonella* effector, SopE, induces IL-1β secretion by GI tract enterocytes through the caspase-1 dependent inflammasome cascade (Muller *et al.*, 2009).

Epithelial cell-dependent immunity to *Salmonella*

Intestinal epithelial cells (IECs) are the primary interface between the microbiotia and potential pathogens located in the GI tract lumen and the underlying immune tissues. As mentioned earlier, IECs provide a physical barrier to commensal organisms and pathogens and contribute to antigenic ignorance via the production of antimicrobial peptides, mucins, and tight junctions that mediate selective permeability of the membrane. However, IECs are a critical component of the GI tract immune system as well, since they not only express a number of pattern recognition receptors (PRRs) and antigen presentation machinery common to immune cells, but they also produce a multitude of cytokines, chemokines, and metabolites that maintain intestinal homeostasis and can drive or prevent inflammatory responses from macrophages, DCs, and T and B cells of the GALT (Artis, 2008).

Because IECs are in constant contact with foreign antigens, expression and signalling through PRRs is tightly regulated. For example, there is a specialized compartmentalization and distribution of toll-like receptors (TLR) in IECs. TLR 4 expression is low on the apical cell surface; TLR 5 is restricted to the basolateral side of the cells, and TLRs 3, 7, and 8 are restricted to endosomes (Shibolet and Podolsky, 2007). Interestingly, TLR 9, the receptor for bacterial DNA, is expressed on both the apical and basolateral sides of IECs. While signals from apical TLR 9 induce only partial activation of NF-κB that does not promote inflammation, basolateral signalling induces a strong inflammatory response (Lee *et al.*, 2006). In fact, NF-κB signalling in IECs is a global regulator of inflammatory responses, and both commensals and non-invasive *Salmonella* species restrict NF-κB activation to tolerize GI tract immunity (Neish *et al.*, 2000; Zaph *et al.*, 2007). TLR signalling within IEC's is especially important during pathogenic *Salmonella* infections, as CX3CR1$^+$ DCs in the LP are able to project balloon shaped extensions between epithelial cells to phagocytose luminal *Salmonella* in a TLR 2 and MyD88 dependent fashion (Chieppa *et al.*, 2006). These 'sampling' DCs also cross the epithelial barrier into the intestinal lumen during *Salmonella* infection, and this is dependent not only on MyD88, but also the presence of bacterial flagellin. While the mechanism of protection is unknown, it is believed that these luminal DCs limit IEC exposure and damage induced by *Salmonella* (Arques et *al.*, 2009). *Salmonella* flagellin and NF-κB signalling through TLR 5 on IECs is also required for the induction of heat shock protein (HSP) 25, which protects IECs against oxidative stress (Petrof *et al.*, 2008).

Signalling through TLRs and NF-κB in IECs can induce the production of a wide range of cytokines and chemokines that are required to either drive immune responses or induce tolerance in the GI tract. T helper cell responses and homeostasis are maintained through the production of IL-25, IL-33, thymic stromal lymphopoietin (TSLP), TGF-β, IL-10, prostaglandin E2 and retinoic acid (RA). IL-25 and IL-33 production by IECs act in a positive feedback loop, inducing the upregulation of their receptors on IECs and innate immune cells. IL-25 and 33 skew naïve T cells towards a Th2 phenotype, and they induce the production of TSLP from IECs. TLSP is a cytokine that drives the positive selection of T cells in the thymus, but it also plays a key role in controlling DC responses in the LP (Saenz *et al.*, 2008). For example, TSLP reduces secretion of IL-12 by DCs and increases IL-10 production; these changes induce DCs to skew T cell responses towards a Th2 phenotype (Rimoldi *et al.*, 2005). In addition, IL-25, IL-33 and TSLP decrease the production of IFNγ and IL-17, reducing gastrointestinal inflammatory responses (Caruso *et al.*, 2009). IEC derived TGF-β and retinoic acid (RA) condition DCs towards a tolerogenic phenotype via the upregulation of CD103, and these cells regulate Treg cell populations in the mucosa (Iliev *et al.*, 2009). Furthermore, TGF-β can directly induce the expression of the transcription factor FoxP3 in Tregs (Chen *et al.*, 2003).

Besides cytokines, epithelial cells produce many other factors involved in bacterial defence and interaction with *Salmonella* induces various changes in IEC secretion of these products. For example, lipocalin-2, an antimicrobial protein that limits bacterial iron acquisition, is produced and secreted by IECs during *Salmonella* infection in response to the production of IL-17 and IL-22 by Th17 cells in the LP. While *Salmonella* infection induces the production of lipocalin-2, resistance to this antimicrobial is conferred in *Salmonella* via the *iroBCDE* operon, which is common to pathogenic organisms (Raffatellu *et al.*, 2009). *Salmonella* also induces the expression of platelet activating factor (PAF) and phospholipase A2, an enzyme required in the PAF formation pathway, from IECs during infection (Egea *et al.*, 2008; Mumy *et al.*, 2008). In addition to initiating the production of PAF, phospholipase A2 induction in IECs plays a role in the eicosanoid hepoxilin A3 (HXA3) mediated neutrophil recruitment that is initiated by the *Salmonella* effector SipA (Lee *et al.*, 2000; Mrsny *et al.*, 2004). *Salmonella* can further alter the inflammatory state of IECs by inducing changes in expression of matrix metalloproteases (MMP) that regulate epithelial barrier permeability. For example, expression of MMP-2 and MMP-9 is increased upon infection with *Salmonella* in the murine model of gastroenteritis. While upregulation of MMP-2 is protective via the negative regulation of neutrophil influx (Garg *et al.*, 2006), MMP-9 upregulation induced by *Salmonella* seems to be a virulence mechanism the bacteria uses to increase the permeability of the epithelial barrier and cause inflammation (Castaneda *et al.*, 2005).

Salmonella flagellin is an important modulator of IEC immune responses in the GI tract. For example, flagellin is required for the release of nutrients such as mucin and glycoconjugates that are produced by IECs as part of mucosal barrier defences during infection (Stecher *et al.*, 2008). While flagellin provides an advantage for pathogen growth in this case, IECs can also exploit flagellin production for defence purposes. IECs stimulate flagellin production from *Salmonella* via the production of lysophosphatidic acid which may serve to stimulate rapid pro-inflammatory responses from TLR 5 on the basolateral side of IECs (Subramanian and Qadri, 2006). Flagellin dependent *Salmonella* motility and invasion of IECs is also inhibited via the production of the monoclonal IgA, Sal-4 (Forbes *et al.*, 2008).

Anti-*Salmonella* immunity in the GALT

Breaching the epithelial barrier

Interactions with commensal organisms and IECs are only the tip of the iceberg in terms of how *Salmonella* defines immune responses in the GI tract. *Salmonella* use a number of mechanisms to cross the epithelial barrier and cause either gastroenteritis or systemic disease. Penetration of the GALT via M cells located in the PP or ILF is the primary way *Salmonella* enter the GI tract; evidence for this is seen in immunofluorescence studies that show bacteria located in highest numbers in lymphoid tissue beneath these cells (Tam *et al.*, 2008a). However, while entry through M cells is common, it is not required. As aforementioned, CX3CR1$^+$ DCs in the FAE can sample luminal *Salmonella,* and these cells are required for non-invasive *Salmonella* strains to breach the epithelial barrier in models of murine colitis (Hapfelmeier *et al.*, 2008). However, in cases of infection with invasive *Salmonella* strains, DC projections are not found in high numbers and are not required for bacterial translocation (Vallon-Eberhard *et al.*, 2006). In addition to penetration through M cells, invasive bacteria utilize a variety of SPI-1 encoded TTSS effectors to disrupt tight junctions between epithelial cells, allowing for paracellular transit into GI tract tissue (Boyle *et al.*, 2006).

The mode of entry that *Salmonella* uses to cross the epithelial barrier is a critical step in governing the immune responses generated in the GALT. Invasive *Salmonella* that cross through M cells activate B2 B cells present in PP and these cells then move to the LP to further differentiate into *Salmonella*-specific IgA secreting cells (Martinoli *et al.*, 2007). The production of anti-*Salmonella* faecal IgA is dependent upon the presence of functional PP, as mice lacking these lymphoid structures fail to produce faecal IgA and have lowered IFNγ production in the LP (Hashizume *et al.*, 2008). In contrast, while non-invasive *Salmonella* strains can cross the epithelium (using methods described above) and traffic to MLN and induce systemic IgG responses as efficiently as invasive strains, they do not elicit the production of sIgA (Martinoli *et al.*, 2007).

Activation of innate responses

Once the epithelial barrier is breached, *Salmonella* encounter a number of innate and adaptive immune cells that mount non-specific inflammatory and antigen-driven immunity to control infection. Ligation of pattern recognition receptors (PRRs) on macrophages, DCs and monocytes with their ligands serves as the initial stimulus, providing a stereotypical cytokine storm required to initiate immune responses. *Salmonella* express ligands for most of the TLRs, and sensing by these molecules is essential to limit infection. TLR 5 ligation is an important signal during *Salmonella* infection because it can stimulate cytokine production via activation of the inflammasome, as well as classical NF-κB transcription. Sensing of *Salmonella* flagellin in the PP by TLR 5 activates the nod-like receptor protein Nalp3 and caspase-1, leading to the production of IL-18 and IL-1β, two cytokines that are key for the maturation of DCs (Fink and Cookson, 2007). While these cytokines are considered pro-inflammatory, caspase-1 is not required for the induction of colitis during *Salmonella* infection, and IL-12 production by flagellin activated cells in the PP can still drive Th1 responses in the absence of inflammasome activation (Lara-Tejero *et al.*, 2006). In addition, TLR stimulation is important for B cell responses, since transfer of B cells from mice lacking TLR 4 or 2 does not produce IgM or IgG in response to *Salmonella* Typhi (Cervantes-Barragan *et al.*, 2009). In addition to TLR signals, functional NOD1 (an intracellular PRR that recognizes bacterial peptidoglycan) in LP DCs is essential for the control of *Salmonella* infection (Le Bourhis *et al.*, 2009).

Function and recruitment of monocytes and neutrophils

PRR stimulation occurs in a variety of cell types in the GALT. The key innate immune cells present in the GALT during *Salmonella* infection are DCs, monocytes, macrophages and neutrophils. After oral *Salmonella* infection, macrophages localize to interfollicular regions of the epithelium, while neutrophils and monocytes accumulate in the PP and MLN (Tam *et al.*, 2008a). Neutrophils and monocytes contain more intracellular bacteria than macrophages or DCs; over 80% of bacteria seen in GI tract tissue are found in these cell types. In the GALT, monocytes outnumber neutrophils during early time points of infection and these cells are the primary producers of the pro-inflammatory mediators TNFα, iNOS, and IL-1β (Rydstrom and Wick, 2007). Both inflammatory and patrolling monocyte subsets are present in the GI tract during *Salmonella* infection. However, it is unknown whether monocytes recruited to the GALT further differentiate into macrophages or DCs in response to *Salmonella* infection. While some monocytes do express low/intermediate levels of CD11c and these cells phagocytose *Salmonella* in the PP and MLN during infection, they express lower levels of co-stimulatory molecules than DCs and cannot present *Salmonella* antigens to induce $CD4^+$ T cell proliferation, suggesting that these cells are more macrophage-like than DC-like in both phenotype and function (Rydstrom and Wick, 2007). In addition, *in vitro* studies showed that heat-killed *Salmonella* and LPS stimulated monocytes are skewed towards a macrophage differentiation program (Palucka *et al.*, 1999).

Monocytes and neutrophils are recruited to the GALT from the blood initially via the production of chemokines from IECs, and populations are maintained during infection via a positive feedback mechanism whereby recruited cells produce chemokines themselves. CCL2, or monocyte chemotactic protein (MCP-1), is highly expressed by cells in the FAE during *Salmonella* infection, and neutrophils and monocytes recruited to the GALT express high levels of the CCL2 receptor (Rydstrom and Wick, 2007). Expression of molecules required for extravasation of these cells from blood vessel endothelium is also required for recruitment to GI tract sites. For example, expression of either *p*-selectin, or its ligand, PSGL-1, is required for the appropriate homing of neutrophils to the GI tract during *Salmonella*-induced colitis. Without the expression of these molecules, ceca of infected mice are filled with $MAdCAM\text{-}1^+$ cells, leading to more severe inflammatory pathology (Kum *et al.*, 2009). Although MAdCAM-1 alone is not required for neutrophil recruitment to the PP or MLN, this study suggests that MAdCAM-1 expression is a compensatory mechanism for the absence of other cell adhesion molecules and its expression under these conditions leads to hyper-recruitment of cells to the GI tract during inflammation (Rydstrom and Wick, 2007).

The role of specialized DC subsets

DCs are key modulators of immune responses to *Salmonella* and the phenotype and function of DCs in the GI tract during infection has been well characterized. After crossing through M cells, *Salmonella* associate with resident $CD11c^+$ $CD8\alpha^-$ or $CD8\alpha^+$ DCs localized in the SED of the PP (Iwasaki and Kelsall, 2000). Death of $CD8\alpha^+$ DCs located in the MLN is induced by *Salmonella* during infection, and this is mediated through TNFα and MyD88 dependent signalling in DCs, suggesting that *Salmonella* has evolved this defence mechanism to partially overcome TNFα driven innate immune responses (Sundquist and Wick, 2009). GI tract DCs mature during *Salmonella* infection either by direct association with the bacteria or by the production of cytokines by macrophages and monocytes. TNFα and IL-1β are sufficient for the upregulation of co-stimulatory markers on the DC surface, and this can occur in either a MyD88 dependent or type 1 IFN-dependent fashion (Sundquist and Wick, 2005; Tam *et al.*, 2008b). Interestingly, while DCs are famous for the production of IL-12, indirect maturation of these cells by TNFα or IL-1β results in lower levels of IL-12 production, leading to decreased activation of T cells to produce IFNγ. In addition, the constitutively high expression of MHCII and the upregulation of co-stimulatory molecules on GI tract DCs is not affected by IFNγ or TNFα levels, suggesting that DCs are primed for antigen presentation regardless of their cytokine environment (Sporri and Reis e Sousa, 2005).

DCs can also be recruited to the GALT during *Salmonella* infection. The FAE preferentially expresses the chemokine CCL20, and DCs expressing its receptor, CCR6, are found in PP (Tam *et al.*, 2008a). Recruited DCs, more so than DCs resident in the GALT, may be more effective in mounting immune responses against *Salmonella*, since they are not exposed to IEC products that can drive DC-dependent anti-inflammatory or Th2 responses. The expression of CCR6 is also important for the expansion of $CD4^{+}$T cells in the PP (Salazar-Gonzalez *et al.*, 2006), and expression of TLR 5 also governs the trafficking of *Salmonella* infected DCs to the MLN (Uematsu and Akira, 2009).

The IL-23/IL-17 cytokine axis

While macrophages, monocytes and DCs in the GALT all contribute to the induction of classic Th1 immunity, cytokines other than IL-12, IFNγ, and TNFα also play an essential role in determining the outcome of *Salmonella* pathogenesis in the GI tract. Recently, the importance of Th17 cytokines has been highlighted in various models of *Salmonella* infection. The expression of IL-23 is upregulated by DCs infected with *Salmonella in vitro* and it is also produced by peritoneal DCs after intraperitoneal *Salmonella* infection, as well as by γδ T cells in the LP during *Salmonella* induced colitis (Godinez *et al.*, 2009). The production of IL-23 during *Salmonella* infection is the key initial step in the production of IL-17 from Th17 cells. Indeed, IL-17, as well as IL-22, are upregulated in colitis models of *Salmonella* infection in mice and macaques, and they are required for control of infection (Raffatellu *et al.*, 2008). Systemic immunity to *Salmonella* is also dependent on IL-23 and IL-22, but not IL-17 (Schulz *et al.*, 2008). Interestingly, IL-17 production during gastroenteritis may be regulated by NK cells and/or the production of IL-15. Support for this comes from a recent report that showed that $IL\text{-}15^{-/-}$ or NK1.1 cell depleted mice are readily colonized in the GI tract by *Salmonella* without the need for antibiotic pre-treatment, and this is associated with increased levels of IL-17, but little inflammation (Ashkar *et al.*, 2009). The requirement for these cytokines to prevent *Salmonella* pathogenesis is not surprising, since both IL-17 and IL-22 stimulate stereotypic innate immune responses, such as c-type lectin and chemokine expression, as well as antimicrobial peptide and mucin secretion from IECs and cytokine production from macrophages, monocytes, and DCs (Ahern *et al.*, 2008). A summary of cytokines and chemokines with a role in controlling *Salmonella* infections is presented in Table 12.2.

Spread and maintenance of infection in systemic locations

Mechanisms of systemic infection

The severity and the outcome of a systemic *Salmonella* infection depends on the 'virulence' of the bacteria, on the infectious dose, as well as on the genetic makeup and immunological status of the host (Mastroeni, 2002). Lethal systemic infections can develop when a susceptible host meets particularly virulent *Salmonellae* or a high bacterial inoculum; in such cases, death can occur quite rapidly, as bacterial numbers in the liver and spleen may reach 10^{8}–10^{9} cfu (Collins, 1974). Given the right host–pathogen interaction, a sublethal infection can be established where bacterial growth is suppressed early on in the reticuloendothelial system (RES). Bacteria are quickly cleared from the blood (within two hours) (Biozzi *et al.*, 1960) by complement activation (Liang-Takasaki *et al.*, 1983). Bacteria are opsonized by complement factors via the alternate pathway and by a C1q-dependent pathway before their rapid clearance (Warren *et al.*, 2002). Once cleared from the blood stream, *Salmonella* remain within an intracellular location; they are recognized by phagocytes, mainly macrophages and DCs, via interactions between their surface PRRs and pathogen-associated molecular patterns (PAMPs) on the bacterial surface. The interaction between CD14/TLR4 complexes on the host cell surface and bacterial lipopolysaccharide (LPS) is crucial for *Salmonella* recognition, as this induces bacterial phagocytosis into the cell and production

Table 12.2 Cytokines and Chemokines required for controlling *Salmonella* infection in mice

Cytokine	Producing cell	Function	Reference
TNFα	Monocytes, MΦ	Inflammation Legion formation	Everest (1998); Mastroeni (1995); Tam (2008)
IFNγ	NK, neutrophils, T cells	Activation ROS, RNS production Lysosomal fusion	Billiau (1996); Foster *et al.* (2003); Jouanguy *et al.* (1999); Kagaya *et al.* (1989)
IL-12	MΦ, DC	IFNγ production RNS production	Koch F (1996); Mastroeni *et al.* (1996); Mastroeni *et al.* (1998)
IL-18	MΦ	IFNγ production RNS production	Mastroeni *et al.* (1999)
IL-15	DC, NK?	IFNγ production Regulation of colitis Contributes to colonization resistance	Eckmann and Kagnoff (2001); Ferlazzo *et al.* (2004); Lucas *et al.* (2007)
IL-4	NK1.1$^+$, regulatory T cells, basophils, mast cells, eosinophils?	Decreases MΦ bactericidal and cytokine responses	Denich *et al.* (1993); Enomoto *et al.* (1997)
IL-10	MΦ, DC	Decreases MΦ function	Arai (1995)
IL-6	MΦ, DC	Enhances PMN killing of bacteria	Nadeau *et al.* (2002)
IL-2	T cells	Final clearance from organs	al-Ramadi *et al.* (2001)
IL-17	Th 17 cells, γδT cells	Regulates colitis and GI colonization	Raffatellu *et al.* (2008); Schulz *et al.* (2008); Siegemund *et al.* (2009)
IL-23	DC, MΦ	Maintenance of IL-17 production Upregulation of IL-22 Required for control of systemic infection	Godinez *et al.* (2009); Siegemund *et al.* (2009)
IL-22	MΦ, NK, DC	Regulates colitis and GI colonization	Raffatellu *et al.* (2008); Siegemund *et al.* (2009)
IL-1β	Monocytes, MΦ	Maturation of DC	Rydstrom and Wick (2007)
IL-8	IECs	Neutrophil recruitment	Zhang *et al.* (2003a)
Fractalkine	IECs	Recruitment of CX3CR1+DCs	Niess *et al.* (2005)
MIP-1α	IECs	Neutrophil, monocyte and MΦ recruitment	Rydstrom and Wick (2007)

of various microbicides, such as acids, enzymes and reactive oxygen species (ROS), which degrade bacteria within the endocytic pathway (Rosenberger and Finlay, 2002; Vazquez-Torres *et al.*, 2000a). Phagocytic cells use these ROS and nitrogen intermediates (reactive nitrogen species, RNS) to control the growth of virulent *Salmonella* with the presence of an intact NADPH oxidase (which generates ROS) being most essential (De Groote *et al.*, 1997). *Salmonella* have SPI-2 encoded mechanisms to defend against ROS, as well as phagolysosome fusion; however, these are not perfect and ROS molecules remain essential for the rapid killing of bacteria by the cell, protecting the host from infection *in vivo* (Mastroeni *et al.*, 2000b; Vazquez-Torres *et al.*, 2000b).

Salmonella growth in tissues results in distribution of the bacteria to uninfected cells and in spread of the bacteria to new infection foci that eventually become pathological lesions. This escape from infected cells is essential for the infection to progress and allows for the continued evasion of local activation of the host immune response (Mastroeni and Sheppard, 2004). In addition to resisting intracellular killing by infected phagocytes, *Salmonella* must escape and disseminate. This process was previously believed to occur via random necrotic cell death, and subsequent release of individual bacteria into the extracellular space (Brown *et al.*, 2006). However, it was recently reported that liver phagocytes can undergo apoptotic caspase-3-mediated cell death *in vivo*, with apoptosis being more prevalent in heavily infected cells and a proportion being dependent on LPS-induced activation of TLR4 (Grant *et al.*, 2008). This density-dependent mechanism of bacterial spread may help to ensure *Salmonella*'s continued occupation of a protective intracellular environment. These findings support earlier studies which showed that *Salmonella* growth results in an increase in number of infected cells rather than an increase of bacteria per cell. High numbers of bacteria are rarely seen in the majority of infected cells; the majority of phagocytes contain only a single bacterium (Sheppard *et al.*, 2003). Essentially, pathological lesions contain a population of cells of which only a small percentage are permissive to bacterial growth.

Intracellular *Salmonella* in systemic sites

Salmonella spread past the GALT in infected $CD18^+$ phagocytes that enter the lymphatic system and bloodstream (Vazquez-Torres *et al.*, 1999) and spread systemically to the MLN, followed by the liver, and spleen. In the MLN, *Salmonella* Typhimurium may persist in macrophages (Monack *et al.*, 2004a), however, Voedisch *et al.* have recently confirmed that the majority of *Salmonella*-containing cells in the MLN show dendritic-cell morphology and express the DC marker CD11c. Furthermore, *Salmonella* transport from the intestine to the MLN is limited by the number of migratory DCs carrying bacteria. Also, modulation of DC migration had no effect on the number of *Salmonella* reaching systemic sites, revealing the vital role of GI tract-associated lymphoid tissue and MLNs as barriers in the systemic dissemination of *Salmonella*. These delay spread and allow for the establishment of protective adaptive immune responses (Voedisch *et al.*, 2009).

Once *Salmonella* have breached the confinement of the MLN, they mainly reside in the liver and spleen. Early studies showed that *Salmonella* reside within red pulp macrophages ($F4/80^+$, $MSR\text{-}A^{low}$) and marginal zone macrophages ($MSR\text{-}A^+$) of the spleen (Salcedo *et al.*, 2001). However, more recent studies using a reporter system to detect secreted bacterial effectors in targeted host cells showed that, despite being capable of targeting both phagocytic and non-phagocytic cells, most viable *Salmonella* preferentially reside within splenic neutrophils (GR-1+/CD11b+) (Geddes *et al.*, 2007). In the liver, specialized macrophages known as Kupffer cells are the main target for *Salmonella* early after infection; however, in later stages of infection, bacteria are seen in polymorphonuclear cell-rich lesions and macrophage-rich granulomas (Richter-Dahlfors *et al.*, 1997). *Salmonella* have also been found extracellularly within sinusoids and tissue debris of lesions of the liver, as well as intracellularly within hepatocytes (Hsu, 1989).

More recently, the nature of the phagocytes that contain *Salmonella* in systemic sites has been further characterized. Nix *et al.* have shown that during the establishment of persistent murine infections, *Salmonella* Typhimurium preferentially replicates in haemophagocytic macrophages (Nix *et al.*, 2007), cells that are characterized by ingestion of non-apoptotic cells of the haematopoietic lineage. Hemophagocytic macrophages represent a specific subset of cells that provide a survival niche to which *Salmonella* appears to have preferentially adapted. In addition, *Salmonella* is also known to survive within DCs (Richter-Dahlfors *et al.*, 1997), polymorphonuclear cells (PMNs) (Sheppard *et al.*, 2003), and B cells (Yrlid *et al.*, 2001) in various systemic organs. It is known that *Salmonella* Typhi can persist in the bone marrow and gallbladder for life (Park; Sinnott and Teall, 1987; Wain *et al.*, 2001). It is suggested that periodic reseeding of the mucosal surface of the small intestine via the bile ducts (and possibly the MLNs) occurs (Gaines *et al.*, 1968), resulting in subsequent shedding of bacteria from the mucosal surface (Monack *et al.*, 2004b). This process allows for re-introduction of *Salmonellae* into the environment and exposure to new hosts.

Other large populations of cells that are permissive to *Salmonella* replication and allow for rapid growth of bacteria do in fact exist, one such population being gallbladder epithelial cells. Menendez *et al.* recently showed that *Salmonella* infect gallbladder epithelial cells in a mouse model of acute typhoid. The bacteria remained confined to these epithelial cells, replicating to significantly high numbers within single cells; cells harbouring more than 30 bacteria per cell were observed (Menendez *et al.*, 2009). *Salmonella* shows itself to be a highly versatile pathogen, failing to limit itself to the canonical-view associated with intracellular replication within macrophages.

Systemic immune responses

Key to phagocytes mounting innate defences against *Salmonella* in mice is the expression of the *Slc11a1* gene (previously known as *Nramp1*) that codes for a divalent cation transporter that localizes to the membrane of bacteria-containing phagosomes within macrophages, DCs and PMNs during *Salmonella* infections (Bellamy, 1999). Susceptibility of mice to typhoid infection is dependent upon the expression of the dominant *Slc11a1* resistant allele. It is not clear whether Slc11a1 is responsible for limiting the movement of essential ions (such as iron and manganese) into, or out of, the phagosome (Forbes and Gros, 2003; Goswami *et al.*, 2001; Kuhn *et al.*, 1999). The expression of this transporter is essential for preventing the replication of *Salmonella in vitro*, as well as in PMNs of the liver and spleen in the mouse model (Forbes and Gros, 2003; Richter-Dahlfors *et al.*, 1997). Expression of the human homologue does not appear to control susceptibility to typhoid infections (Dunstan *et al.*, 2001); however, it is required to combat other intracellular pathogens, including *Mycobacterium tuberculosis* and *Leishmania* (Blackwell *et al.*, 2001).

In order to ensure survival of the host, suppression of exponential bacterial growth in the RES is necessary, thus establishing a 'plateau' phase of infection. This is a crucial turning point during the infection and requires recognition of *Salmonella* by macrophages and other professional phagocytes for induction of a complex adaptive/innate immune response (Table 12.3). During this phase of infection, bone-marrow derived macrophages infiltrate the spleen and liver, leading to hepatosplenomegaly (enlargement of both the spleen and the liver), as well as the formation of macrophage-rich granulomas (Mastroeni, 2002). Significant production of several cytokines by macrophages, DCs, PMNs, and NK cells is required for controlling systemic bacterial replication, as well as initiating later antigen-specific responses (Table 12.2). For example, the classic pro-inflammatory mediator TNFα is required in the formation and persistence of granulomas, as well as in the regulation of NADPH oxidase-mediated killing of *Salmonella* by macrophages (Everest, 1998; Mastroeni, 1995). Interleukins IL-12, IL-15, and IL-18 are essential for the production of IFNγ and the subsequent activation of macrophages (Kagaya *et al.*, 1989). Elimination of any of these cytokines from the mouse during typhoid infection greatly increases susceptibility to *Salmonella* (Stoycheva and Murdjeva, 2004).

Macrophage recruitment and activation, as well as the formation of granulomas, are essential during this phase of infection. The early bactericidal role of ROS is gradually replaced by the bacteriostatic role of *NOS2*-dependent reactive nitrogen species (RNS). *NOS2* expression occurs within macrophages found in foci of infected tissues and is upregulated by the production of IFNγ, IL-12 and IL-18 (Khan *et al.*, 2001; Mastroeni *et al.*, 1998). The host's response to LPS is another key factor in the establishment of the plateau phase of *Salmonella* infection. For example, C3H/HeJ mice and C57BL10ScCr mice are homozygous for a mutant *Lps* allele ($Lps^{d/d}$) that results in an inability to respond to LPS

Table 12.3 Immune responses to *Salmonella* Typhimurium infection *in vivo*

Innate responses – bactericidal (0–5 days)	Adaptive/innate responses – bacteriostatic (5–20 days)	Antigen-specific responses – bactericidal, bacteriostatic (20+ days)
Complement (blood)	TNFα, IFNγ, IL12, IL15, and IL18 (Mϕ, DCs, NK cells, etc.)	CD4+ T cells
Slc11a1 (Mϕ)	Granuloma formation	CD8+ T cells
ROS/RNS(Mϕ)	RNS and TLR4 (Mϕ)	B cells: antibody production and APC function

Modified from Mastroeni (2002).

(Poltorak *et al.*, 1998). This hyporesponsiveness is due to a point mutation in the *Tlr4* gene that encodes TLR 4, a protein primarily adapted for the recognition and signalling of LPS. Neither of these mouse strains' macrophages, defective in TLR 4 signalling and function, can effectively suppress the growth of virulent *Salmonella* in the spleen and liver (Hormaeche, 1990; O'Brien *et al.*, 1982).

Generation of protective immunity in systemic sites

Establishment and maintenance of the 'plateau' phase of *Salmonella* infection is not sufficient for survival of the host. Relapses of bacterial growth or persistence in the RES can occur, leading to death or the establishment of a chronic carrier state (Table 12.3). Importantly, clearance of *Salmonella* from the host depends on the involvement of T-cells and adaptive immune mechanisms.

The principal cells involved in the clearance of *Salmonella* in later stages of infection are $CD4^+$ TCR-$\alpha\beta^+$ T-cells, as mouse strains lacking TCR-β^+ T-cells and mature $CD4^+$ TCR-$\alpha\beta^+$ T-cells (H2I-$A\beta^{-/-}$ and TCR-$\beta^{-/-}$ mice, respectively) cannot suppress the progressive growth of *Salmonella* in the weeks following the 'plateau' phase (Hess *et al.*, 1996). The exact mechanisms by which $CD4^+$ T-cells control bacterial growth in late stages of infection remain unknown. For example, IFNγ that is produced by T-cells does not affect the elimination of bacteria from tissues during later phases of infection (Muotiala and Makela, 1993); TNFα is also dispensable for the clearance of *Salmonella* (Fangtham and Wilde, 2008; Levine *et al.*, 2007).

Antibody responses to a large number of *Salmonella* antigens can be detected following exposure to infection. Anti-*Salmonella* IgM appears in serum early on after infection, which is then followed by IgG; IgA is detected in animals and humans exposed to live *Salmonella* (Forrest *et al.*, 1991; Harrison *et al.*, 1997; Metcalf and O'Brien, 1981). T-cells predominantly control antibody production. For example, athymic *nu/nu* mice, which are T-cell deficient, produce low levels of IgM and IgG3, with no other IgG subclasses detected after exposure to live attenuated strains of *Salmonella* (Sinha *et al.*, 1997). $CD28^{-/-}$ mice, which have impaired T-cell activation and lack proper T-B-cell communication, are highly susceptible to wild-type *Salmonella* infection and completely lack specific IgG1 and IgG2a antibodies (Mittrucker *et al.*, 1999).

The cytokine profile produced during *Salmonella* infection skews $CD4^+$ T helper cells to a Th1 response; IL-2 and IFNγ production by these cells is critical for rapid clearance of the pathogen by the host (al-Ramadi *et al.*, 2001; Mittrucker *et al.*, 2002; Srinivasan *et al.*, 2004). Th1 cells, along with $CD8^+$ cytotoxic lymphocytes, also modulate a variety of antigen-specific immune events central to the clearance of *Salmonella* from systemic sites, including B cell antibody production and isotype profile. B cells produce antibodies against various *Salmonella* antigens, including LPS, the capsular Vi polysaccharide, outer membrane proteins, flagellin and heat shock proteins (Mastroeni, 2002). B cells are classically known to play an important role in the memory response to secondary *Salmonella* infections; however, they also act during initial infections as APCs, which is required for the eventual development of Th1 responses (Mastroeni *et al.*, 2000a).

Continuous activation of immune mechanisms over the course of *Salmonella* infection adds to the control and eventual clearance of the infection as well as to the development of acquired immunity to re-infection (Mastroeni, 2002). Host resistance increases over the course of infection as new immunological mechanisms are activated with the progression of disease. Expression of the highest level of host protection, the production of *Salmonella*-specific antibodies, depends on the collective action of all preceding immune mechanisms, including the innate, adaptive and antigen specific immune mechanisms (Table 12.3). This progressive reinforcement provided by the host immune system at every turn over the course of infection is the only means of combat against a most ruthless opponent, *Salmonella*.

Importance of systemic sites in carriage and transmission

A bacterial pathogen is a microbe that causes disease in another organism. As the pathogen enters and moves through its animal host, it encounters a number of unique environments, each of which possess distinct chemical and physical properties and immune responses (Ohl and Miller, 2001). As a result, it must react quickly, adapting to each new environment. Doing so successfully contributes to the virulence of the bacterial pathogen, i.e. the ability of a pathogen to cause disease within its host. This is a complex process that involves coordinated expression of virulence factors, bacterial gene products that aid in the pathogen's continued invasion and colonization of its host. Exit from the host and transmission to new hosts are subsequent stages in the infectious cycle that also contribute to the virulence of the pathogen (Chiang *et al.*, 1999). Evolutionary success of a bacterial pathogen, such as *Salmonella*, is therefore dependent upon its ability to replicate within a host and its spread to new hosts.

The fitness of pathogenic microorganisms depends on their ability to complete the transmission process that consists of multiplication within a host, dissemination from that host, and translocation to and colonization of a new host (Lipsitch and Moxon, 1997). For the majority of bacterial pathogens, including *Salmonella*, very little is known regarding the genes that contribute to the success of transmission. Generally, *Salmonellae* have an enhanced survival compared to other enteric pathogens in the external environment; this ensures its passage to the next host and is a strategy employed by many *Salmonella* serovars to successfully transmit to new hosts in the absence of direct host-to-host contact (Winfield and Groisman, 2003). Shedding of transmissive bacteria into faeces plays an obvious role in this process, allowing for a route of passage into the environment, and other animals (Kingsley *et al.*, 2000). Importantly, with the exception of host-restricted serovars such as *Salmonella* Typhi, *S. enterica* present in faeces of infected hosts can be infectious for other organisms besides the host species (Hornick *et al.*, 1970a), allowing the continual dissemination of this pathogen to various hosts.

The maintenance of a chronic carrier states is also crucial for pathogenic fitness and transmission to new hosts. A relationship between biliary disease and chronic *Salmonella* Typhi carriage has been recognized for years, though the presence of biliary tract disease is not an absolute requirement for the development of the carrier state. This relationship has lead researchers to examine the exact role of the gallbladder and associated biliary ducts in *Salmonella* infection. Much work was carried out in the early 1900's, when gallbladder removal was commonly used as a possible cure for *Salmonella* Typhi carriage (Dinbar *et al.*, 1969; Leary, 1913); cholecystectomy (gallbladder removal) was first advocated as treatment for the typhoid carrier by Dehler in 1907 (Dehler, 1907). However, results were variable, as some patients ceased excreting *Salmonella* Typhi in their faeces permanently, whereas others continued shedding after the procedure. As Wilson and Miles explained of such failures (Wilson and Miles, 1964): 'It is not clear in what part of the intestinal tract the focus of infection persists, though there is reason to believe that it is in the biliary passages of the liver.' However, the presence of *Salmonella* Typhi in the gallbladder was not simply restricted to persistent infections. *Salmonella* Typhi could be isolated from the gallbladder in fatal (*i.e.* acute) cases of typhoid fever (Anton and Fütterer, 1888) and was in fact present in the majority of cases of typhoid fever (Ledingham and Arkwright, 1912).

In order to study this particular disease in depth, a model for *Salmonella* Typhi infections in humans was needed. *Salmonella* Typhi appeared to form bacterial 'nests' in the gallbladder of humans (Koch, 1909). Direct injection into the gallbladder of rabbits was the first model employed in early *Salmonella* pathogenesis studies (Hailer and Ungermann, 1914); the infection resulted in early invasion of the gallbladder, where bacilli were able to survive and replicate for a long period of time. Ultimately, *Salmonella* Typhi was cleared from other parts of the rabbits, including the blood, urine, liver, spleen, and bone marrow, yet remained within the gallbladder (Doerr, 1905).

As a potential reservoir for *Salmonella* Typhi in chronic cases of typhoid, the gallbladder (via bile) has been suggested to be a main route of passage for bacilli back to the gastrointestinal tract for eventual reseeding and faecal shedding (Everest *et al.*, 2001). This process has not been restricted to *Salmonella* Typhi in humans. Extracellular replication, biliary excretion and subsequent faecal shedding of *Listeria monocytogenes* have been documented in an animal model of listeriosis (Hardy *et al.*, 2004; Hardy *et al.*, 2006; Nichterlein *et al.*, 1994). To date, no animal infection (chronic or acute) of *Salmonella* Typhimurium has been developed to model the above process. However, the effects of bile salts *in vitro* on *Salmonella* Typhimurium have been examined, as this organism is exposed to harsh environments found in the small intestine and gallbladder. Whereas bile (a detergent-like substance) is toxic to most organisms, most enteric bacteria and bacterial pathogens, including *L. monocytogenes*, *Salmonella* Typhimurium and *Salmonella* Typhi, have evolved resistance to bile toxicity (Dussurget *et al.*, 2002; Prieto *et al.*, 2007; Prouty *et al.*, 2004a; Prouty *et al.*, 2004b; Prouty *et al.*, 2002; Sleator *et al.*, 2005). In fact, some bacteria (including *Salmonella* Typhimurium) utilize bile as a regulatory environmental signal, affecting expression of bile-resistance genes, as well as genes within important virulence pathways (Chatterjee *et al.*, 2007; Prouty *et al.*, 2004a; Prouty *et al.*, 2004b). An array of studies have shown that *Salmonella* responds to bile by inducing bile resistance genes (*i.e.* the *mar* operon and the *acrAB* efflux pump) and down-regulating genes involved in invasion of epithelial cells (*i.e.* genes within the SPI-1 invasion pathway) (Prouty *et al.*, 2004a; Prouty *et al.*, 2004b; Prouty and Gunn, 2000; Prouty *et al.*, 2002). Recently, Menendez *et al.* showed that not only does *Salmonella* invade epithelial cells in the presence of bile in the gallbladder, it can survive and actively replicate within the harsh environment of the gallbladder lumen that is made up of 100% bile (Menendez, 2009). Depending on its location, whether the intestine or gallbladder, *Salmonella* appears to be able to sense and respond to its environment, facilitating its current stage of infection.

Conclusion

Studying *Salmonella* infections, whether systemic or localized to the GI tract, is extremely complex, and new advancements are made daily in our understanding of how this pathogen interacts with every organ system. Recently, mucosal immunology studies that have delineated the specific cell types that maintain homeostasis in the gut have been a critical breakthrough in our knowledge about the way *Salmonella* establishes infection and disseminates throughout the body. The field of mucosal immunology is at the forefront of host-pathogen studies and is rapidly expanding. For example, a recent issue of Immunity (September 2009) was dedicated to interactions between the mucosa and commensal as well as pathogenic microbes, and two exciting new papers not only established the origins of dendritic cell subsets in the LP, but also further narrowed down the specific $CD103^{-}CX3CR1^{+}CCR7^{+}$ DC that are the first to transport *Salmonella* to the MLN (Bogunovic *et al.*, 2009; Varol *et al.*, 2009). In addition, unpublished studies featured in a recent Nature article showed how new imaging techniques, such as multiphoton microscopy, can elucidate how *Salmonella* interacts with host cells at extremely early time points after infection. In the future, the ability to combine this method with biochemical assays that can track changes in cell signalling or bacterial virulence genes, will lay the groundwork for highly specialized studies of *Salmonella* pathogenesis (Erdmann, 2009). In the wake of this constantly evolving body of knowledge, we hope that great strides will be made towards the prevention and treatment of enteric fevers and NTS infections worldwide.

List of abbreviations

APC	antigen presenting cell
CCL2	chemokine c-c motif ligand 2
CTL	cytotoxic T lymphocyte
CVD	Center for Vaccine Development
CX3CR1	CX3-motif chemokine receptor 1
DC	dendritic cell
FAE	follicle-associated epithelium
GALT	gut-associated lymphoid tissue
GI	gastrointestinal
HXA3	eicosanoid hepoxilin A_3
IFNγR	interferon gamma receptor
IFNγ	interferon gamma
IBD	inflammatory bowel disease
IEC	intestinal epithelial cell
IL	interleukin
ILF	isolated lymphoid follicles
iNOS	inducible nitric oxide synthase
LP	lamina propria
LPS	lipopolysaccharide
Mϕ	macrophage
MAdCAM-1	mucosal addressin cell adhesion molecule 1
M cell	microfold cell
MCP-1	monocyte chemotactic protein 1
MDR	multi-drug resistance
MHC	major histocompatibility complex
MLN	mesenteric lymph node
MMP	matrix metalloprotease
MSR-A	macrophage scavenger receptor A
MyD88	Myeloid differentiation primary response 88
NF-κB	nuclear factor kappa B
NK	natural killer
NLR	nod-like receptor
NTS	non-typhoidal salmonellosis
PAF	platelet activating factor
PAMP	pathogen associated molecular pattern
PMN	polymorphonuclear
PP	Peyer's patch
PRR	pattern recognition receptor
PSGL-1	p-selectin glycoprotein ligand 1
RA	retinoic acid
RES	reticulo-endothelial system
ROS	reactive oxygen species
RNS	reactive nitrogen species
SCV	*Salmonella* containing vacuole
SED	sub-epithelial dome
SPI	*Salmonella* pathogenicity island
TGF-β	transforming growth factor beta
Th1	T helper
TLR	toll-like receptor
TNFα	tumor necrosis factor alpha
TSLP	thymic stromal lymphopoietin
Treg	T regulatory
TTSS	type three secretion system
Vi	virulence

References

Ahern, P.P., Izcue, A., Maloy, K.J., and Powrie, F. (2008). The interleukin-23 axis in intestinal inflammation. Immunol. Rev. *226*, 147–159.

al-Ramadi, B.K., Al-Dhaheri, M.H., Mustafa, N., Abouhaidar, M., Xu, D., Liew, F.Y., Lukic, M.L., and Fernandez-Cabezudo, M.J. (2001). Influence of vector-encoded cytokines on anti-*Salmonella* immunity: divergent effects of interleukin-2 and tumor necrosis factor alpha. Infect. Immun. *69*, 3980–3988.

Ali, S., Vollaard, A.M., Widjaja, S., Surjadi, C., van de Vosse, E., and van Dissel, J.T. (2006). PARK2/PACRG polymorphisms and susceptibility to typhoid and paratyphoid fever. Clin. Exp. Immunol. *144*, 425–431.

Anton, B., and Fütterer, G. (1888). MMW Münch. Med. Wochenschr. 35, 316.

Arai, T., Hiromatsu, K., Nishimura, H., Kimura, Y., Kobayashi, N., Ishida, H., Nimura, Y., and Yashikai, Y. (1995). Effects of *in vivo* administration of anti-IL-10 monoclonal antibody on the host defence mechanism in Samonella Typhimurium infection in mice. Infect. Immun. *66*, 1432–1438.

Arques, J.L., Hautefort, I., Ivory, K., Bertelli, E., Regoli, M., Clare, S., Hinton, J.C., and Nicoletti, C. (2009). *Salmonella* induces flagellin- and MyD88-dependent migration of bacteria-capturing dendritic cells into the gut lumen. Gastroenterology 137, 579–587, 587, e571–572.

Artis, D. (2008). Epithelial-cell recognition of commensal bacteria and maintenance of immune homeostasis in the gut. Nat. Rev. Immunol. *8*, 411–420.

Ashkar, A.A., Reid, S., Verdu, E.F., Zhang, K., and Coombes, B.K. (2009). Interleukin-15 and NK1.1+ cells provide innate protection against acute *Salmonella enterica* serovar Typhimurium infection in the gut and in systemic tissues. Infect. Immun. *77*, 214–222.

Atarashi, K., Nishimura, J., Shima, T., Umesaki, Y., Yamamoto, M., Onoue, M., Yagita, H., Ishii, N., Evans, R., Honda, K., *et al.* (2008). ATP drives lamina propria T(H)17 cell differentiation. Nature *455*, 808–812.

Awasthi, A., and Kuchroo, V.K. (2009). Th17 cells: from precursors to players in inflammation and infection. Int. Immunol. *21*, 489–498.

Bakshi, C.S., Singh, V.P., Wood, M.W., Jones, P.W., Wallis, T.S., and Galyov, E.E. (2000). Identification of SopE2, a *Salmonella* secreted protein which is highly homologous to SopE and involved in bacterial invasion of epithelial cells. J. Bacteriol. *182*, 2341–2344.

Barman, M., Unold, D., Shifley, K., Amir, E., Hung, K., Bos, N., and Salzman, N. (2008). Enteric salmonellosis disrupts the microbial ecology of the murine gastrointestinal tract. Infect. Immun. *76*, 907–915.

Barthel, M., Hapfelmeier, S., Quintanilla-Martinez, L., Kremer, M., Rohde, M., Hogardt, M., Pfeffer, K., Russmann, H., and Hardt, W.D. (2003). Pretreatment of mice with streptomycin provides a *Salmonella enterica* serovar Typhimurium colitis model that allows analysis of both pathogen and host. Infect. Immun. *71*, 2839–2858.

Bellamy, R. (1999). The natural resistance-associated macrophage protein and susceptibility to intracellular pathogens. Microbes Infect *1*, 23–27.

Beuzon, C.R., Meresse, S., Unsworth, K.E., Ruiz-Albert, J., Garvis, S., Waterman, S.R., Ryder, T.A., Boucrot, E., and Holden, D.W. (2000). *Salmonella* maintains the integrity of its intracellular vacuole through the action of SifA. The EMBO journal *19*, 3235–3249.

Billiau, A. (1996). Interferon-gamma: biology and role in pathogenesis. Adv Immunol. *62*, 61–130.

Biozzi, G., Howard, J.G., Halpern, B.N., Stiffel, C., and Mouton, D. (1960). The kinetics of blood clearance o isotopically labelled *Salmonella entertidis* by the reticulo-endothelial system in mice. Immunology *3*, 74–89.

Bishop, J.L. (2008). Characterizing the role of the LYN tyrosine kinase in innate immune responses to *Salmonella* infection. PhD Thesis, University of British Columbia, Vancouver, BC.

Blackwell, J.M., Goswami, T., Evans, C.A., Sibthorpe, D., Papo, N., White, J.K., Searle, S., Miller, E.N., Peacock, C.S., Mohammed, H., *et al.* (2001). SLC11A1 (formerly NRAMP1) and disease resistance. Cell Microbiol. *3*, 773–784.

Bogunovic, M., Ginhoux, F., Helft, J., Shang, L., Hashimoto, D., Greter, M., Liu, K., Jakubzick, C., Ingersoll, M.A., Leboef, M., *et al.* (2009). Origin of the Lamina Propria Dendritic Cell Network. Immunity *31*, 513–525.

Boyle, E.C., Bishop, J.L., Grassl, G.A., and Finlay, B.B. (2007). *Salmonella*: from pathogenesis to therapeutics. J. Bacteriol. *189*, 1489–1149.

Boyle, E.C., Brown, N.F., and Finlay, B.B. (2006). *Salmonella enterica* serovar Typhimurium effectors SopB, SopE, SopE2 and SipA disrupt tight junction structure and function. Cell Microbiol. *8*, 1946–1957.

Brown, S.P., Cornell, S.J., Sheppard, M., Grant, A.J., Maskell, D.J., Grenfell, B.T., and Mastroeni, P. (2006). Intracellular demography and the dynamics of *Salmonella enterica* infections. PLoS Biol. *4*, e349.

Bruno, V.M., Hannemann, S., Lara-Tejero, M., Flavell, R.A., Kleinstein, S.H., and Galan, J.E. (2009). *Salmonella* Typhimurium type III secretion effectors stimulate innate immune responses in cultured epithelial cells. PLoS Pathog *5*, e1000538.

Butler, T., Bell, W.R., Levin, J., Linh, N.N., and Arnold, K. (1978). Typhoid fever. Studies of blood coagulation, bacteremia, and endotoxemia. Arch. Int. Med. *138*, 407–410.

Butler, T., Knight, J., Nath, S.K., Speelman, P., Roy, S.K., and Azad, M.A. (1985). Typhoid fever complicated by intestinal perforation: a persisting fatal disease requiring surgical management. Rev. Infect Dis. *7*, 244–256.

Caruso, R., Sarra, M., Stolfi, C., Rizzo, A., Fina, D., Fantini, M.C., Pallone, F., MacDonald, T.T., and Monteleone, G. (2009). Interleukin-25 inhibits interleukin-12 production and Th1 cell-driven inflammation in the gut. Gastroenterology *136*, 2270–2279.

Castaneda, F.E., Walia, B., Vijay-Kumar, M., Patel, N.R., Roser, S., Kolachala, V.L., Rojas, M., Wang, L., Oprea, G., Garg, P., *et al.* (2005). Targeted deletion of metalloproteinase 9 attenuates experimental colitis in mice: central role of epithelial-derived MMP. Gastroenterology *129*, 1991–2008.

Caygill, C.P., Hill, M.J., Braddick, M., and Sharp, J.C. (1994). Cancer mortality in chronic typhoid and paratyphoid carriers. Lancet *343*, 83–84.

Cella, M., Fuchs, A., Vermi, W., Facchetti, F., Otero, K., Lennerz, J.K., Doherty, J.M., Mills, J.C., and Colonna, M. (2009). A human natural killer cell subset provides an innate source of IL-22 for mucosal immunity. Nature *457*, 722–725.

Cervantes-Barragan, L., Gil-Cruz, C., Pastelin-Palacios, R., Lang, K.S., Isibasi, A., Ludewig, B., and Lopez-Macias, C. (2009). TLR2 and TLR4 signaling shapes specific antibody responses to *Salmonella* Typhi antigens. Eur. J. Immunol. *39*, 126–135.

Chatterjee, A., Dutta, P.K., and Chowdhury, R. (2007). Effect of fatty acids and cholesterol present in bile on expression of virulence factors and motility of Vibrio cholerae. Infect. Immun. *75*, 1946–1953.

Chen, W., Jin, W., Hardegen, N., Lei, K.J., Li, L., Marinos, N., McGrady, G., and Wahl, S.M. (2003). Conversion of peripheral CD4+CD25- naive T cells to CD4+CD25+ regulatory T cells by TGF-beta induction of transcription factor Foxp3. J. Exp. Med. *198*, 1875–1886.

Chiang, S.L., Mekalanos, J.J., and Holden, D.W. (1999). *In vivo* genetic analysis of bacterial virulence. Annu. Rev. Microbiol. *53*, 129–154.

Chieppa, M., Rescigno, M., Huang, A.Y., and Germain, R.N. (2006). Dynamic imaging of dendritic cell extension into the small bowel lumen in response to epithelial cell TLR engagement. J. Exp. Med. *203*, 2841–2852.

Collins, F.M. (1974). Vaccines and cell-mediated immunity. Bacteriol. Rev. *38*, 371–402.

Colonna, M. (2009). Interleukin-22-producing natural killer cells and lymphoid tissue inducer-like cells in mucosal immunity. Immunity *31*, 15–23.

Connor, B.A., and Schwartz, E. (2005). Typhoid and paratyphoid fever in travellers. Lancet Infect Dis. *5*, 623–628.

Coombes, J.L., Siddiqui, K.R., Arancibia-Carcamo, C.V., Hall, J., Sun, C.M., Belkaid, Y., and Powrie, F. (2007). A functionally specialized population of mucosal CD103+ DCs induces Foxp3+ regulatory T cells via a TGF-beta and retinoic acid-dependent mechanism. J. Exp. Med. *204*, 1757–1764.

Croswell, A., Amir, E., Teggatz, P., Barman, M., and Salzman, N.H. (2009). Prolonged impact of antibiotics on intestinal microbial ecology and susceptibility to enteric *Salmonella* infection. Infect. Immun. *77*, 2741–2753.

Crump, J.A., Luby, S.P., and Mintz, E.D. (2004). The global burden of typhoid fever. Bull World Health Organ *82*, 346–353.

Crump, J.A., Ram, P.K., Gupta, S.K., Miller, M.A., and Mintz, E.D. (2008). Part I. Analysis of data gaps pertaining to *Salmonella enterica* serotype Typhi infections in low and medium human development index countries, 1984–2005. Epidemiol. Infect. *136*, 436–448.

Curtiss, R., 3rd, Wanda, S.Y., Gunn, B.M., Zhang, X., Tinge, S.A., Ananthnarayan, V., Mo, H., Wang, S., and Kong, W. (2009). *Salmonella enterica* serovar Typhimurium strains with regulated delayed attenuation *in vivo*. Infect. Immun. *77*, 1071–1082.

De Groote, M.A., Ochsner, U.A., Shiloh, M.U., Nathan, C., McCord, J.M., Dinauer, M.C., Libby, S.J., Vazquez-Torres, A., Xu, Y., and Fang, F.C. (1997). Periplasmic superoxide dismutase protects *Salmonella* from products of phagocyte NADPH-oxidase and nitric oxide synthase. Proc. Natl. Acad. Sci. U.S.A. *94*, 13997–14001.

Dehler, H. (1907). MMW Münch Med Wochenschr *54*, 779.

Denich, K., Borlin, P., O'Hanley, P.D., Howard, M., and Heath, A.W. (1993). Expression of the murine interleukin-4 gene in an attenuated aroA strain of *Salmonella* Typhimurium: persistence and immune response in BALB/c mice and susceptibility to macrophage killing. Infect. Immun. *61*, 4818–4827.

Dhanoa, A., and Fatt, Q.K. (2009). Non-typhoidal *Salmonella* bacteraemia: epidemiology, clinical characteristics and its' association with severe immunosuppression. Ann Clin. Microbiol. Antimicrob *8*, 15.

Dinbar, A., Altmann, G., and Tulcinsky, D.B. (1969). The treatment of chronic biliary *Salmonella* carriers. Am. J. Med. *47*, 236–242.

Doerr, R. (1905). Experimentelle untersuchungen über das fortwuchern von typhusbacillen in der gallenblase. Centralbl. f. Bakt. *39*, 624–634.

Doorduyn, Y., Van Pelt, W., Siezen, C.L., Van Der Horst, F., Van Duynhoven, Y.T., Hoebee, B., and Janssen, R. (2008). Novel insight in the association between salmonellosis or campylobacteriosis and chronic illness, and the role of host genetics in susceptibility to these diseases. Epidemiol. Infect. *136*, 1225–1234.

Dukes, J.D., Lee, H., Hagen, R., Reaves, B.J., Layton, A.N., Galyov, E.E., and Whitley, P. (2006). The secreted *Salmonella* dublin phosphoinositide phosphatase, SopB, localizes to PtdIns(3)P-containing endosomes and perturbs normal endosome to lysosome trafficking. The Biochemical journal *395*, 239–247.

Dunstan, S.J., Ho, V.A., Duc, C.M., Lanh, M.N., Phuong, C.X., Luxemburger, C., Wain, J., Dudbridge, F., Peacock, C.S., House, D., *et al.* (2001). Typhoid fever and genetic polymorphisms at the natural resistance-associated macrophage protein 1. J. Infect Dis. *183*, 1156–1160.

Dussurget, O., Cabanes, D., Dehoux, P., Lecuit, M., Buchrieser, C., Glaser, P., and Cossart, P. (2002). Listeria monocytogenes bile salt hydrolase is a PrfA-regulated virulence factor involved in the intestinal and hepatic phases of listeriosis. Mol. Microbiol. *45*, 1095–1106.

Eckmann, L., and Kagnoff, M.F. (2001). Cytokines in host defence against *Salmonella*. Microbes Infect. *3*, 1191–1200.

Egea, L., Gimenez, R., Lucia, D., Modolell, I., Badia, J., Baldoma, L., and Aguilar, J. (2008). Increased production of the ether-lipid platelet-activating factor in intestinal epithelial cells infected by *Salmonella* Enteritidis. Biochim. Biophys. Acta. *1781*, 270–276.

Engels, E.A., Falagas, M.E., Lau, J., and Bennish, M.L. (1998). Typhoid fever vaccines: a meta-analysis of studies on efficacy and toxicity. BMJ (Clinical research ed *316*, 110–116.

Enomoto, A., Nishimura, H., and Yoshikai, Y. (1997). Predominant appearance of NK1.1+ T cells producing IL-4 may be involved in the increased susceptibility of mice with the beige mutation during *Salmonella* infection. J. Immunol. *158*, 2268–2277.

Erdmann, J. (2009). Immunology: Lights, Camera, Infection. Nature *460*, 568–570.

Everest, P., Roberts, M., and Dougan, G. (1998). Susceptibility to Samonella Typhimurium infection and effectiveness of vaccination in mice deficient in the tumor necrosis factor alpha p55 receptor. Infect. Immun. *66*, 3355–3364.

Everest, P., Wain, J., Roberts, M., Rook, G., and Dougan, G. (2001). The molecular mechanisms of severe typhoid fever. Trends Microbiol. *9*, 316–320.

Fangtham, M., and Wilde, H. (2008). Emergence of *Salmonella* Paratyphi A as a major cause of enteric fever: need for early detection, preventive measures, and effective vaccines. J. Travel Med. *15*, 344–350.

Ferlazzo, G., Pack, M., Thomas, D., Paludan, C., Schmid, D., Strowig, T., Bougras, G., Muller, W.A., Moretta, L., and Munz, C. (2004). Distinct roles of IL-12 and IL-15 in human natural killer cell activation by dendritic cells from secondary lymphoid organs. Proceedings of the National Academy of Sciences of the United States of America *101*, 16606–16611.

Fink, S.L., and Cookson, B.T. (2007). Pyroptosis and host cell death responses during *Salmonella* infection. Cell Microbiol. *9*, 2562–2570.

Forbes, J.R., and Gros, P. (2003). Iron, manganese, and cobalt transport by Nramp1 (Slc11a1) and Nramp2 (Slc11a2) expressed at the plasma membrane. Blood *102*, 1884–1892.

Forbes, S.J., Eschmann, M., and Mantis, N.J. (2008). Inhibition of *Salmonella enterica* serovar Typhimurium motility and entry into epithelial cells by a protective antilipopolysaccharide monoclonal immunoglobulin A antibody. Infect. Immun. *76,* 4137–4144.

Forrest, B.D., LaBrooy, J.T., Beyer, L., Dearlove, C.E., and Shearman, D.J. (1991). The human humoral immune response to *Salmonella* Typhi Ty21a. J. Infect Dis. *163,* 336–345.

Fortin, G., Raymond, M., Van, V.Q., Rubio, M., Gautier, P., Sarfati, M., and Franchimont, D. (2009). A role for CD47 in the development of experimental colitis mediated by SIRP{alpha}+CD103- dendritic cells. J. Exp. Med. *206,* 1995–2011.

Foster, N., Hulme, S.D., and Barrow, P.A. (2003). Induction of antimicrobial pathways during early-phase immune response to *Salmonella* spp. in murine macrophages: gamma interferon (IFN-gamma) and upregulation of IFN-gamma receptor alpha expression are required for NADPH phagocytic oxidase gp91-stimulated oxidative burst and control of virulent *Salmonella* spp. Infect. Immun. *71,* 4733–4741.

Fraser, A., Paul, M., Goldberg, E., Acosta, C.J., and Leibovici, L. (2007). Typhoid fever vaccines: systematic review and meta-analysis of randomised controlled trials. Vaccine *25,* 7848–7857.

Gaines, S., Sprinz, H., Tully, J.G., and Tigertt, W.D. (1968). Studies on infection and immunity in experimental typhoid fever. VII. The distribution of *Salmonella* Typhi in chimpanzee tissue following oral challenge, and the relationship between the numbers of bacilli and morphologic lesions. J. Infect Dis. *118,* 293–306.

Gal-Mor, O., and Finlay, B.B. (2006). Pathogenicity islands: a molecular toolbox for bacterial virulence. Cellular microbiology *8,* 1707–1719.

Galan, J.E. (2001). *Salmonella* interactions with host cells: type III secretion at work. Annu Rev. Cell Dev Biol. *17,* 53–86.

Galan, J.E., and Curtiss, R., 3rd (1989). Cloning and molecular characterization of genes whose products allow *Salmonella* Typhimurium to penetrate tissue culture cells. Proceedings of the National Academy of Sciences of the United States of America *86,* 6383–6387.

Garg, P., Rojas, M., Ravi, A., Bockbrader, K., Epstein, S., Vijay-Kumar, M., Gewirtz, A.T., Merlin, D., and Sitaraman, S.V. (2006). Selective ablation of matrix metalloproteinase-2 exacerbates experimental colitis: contrasting role of gelatinases in the pathogenesis of colitis. J. Immunol. *177,* 4103–4112.

Garmory, H.S., Brown, K.A., and Titball, R.W. (2002). *Salmonella* vaccines for use in humans: present and future perspectives. FEMS Microbiol. Rev. *26,* 339–353.

Garner, C.D., Antonopoulos, D.A., Wagner, B., Duhamel, G.E., Keresztes, I., Ross, D.A., Young, V.B., and Altier, C. (2009). Perturbation of the small intestine microbial ecology by streptomycin alters pathology in a *Salmonella enterica* serovar Typhimurium murine model of infection. Infect. Immun. *77,* 2691–2702.

Gasem, M.H., Dolmans, W.M., Keuter, M.M., and Djokomoeljanto, R.R. (2001). Poor food hygiene and housing as risk factors for typhoid fever in Semarang, Indonesia. Trop Med. Int. Health *6,* 484–490.

Geddes, K., Cruz, F., and Heffron, F. (2007). Analysis of cells targeted by *Salmonella* type III secretion *in vivo.* PLoS Pathog *3,* e196.

Germanier, R., and Furer, E. (1983). Characteristics of the attenuated oral vaccine strain '*S.* Typhi' Ty 21a. Dev Biol. Stand *53,* 3–7.

Godinez, I., Raffatellu, M., Chu, H., Paixao, T.A., Haneda, T., Santos, R.L., Bevins, C.L., Tsolis, R.M., and Baumler, A.J. (2009). Interleukin-23 orchestrates mucosal responses to *Salmonella enterica* serotype Typhimurium in the intestine. Infect. Immun. *77,* 387–398.

Gordon, M.A. (2008). *Salmonella* infections in immunocompromised adults. J. Infect *56,* 413–422.

Gordon M.A., Banda, H.T., Gondwe, M., Gordon, S.B., Boeree, M.J., Walsh, A.L., Corkill, J.E., Hart, C.A., Gilks, C.F., and Molyneux, M.E. (2002). Non-typhoidal *Salmonella* bacteraemia among HIV-infected Malawian adults: high mortality and frequent recrudescence. AIDS *16,* 1633–1641.

Goswami, T., Bhattacharjee, A., Babal, P., Searle, S., Moore, E., Li, M., and Blackwell, J.M. (2001). Natural-resistance-associated macrophage protein 1 is an H+/bivalent cation antiporter. Biochem. J. *354,* 511–519.

Graham, S.M. (2002). Salmonellosis in children in developing and developed countries and populations. Curr. Opin Infect Dis. *15,* 507–512.

Grant, A.J., Sheppard, M., Deardon, R., Brown, S.P., Foster, G., Bryant, C.E., Maskell, D.J., and Mastroeni, P. (2008). Caspase-3-dependent phagocyte death during systemic *Salmonella enterica* serovar Typhimurium infection of mice. Immunology *125,* 28–37.

Gulig, P.A., Danbara, H., Guiney, D.G., Lax, A.J., Norel, F., and Rhen, M. (1993). Molecular analysis of spv virulence genes of the *Salmonella* virulence plasmids. Mol. Microbiol. *7,* 825–830.

Guzman, C.A., Borsutzky, S., Griot-Wenk, M., Metcalfe, I.C., Pearman, J., Collioud, A., Favre, D., and Dietrich, G. (2006). Vaccines against typhoid fever. Vaccine *24,* 3804–3811.

Hailer, E., and Ungermann, E. (1914). Zur technick der experimentellen typusinfektion. Arbeiten aus dem Kaiserlichen Gesundheitsamte *47,* 451–469.

Hapfelmeier, S., Muller, A.J., Stecher, B., Kaiser, P., Barthel, M., Endt, K., Eberhard, M., Robbiani, R., Jacobi, C.A., Heikenwalder, M., *et al.* (2008). Microbe sampling by mucosal dendritic cells is a discrete, MyD88-independent step in Deltainv G S. Typhimurium colitis. J. Exp. Med. *205,* 437–450.

Haraga, A., and Miller, S.I. (2003). A *Salmonella enterica* serovar Typhimurium translocated leucine-rich repeat effector protein inhibits NF-kappa B-dependent gene expression. Infection and immunity *71,* 4052–4058.

Hardy, J., Francis, K.P., DeBoer, M., Chu, P., Gibbs, K., and Contag, C.H. (2004). Extracellular replication of Listeria monocytogenes in the murine gall bladder. Science *303,* 851–853.

Hardy, J., Margolis, J.J., and Contag, C.H. (2006). Induced biliary excretion of Listeria monocytogenes. Infect. Immun. *74,* 1819–1827.

Harrison, J.A., Villarreal-Ramos, B., Mastroeni, P., Demarco de Hormaeche, R., and Hormaeche, C.E. (1997). Correlates of protection induced by live Aro- *Salmonella* Typhimurium vaccines in the murine typhoid model. Immunology *90*, 618–625.

Hashizume, T., Togawa, A., Nochi, T., Igarashi, O., Kweon, M.N., Kiyono, H., and Yamamoto, M. (2008). Peyer's patches are required for intestinal immunoglobulin A responses to *Salmonella* spp. Infect. Immun. *76*, 927–934.

Hensel, M. (2004). Evolution of pathogenicity islands of *Salmonella enterica*. Int. J. Med. Microbiol. *294*, 95–102.

Hess, J., Ladel, C., Miko, D., and Kaufmann, S.H. (1996). *Salmonella* Typhimurium aroA- infection in gene-targeted immunodeficient mice: major role of CD4+ TCR-alpha beta cells and IFN-gamma in bacterial clearance independent of intracellular location. J. Immunol. *156*, 3321–3326.

Hohmann, E.L. (2001). Nontyphoidal salmonellosis. Clin. Infect. Dis. *32*, 263–269.

Hohmann, E.L., Oletta, C.A., Killeen, K.P., and Miller, S.I. (1996). phoP/phoQ-deleted *Salmonella* Typhi (Ty800) is a safe and immunogenic single-dose typhoid fever vaccine in volunteers. J. Infect Dis. *173*, 1408–1414.

Hormaeche, C.E. (1990). Dead *Salmonellae* or their endotoxin accelerate the early course of a *Salmonella* infection in mice. Microb Pathog *9*, 213–218.

Hornick, R.B., Greisman, S.E., Woodward, T.E., DuPont, H.L., Dawkins, A.T., and Snyder, M.J. (1970a). Typhoid fever: pathogenesis and immunologic control. N. Engl. J. Med. *283*, 686–691.

Hornick, R.B., Greisman, S.E., Woodward, T.E., DuPont, H.L., Dawkins, A.T., and Snyder, M.J. (1970b). Typhoid fever: pathogenesis and immunologic control. 2. N. Engl. J. Med. *283*, 739–746.

House, D., Bishop, A., Parry, C., Dougan, G., and Wain, J. (2001). Typhoid fever: pathogenesis and disease. Current opinion in infectious diseases *14*, 573–578.

Hsu, H.S. (1989). Pathogenesis and immunity in murine salmonellosis. Microbiol. Rev. *53*, 390–409.

Huang, D.B., and DuPont, H.L. (2005). Problem pathogens: extra-intestinal complications of *Salmonella enterica* serotype Typhi infection. Lancet Infect Dis. *5*, 341–348.

Iliev, I.D., Mileti, E., Matteoli, G., Chieppa, M., and Rescigno, M. (2009). Intestinal epithelial cells promote colitis-protective regulatory T-cell differentiation through dendritic cell conditioning. Mucosal Immunol. *2*, 340–350.

Iwasaki, A., and Kelsall, B.L. (1999). Freshly isolated Peyer's patch, but not spleen, dendritic cells produce interleukin 10 and induce the differentiation of T helper type 2 cells. J. Exp. Med. *190*, 229–239.

Iwasaki, A., and Kelsall, B.L. (2000). Localization of distinct Peyer's patch dendritic cell subsets and their recruitment by chemokines macrophage inflammatory protein (MIP)–3alpha, MIP-3beta, and secondary lymphoid organ chemokine. J. Exp. Med. *191*, 1381–1394.

Iwasaki, A., and Kelsall, B.L. (2001). Unique functions of CD11b+, CD8 alpha+, and double-negative Peyer's patch dendritic cells. J. Immunol. *166*, 4884–4890.

Jouanguy, E., Doffinger, R., Dupuis, S., Pallier, A., Altare, F., and Casanova, J.L. (1999). IL-12 and IFN-gamma in host defence against mycobacteria and *Salmonella* in mice and men. Curr. Opin Immunol. *11*, 346–351.

Kagaya, K., Watanabe, K., and Fukazawa, Y. (1989). Capacity of recombinant gamma interferon to activate macrophages for *Salmonella*-killing activity. Infect. Immun. *57*, 609–615.

Khan, S.A., Strijbos, P.J., Everest, P., Moss, D., Stratford, R., Mastroeni, P., Allen, J., Servos, S., Charles, I.G., Dougan, G., *et al.* (2001). Early responses to *Salmonella* Typhimurium infection in mice occur at focal lesions in infected organs. Microb Pathog *30*, 29–38.

Kingsley, R.A., van Amsterdam, K., Kramer, N., and Baumler, A.J. (2000). The shdA gene is restricted to serotypes of *Salmonella enterica* subspecies I and contributes to efficient and prolonged fecal shedding. Infect. Immun. *68*, 2720–2727.

Knodler, L.A., and Steele-Mortimer, O. (2005). The *Salmonella* effector PipB2 affects late endosome/lysosome distribution to mediate Sif extension. Molecular biology of the cell *16*, 4108–4123.

Koch F., Stanzl, U., Jennewein, P., Janke, K., Heufler, C., Kampgen, E., Romani, N., and Schuler, G., (1996). High level IL-12 production by murine dendritic cells: upregulation via MHC class II and CD40 molecules and downregulation by IL-4 and IL-10. J. Exp. Med. *184*, 741–746.

Koch, J. (1909). Typhusbazillen und gallenblase. Ztschr f Hyg, 1–11.

Konadu, E.Y., Lin, F.Y., Ho, V.A., Thuy, N.T., Van Bay, P., Thanh, T.C., Khiem, H.B., Trach, D.D., Karpas, A.B., Li, J., *et al.* (2000). Phase 1 and phase 2 studies of *Salmonella enterica* serovar Paratyphi A O-specific polysaccharide-tetanus toxoid conjugates in adults, teenagers, and 2- to 4-year-old children in Vietnam. Infect. Immun. *68*, 1529–1534.

Kopecko, D.J., Sieber, H., Ures, J.A., Furer, A., Schlup, J., Knof, U., Collioud, A., Xu, D., Colburn, K., and Dietrich, G. (2009). Genetic stability of vaccine strain *Salmonella* Typhi Ty21a over 25 years. Int. J. Med. Microbiol. *299*, 233–246.

Kuhle, V., and Hensel, M. (2002). SseF and SseG are translocated effectors of the type III secretion system of *Salmonella* pathogenicity island 2 that modulate aggregation of endosomal compartments. Cellular microbiology *4*, 813–824.

Kuhn, D.E., Baker, B.D., Lafuse, W.P., and Zwilling, B.S. (1999). Differential iron transport into phagosomes isolated from the RAW264.7 macrophage cell lines transfected with Nramp1Gly169 or Nramp1Asp169. J. Leukoc. Biol. *66*, 113–119.

Kum, W.W., Lee, S., Grassl, G.A., Bidshahri, R., Hsu, K., Ziltener, H.J., and Finlay, B.B. (2009). Lack of functional P-selectin ligand exacerbates *Salmonella* serovar Typhimurium infection. J. Immunol. *182*, 6550–6561.

Kuvandik, C., Karaoglan, I., Namiduru, M., and Baydar, I. (2009). Predictive value of clinical and laboratory

findings in the diagnosis of the enteric fever. New Microbiol. *32*, 25–30.
Lara-Tejero, M., Sutterwala, F.S., Ogura, Y., Grant, E.P., Bertin, J., Coyle, A.J., Flavell, R.A., and Galan, J.E. (2006). Role of the caspase-1 inflammasome in *Salmonella* Typhimurium pathogenesis. J. Exp. Med. *203*, 1407–1412.
Lawley, T.D., Bouley, D.M., Hoy, Y.E., Gerke, C., Relman, D.A., and Monack, D.M. (2008). Host transmission of *Salmonella enterica* serovar Typhimurium is controlled by virulence factors and indigenous intestinal microbiota. Infect. Immun. *76*, 403–416.
Le Bourhis, L., Magalhaes, J.G., Selvanantham, T., Travassos, L.H., Geddes, K., Fritz, J.H., Viala, J., Tedin, K., Girardin, S.E., and Philpott, D.J. (2009). Role of Nod1 in mucosal dendritic cells during Spi-1 independent *Salmonella* infection. Infect. Immun. 77, 4480–4486.
Leary, T.J. (1913). Surgical method of clearing up chronic typhoid carriers. JAMA *LX*, 1293–1294.
Ledingham, J.C., and Arkwright, J.A. (1912). The carrier problem in infectious disease (London, Edward Arnold).
Lee, C.A., Silva, M., Siber, A.M., Kelly, A.J., Galyov, E., and McCormick, B.A. (2000). A secreted *Salmonella* protein induces a proinflammatory response in epithelial cells, which promotes neutrophil migration. Proc. Natl. Acad. Sci. U.S.A. *97*, 12283–12288.
Lee, J., Mo, J.H., Katakura, K., Alkalay, I., Rucker, A.N., Liu, Y.T., Lee, H.K., Shen, C., Cojocaru, G., Shenouda, S., *et al.* (2006). Maintenance of colonic homeostasis by distinctive apical TLR9 signalling in intestinal epithelial cells. Nat. Cell Biol. *8*, 1327–1336.
Levine, M.M., Ferreccio, C., Black, R.E., Lagos, R., San Martin, O., and Blackwelder, W.C. (2007). Ty21a live oral typhoid vaccine and prevention of paratyphoid fever caused by *Salmonella enterica* Serovar Paratyphi B. Clin. Infect. Dis. *45 Suppl 1*, S24–28.
Liang-Takasaki, C.J., Saxen, H., Makela, P.H., and Leive, L. (1983). Complement activation by polysaccharide of lipopolysaccharide: an important virulence determinant of *Salmonellae*. Infect. Immun. *41*, 563–569.
Lipsitch, M., and Moxon, E.R. (1997). Virulence and transmissibility of pathogens: what is the relationship? Trends Microbiol. *5*, 31–37.
Lu, S., Wang, S., and Grimes-Serrano, J.M. (2008). Current progress of DNA vaccine studies in humans. Expert Rev. Vaccines *7*, 175–191.
Luby, S.P., Faizan, M.K., Fisher-Hoch, S.P., Syed, A., Mintz, E.D., Bhutta, Z.A., and McCormick, J.B. (1998). Risk factors for typhoid fever in an endemic setting, Karachi, Pakistan. Epidemiology and infection *120*, 129–138.
Lucas, M., Schachterle, W., Oberle, K., Aichele, P., and Diefenbach, A. (2007). Dendritic cells prime natural killer cells by trans-presenting interleukin 15. Immunity *26*, 503–517.
Marmion, D.E., Naylor, G.R., and Stewart, I.O. (1953). Second attacks of typhoid fever. The Journal of hygiene *51*, 260–267.
Martinoli, C., Chiavelli, A., and Rescigno, M. (2007). Entry route of *Salmonella* Typhimurium directs the type of induced immune response. Immunity *27*, 975–984.
Mason, K.L., Huffnagle, G.B., Noverr, M.C., and Kao, J.Y. (2008). Overview of gut immunology. Adv Exp. Med. Biol. *635*, 1–14.
Mastroeni, P. (2002). Immunity to systemic *Salmonella* infections. Curr. Mol. Med. *2*, 393–406.
Mastroeni, P., Clare, S., Khan, S., Harrison, J.A., Hormaeche, C.E., Okamura, H., Kurimoto, M., and Dougan, G. (1999). Interleukin 18 contributes to host resistance and gamma interferon production in mice infected with virulent *Salmonella* Typhimurium. Infect. Immun. *67*, 478–483.
Mastroeni, P., Harrison, J.A., Chabalgoity, J.A., and Hormaeche, C.E. (1996). Effect of interleukin 12 neutralization on host resistance and gamma interferon production in mouse typhoid. Infect. Immun. *64*, 189–196.
Mastroeni, P., Harrison, J.A., Robinson, J.H., Clare, S., Khan, S., Maskell, D.J., Dougan, G., and Hormaeche, C.E. (1998). Interleukin-12 is required for control of the growth of attenuated aromatic-compound-dependent *Salmonellae* in BALB/c mice: role of gamma interferon and macrophage activation. Infect. Immun. *66*, 4767–4776.
Mastroeni, P., Skepper, J.N., and Hormaeche, C.E. (1995). Effect of anti-tumor necrosis factor alpha antibodies on histopathology of primary Samonella infections. Infect. Immun. *63*, 3674–3682.
Mastroeni, P., and Maskell, D. (2006). *Salmonella* infections: clinical, immunological and molecular aspects (Cambridge, UK; New York, Cambridge University Press).
Mastroeni, P., and Sheppard, M. (2004). *Salmonella* infections in the mouse model: host resistance factors and *in vivo* dynamics of bacterial spread and distribution in the tissues. Microbes Infect *6*, 398–405.
Mastroeni, P., Simmons, C., Fowler, R., Hormaeche, C.E., and Dougan, G. (2000a). Igh-6(-/-) (B-cell-deficient) mice fail to mount solid acquired resistance to oral challenge with virulent *Salmonella enterica* serovar Typhimurium and show impaired Th1 T-cell responses to *Salmonella* antigens. Infection and immunity *68*, 46–53.
Mastroeni, P., Vazquez-Torres, A., Fang, F.C., Xu, Y., Khan, S., Hormaeche, C.E., and Dougan, G. (2000b). Antimicrobial actions of the NADPH phagocyte oxidase and inducible nitric oxide synthase in experimental salmonellosis. II. Effects on microbial proliferation and host survival *in vivo*. J. Exp. Med. *192*, 237–248.
Menendez, A., Arena, E.T.A, Gutman, J.A., Thorson, L., Vallance, B.A., Vogl, W., Finlay, B.B.F. *Salmonella* infection of gallbladder epithelial cells drives local inflammation and injury in a model of acute typhoid fever. J. Infect. Dis. *200*, 1703–1713.
Mermin, J.H., Villar, R., Carpenter, J., Roberts, L., Samaridden, A., Gasanova, L., Lomakina, S., Bopp, C., Hutwagner, L., Mead, P., *et al.* (1999). A massive epidemic of multidrug-resistant typhoid fever in Tajikistan associated with consumption of municipal water. J. Infect Dis. *179*, 1416–1422.

Metcalf, E.S., and O'Brien, A.D. (1981). Characterization of murine antibody response to *Salmonella* Typhimurium by a class-specific solid-phase radioimmunoassay. Infect. Immun. *31*, 33–41.

Mittrucker, H.W., Kohler, A., and Kaufmann, S.H. (2002). Characterization of the murine T-lymphocyte response to *Salmonella enterica* serovar Typhimurium infection. Infection and immunity *70*, 199–203.

Mittrucker, H.W., Kohler, A., Mak, T.W., and Kaufmann, S.H. (1999). Critical role of CD28 in protective immunity against *Salmonella* Typhimurium. J. Immunol. *163*, 6769–6776.

Monack, D.M., Bouley, D.M., and Falkow, S. (2004a). *Salmonella* Typhimurium persists within macrophages in the mesenteric lymph nodes of chronically infected Nramp1+/+ mice and can be reactivated by IFNgamma neutralization. J. Exp. Med. *199*, 231–241.

Monack, D.M., Hersh, D., Ghori, N., Bouley, D., Zychlinsky, A., and Falkow, S. (2000). *Salmonella* exploits caspase-1 to colonize Peyer's patches in a murine typhoid model. The Journal of experimental medicine *192*, 249–258.

Monack, D.M., Mueller, A., and Falkow, S. (2004b). Persistent bacterial infections: the interface of the pathogen and the host immune system. Nat. Rev. Microbiol. *2*, 747–765.

Morpeth, S.C., Ramadhani, H.O., and Crump, J.A. (2009). Invasive non-Typhi *Salmonella* disease in Africa. Clin. Infect. Dis. *49*, 606–611.

Mrsny, R.J., Gewirtz, A.T., Siccardi, D., Savidge, T., Hurley, B.P., Madara, J.L., and McCormick, B.A. (2004). Identification of hepoxilin A3 in inflammatory events: a required role in neutrophil migration across intestinal epithelia. Proc. Natl. Acad. Sci. U.S.A. *101*, 7421–7426.

Muller, A.J., Hoffmann, C., Galle, M., Van Den Broeke, A., Heikenwalder, M., Falter, L., Misselwitz, B., Kremer, M., Beyaert, R., and Hardt, W.D. (2009). The S. Typhimurium Effector SopE Induces Caspase-1 Activation in Stromal Cells to Initiate Gut Inflammation. Cell Host Microbe *6*, 125–136.

Mumy, K.L., Bien, J.D., Pazos, M.A., Gronert, K., Hurley, B.P., and McCormick, B.A. (2008). Distinct isoforms of phospholipase A2 mediate the ability of *Salmonella enterica* serotype Typhimurium and Shigella flexneri to induce the transepithelial migration of neutrophils. Infect. Immun. *76*, 3614–3627.

Muotiala, A., and Makela, P.H. (1993). Role of gamma interferon in late stages of murine salmonellosis. Infect. Immun. *61*, 4248–4253.

Nagarajan, A.G., Balasundaram, S.V., Janice, J., Karnam, G., Eswarappa, S.M., and Chakravortty, D. (2009a). SopB of *Salmonella enterica* serovar Typhimurium is a potential DNA vaccine candidate in conjugation with live attenuated bacteria. Vaccine *27*, 2804–2811.

Nagarajan, A.G., Karnam, G., Lahiri, A., Allam, U.S., and Chakravortty, D. (2009b). Reliable means of diagnosis and serovar determination of blood-borne *Salmonella* strains: quick PCR amplification of unique genomic loci by novel primer sets. J. Clin. Microbiol. *47*, 2435–2441.

Nagy, G., Dobrindt, U., Hacker, J., and Emody, L. (2004). Oral immunization with an rfaH mutant elicits protection against salmonellosis in mice. Infect. Immun. *72*, 4297–4301.

Nagy, G., Palkovics, T., Otto, A., Kusch, H., Kocsis, B., Dobrindt, U., Engelmann, S., Hecker, M., Emody, L., Pal, T., *et al.* (2008). 'Gently rough': the vaccine potential of a *Salmonella enterica* regulatory lipopolysaccharide mutant. J. Infect Dis. *198*, 1699–1706.

Negi, V.D., Singhamahapatra, S., and Chakravortty, D. (2007). *Salmonella enterica* serovar Typhimurium strain lacking pmrG-HM-D provides excellent protection against salmonellosis in murine typhoid model. Vaccine *25*, 5315–5323.

Neish, A.S., Gewirtz, A.T., Zeng, H., Young, A.N., Hobert, M.E., Karmali, V., Rao, A.S., and Madara, J.L. (2000). Prokaryotic regulation of epithelial responses by inhibition of IkappaB-alpha ubiquitination. Science *289*, 1560–1563.

Nichterlein, T., Kretschmar, M., and Hof, H. (1994). Fecal shedding of Listeria monocytogenes during murine listeriosis after intravenous infection. Zentralbl Bakteriol *281*, 192–195.

Niess, J.H., Brand, S., Gu, X., Landsman, L., Jung, S., McCormick, B.A., Vyas, J.M., Boes, M., Ploegh, H.L., Fox, J.G., *et al.* (2005). CX3CR1-mediated dendritic cell access to the intestinal lumen and bacterial clearance. Science *307*, 254–258.

Nix, R.N., Altschuler, S.E., Henson, P.M., and Detweiler, C.S. (2007). Hemophagocytic macrophages harbor *Salmonella enterica* during persistent infection. PLoS Pathog *3*, e193.

O'Brien, A.D., Metcalf, E.S., and Rosenstreich, D.L. (1982). Defect in macrophage effector function confers *Salmonella* Typhimurium susceptibility on C3H/HeJ. mice. Cell Immunol. *67*, 325–333.

O'Hara, A.M., and Shanahan, F. (2006). The gut flora as a forgotten organ. EMBO Rep *7*, 688–693.

Ohl, M.E., and Miller, S.I. (2001). *Salmonella*: a model for bacterial pathogenesis. Annu Rev. Med. *52*, 259–274.

Olsen, S.J., Pruckler, J., Bibb, W., Nguyen, T.M., Tran, M.T., Sivapalasingam, S., Gupta, A., Phan, T.P., Nguyen, T.C., Nguyen, V.C., *et al.* (2004). Evaluation of rapid diagnostic tests for typhoid fever. J. Clin. Microbiol. *42*, 1885–1889.

Palucka, K.A., Taquet, N., Sanchez-Chapuis, F., and Gluckman, J.C. (1999). Lipopolysaccharide can block the potential of monocytes to differentiate into dendritic cells. J. Leukoc. Biol. *65*, 232–240.

Pang, T., Bhutta, Z.A., Finlay, B.B., and Altwegg, M. (1995). Typhoid fever and other salmonellosis: a continuing challenge. Trends in microbiology *3*, 253–255.

Park, Typhoid, W.H. bacilli carriers (Primary Sources, Workshops in American History).

Parry, C.M. (2004). Typhoid Fever. Curr. Infect Dis. Rep *6*, 27–33.

White, N.J., and Farrar, J.J. (1999). Value of a single-tube Widal test in diagnosis of typhoid fever in Vietnam. J. Clin. Microbiol. *37*, 2882–2886.

Parry, C.M., and Threlfall, E.J. (2008). Antimicrobial resistance in typhoidal and nontyphoidal *Salmonellae*. Curr. Opin Infect Dis. *21*, 531–538.

Paterson, G.K., Northen, H., Cone, D.B., Willers, C., Peters, S.E., and Maskell, D.J. (2009). Deletion of tolA in *Salmonella* Typhimurium generates an attenuated strain with vaccine potential. Microbiology *155*, 220–228.

Petrof, E.O., Musch, M.W., Ciancio, M., Sun, J., Hobert, M.E., Claud, E.C., Gewirtz, A., and Chang, E.B. (2008). Flagellin is required for *Salmonella*-induced expression of heat shock protein Hsp25 in intestinal epithelium. Am. J. Physiol. Gastrointest. Liver Physiol. *294*, G808–818.

Poltorak, A., He, X., Smirnova, I., Liu, M.Y., Van Huffel, C., Du, X., Birdwell, D., Alejos, E., Silva, M., Galanos, C., *et al.* (1998). Defective LPS signaling in C3H/HeJ. and C57BL/10ScCr mice: mutations in Tlr4 gene. Science *282*, 2085–2088.

Prieto, A.I., Jakomin, M., Segura, I., Pucciarelli, M.G., Ramos-Morales, F., Garcia-Del Portillo, F., and Casadesus, J. (2007). The GATC-binding protein SeqA is required for bile resistance and virulence in *Salmonella enterica* serovar Typhimurium. J. Bacteriol. *189*, 8496–8502.

Prouty, A.M., Brodsky, I.E., Falkow, S., and Gunn, J.S. (2004a). Bile-salt-mediated induction of antimicrobial and bile resistance in *Salmonella* Typhimurium. Microbiology *150*, 775–783.

Prouty, A.M., Brodsky, I.E., Manos, J., Belas, R., Falkow, S., and Gunn, J.S. (2004b). Transcriptional regulation of *Salmonella enterica* serovar Typhimurium genes by bile. FEMS Immunol. Med. Microbiol. *41*, 177–185.

Prouty, A.M., and Gunn, J.S. (2000). *Salmonella enterica* serovar Typhimurium invasion is repressed in the presence of bile. Infect. Immun. *68*, 6763–6769.

Prouty, A.M., Van Velkinburgh, J.C., and Gunn, J.S. (2002). *Salmonella enterica* serovar Typhimurium resistance to bile: identification and characterization of the tolQRA cluster. J. Bacteriol. *184*, 1270–1276.

Rabsch, W., Tschape, H., and Baumler, A.J. (2001). Non-typhoidal salmonellosis: emerging problems. Microbes Infect *3*, 237–247.

Raffatellu, M., George, M.D., Akiyama, Y., Hornsby, M.J., Nuccio, S.P., Paixao, T.A., Butler, B.P., Chu, H., Santos, R.L., Berger, T., *et al.* (2009). Lipocalin-2 resistance confers an advantage to *Salmonella enterica* serotype Typhimurium for growth and survival in the inflamed intestine. Cell Host Microbe *5*, 476–486.

Raffatellu, M., Santos, R.L., Verhoeven, D.E., George, M.D., Wilson, R.P., Winter, S.E., Godinez, I., Sankaran, S., Paixao, T.A., Gordon, M.A., *et al.* (2008). Simian immunodeficiency virus-induced mucosal interleukin-17 deficiency promotes *Salmonella* dissemination from the gut. Nat. Med. *14*, 421–428.

Raffatellu, M., Wilson, R.P., Chessa, D., Andrews-Polymenis, H., Tran, Q.T., Lawhon, S., Khare, S., Adams, L.G., and Baumler, A.J. (2005). SipA, SopA, SopB, SopD, and SopE2 contribute to *Salmonella enterica* serotype Typhimurium invasion of epithelial cells. Infection and immunity *73*, 146–154.

Richter-Dahlfors, A., Buchan, A.M., and Finlay, B.B. (1997). Murine salmonellosis studied by confocal microscopy: *Salmonella* Typhimurium resides intracellularly inside macrophages and exerts a cytotoxic effect on phagocytes *in vivo*. J. Exp. Med. *186*, 569–580.

Rimoldi, M., Chieppa, M., Salucci, V., Avogadri, F., Sonzogni, A., Sampietro, G.M., Nespoli, A., Viale, G., Allavena, P., and Rescigno, M. (2005). Intestinal immune homeostasis is regulated by the crosstalk between epithelial cells and dendritic cells. Nat. Immunol. *6*, 507–514.

Round, J.L., and Mazmanian, S.K. (2009). The gut microbiota shapes intestinal immune responses during health and disease. Nat. Rev. Immunol. *9*, 313–323.

Rosenberger, C.M., and Finlay, B.B. (2002). Macrophages inhibit *Salmonella* Typhimurium replication through MEK/ERK kinase and phagocyte NADPH oxidase activities. J. Biol. Chem. *277*, 18753–18762.

Round, J.L., and Mazmanian, S.K. (2009). The gut microbiota shapes intestinal immune responses during health and disease. Nat. Rev. Immunol. *9*, 313–323.

Rydstrom, A., and Wick, M.J. (2007). Monocyte recruitment, activation, and function in the gut-associated lymphoid tissue during oral *Salmonella* infection. J. Immunol. *178*, 5789–5801.

Saemann, M.D., Bohmig, G.A., Osterreicher, C.H., Burtscher, H., Parolini, O., Diakos, C., Stockl, J., Horl, W.H., and Zlabinger, G.J. (2000). Anti-inflammatory effects of sodium butyrate on human monocytes: potent inhibition of IL-12 and up-regulation of IL-10 production. FASEB J. *14*, 2380–2382.

Saenz, S.A., Taylor, B.C., and Artis, D. (2008). Welcome to the neighborhood: epithelial cell-derived cytokines license innate and adaptive immune responses at mucosal sites. Immunol. Rev. *226*, 172–190.

Salazar-Gonzalez, R.M., Niess, J.H., Zammit, D.J., Ravindran, R., Srinivasan, A., Maxwell, J.R., Stoklasek, T., Yadav, R., Williams, I.R., Gu, X., *et al.* (2006). CCR6-mediated dendritic cell activation of pathogen-specific T cells in Peyer's patches. Immunity *24*, 623–632.

Salcedo, S.P., Noursadeghi, M., Cohen, J., and Holden, D.W. (2001). Intracellular replication of *Salmonella* Typhimurium strains in specific subsets of splenic macrophages *in vivo*. Cell Microbiol. 3, 587–597.

Santos, R.L., Zhang, S., Tsolis, R.M., Kingsley, R.A., Adams, L.G., and Baumler, A.J. (2001). Animal models of *Salmonella* infections: enteritis versus typhoid fever. Microbes and infection/Institut Pasteur *3*, 1335–1344.

Schlumberger, M.C., and Hardt, W.D. (2006). *Salmonella* type III secretion effectors: pulling the host cell's strings. Current opinion in microbiology *9*, 46–54.

Schulz, S.M., Kohler, G., Holscher, C., Iwakura, Y., and Alber, G. (2008). IL-17A is produced by Th17, gammadelta T cells and other CD4- lymphocytes during infection with *Salmonella enterica* serovar Enteritidis and has a mild effect in bacterial clearance. Int. Immunol. *20*, 1129–1138.

Searle, L.E., Best, A., Nunez, A., Salguero, F.J., Johnson, L., Weyer, U., Dugdale, A.H., Cooley, W.A., Carter, B., Jones, G., *et al.* (2009). A mixture containing galactooligosaccharide, produced by the enzymic activity of *Bifidobacterium bifidum*, reduces *Salmonella enterica* serovar Typhimurium infection in mice. J. Med. Microbiol. *58*, 37–48.

Sekirov, I., and Finlay, B.B. (2009). The role of the intestinal microbiota in enteric infection. J. Physiol. 587 (pt 17), 4159–4167.

Sekirov, I., Tam, N.M., Jogova, M., Robertson, M.L., Li, Y., Lupp, C., and Finlay, B.B. (2008). Antibiotic-induced perturbations of the intestinal microbiota alter host susceptibility to enteric infection. Infect. Immun. *76*, 4726–4736.

Sheppard, M., Webb, C., Heath, F., Mallows, V., Emilianus, R., Maskell, D., and Mastroeni, P. (2003). Dynamics of bacterial growth and distribution within the liver during *Salmonella* infection. Cell Microbiol. *5*, 593–600.

Shibolet, O., and Podolsky, D.K. (2007). TLRs in the Gut. IV. Negative regulation of Toll-like receptors and intestinal homeostasis: addition by subtraction. Am. J. Physiol. Gastrointest. Liver Physiol. *292*, G1469–1473.

Siegemund, S., Schutze, N., Schulz, S., Wolk, K., Nasilowska, K., Straubinger, R.K., Sabat, R., and Alber, G. (2009). Differential IL-23 requirement for IL-22 and IL-17A production during innate immunity against *Salmonella enterica* serovar Enteritidis. Int. Immunol. *21*, 555–565.

Sinha, K., Mastroeni, P., Harrison, J., de Hormaeche, R.D., and Hormaeche, C.E. (1997). *Salmonella* Typhimurium aroA, htrA, and aroD htrA mutants cause progressive infections in athymic (nu/nu) BALB/c mice. Infect. Immun. *65*, 1566–1569.

Sinnott, C.R., and Teall, A.J. (1987). Persistent gallbladder carriage of *Salmonella* Typhi. Lancet *329*, 976.

Sirinavin, S., and Garner, P. (2000). Antibiotics for treating *Salmonella* gut infections. Cochrane Database Syst. Rev., CD001167.

Sleator, R.D., Wemekamp-Kamphuis, H.H., Gahan, C.G., Abee, T., and Hill, C. (2005). A PrfA-regulated bile exclusion system (BilE) is a novel virulence factor in Listeria monocytogenes. Mol. Microbiol. *55*, 1183–1195.

Sporri, R., and Reis e Sousa, C. (2005). Inflammatory mediators are insufficient for full dendritic cell activation and promote expansion of CD4+ T cell populations lacking helper function. Nat. Immunol. *6*, 163–170.

Srinivasan, A., Foley, J., and McSorley, S.J. (2004). Massive number of antigen-specific CD4 T cells during vaccination with live attenuated *Salmonella* causes interclonal competition. J. Immunol. *172*, 6884–6893.

Stecher, B., Barthel, M., Schlumberger, M.C., Haberli, L., Rabsch, W., Kremer, M., and Hardt, W.D. (2008). Motility allows *S.* Typhimurium to benefit from the mucosal defence. Cell Microbiol. *10*, 1166–1180.

Stecher, B., and Hardt, W.D. (2008). The role of microbiota in infectious disease. Trends Microbiol. *16*, 107–114.

Stecher, B., Macpherson, A.J., Hapfelmeier, S., Kremer, M., Stallmach, T., and Hardt, W.D. (2005). Comparison of *Salmonella enterica* serovar Typhimurium colitis in germfree mice and mice pretreated with streptomycin. Infect. Immun. 73, 3228–3241.

Stein, M.A., Leung, K.Y., Zwick, M., Garcia-del Portillo, F., and Finlay, B.B. (1996). Identification of a *Salmonella* virulence gene required for formation of filamentous structures containing lysosomal membrane glycoproteins within epithelial cells. Molecular microbiology *20*, 151–164.

Stoycheva, M.V., and Murdjeva, M.A. (2004). Cytokines in *Salmonella* infections. Folia Medica *46*, 5–10.

Strauch, U.G., Obermeier, F., Grunwald, N., Gurster, S., Dunger, N., Schultz, M., Griese, D.P., Mahler, M., Scholmerich, J., and Rath, H.C. (2005). Influence of intestinal bacteria on induction of regulatory T cells: lessons from a transfer model of colitis. Gut *54*, 1546–1552.

Subramanian, N., and Qadri, A. (2006). Lysophospholipid sensing triggers secretion of flagellin from pathogenic *Salmonella*. Nat. Immunol. *7*, 583–589.

Sundquist, M., and Wick, M.J. (2005). TNF-alpha-dependent and -independent maturation of dendritic cells and recruited CD11c(int)CD11b+ Cells during oral *Salmonella* infection. J. Immunol. *175*, 3287–3298.

Sundquist, M., and Wick, M.J. (2009). *Salmonella* induces death of CD8alpha(+) dendritic cells but not CD11c(int)CD11b(+) inflammatory cells *in vivo* via MyD88 and TNFR1. J. Leukoc. Biol. *85*, 225–234.

Tacket, C.O., and Levine, M.M. (2007). CVD 908, CVD 908-htrA, and CVD 909 live oral typhoid vaccines: a logical progression. Clin. Infect. Dis. *45 Suppl 1*, S20–23.

Tacket, C.O., Pasetti, M.F., Sztein, M.B., Livio, S., and Levine, M.M. (2004). Immune responses to an oral typhoid vaccine strain that is modified to constitutively express Vi capsular polysaccharide. J. Infect Dis. *190*, 565–570.

Tam, M.A., Rydstrom, A., Sundquist, M., and Wick, M.J. (2008a). Early cellular responses to *Salmonella* infection: dendritic cells, monocytes, and. more. Immunol. Rev. *225*, 140–162.

Tam, M.A., Sundquist, M., and Wick, M.J. (2008b). MyD88 and IFN-alphabeta differentially control maturation of bystander but not *Salmonella*-associated dendritic cells or CD11cintCD11b+ cells during infection. Cell Microbiol. *10*, 1517–1529.

Troy, A.E., Zaph, C., Du, Y., Taylor, B.C., Guild, K.J., Hunter, C.A., Saris, C.J., and Artis, D. (2009). IL-27 regulates homeostasis of the intestinal CD4+ effector T cell pool and limits intestinal inflammation in a murine model of colitis. J. Immunol. *183*, 2037–2044.

Truusalu, K., Mikelsaar, R.H., Naaber, P., Karki, T., Kullisaar, T., Zilmer, M., and Mikelsaar, M. (2008). Eradication of *Salmonella* Typhimurium infection in a murine model of typhoid fever with the combination of probiotic Lactobacillus fermentum ME-3 and ofloxacin. BMC Microbiol. *8*, 132.

Uchiya, K., Barbieri, M.A., Funato, K., Shah, A.H., Stahl, P.D., and Groisman, E.A. (1999). A *Salmonella* virulence protein that inhibits cellular trafficking. The EMBO journal *18*, 3924–3933.

Uematsu, S., and Akira, S. (2009). Immune responses of TLR5(+) lamina propria dendritic cells in enterobacterial infection. J. Gastroenterol. *44*, 803–811.

Vallon-Eberhard, A., Landsman, L., Yogev, N., Verrier, B., and Jung, S. (2006). Transepithelial pathogen uptake into the small intestinal lamina propria. J. Immunol. *176*, 2465–2469.

van de Vosse, E., Hoeve, M.A., and Ottenhoff, T.H. (2004). Human genetics of intracellular infectious diseases: molecular and cellular immunity against mycobacteria and *Salmonellae*. Lancet Infect Dis. *4*, 739–749.

Varol, C., Vallon-Eberhard, A., Elinav, E., Aycheck, T., Sapira, Y., Luche, H., Fehling, H.J., Hardt, W.D., Shakar, G., and Jung, S. (2009). Intestinal lamina propria dendritic cell subsets have different origin and functions. Immunity *31*, 502–512.

Vazquez-Torres, A., Jones-Carson, J., Baumler, A.J., Falkow, S., Valdivia, R., Brown, W., Le, M., Berggren, R., Parks, W.T., and Fang, F.C. (1999). Extraintestinal dissemination of *Salmonella* by CD18-expressing phagocytes. Nature *401*, 804–808.

Vazquez-Torres, A., Jones-Carson, J., Mastroeni, P., Ischiropoulos, H., and Fang, F.C. (2000a). Antimicrobial actions of the NADPH phagocyte oxidase and inducible nitric oxide synthase in experimental salmonellosis. I. Effects on microbial killing by activated peritoneal macrophages *in vitro*. J. Exp. Med. *192*, 227–236.

Vazquez-Torres, A., Xu, Y., Jones-Carson, J., Holden, D.W., Lucia, S.M., Dinauer, M.C., Mastroeni, P., and Fang, F.C. (2000b). *Salmonella* pathogenicity island 2-dependent evasion of the phagocyte NADPH oxidase. Science *287*, 1655–1658.

Voedisch, S., Koenecke, C., David, S., Herbrand, H., Forster, R., Rhen, M., and Pabst, O. (2009). Mesenteric lymph nodes confine dendritic cell-mediated dissemination of *Salmonella enterica* serovar Typhimurium and limit systemic disease in mice. Infect. Immun. *77*, 3170–3180.

Vollaard, A.M., Ali, S., van Asten, H.A., Widjaja, S., Visser, L.G., Surjadi, C., and van Dissel, J.T. (2004). Risk factors for typhoid and paratyphoid fever in Jakarta, Indonesia. JAMA *291*, 2607–2615.

Wahid, R., Salerno-Goncalves, R., Tacket, C.O., Levine, M.M., and Sztein, M.B. (2008). Generation of specific effector and memory T cells with gut- and secondary lymphoid tissue-homing potential by oral attenuated CVD 909 typhoid vaccine in humans. Mucosal Immunol. *1*, 389–398.

Wain, J., Pham, V.B., Ha, V., Nguyen, N.M., To, S.D., Walsh, A.L., Parry, C.M., Hasserjian, R.P., HoHo, V.A., Tran, T.H., *et al.* (2001). Quantitation of bacteria in bone marrow from patients with typhoid fever: relationship between counts and clinical features. J. Clin. Microbiol. *39*, 1571–1576.

Warren, J., Mastroeni, P., Dougan, G., Noursadeghi, M., Cohen, J., Walport, M.J., and Botto, M. (2002). Increased susceptibility of C1q-deficient mice to *Salmonella enterica* serovar Typhimurium infection. Infect. Immun. *70*, 551–557.

Whitaker, J.A., Franco-Paredes, C., del Rio, C., and Edupuganti, S. (2009). Rethinking typhoid fever vaccines: implications for travelers and people living in highly endemic areas. J. Travel Med. *16*, 46–52.

Wilson, G.S., and Miles, A.A. (1964). Topley and Wilson's principles of bacteriology and immunity, 5th ed. (London, Arnold).

Winfield, M.D., and Groisman, E.A. (2003). Role of nonhost environments in the lifestyles of *Salmonella* and Escherichia coli. Appl. Environ. Microbiol. *69*, 3687–3694.

Wood, M.W., Jones, M.A., Watson, P.R., Siber, A.M., McCormick, B.A., Hedges, S., Rosqvist, R., Wallis, T.S., and Galyov, E.E. (2000). The secreted effector protein of *Salmonella* dublin, SopA, is translocated into eukaryotic cells and influences the induction of enteritis. Cellular microbiology *2*, 293–303.

Yrlid, U., Svensson, M., Hakansson, A., Chambers, B.J., Ljunggren, H.G., and Wick, M.J. (2001). *In vivo* activation of dendritic cells and T cells during *Salmonella enterica* serovar Typhimurium infection. Infect. Immun. *69*, 5726–5735.

Zaph, C., Troy, A.E., Taylor, B.C., Berman-Booty, L.D., Guild, K.J., Du, Y., Yost, E.A., Gruber, A.D., May, M.J., Greten, F.R., *et al.* (2007). Epithelial-cell-intrinsic IKK-beta expression regulates intestinal immune homeostasis. Nature *446*, 552–556.

Zhang, S., Adams, L.G., Nunes, J., Khare, S., Tsolis, R.M., and Baumler, A.J. (2003a). Secreted effector proteins of *Salmonella enterica* serotype Typhimurium elicit host-specific chemokine profiles in animal models of typhoid fever and enterocolitis. Infect. Immun. *71*, 4795–4803.

Zhang, S., Kingsley, R.A., Santos, R.L., Andrews-Polymenis, H., Raffatellu, M., Figueiredo, J., Nunes, J., Tsolis, R.M., Adams, L.G., and Baumler, A.J. (2003b). Molecular pathogenesis of *Salmonella enterica* serotype Typhimurium-induced diarrhea. Infection and immunity *71*, 1–12.

Zhang, S., Santos, R.L., Tsolis, R.M., Stender, S., Hardt, W.D., Baumler, A.J., and Adams, L.G. (2002). The *Salmonella enterica* serotype Typhimurium effector proteins SipA, SopA, SopB, SopD, and SopE2 act in concert to induce diarrhea in calves. Infection and immunity *70*, 3843–3855.

Zhang, Y., Higashide, W.M., McCormick, B.A., Chen, J., and Zhou, D. (2006). The inflammation-associated *Salmonella* SopA is a HECT-like E3 ubiquitin ligase. Molecular microbiology *62*, 786–793.

Zhou, D., L.M Chen, L Hernandez, S.B Shears and J.E Galan (2001). A *Salmonella* inositol polyphosphatase acts in conjunction with other bacterial effectors to promote host cell actin cytoskeleton rearrangements and bacterial internalization. Molecular microbiology *39*, 248–259.

Salmonella Biofilms: From Food to Human Disease

13

Robert W. Crawford, Geoffrey Gonzalez-Escobedo and John S. Gunn*

Abstract

Bacterial biofilms are increasingly implicated as burdens to food and public safety. Over the past few decades, we have learned that this sessile environment provides diverse species of bacteria selective advantages in natural, medical, and industrial ecosystems, as well as resistance to commonly administered antibiotics and protection from host immune responses during chronic infection of humans and animals. *Salmonella* spp. are food-borne pathogens that remain a critical health concern in impoverished and industrialized nations. In the laboratory, *Salmonellae* have been shown to form biofilms on a variety of surfaces. These *Salmonella* spp. biofilms have been found to contaminate plant and animal food sources to cause human disease upon consumption, and/or to enhance *Salmonella* colonization of and persistence at sites of infection.

Introduction

Over the past few decades, bacterial biofilms have been increasingly implicated as burdens to food and public safety worldwide, and are broadly defined as heterogeneous communities of microorganisms that adhere to each other and to inert or live substrates (Costerton *et al.*, 1995; Davey and O'Toole, 2000; Watnick and Kolter, 2000). Biofilm formation occurs in sequential, highly regulated stages that begin with adherence of free-swimming, planktonic bacteria to a surface. Subsequent biofilm maturation is characterized by the production of a self-initiated extracellular matrix (ECM) composed of nucleic acid, proteins, or exopolysaccharides (EPS) that encase the community of microorganisms and provide structure and protection. Planktonic cells from this sessile, matrix-bound population are continuously shed, which can result in reattachment and fortification of the biofilm, or release of the organism into the environment (Fig. 13.1).

Biofilm formation was found to be clinically relevant in 1978 by Dr. R.J. Gibbons (Jefferson, 2004) and is currently thought to characterize 65% of chronic infections treated by physicians (Schaechter, 1987). Bacterial biofilm infections develop on tissues and medical devices. They are generally slow to produce overt symptoms, demonstrate heightened resistance to cellular and humoral immune host responses, antibiotics and household sanitizers, and persist for long periods of time with intermittent recurrence of symptoms until the community is physically removed (for a review see Costerton *et al.*, 1999).

Wild-type and mutant bacteria strains have been investigated to determine important ligands mediating biofilm development using static microtitre plates, dynamic flow cell apparati, tissue cultured cells, and human gallstones (Stoodley *et al.*, 2002; Stewart and Costerton, 2001; Prouty and Gunn, 2003; Ma *et al.*, 2009). This includes diverse species of commensal and pathogenic microorganisms such as *Streptococcus mutans* (Wen and Burne, 2002; Svensater *et al.*, 2001), *Staphylococcus aureus* (Vaudaux *et al.*, 1995; Gross *et al.*, 2001), non-typeable *Haemophilus influenzae* (NTHI), *Escherichia coli* (Danese *et al.*, 2000), *Vibrio cholerae* (Lejeune, 2003; Yildiz and Schoolnik, 1999), and *Pseudomonas aeruginosa* (Stewart and Franklin, 2008). Interestingly, bacterial gene

*Corresponding author

expression profiles differ between planktonic and biofilm phenotypes (Stoodley *et al.*, 2002; Jefferson, 2004), and these changes are probably regulated by external stimuli including nutrient availability, oxygen levels, pH, the presence or absence of antimicrobials, and the composition of binding substrates.

Bacterial biofilms resist the deleterious effects of antibiotics whereas planktonic cells dispersed from the biofilm microenvironment quickly regain susceptibility to those antibiotics, suggesting that biofilm-mediated resistance is not acquired through mutations or horizontal transfer of genes (Stewart and Costerton, 2001; Anwar *et al.*, 1989; Williams *et al.*, 1997). For example, a *Klebsiella pneumoniae* strain lacking β-lactamase production was killed in aqueous solution by ampicillin at concentrations as low as 2 μg/ml, but, when grown as a biofilm, was unaffected by 5000 μg/ml ampicillin (Anderl *et al.*, 2000). It was also demonstrated that established mechanisms of resistance to antimicrobials and bile salts including efflux pumps and the multiple antibiotic resistance (*mar*) locus of *E. coli* did not provide protection to bacteria in biofilms (Williams *et al.*, 1997; Maira-Litran *et al.*, 2000), suggesting that biofilm-mediated mechanisms of resistance may differ from those conventionally employed by planktonic cells.

Biofilm formation by *P. aeruginosa* has been hypothesized to mediate colonization and antibiotic resistance in the context of cystic fibrosis, a lung disease that features persistent airway infiltration by neutrophils. Interestingly, human neutrophils were shown *in vitro* to enhance *P. aeruginosa* biofilm development and even settle on the mature biofilm to become phagocytically engorged, degranulated, immobilized and rounded, contributing to a compromised host defence (Walker *et al.*, 2005; Jesaitis *et al.*, 2003). Similarly, *Staphylococcus epidermidis* biofilm formation mediates a protective response *in vitro* against complement deposition and neutrophil-dependent killing (Kristian *et al.*, 2008).

Mathematical approaches and empirical observations are constantly challenging and redefining the developmental model of microbial biofilm formation (Monds and O'Toole, 2009; van Loosdrecht *et al.*, 2002; Picioreanu *et al.*, 2007; Saravanan and Sreekrishnan, 2006). Regarding *Salmonella* spp., current work addresses the molecular details pertaining to *Salmonella* spp. biofilm formation including important structural and protective constituents of the extracellular matrix, adherence factors, quorum sensing molecules, outer membrane proteins, environmental signals that induce biofilm formation, and the role of enzymes controlling cellular c-di-GMP levels such as diguanylate cyclases containing GGDEF as well as phosphodiesterases containing EAL domains. A better understanding of the regulatory mechanisms that mediate *Salmonella* spp. biofilm formation and maintenance promises to elucidate important targets for the dissolution of these microbial communities and the risks they pose to human health. Here we review significant findings that have been uncovered to date.

Salmonella biofilm formation in the environment and human host

Plants

Salmonella contamination of produce is a common cause of human infection and has been linked to several recent outbreaks of salmonellosis (Heaton and Jones, 2008). Such outbreaks, if from commercial food sources, negatively affect consumer confidence, have a high economic burden, and point to a need for increased microbial surveillance. A key factor in produce contamination probably initiates from the ability of *Salmonella* spp. to form a biofilm on or within fruits and vegetables. It has been stated that 30–80% of the total bacterial population on a leaf surface will be biofilm in nature (Morris and Monier, 2003). This has been studied extensively in the laboratory environment and documented in nature.

The nature of the exopolysaccharide cellulose and curli (Tafi) pili in biofilms has been investigated on parsley. From water sources, bacteria were found to aggregate below the surface of the leaf with both curli and cellulose playing a significant role in the leaf penetration or survival within the leaves (Lapidot and Yaron, 2009). *Salmonella* biofilms were also observed on lettuce, and such biofilm formation increased acid tolerance and enhanced persistence during storage (Kroupitski

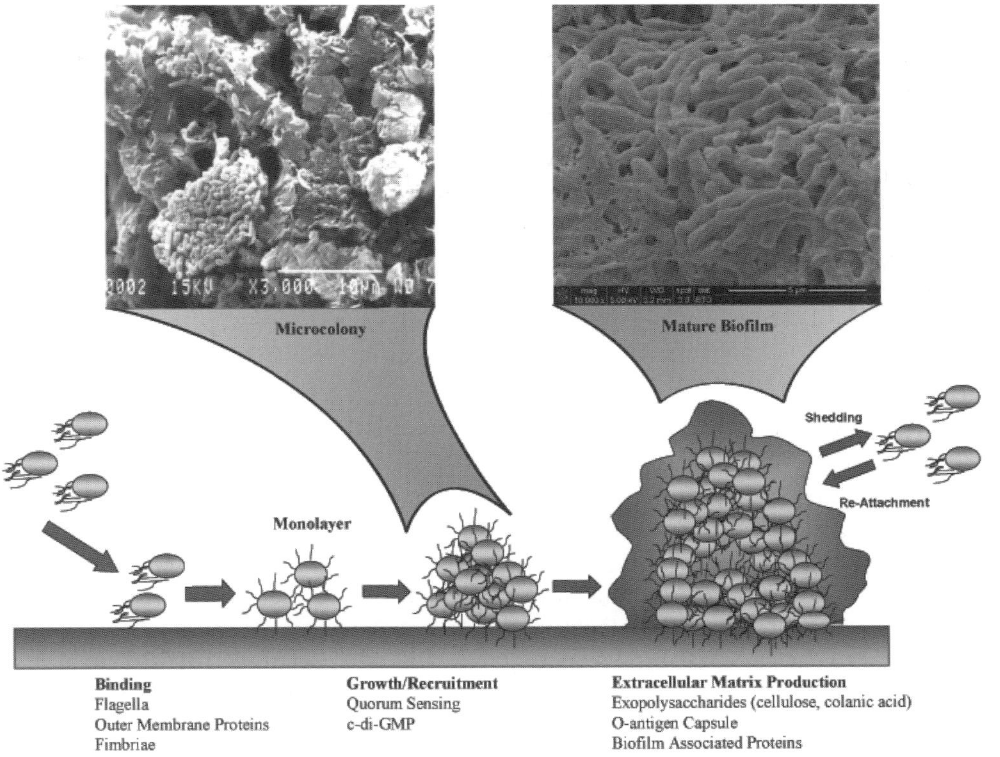

Figure 13.1 The sequential, highly regulated stages of biofilm development. Planktonic, free-swimming bacteria adhere to biotic or abiotic surfaces as mediated by environmental conditions and swim along the substrate to facilitate monolayer formation. The growing community is enlarged and fortified by replication and recruitment of new cells. Subsequent biofilm formation is characterized by the production of a self-initiated extracellular matrix composed of nucleic acid, proteins, and/or exopolysaccharides that encase the microorganisms and provide structure and protection. Planktonic cells from this sessile, matrix-bound population are continuously shed and can reattach or spread to distal sites.

et al., 2009). Numerous other examples exist and are described in a review by Heaton and Jones (Heaton and Jones, 2008). In addition to aiding bacterial adherence and persistence on plants and fruits, biofilms are protective against disinfection procedures including chemicals and UV irradiation. Thus, biofilm formation on contaminated produce is a major contributor to the incidence of *Salmonella* spp. foodborne infections globally.

Typhoid fever and chronic gallbladder carriage

During systemic typhoid infections, *Salmonella* Typhi can reach the gallbladder and establish an acute, active infection (cholecystitis) or persist in this organ relatively asymptomatically long after overt symptoms subside (Vaishnavi *et al.*, 2005; Sinnott and Teall, 1987). A study by Levine *et al.* in 1982 demonstrated that serovar Typhi could be recovered in stool samples for over 1 year from at least 2 to 5% of those suffering acute typhoid illnesses (Levine *et al.*, 1982), yet bacterial recovery in faeces is sporadic and as such a poor diagnostic tool. Shedding of *Salmonella* Typhi into the environment by chronic carriers accelerates the faecal–oral route of person-to-person transmission in endemic areas, suggesting the need for improved food and water sanitation, awareness of personal hygiene, and the development of novel therapeutic strategies.

An estimated 25% of asymptomatic carriers have no history of typhoid fever (Parry *et al.*,

2002), making the chronic carrier state difficult to confirm and an understudied aspect of global human health. Despite difficulties in diagnosis, it has been demonstrated that the asymptomatic carrier state is frequently associated with gallbladder abnormalities, particularly gallstones (Lai *et al.*, 1992). Gallstone formation (cholelithogenesis) is attributed to a combination of environmental and genetic causes including obesity, female gender, multiple pregnancies, nutrition, old age, and Crohn's disease, and can persist in patients for many years without symptoms (Shoheiber *et al.*, 1997; Afdhal, 1999). The primary constituent of gallbladder stones is cholesterol, whereas calcium bilirubinate predominates in bile duct stones in certain parts of the world (Hofmann, 2005).

In patients carrying both *Salmonella* Typhi and cholesterol gallstones in the gallbladder, clinically administered antibiotics are typically ineffective against dissolution of the persistent bacterial infection (Lai *et al.*, 1992). Removal of the gallbladder (cholecystectomy) remains the most effective treatment option for these individuals, but this procedure is cost prohibitive in typhoid endemic areas. Chronic typhoid carriers with gallstones also experience an 8.47-fold-higher risk for developing hepatobiliary diseases, including carcinoma of the gallbladder (Dutta *et al.*, 2000; Kumar *et al.*, 2006; Surette and Bassler, 1999). In 1994; Caygill *et al.* demonstrated that gallbladder cancers were associated with long-term typhoid carriage, but not with acute infections (Caygill *et al.*, 1994; Nath *et al.*, 1997).

Because gallbladder carriage of *Salmonella* Typhi is basically confined to this organ with intestinal shedding, difficult to treat with antibiotics, associated with the presence of gallstones, and cured most effectively by removal, it was hypothesized that biofilm formation on gallstones aided in the establishment and maintenance of this disease. To examine the role that gallstone biofilms may play in the establishment and maintenance of *Salmonella* Typhi chronic carriage, our laboratory utilized a number of methods to visualize biofilm formation on human gallstones *in vitro*, including confocal microscopy and scanning electron microscopy. A culture procedure was developed to try to mimic the emptying and re-filling of the gallbladder that occurs multiple times per day. Bacteria are grown in a test tube with gallstone material in 1 ml of media (LB or LB + 3% bile) on a roller drum at 37 °C. Every 24 hours, the liquid culture is removed and the gallstones washed. Using this assay, *Salmonellae* form a mature biofilm on the surface of human gallstones and excrete an exopolysaccharide that binds the bacteria to the surface and to each other. Efficient biofilm formation on gallstones is dependent upon the presence of bile in the growth medium. Bacteria such as *E. coli* or *Klebsiella* that can cause acute gallbladder infections do not form significant biofilms on gallstone surfaces.

Using this assay, numerous genetic factors have been investigated for a role in gallstone biofilm formation, including fimbriae, flagella, LPS, RpoS, PhoPQ, and exopolysaccharide (Prouty and Gunn, 2003). Many of these factors will be discussed in more detail in proceeding sections of this review. To eliminate dependence on human gallstone samples and to allow gallstone constituents to be analysed individually for their abilities to interact with *Salmonellae*, siliconized Eppendorff tubes were coated with cholesterol and shown to support *Salmonella* biofilm formation in a Tube Biofilm Assay (TBA). This assay has proved to mimic results seen with *Salmonella* and human gallstones, and will be a useful tool going forward to study gallstone biofilm formation (Crawford *et al.*, 2008).

The linkage of gallstone biofilms to *Salmonella* carriage has also been studied in animal models and in humans. Naturally resistant 129X1/SvJ mice fed a lithogenic diet developed cholesterol gallstones and, as a result, demonstrated enhanced colonization of gallbladder tissue and bile by Typhimurium (Crawford *et al.*, 2010). These mice also showed pronounced faecal shedding during persistent Typhimurium infection. Furthermore, cholesterol gallstones were shown to associate with significant amounts of *Salmonellae* that increased with time post infection. In a human study conducted in typhoid endemic Mexico City, individuals seeking cholecystectomy due to gallstones were analysed for asymptomatic Typhi carriage (Crawford *et al.*, 2010). Results indicated that 5% of patients with cholelithiasis carried Typhi and that large gallstones encompassing bacterial biofilms could be visualized in 3 of 4 of these carriers. Significant gallstone biofilm formation was not observed in 8 patients with *E. coli/Shigella* present in the gallbladder, suggesting that gallstone biofilms are a typhoid specific phenomenon in humans. These findings offer direct evidence that gallstone biofilms occur in humans and mice carrying *Salmonellae* asymptomatically in their gallbladders, which facilitates gallbladder colonization and shedding. By understanding the basis of carrier development, it may be possible to identify effective strategies to prevent/treat this chronic, apparently biofilm-mediated infection.

Surface appendages that may initiate *Salmonella* biofilm formation

Flagella

The molecular genetic mechanisms mediating adherence and microcolony formation during biofilm growth have recently begun to be elucidated for many microorganisms. Flagella, for example, promote surface binding during the initial stages of biofilm development (Pratt and Kolter, 1998; O'Toole and Kolter, 1998) as well as adhesion during biofilm-mediated colonization of tissue cultured cells and mucus by *P. aeruginosa* (Rudner *et al.*, 1992; Lillehoj *et al.*, 2002), *Campylobacter jejuni* (McSweegan and Walker, 1986), and *E. coli* (Giron *et al.*, 2002). Similarly, flagella were shown to mediate adherence of *Salmonella* Enteritidis to cultured epithelial cells (Dibb-Fuller *et al.*, 1999) and *Salmonella* Derby diarrhea isolates to HEp-2 cell monolayers (Budiarti *et al.*, 1991). However, expression of Typhimurium flagella has been shown to inhibit biofilm formation on polystyrene wells and glass surfaces (Teplitski *et al.*, 2006; Prouty and Gunn, 2003), but directly mediates biofilm development on cholesterol-coated surfaces (Crawford *et al.*, 2010), suggesting that the contribution of this surface appendage to biofilm formation varies with binding substrates.

Interestingly, flagellar-mediated phenotypes including swimming motility and chemotaxis differentially contributed to colonization and biofilm formation depending on environmental cues (O'Toole and Kolter, 1998; Pratt and Kolter, 1998; Prouty and Gunn, 2003; Huber *et al.*, 2002; Mey *et al.*, 2005). Bile, for example, is a caustic digestive secretion encountered by pathogenic Enterobacteriaceae during infection of the mammalian gastrointestinal tract. *In vitro*, bile increases motility and is a chemical attractant at physiological concentrations for *V. cholerae* (Butler and Camilli, 2005), acts as a repellent for *Helicobacter pylori* (Worku *et al.*, 2004), and enhances tumbling frequency while downregulating motility in *S. enterica* (Prouty *et al.*, 2004). Bacterial chemotaxis (tumbling motility) is controlled by clockwise flagellar rotation and has been shown to be dispensable for *E. coli* biofilms in a PVC microtitre dish assay (Pratt and Kolter, 1998), but important for *V. cholerae* biofilms on polystyrene wells (Mey *et al.*, 2005). Swarming motility is inversely correlated with Typhimurium biofilm formation on microtitre plates (Mireles *et al.*, 2001).

Fimbriae

Fimbriae have been shown to mediate adherence during biofilm initiation and cell–cell interactions during biofilm growth for a variety of microorganisms (Davey and O'Toole, 2000; O'Toole and Kolter, 1998). The Typhimurium LT2 genome encodes 13 putative fimbrial operons, some of which are not expressed *in vitro* (Nuccio *et al.*, 2007). Type 1 fimbriae are produced by numerous members of the family Enterobacteriaceae, mediate D-mannose-sensitive (MS) agglutination of guinea pig erythrocytes (Clegg *et al.*, 1987), and have been shown to be important for biofilm formation on HEp-2 tissue culture cells, the murine intestinal epithelium, and chicken intestinal epithelium (Boddicker *et al.*, 2002; Ledeboer and Jones, 2005), but not on human gallstones incubated with bile (Prouty *et al.*, 2002) or chick gut explant (Allen-Vercoe and Woodward, 1999). Over-expression of type 1 fimbriae inhibited *Salmonella* biofilm formation on cholesterol-coated surfaces but not to glass or plastic (Crawford *et al.*, 2010). These data again suggest that there is variable reliance upon certain adherence factors from one set of environmental conditions to the next.

Development of a self-initiated extracellular matrix

Cellulose, colonic acid and curli

The major exopolysaccharide (EPS) components in *Salmonella spp.* biofilms include cellulose, colanic acid, O-antigen capsule, Vi-antigen (*Salmonella* Typhi only), curli pili, biofilm-related proteins and possibly nucleic acids. Extracellular matrix (ECM) material within bacterial biofilms is thus heterogeneous and provides both rigidity to the biofilm and protection to bacteria embedded within established biofilms. It has previously been shown that *Salmonella* biofilm-associated ECM factors are not uniformly required for biofilm formation on all surfaces (Prouty and Gunn, 2003). Cellulose and colonic acids have been shown to play a role in biofilm formation on biotic as well as abiotic surfaces, as mutations in either *wcaM* (colanic acid) or *yhjN* (cellulose) have been shown to disrupt Typhimurium biofilm formation on HEp-2 cells in flow through chambers (Ledeboer and Jones, 2005). However, in the presence of bile, cellulose and colanic acid are not a factor in biofilm formation on human gallstones (Prouty and Gunn, 2003). Finally, atomic force microscopy of Typhimurium biofilms grown on agar plates or mica slides demonstrated that curli and cellulose were important for ECM production and suggested that expression of surface appendages may be co-regulated during biofilm formation (Jonas *et al.*, 2007).

CsgD (AgfD) is an important regulator of *Salmonella* biofilm formation (Brombacher *et al.*, 2003). CsgD coordinately regulates the biofilm-associated protein BapA and curli expression (directly) and cellulose production (indirectly). CsgD activates the GGDEF-containing protein AdrA, which is involved in the activation of cellulose expression (Zogaj *et al.*, 2001). Transcription of *csgD* is itself regulated by at least six different transcription factors (Gerstel and Romling, 2003). The details of all of the above regulatory functions have yet to be clearly elucidated.

On the basis of colony morphology, bacteria expressing cellulose and curli pili have a red, dry and rough (rdar) appearance on Congo red agar plates. This phenotype is dependent upon both curli and cellulose (Romling, 2005). Competition experiments were performed to determine if aggregation via the rdar morphotype was a virulence adaptation in *Salmonella* Typhimurium. It was shown that the aggregative rdar phenotype was outcompeted, and furthermore, that curli pili gene expression was very low during mouse infection (White *et al.*, 2008). Thus, it appears that the primary role of the rdar morphotype is to foster survival of *Salmonella* in the environment, probably via biofilms (Jain and Chen, 2007), thereby aiding in transmission.

O-antigen capsule

CsgD has also been demonstrated to control biosynthesis of the recently identified O-antigen (O-ag) capsule (Gibson *et al.*, 2006). This exopolysaccharide is membrane linked by a 2-hydroxymyristic acid residue (like a capsule), similar in structure to the LPS O-ag of serovar Enteritidis, conserved amongst *Salmonellae*, and encoded by two divergent operons (*yihU-yshA* and *yihV-yihW*) (Gibson *et al.*, 2006). Despite having similar repeating sugar units, the O-ag capsule and the LPS O-ag differ in size, charge, substitution patterns, and immunoreactivity (White *et al.*, 2003).

O-ag capsule expression was found to be necessary for Enteritidis biofilm formation leading to enhanced plant colonization (Barak *et al.*, 2007). Furthermore, production of the O-ag capsule in Typhimurium was shown to be transcriptionally induced during murine infection (White *et al.*, 2008), suggesting involvement during systemic infection and the development of typhoid fever. The O-antigen capsule was also shown to be a crucial determinant in gallstone and cholesterol biofilms, but the expression of this exopolysaccharide is not necessary for binding of *Salmonella* to glass or plastic. Growth in bile resulted in upregulation of the O-antigen capsule-encoding operon in a *csgD*-independent manner, and enhanced capsule expression in biofilms on gallstone/cholesterol-coated surfaces. Thus, the O-antigen capsule is bile induced and specifically required for binding to cholesterol-coated surfaces (Crawford *et al.*, 2008).

Biofilm-associated proteins

In addition to polysaccharides, biofilm-associated proteins (BAPs) contribute to ECM production in *S. aureus* and *E. faecalis* biofilms (Latasa *et al.*, 2006). *Salmonellae* express a large cell surface protein encoded by *bapA* that was shown to be important for Enteritidis biofilm formation but not for Typhimurium (Latasa *et al.*, 2005; Jonas *et al.*, 2007). BapA is coordinately regulated by CsgD at 30°C with curli pili, cellulose and O-antigen capsule, and it is secreted by proteins produced from the genetically linked BapBCD Type I secretion system. Perhaps not unexpectedly, BapA had only modest effects on virulence in the mouse model (Latasa *et al.*, 2005).

Cell-to-cell communication and *Salmonella* biofilms

Quorum sensing

Quorum sensing (QS) is a mechanism of cell–cell communication that enables bacteria to regulate population behaviour by monitoring the local environment with diffusible signalling molecules known as autoinducers (Bassler, 2002). In Gram negative cell populations, this autoinduced positive feedback system is mediated by acyl-homoserine lactones (AHLs) and/or autoinducers 2 and 3 (AI-2, AI-3) (Irie and Parsek, 2008). AHLs provide bacteria with a measuring stick for detection of its own species and, as such, have been shown to mediate the formation of differentiated multicellular microcolonies in *P. aeruginosa* biofilms on both live and inert surfaces (reviewed in Kjelleberg and Molin, 2002). *Salmonellae* do not synthesize AHLs, but can sense those produced by other species using SdiA, a LuxR homologue that detects mixed microbial communities and has been implicated in *E. coli* biofilm formation (Michael *et al.*, 2001; Lee *et al.*, 2009). To date, no role for SdiA in mixed biofilm formation involving *Salmonella* has been elucidated.

In *Salmonella* Typhimurium, AI-2 molecules are synthesized by the highly conserved LuxS enzyme and are derived from the same compounds as those for *Vibrio harveyi*, a model QS microorganism (Xavier and Bassler, 2003). The *lsr* operon of *Salmonella* encodes a receptor, LsrB, that, when activated, transports AI-2 molecules from the extracellular space into the cell (Taga *et al.*, 2003). This operon is thought to be upregulated in the intestinal microenvironment (Sonck *et al.*, 2009). It has been reported that expression of LuxS and AI-2 is necessary for *Salmonella* virulence (Choi *et al.*, 2007) and mutations in *luxS* inhibited *Salmonella* biofilm formation on polystyrene and gallstones (Prouty *et al.*, 2002; De Keersmaecker *et al.*, 2005), yet it remains unclear whether these phenotypes are directly mediated by QS. New strategies that more accurately mimic the natural microbe environment may be necessary to provide significant additional insights as to how QS pathways mediate biofilm development and maintenance.

Cyclic diguanylic acid

The novel second messenger cyclic-3′5′-diguanylic acid (c-di-GMP) has been demonstrated to mediate the transition from motility to sessility during biofilm formation for a variety of microorganisms including *Salmonella*, *Vibrio* and *Pseudomonas* (Romling and Amikam, 2006; Tischler and Camilli, 2004; Ueda and Wood, 2009). This bacterial signal transduction molecule is synthesized and degraded by two protein domain superfamilies, diguanylate cyclases GGDEF and glutamate-alanine-leucine (EAL), although the molecular details and interplay of these complicated transcriptional cascades remain unclear (Simm *et al.*, 2004).

Elevated concentrations of c-di-GMP have been shown to activate CsgD, a biofilm-associated protein that, as mentioned above, coordinately regulates expression of cellulose, curli fimbriae, BapA, and the O-ag capsule in *Salmonella* Typhimurium (Zogaj *et al.*, 2001; Gibson *et al.*, 2006). Interestingly, cellulose biosynthesis requires c-di-GMP availability (Solano *et al.*, 2009) and can occur independently of CsgD (Zogaj *et al.*, 2001), suggesting that CsgD activity is not a critical component of cellulose-mediated biofilm formation in every experimental condition. *Salmonella* Typhimurium also regulates GGDEF-EAL domain proteins and the transition from motility to sessility via carbon storage regulator A (CsrA) (Teplitski *et al.*, 2006; Jonas *et al.*, 2007), suggesting a potential role for this protein during *Salmonella* biofilm development. Recent work suggests that high c-di-GMP levels negatively regulate virulence properties of *Salmonella* Typhimurium during epithelial cell invasion at the mucosal lining, suggesting an important role for c-di-GMP during intestinal colonization (Lamprokostopoulou *et al.*, 2009), probably independent of biofilm formation.

Perspectives

A better understanding of the regulatory mechanisms that mediate *Salmonella* spp. biofilm formation and maintenance promises to elucidate important targets for the dissolution of these microbial communities and the risks they pose to human health. Although much progress has been made in this characterization, complicated issues remain. The factors mediating attachment and biofilm development, including membrane-bound capsules and appendages and secreted enzymes, are often uniquely required depending on environmental conditions such as osmolarity, temperature, and composition of the binding substrate in question, and can vary greatly from one experimental model to the next. Despite this seeming need for specificity, bacteria also rely on functional redundancy and connectivity of pathways during biofilm formation. Recent investigations suggest that quorum sensing, polysaccharide production, and c-di-GMP expression are inseparable components of biofilm formation in *P. aeruginosa* (Ueda and Wood, 2009), and that motility, anti-microbial resistance mechanisms and c-di-GMP activity are co-regulated to achieve biofilms in *V. cholerae* (Pratt *et al.*, 2009). Although *Salmonella* biofilm research is still a relatively recent endeavour, data presented in this review suggests a familiar interdependence and coordination of biofilm-related behaviours. It is hoped that better diagnostic tools and continued research efforts will reveal the genetic and phenotypic profiles of these microbial communities that continue to burden mankind.

References

Afdhal, N.H. (1999). Crohn's and stones. Am. J. Gastroenterol. *94*, 1130–1132.

Allen-Vercoe, E., and Woodward, M.J. (1999). The role of flagella, but not fimbriae, in the adherence of *Salmonella enterica* serotype Enteritidis to chick gut explant. J. Med. Microbiol. *48*, 771–780.

Anderl, J.N., Franklin, M.J., and Stewart, P.S. (2000). Role of antibiotic penetration limitation in Klebsiella pneumoniae biofilm resistance to ampicillin and ciprofloxacin. Antimicrob. Agents Chemother. *44*, 1818–1824.

Anwar, H., van Biesen, T., Dasgupta, M., Lam, K., and Costerton, J.W. (1989). Interaction of biofilm bacteria with antibiotics in a novel *in vitro* chemostat system. Antimicrob. Agents Chemother. *33*, 1824–1826.

Barak, J.D., Jahn, C.E., Gibson, D.L., and Charkowski, A.O. (2007). The role of cellulose and O-antigen capsule in the colonization of plants by *Salmonella enterica*. Mol. Plant Microbe Interact *20*, 1083–1091.

Bassler, B.L. (2002). Small talk. Cell-to-cell communication in bacteria. Cell *109*, 421–424.

Boddicker, J.D., Ledeboer, N.A., Jagnow, J., Jones, B.D., and Clegg, S. (2002). Differential binding to and biofilm formation on, HEp-2 cells by *Salmonella enterica* serovar Typhimurium is dependent upon allelic variation in the fimH gene of the fim gene cluster. Mol. Microbiol. *45*, 1255–1265.

Brombacher, E., Dorel, C., Zehnder, A.J., and Landini, P. (2003). The curli biosynthesis regulator CsgD coordinates the expression of both positive and negative determinants for biofilm formation in Escherichia coli. Microbiology *149*, 2847–2857.

Budiarti, S., Hirai, Y., Minami, J., Katayama, S., Shimizu, T., and Okabe, A. (1991). Adherence to HEp-2 cells and replication in macrophages of *Salmonella* derby of human origin. Microbiol. Immunol. *35*, 111–123.

Butler, S.M., and Camilli, A. (2005). Going against the grain: chemotaxis and infection in Vibrio cholerae. Nat. Rev. Microbiol. *3*, 611–620.

Caygill, C.P., Hill, M.J., Braddick, M., and Sharp, J.C. (1994). Cancer mortality in chronic typhoid and paratyphoid carriers. Lancet *343*, 83–84.

Choi, J., Shin, D., and Ryu, S. (2007). Implication of quorum sensing in *Salmonella enterica* serovar Typhimurium virulence: the luxS gene is necessary for expression of genes in pathogenicity island 1. Infect. Immun. *75*, 4885–4890.

Clegg, S., Purcell, B.K., and Pruckler, J. (1987). Characterization of genes encoding type 1 fimbriae of Klebsiella pneumoniae, *Salmonella* Typhimurium, and Serratia marcescens. Infect. Immun. *55*, 281–287.

Costerton, J.W., Lewandowski, Z., Caldwell, D.E., Korber, D.R., and Lappin-Scott, H.M. (1995). Microbial biofilms. Annu. Rev. Microbiol. *49*, 711–745.

Costerton, J.W., Stewart, P.S., and Greenberg, E.P. (1999). Bacterial biofilms: a common cause of persistent infections. Science *284*, 1318–1322.

Crawford, R.W., Gibson, D.L., Kay, W.W., and Gunn, J.S. (2008). Identification of a bile-induced exopolysaccharide required for *Salmonella* biofilm formation on gallstone surfaces. Infect. Immun. *76*, 5341–5349.

Crawford, R.W., Rosales-Reyes, R., Ramírez-Aguilar, M., Chapa-Azuela, O., Alpuche-Aranda, C., and Gunn, J.S. (2010). Gallstones play a significant role in *Salmonella* spp. gallbladder colonization and carriage. Proc. Natl. Acad. Sci. U.S.A. *107*, 4353–4358.

Crawford, R.W., Reeve, K.E., and Gunn, J.S. (2010). Flagellated but not hyperfimbrinted *Salmonella*

Enterica serovar Typhimurium attaches to and forms biofilm: cholesterol-coated surfaces. J. Bacteriol. *192*, 2981–2929.

Danese, P.N., Pratt, L.A., and Kolter, R. (2000). Exopolysaccharide production is required for development of Escherichia coli K-12 biofilm architecture. J. Bacteriol. *182*, 3593–3596.

Davey, M.E., and O'Toole, G.A. (2000). Microbial biofilms: from ecology to molecular genetics. Microbiol. Mol. Biol. Rev. *64*, 847–867.

De Keersmaecker, S.C., Varszegi, C., van Boxel, N., Habel, L.W., Metzger, K., Daniels, R., Marchal, K., De Vos, D., and Vanderleyden, J. (2005). Chemical synthesis of (S)-4,5-dihydroxy-2,3-pentanedione, a bacterial signal molecule precursor, and validation of its activity in *Salmonella* Typhimurium. J. Biol. Chem. *280*, 19563–19568.

Dibb-Fuller, M.P., Allen-Vercoe, E., Thorns, C.J., and Woodward, M.J. (1999). Fimbriae- and flagella-mediated association with and invasion of cultured epithelial cells by *Salmonella* Enteritidis. Microbiology *145*, 1023–1031.

Dutta, U., Garg, P.K., Kumar, R., and Tandon, R.K. (2000). Typhoid carriers among patients with gallstones are at increased risk for carcinoma of the gallbladder. Am. J. Gastroenterol. *95*, 784–787.

Gerstel, U., and Romling, U. (2003). The csgD promoter, a control unit for biofilm formation in *Salmonella* Typhimurium. Res Microbiol. *154*, 659–667.

Gibson, D.L., White, A.P., Snyder, S.D., Martin, S., Heiss, C., Azadi, P., Surette, M., and Kay, W.W. (2006). *Salmonella* produces an O-antigen capsule regulated by AgfD and important for environmental persistence. J. Bacteriol. *188*, 7722–7730.

Giron, J.A., Torres, A.G., Freer, E., and Kaper, J.B. (2002). The flagella of enteropathogenic Escherichia coli mediate adherence to epithelial cells. Mol. Microbiol. *44*, 361–379.

Gross, M., Cramton, S.E., Gotz, F., and Peschel, A. (2001). Key role of teichoic acid net charge in Staphylococcus aureus colonization of artificial surfaces. Infect. Immun. *69*, 3423–3426.

Heaton, J.C., and Jones, K. (2008). Microbial contamination of fruit and vegetables and the behaviour of enteropathogens in the phyllosphere: a review. J. Appl. Microbiol. *104*, 613–626.

Hofmann, A.F. (2005). Helicobacter and cholesterol gallstones: do findings in the mouse apply to man? Gastroenterology *128*, 1126–1129.

Huber, B., Riedel, K., Kothe, M., Givskov, M., Molin, S., and Eberl, L. (2002). Genetic analysis of functions involved in the late stages of biofilm development in *Burkholderia cepacia* H111. Mol. Microbiol. *46*, 411–426.

Irie, Y., and Parsek, M.R. (2008). Quorum sensing and microbial biofilms. Curr. Top Microbiol. Immunol. *322*, 67–84.

Jain, S., and Chen, J. (2007). Attachment and biofilm formation by various serotypes of *Salmonella* as influenced by cellulose production and thin aggregative fimbriae biosynthesis. J. Food Prot. *70*, 2473–2479.

Jefferson, K.K. (2004). What drives bacteria to produce a biofilm? FEMS Microbiol. Lett *236*, 163–173.

Jesaitis, A.J., Franklin, M.J., Berglund, D., Sasaki, M., Lord, C.I., Bleazard, J.B., Duffy, J.E., Beyenal, H., and Lewandowski, Z. (2003). Compromised host defence on Pseudomonas aeruginosa biofilms: characterization of neutrophil and biofilm interactions. J. Immunol. *171*, 4329–4339.

Jonas, K., Tomenius, H., Kader, A., Normark, S., Romling, U., Belova, L.M., and Melefors, O. (2007). Roles of curli, cellulose and BapA in *Salmonella* biofilm morphology studied by atomic force microscopy. BMC Microbiol. *7*, 70.

Kjelleberg, S., and Molin, S. (2002). Is there a role for quorum sensing signals in bacterial biofilms? Curr. Opin Microbiol. *5*, 254–258.

Kristian, S.A., Birkenstock, T.A., Sauder, U., Mack, D., Gotz, F., and Landmann, R. (2008). Biofilm formation induces C3a release and protects Staphylococcus epidermidis from IgG and complement deposition and from neutrophil-dependent killing. J. Infect Dis. *197*, 1028–1035.

Kroupitski, Y., Pinto, R., Brandl, M.T., Belausov, E., and Sela, S. (2009). Interactions of *Salmonella enterica* with lettuce leaves. J. Appl. Microbiol. *106*, 1876–1885.

Kumar, S., Kumar, S., and Kumar, S. (2006). Infection as a risk factor for gallbladder cancer. J. Surg. Oncol. *93*, 633–639.

Lai, C.W., Chan, R.C., Cheng, A.F., Sung, J.Y., and Leung, J.W. (1992). Common bile duct stones: a cause of chronic salmonellosis. Am. J. Gastroenterol. *87*, 1198–1199.

Lamprokostopoulou, A., Monteiro, C., Rhen, M., and Romling, U. (2010). Cyclic di-GMP signalling controls virulence properties of *Salmonella enterica* serovar Typhimurium at the mucosal lining. Environ. Microbiol. *12*, 40–53.

Lapidot, A., and Yaron, S. (2009). Transfer of *Salmonella enterica* serovar Typhimurium from contaminated irrigation water to parsley is dependent on curli and cellulose, the biofilm matrix components. J. Food Prot. *72*, 618–623.

Latasa, C., Roux, A., Toledo-Arana, A., Ghigo, J.M., Gamazo, C., Penades, J.R., and Lasa, I. (2005). BapA, a large secreted protein required for biofilm formation and host colonization of *Salmonella enterica* serovar Enteritidis. Mol. Microbiol. *58*, 1322–1339.

Latasa, C., Solano, C., Penades, J.R., and Lasa, I. (2006). Biofilm-associated proteins. C R Biol. *329*, 849–857.

Ledeboer, N.A., and Jones, B.D. (2005). Exopolysaccharide sugars contribute to biofilm formation by *Salmonella enterica* serovar Typhimurium on HEp-2 cells and chicken intestinal epithelium. J. Bacteriol. *187*, 3214–3226.

Lee, J., Maeda, T., Hong, S.H., and Wood, T.K. (2009). Reconfiguring the quorum-sensing regulator SdiA of Escherichia coli to control biofilm formation via indole and N-acylhomoserine lactones. Appl. Environ. Microbiol. *75*, 1703–1716.

Lejeune, P. (2003). Contamination of abiotic surfaces: what a colonizing bacterium sees and how to blur it. Trends Microbiol. *11*, 179–184.

Levine, M.M., Black, R.E., and Lanata, C. (1982). Precise estimation of the numbers of chronic carriers of *Salmonella* Typhi in Santiago, Chile, an endemic area. J. Infect Dis. *146*, 724–726.

Lillehoj, E.P., Kim, B.T., and Kim, K.C. (2002). Identification of Pseudomonas aeruginosa flagellin as an adhesin for Muc1 mucin. Am. J. Physiol. Lung Cell Mol. Physiol. *282*, L751–L756.

Ma, L., Conover, M., Lu, H., Parsek, M.R., Bayles, K., and Wozniak, D.J. (2009). Assembly and Development of the Pseudomonas aeruginosa Biofilm Matrix. PLoS Pathog *5*, e1000354.

Maira-Litran, T., Allison, D.G., and Gilbert, P. (2000). Expression of the multiple antibiotic resistance operon (mar) during growth of Escherichia coli as a biofilm. J. Appl. Microbiol. *88*, 243–247.

McSweegan, E., and Walker, R.I. (1986). Identification and characterization of two Campylobacter jejuni adhesins for cellular and mucous substrates. Infect. Immun. *53*, 141–148.

Mey, A.R., Craig, S.A., and Payne, S.M. (2005). Characterization of Vibrio cholerae RyhB: the RyhB regulon and role of ryhB in biofilm formation. Infect. Immun. *73*, 5706–5719.

Michael, B., Smith, J.N., Swift, S., Heffron, F., and Ahmer, B.M. (2001). SdiA of *Salmonella enterica* is a LuxR homolog that detects mixed microbial communities. J. Bacteriol. *183*, 5733–5742.

Mireles, J.R. 2nd, Toguchi, A., and Harshey, R.M. (2001). *Salmonella enterica* serovar Typhimurium swarming mutants with altered biofilm-forming abilities: surfactin inhibits biofilm formation. J. Bacteriol. *183*, 5848–5854.

Monds, R.D., and O'Toole, G.A. (2009). The developmental model of microbial biofilms: ten years of a paradigm up for review. Trends Microbiol. *17*, 73–87.

Morris, C.E., and Monier, J.M. (2003). The ecological significance of biofilm formation by plant-associated bacteria. Annu Rev. Phytopathol. *41*, 429–453.

Nath, G., Singh, H., and Shukla, V.K. (1997). Chronic typhoid carriage and carcinoma of the gallbladder. Eur. J. Cancer Prev. *6*, 557–559.

Nuccio, S.P., Chessa, D., Weening, E.H., Raffatellu, M., Clegg, S., and Baumler, A.J. (2007). SIMPLE approach for isolating mutants expressing fimbriae. Appl. Environ. Microbiol. *73*, 4455–4462.

O'Toole, G.A., and Kolter, R. (1998). Flagellar and twitching motility are necessary for Pseudomonas aeruginosa biofilm development. Mol. Microbiol. *30*, 295–304.

Parry, C.M., Hien, T.T., Dougan, G., White, N.J., and Farrar, J.J. (2002). Typhoid fever. N. Engl. J. Med. *347*, 1770–1782.

Picioreanu, C., Kreft, J.U., Klausen, M., Haagensen, J.A., Tolker-Nielsen, T., and Molin, S. (2007). Microbial motility involvement in biofilm structure formation–a 3D modelling study. Water Sci. Technol. *55*, 337–343.

Pratt, J.T., McDonough, E., and Camilli, A. (2009). PhoB regulates motility, biofilm and c-di-GMP in Vibrio cholerae. J. Bacteriol. *191*, 6632–6642.

Pratt, L.A., and Kolter, R. (1998). Genetic analysis of Escherichia coli biofilm formation: roles of flagella, motility, chemotaxis and type I pili. Mol. Microbiol. *30*, 285–293.

Prouty, A.M., Brodsky, I.E., Manos, J., Belas, R., Falkow, S., and Gunn, J.S. (2004). Transcriptional regulation of *Salmonella enterica* serovar Typhimurium genes by bile. FEMS Immunol. Med. Microbiol. *41*, 177–185.

Prouty, A.M., and Gunn, J.S. (2003). Comparative analysis of *Salmonella enterica* serovar Typhimurium biofilm formation on gallstones and on glass. Infect. Immun. *71*, 7154–7158.

Prouty, A.M., Schwesinger, W.H., and Gunn, J.S. (2002). Biofilm formation and interaction with the surfaces of gallstones by *Salmonella* spp. Infect. Immun. *70*, 2640–2649.

Romling, U. (2005). Characterization of the rdar morphotype, a multicellular behaviour in Enterobacteriaceae. Cell Mol. Life Sci *62*, 1234–1246.

Romling, U., and Amikam, D. (2006). Cyclic di-GMP as a second messenger. Curr. Opin Microbiol. *9*, 218–228.

Rudner, X.L., Hazlett, L.D., and Berk, R.S. (1992). Systemic and topical protection studies using Pseudomonas aeruginosa flagella in an ocular model of infection. Curr. Eye Res *11*, 727–738.

Saravanan, V., and Sreekrishnan, T.R. (2006). Modelling anaerobic biofilm reactors–a review. J. Environ. Manage *81*, 1–18.

Schaechter, M.a.N.F.C. (1987). Introduction. In *Escherichia coli* and *Salmonella* Typhimurium Cellular and Molecular Biology. Anonymous. (Washington, D.C., American Society for Microbiology), pp. 1–2.

Shoheiber, O., Biskupiak, J.E., and Nash, D.B. (1997). Estimation of the cost savings resulting from the use of ursodiol for the prevention of gallstones in obese patients undergoing rapid weight reduction. Int. J. Obes. Relat. Metab. Disord. *21*, 1038–1045.

Simm, R., Morr, M., Kader, A., Nimtz, M., and Romling, U. (2004). GGDEF and EAL domains inversely regulate cyclic di-GMP levels and transition from sessility to motility. Mol. Microbiol. *53*, 1123–1134.

Sinnott, C.R., and Teall, A.J. (1987). Persistent gallbladder carriage of *Salmonella* Typhi. Lancet *1*, 976.

Solano, C., Garcia, B., Latasa, C., Toledo-Arana, A., Zorraquino, V., Valle, J., Casals, J., Pedroso, E., and Lasa, I. (2009). Genetic reductionist approach for dissecting individual roles of GGDEF proteins within the c-di-GMP signaling network in *Salmonella*. Proc. Natl. Acad. Sci. U.S.A. *106*, 7997–8002.

Sonck, K.A., Kint, G., Schoofs, G., Vander Wauven, C., Vanderleyden, J., and De Keersmaecker, S.C. (2009). The proteome of *Salmonella* Typhimurium grown under *in vivo*-mimicking conditions. Proteomics *9*, 565–579.

Stewart, P.S., and Costerton, J.W. (2001). Antibiotic resistance of bacteria in biofilms. Lancet *358*, 135–138.

Stewart, P.S., and Franklin, M.J. (2008). Physiological heterogeneity in biofilms. Nat. Rev. Microbiol. *6*, 199–210.

Stoodley, P., Sauer, K., Davies, D.G., and Costerton, J.W. (2002). Biofilms as complex differentiated communities. Annu. Rev. Microbiol. *56*, 187–209.

Surette, M.G., and Bassler, B.L. (1999). Regulation of autoinducer production in *Salmonella* Typhimurium. Mol. Microbiol. *31*, 585–595.

Svensater, G., Welin, J., Wilkins, J.C., Beighton, D., and Hamilton, I.R. (2001). Protein expression by planktonic and biofilm cells of Streptococcus mutans. FEMS Microbiol. Lett *205*, 139–146.

Taga, M.E., Miller, S.T., and Bassler, B.L. (2003). Lsr-mediated transport and processing of AI-2 in *Salmonella* Typhimurium. Mol. Microbiol. *50*, 1411–1427.

Teplitski, M., Al-Agely, A., and Ahmer, B.M. (2006). Contribution of the SirA regulon to biofilm formation in *Salmonella enterica* serovar Typhimurium. Microbiology *152*, 3411–3424.

Tischler, A.D., and Camilli, A. (2004). Cyclic diguanylate (c-di-GMP) regulates Vibrio cholerae biofilm formation. Mol. Microbiol. *53*, 857–869.

Ueda, A., and Wood, T.K. (2009). Connecting quorum sensing, c-di-GMP, pel polysaccharide, and biofilm formation in Pseudomonas aeruginosa through tyrosine phosphatase TpbA (PA3885). PLoS Pathog *5*, e1000483.

Vaishnavi, C., Singh, S., Kochhar, R., Bhasin, D., Singh, G., and Singh, K. (2005). Prevalence of *Salmonella enterica* serovar Typhi in bile and stool of patients with biliary diseases and those requiring biliary drainage for other purposes. Jpn J. Infect Dis. *58*, 363–365.

van Loosdrecht, M.C., Heijnen, J.J., Eberl, H., Kreft, J., and Picioreanu, C. (2002). Mathematical modelling of biofilm structures. Antonie Van Leeuwenhoek *81*, 245–256.

Vaudaux, P.E., Francois, P., Proctor, R.A., McDevitt, D., Foster, T.J., Albrecht, R.M., Lew, D.P., Wabers, H., and Cooper, S.L. (1995). Use of adhesion-defective mutants of Staphylococcus aureus to define the role of specific plasma proteins in promoting bacterial adhesion to canine arteriovenous shunts. Infect. Immun. *63*, 585–590.

Walker, T.S., Tomlin, K.L., Worthen, G.S., Poch, K.R., Lieber, J.G., Saavedra, M.T., Fessler, M.B., Malcolm, K.C., Vasil, M.L., and Nick, J.A. (2005). Enhanced Pseudomonas aeruginosa biofilm development mediated by human neutrophils. Infect. Immun. *73*, 3693–3701.

Watnick, P., and Kolter, R. (2000). Biofilm, city of microbes. J. Bacteriol. *182*, 2675–2679.

Wen, Z.T., and Burne, R.A. (2002). Functional genomics approach to identifying genes required for biofilm development by Streptococcus mutans. Appl. Environ. Microbiol. *68*, 1196–1203.

White, A.P., Gibson, D.L., Collinson, S.K., Banser, P.A., and Kay, W.W. (2003). Extracellular polysaccharides associated with thin aggregative fimbriae of *Salmonella enterica* serovar Enteritidis. J. Bacteriol. *185*, 5398–5407.

White, A.P., Gibson, D.L., Grassl, G.A., Kay, W.W., Finlay, B.B., Vallance, B.A., and Surette, M.G. (2008). Aggregation via the red, dry, and rough morphotype is not a virulence adaptation in *Salmonella enterica* serovar Typhimurium. Infect. Immun. *76*, 1048–1058.

Williams, I., Venables, W.A., Lloyd, D., Paul, F., and Critchley, I. (1997). The effects of adherence to silicone surfaces on antibiotic susceptibility in Staphylococcus aureus. Microbiology *143*, 2407–2413.

Worku, M.L., Karim, Q.N., Spencer, J., and Sidebotham, R.L. (2004). Chemotactic response of Helicobacter pylori to human plasma and bile. J. Med. Microbiol. *53*, 807–811.

Xavier, K.B., and Bassler, B.L. (2003). LuxS quorum sensing: more than just a numbers game. Curr. Opin Microbiol. *6*, 191–197.

Yildiz, F.H., and Schoolnik, G.K. (1999). Vibrio cholerae O1 El Tor: identification of a gene cluster required for the rugose colony type, exopolysaccharide production, chlorine resistance, and biofilm formation. Proc. Natl. Acad. Sci. U.S.A. *96*, 4028–4033.

Zogaj, X., Nimtz, M., Rohde, M., Bokranz, W., and Romling, U. (2001). The multicellular morphotypes of *Salmonella* Typhimurium and Escherichia coli produce cellulose as the second component of the extracellular matrix. Mol. Microbiol. *39*, 1452–1463.

Salmonella as the Paradigm for Bacterial Therapy of Cancer: A Progress Report

14

Robert M. Hoffman*

Abstract

For over 300 years it has been observed that cancer patients who became infected with bacteria sometimes experienced spontaneous remission of their cancer. Recently, there have been attempts to develop cancer treatments by using tumour-targeting bacteria. Anaerobic microorganisms, such as *Clostridium,* that preferentially grow in necrotic tumour areas have mostly been used. However, the resulting tumour killing was, at best, limited. *Salmonella* was originally developed as an antitumour agent by attenuating the bacteria with multiple mutations, including auxotrophs. These multiple auxotrophs appeared to direct the bacteria to the necrotic areas of tumours where more nutrients are available. We have developed a more effective bacterial cancer therapy strategy by targeting viable tumour tissue with *Salmonella enterica* serovar Typhimurium containing only two auxotrophic mutations. These auxotrophs grow in viable as well as necrotic areas of tumours. However, the auxotrophy severely restricts growth of these bacteria in normal tissue, making this a safe treatment. The *Salmonella* Typhimurium A1-R mutant, which is auxotrophic for leucine and arginine and had been selected for high antitumour virulence, was effective as monotherapy against human prostate and breast tumours that had been orthotopically implanted in nude mice. The approach described here, where bacterial monotherapy effectively treats primary and metastatic tumours, is a significant improvement over previous bacterial tumour-therapy strategies that require combination with toxic chemotherapy. Exploitation of the tumour-killing capability of *Salmonella* has great potential for a new paradigm of cancer therapy.

Introduction

It has been observed over the past 300 years that cancer patients who became infected with cancer sometimes experienced spontaneous regression of their cancer. Coley observed more than a century ago that some cancer patients were cured of their tumours following post-operative bacterial infection (Coley, 1906). Based on these observations, Coley developed extracts of the bacteria *Streptococcus pyrogenes* to treat cancer patients. These became known as Coley's toxins. In the middle part of the last century, Malmgren *et al.* showed that anaerobic bacteria had the ability to survive and replicate in necrotic tumour tissue with low oxygen content in mouse tumours (Malmgren *et al.,* 1955). Several approaches utilizing bacteria for cancer therapy have subsequently been described (Gericke, *et al.,* 1964; Moese *et al.,* 1964; Thiele, *et al.,* 1964; Kohwi, *et al.,* 1978; Kimura *et al.,* 1980; Fox *et al.,* 1996; Lemmon *et al.,* 1997; Brown *et al.,* 1998; Low *et al.,* 1999; Clairmont *et al.,* 2000; Sznol *et al.,* 2000; Yazawa *et al.,* 2000, 2001).

Bifidobacterium longum has been shown to selectively grow in hypoxic regions of tumours following intravenous administration. This effect was demonstrated in 7,12-dimethylbenzanthracene induced rat mammary tumours by Yazawa *et al.* (Yazawa *et al.,* 2000; Yazawa *et al.,* 2001).

Vogelstein *et al.* created a strain of *Clostridium novyi,* an obligate anaerobe, which was depleted of its lethal toxin (Dang *et al.,* 2001). This strain

*Corresponding author

was termed *C. novyi* NT. Following intravenous administration, the *C. novyi* NT spores germinated in the avascular necrotic regions of tumours in mice, causing damage to the surrounding viable tumour (Dang *et al.*, 2001). Combined with conventional chemotherapy or radiotherapy, intravenous *C. novyi* NT spores caused extensive tumour damage within 24 hours (Dang *et al.*, 2001).

Following attenuation by purine and other auxotrophic mutations, the facultative anaerobe *Salmonella* Typhimurium was also used for cancer therapy (Hoiseth *et al.*, 1981; Pawelek *et al.*, 1997; Low *et al.*, 1999). These genetically modified bacteria replicated in tumours to levels of more than 1000 fold greater than in normal tissue (Low *et al.*, 1999). *Salmonella* Typhimurium was further modified genetically by disrupting the *msbB* gene to reduce the incidence of septic shock (Low *et al.*, 1999). The *msbB* mutant has been tested in a Phase I clinical trial to determine its efficacy on metastatic melanoma and metastatic renal cell carcinoma (Toso *et al.*, 2002). To raise the therapeutic index, *Salmonella* Typhimurium was further attenuated by deletion of both the previously disrupted *msbB* and the novel *purI* genes (Toso *et al.*, 2002). The new strain, termed VNP20009, could then be safely administered to patients (Toso *et al.*, 2002). This multiple auxotrophic mutant of *Salmonella* Typhimurium may have too many nutritional requirements and thereby may be restricted to the necrotic zone of tumours, where more nutrients are available from degrading tissue than in the viable tumour rim. Mengesha *et al.* utilized *Salmonella* Typhimurium as a vector for gene delivery by developing a hypoxia-inducible promoter (HIP-1) to limit gene expression to hypoxic tumours. HIP-1 was able to drive gene expression in bacteria residing in human tumour xenografts implanted in mice (Mengesha *et al.*, 2006). Genes linked to the HIP-1 promoter showed selective expression in tumours (Mengesha *et al.*, 2006).

As a further technological advancement, Yu *et al.* used green fluorescent protein (GFP) labelled bacteria to visualize tumour targeting abilities of three pathogens: *Salmonella* Typhimurium, *Vibrio cholerae*, and *Listeria* monocytogenes (Yu *et al.*, 2003; Yu *et al.*, 2004). Tumor invasion of *Salmonella* Typhimurium is accompanied by strong haemorrhage promoted by TNF-alpha which helps to explain the tumour therapeutic potential of *Salmonella* (Leschner *et al.*, 2009). In the following paragraphs I will outline our own research results, using this promising bacterium for tumour targeting and killing.

Design of a *Salmonella* Typhimurium strain for safe tumour targeting

We initially developed a strain of *Salmonella* Typhimurium, termed A1, which selectively grows in tumours in mouse models (Zhao *et al.*, 2005). In contrast, normal tissue rapidly cleared infecting bacteria, even in immunodeficient athymic mice. *Salmonella* Typhimurium A1 is auxotrophic (leucine/arginine dependent) and receives sufficient support from tumour tissue, but not from normal tissue.

Salmonella Typhimurium A1 was developed as follows: As a first step, *Salmonella* Typhimurium 14028 was transfected with the GFP gene by electroporation. The transformed strain stably expressed GFP over 100 passages. To observe the intracellular replication and virulence of *Salmonella* Typhimurium-GFP in a human prostate cancer cell line *in vitro*, PC-3 human prostate cancer cells were labelled with retroviral red fluorescent protein (RFP) in the cytoplasm, and GFP in the nucleus by means of a fusion of GFP with histone H2B. This allowed the interaction between bacteria and cancer cells to be visualized by dual colour spatial-temporal imaging (Fig. 14.1). The quantitative ability of *Salmonella* Typhimurium to kill prostate cancer cells was determined with the MTT method, a cell viability assay using 3-(4,5-dimethylthiazol-2-yl)-2,5-diphenyl-2H tetrazolium bromide, and observed to be dose dependent (Zhao *et al.*, 2005).

Salmonella Typhimurium-GFP auxotrophic strains were subsequently obtained after nitrosoguanidine (NTG) mutagenesis. Twelve of three hundred isolates tested were identified as auxotrophic mutants (4%), using minimal medium supplemented with various amino acids. Nude mice were inoculated intravenously with 10^7 CFU of each mutant. After inoculation with wild-type *Salmonella* Typhimurium, the mice died within two days. The longest surviving mice were those

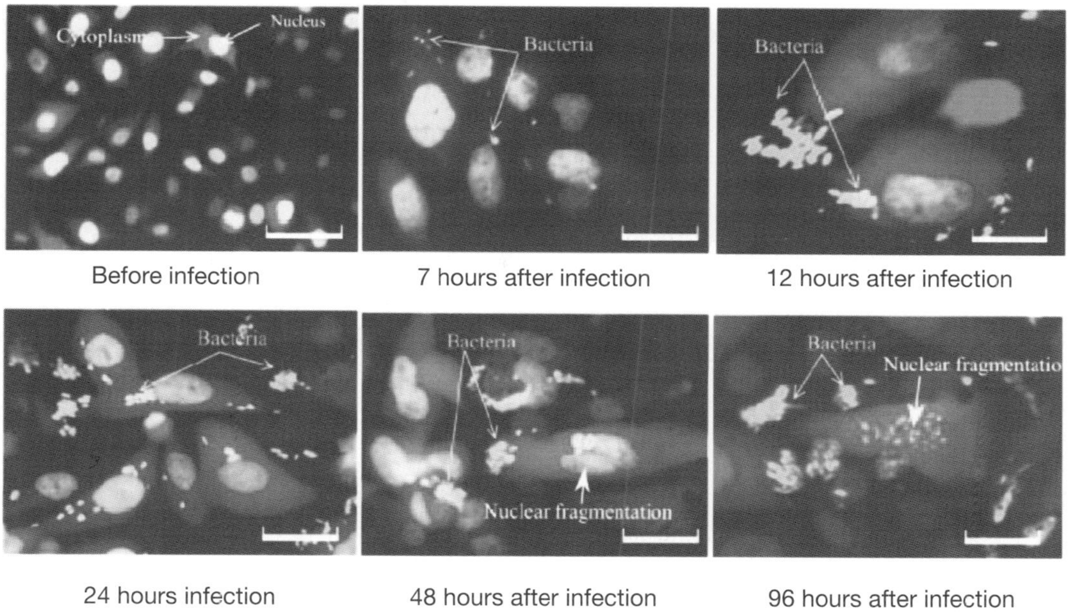

Figure 14.1 Intracellular growth of *Salmonella* Typhimurium A1. Time course of GFP-labelled *Salmonella* Typhimurium A1 growing in GFP–RFP-labelled PC-3 human prostate cells *in vitro*. PC-3 human prostate tumour cells labelled with RFP in the cytoplasm and GFP in the nucleus by means of a fusion with histone H2B were grown in 24-well tissue culture plates to a density of ~10^4 cells per well. Bacteria were grown in LB and harvested at late logarithmic phase, then diluted in cell culture medium and added to the tumour cells (1×10^5 cfu per well). After 1 h of incubation at 37°C, the cells were rinsed and cultured in medium containing gentamicin sulfate (20 µg/ml) to kill external but not internal bacteria. Interaction between bacteria and tumour cells was observed at the indicated time points under fluorescence microscopy magnification. (Bar: 156 µm for upper left; otherwise, bar: 78 µm). From Zhao *et al.* (2005).

inoculated with auxotroph A1, which lived as long as control uninfected mice. A1 required leucine and arginine for growth and was chosen for efficacy studies.

Ten NCR nude mice, 6–8 weeks, were implanted subcutaneously (s.c.) on the mid-right side with 2×10^6 RFP-labelled PC-3 human prostate cancer cells. Bacteria were grown and harvested at late-log phase and then diluted in PBS and injected directly into the tail vein (5×10^7 cfu/100 µl of PBS). Tumor size was determined from fluorescence imaging at each time-point after infection. *Salmonella* Typhimurium A1 selectively colonized the PC-3 tumour and suppressed its growth (Zhao *et al.*, 2005).

To enhance tumour virulence, *Salmonella* Typhimurium A1 was passaged by injection in nude mice transplanted with the HT-29 human colon tumour. Bacteria, expressing GFP, isolated from the infected tumour, were then cultured. The bacterium thus selected was termed A1-R. The ability of A1-R to adhere to tumour cells was evaluated in comparison with the parental A1 strain *in vitro*. The number of A1-R bacteria attached to HT-29 human colon cancer cells was approximately 6 times higher than parental A1 (Zhao *et al.*, 2006).

Using *Salmonella* Typhimurium A1-R to fight cancer

Prostate cancer

Virulence of GFP-labelled *Salmonella* Typhimurium A1 and A1-R bacteria were compared *in vitro* under fluorescence microscopy. Both strains infected dual colour PC-3 prostate cancer cells. Whereas almost all cells are infected and dead after two hours with A1-R, it took 24 hours to get the same result with A1 (Zhao *et al.*, 2007). Thus, the ability of A1-R to kill cancer cells was greatly increased.

GFP-labelled *Salmonella* Typhimurium A1 and A1-R (5×10^7 cfu per 100 μl) were administered intravenously (iv) to PC-3-bearing nude mice. The biodistribution of the bacteria in tumour tissue was determined at day 4. A1-R had a 100× greater CFU in PC-3 tumour tissue than A1. This result suggested that A1-R has greater tumour targeting than A1. A1-R was subsequently used to treat metastatic PC-3 human prostate tumours that had been orthotopically implanted in nude mice. A1-R could effect cures in the orthotopic nude mouse models of PC-3. Of 10 mice with the PC-3 tumours that were injected weekly with A1-R, seven were alive at the time the last untreated mouse died. Four of the tumour-bearing mice were apparently cured by weekly bacterial treatment (Fig. 14.2). In contrast to 12 weekly doses, a regimen of only two doses cured only 1 of 10 mice (Zhao *et al.*, 2007). To compare bacterial infection in the tumour with infection in normal tissue, A1-R bacteria (5 10^7 CFU per 100 μl) were administered intravenously in PC-3-bearing nude mice. On day 4 after injection, the tumour, liver and spleen were removed. The tissues were homogenized and plated on LB agar plates. After overnight growth at 37°C, the CFU were counted. The ratio of tumour to normal tissue was approximately 10^6, indicating a very high degree of tumour targeting by A1-R.

Breast cancer

The tumour-targeting capabilities of *Salmonella* Typhimurium A1-R are not restricted to only one type of cancer. Treatment with A1-R resulted in significant tumour shrinkage in nude mice with subcutaneous MARY-X human xerographs.

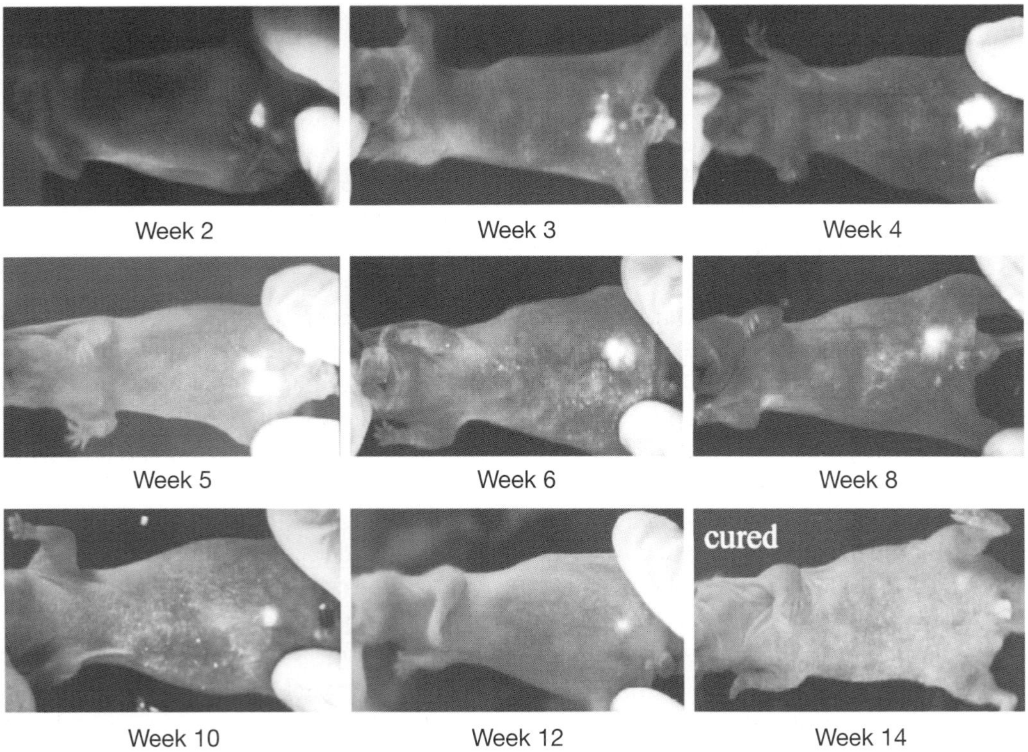

Figure 14.2 Time course of tumour growth and regression of A1-R-treated orthotopic PC-3 human prostate cancer in nude mice. Whole-body images of an A1-R-treated mouse. The mouse was implanted orthotopically with PC-3 tumours expressing GFP. Approximately 2 weeks after implantation, images of the tumours could be obtained externally. The mouse was one of ten mice intravenously injected with *Salmonella* Typhimurium A1-R on a weekly regimen. No tumour regression was observed in 10 untreated control mice. From Zhao *et al.* (2007).

Bacteria (5×10^7CFU/100 μl) were inoculated intravenously in MARY-X-bearing nude mice. Tumour growth was monitored by caliper measurement in two dimensions. The infected tumours regressed by day 5 after infection, and complete regression occurred by day 25. In orthotopic models of MARY-X, A1-R treatment also led to tumour regression following a single intravenous injection of A1-R. The destruction of the tumour in treated mice was visualized by whole-body imaging. The difference in tumour volume between the treated group, which showed quantitative regression, and the control was statistically significant ($P<0.05$) (Zhao *et al.*, 2006). The survival of the A1-R-treated animals was prolonged with a 50% survival time of 13 weeks compared with 5 weeks of control animals. Forty per cent of the mice survived as long as control non–tumour-bearing mice. In the cured animals, tumours were completely eradicated with no regrowth.

Pancreatic cancer

A1-R was able to invade and replicate intracellularly in the XPA-1 human pancreatic cancer cell line expressing GFP in the nucleus and RFP in the cytoplasm. Intracellular bacterial infection led to cell fragmentation and cell death (Nagakura *et al.*, 2009). Before treatment, the average tumour size of the transplanted XPA-1 pancreatic tumour (fluorescent area) on day 7 was 3.2 ± 1.9 mm^2 in the untreated group, 3.1 ± 1.4 in the high-bacteria-dose group (treated with 10^8 CFU *Salmonella* Typhimurium A1-R in 1 ml) and 3.5 ± 0.75 in the low-bacteria-dose group (10^7 CFU in 1 ml). On day 14, after 7 days of treatment, the tumour fluorescence area was 19.9 ± 4.3 mm^2 in the untreated group, 2.2 ± 0.89 in the high-bacteria dose group and 12.7 ± 6.5 in the low-bacteria does group.

Metastasis

We demonstrated the efficacy of locally as well as systemically administered A1-R on liver metastases of XPA-1 human pancreatic cancer cells expressing RFP. Mice treated with A1-R given locally via intrasplenic injection or systemically via tail-vein injection had a much lower hepatic and splenic tumour burden as compared to control mice. Systemic treatment with intravenous A1-R also increased survival time of the nude mice with liver metastases of pancreatic cancer. All results were statistically significant (Yam *et al.*, 2009).

A new experimental model of lymph node metastases was developed to investigate the effect of A1-R on this type of metastases. XPA-1-RFP human pancreatic cancer cells were injected into the inguinal lymph node in nude mice. Just after injection, cancer cells were imaged trafficking in the efferent lymph duct to the axillary lymph node. Metastases in the axillary lymph node were subsequently formed. A1-R bacteria were then injected into the inguinal lymph node to target the axillary lymph node metastases. Just after bacterial injection, a large amount of bacteria was visualized around the axillary lymph node metastases. By day 7, all lymph node metastases had been eradicated in contrast to growing metastases in the control group. There were very few bacteria in the lymph node by day 7, and no bacteria were detected after day 10. This route of administration was therefore able to deliver sufficient bacteria to eradicate the lymph node metastases after which the bacteria became undetectable. The average tumour size (fluorescent area) in the axillary lymph nodes on day 0 was 0.4 ± 0.19 mm^2 in the treatment group and 0.46 ± 0.08 in the untreated group, respectively. On day 7, it was 0 mm^2 in the treatment group and 0.98 ± 0.17 in the untreated group (Hayashi *et al.*, 2009a).

We subsequently tested the *Salmonella* Typhimurium A1-R therapy strategy for spontaneous lymph node metastases from a fibrosarcoma tumour growing in the footpad. At first, only A1-R bacteria were injected in the foot pad in nude mice in order to determine any adverse effects. No infection, skin necrosis, or body weight loss or fatality was detected. Then, HT1080-GFP-RFP human fibrosarcoma cells were injected into the footpad of additional nude mice. The presence of popliteal lymph node metastases was determined by weekly imaging. Once the metastases were detected, A1-R bacteria were injected subcutaneously in the footpad. Bacteria are small particles, and when injected subcutaneously, the lymph system immediately collects them from the site of injection. We observed the injected bacteria trafficking in the lymphatic channel. The popliteal region was exposed just after bacteria injection and a large amount of GFP bacteria targeting the popliteal lymph node metastases was observed by fluorescence imaging. Dual-colour labelling of the cancer cells with RFP in the cytoplasm and GFP in the nucleus distinguished them from the GFP bacteria. After treatment, the popliteal lymph node was observed every week by fluorescence imaging. One mouse was used to image the bacteria by exposing the popliteal lymph node on day 7. GFP bacteria invading the lymph node metastases were observed. All lymph node metastases shrank, and 5 out of 6 were eradicated within 7 to 21 days after treatment, in contrast to growing metastases in the control group.

To study the effect of A1-R on lung metastasis, RFP-GFP HT-1080 cells were injected into the tail vein of nude mice (day 0). On day 4 and day 11, bacteria were injected into the tail vein. On day 16, all animals were sacrificed, and the lungs were imaged to determine the effect of bacterial therapy on lung metastases. To observe the lung metastasis at lower magnification, an RFP filter was used (excitation 545 nm, emission 570–625 nm). In the bacterial treatment group, only a few cancer cells were observed, in contrast to multiple metastases in the control (untreated) group. The number of metastases on the surface of the lung was significantly lower in the treatment group than in the control group ($P<0.005$) (Hayashi *et al.*, 2009a).

Bone tumour

To investigate targeting of *Salmonella* Typhimurium A1-R to primary bone tumour and lung metastases of high-grade osteosarcoma in nude mice, animals were transplanted with 143B-RFP osteosarcoma cells in the tibia, and subsequently developed primary bone tumour and lung metastases. Seven days after tumour injection, RFP tumour was confirmed inside of the tibia by fluorescence imaging. After three weekly injections of bacteria, the bone tumour size and lung metastases were examined on day 28. The bone tumour size (RFP area) was $231.7 \pm 69.7\,\text{mm}^2$ in the untreated group and $94.6 \pm 22.7\,\text{mm}^2$ in the treated group, respectively ($P<0.05$). The lung was excised and the metastases on the surface were counted. The number of metastases was 52 ± 29.6 in the untreated group and 2.3 ± 2.1 in the treated group, respectively ($P<0.05$). Bacterial therapy was therefore effective for primary and metastatic osteosarcoma (Hayashi *et al.*, 2009b).

Spinal cord tumours are highly malignant and often lead to paralysis and death mainly due to their infiltrative nature, high recurrence rate, and limited treatment options. We measured the anti-tumour efficacy of the *Salmonella* Typhimurium A1-R, administered systemically or intrathecally, to spinal cord cancer in orthotopic mouse models. Tumor fragments of U87-RFP human glioma cells were implanted by surgical orthotopic implantation into the dorsal site of the spinal cord. Five and ten days after transplantation, 8 mice in each group were treated with A1-R (2×10^7 cfu by intravenous injection or 2×10^6 cfu by intrathecal injection). The untreated mice showed progressive paralysis beginning at 6 days after tumour transplantation and developed complete paralysis between 18 to 25 days. The mice treated intravenously with A1-R had an onset of paralysis at approximately 11 days and at day 30, 5 mice developed complete paralysis, while other 3 mice had partial paralysis. Mice treated via intrathecal injection of A1-R had an onset of paralysis at approximately 18 days, and one mouse was still not paralysed at day 30. Only one mouse developed complete paralysis at day 30 in this group. The intrathecally treated animals had a significant increase in survival over the intravenously treated group as well as the control group. These results suggest that *Salmonella* Typhimurium A1-R monotherapy can effectively treat spinal cord glioma (Kimura *et al.*, 2009).

Identification of tumour-specific promoters

Salmonella Typhimurium has the ability to preferentially grow in solid tumours as described above. We have used a high-throughput method to screen for *Salmonella* Typhimurium promoters that are selectively activated in tumours in the mouse. A random library of *Salmonella* Typhimurium 14028 with DNA fragments cloned upstream of a promoterless GFP was injected intravenously in nude mice with subcutaneous human PC-3 prostate tumours as well as control nude mice. GFP-positive *Salmonella* Typhimurium bacteria from tumour, spleen, liver and *in vitro* growth in LB medium were isolated by fluorescence activated cell sorting (FACS). Active promoters in all environments were amplified by PCR and identified by DNA–microarray hybridization. Among promoters identified as preferentially induced in tumours, and not induced in any of the other environments (spleen, liver, or *in vitro*) were at least five genes known to be controlled by the fumarate and nitrate reduction global regulator (Fnr). At least five other genes with unknown regulation were also enriched in tumours. The natural tendency of *Salmonella* Typhimurium to target tumours preferentially over other tissues, combined with the use of promoters preferentially induced in the tumour environment versus other environments, may allow the exquisitely tumour-specific expression of fusion proteins on the surface or secreted by *Salmonella* Typhimurium for highly selective tumour therapy (Arrach *et al.*, 2008).

Conclusion

Our goal is to develop tumour-targeting *Salmonella* Typhimurium strains that can kill primary and metastatic cancer without toxic effects to the host and without the need for combination with toxic chemotherapy. Toward this goal, we have developed a new strain of *Salmonella* Typhimurium, A1-R, which has greatly increased anti-tumour efficacy but maintains its original auxotrophy for leucine and arginine that prevents it from mounting a continuous infection in normal tissues. A1-R was able to trigger cures in monotherapy on mouse models of metastatic human cancer. We have also identified candidate *Salmonella* Typhimurium tumour-specific promoters that may enhance the anti-tumour efficacy of A1-R by driving expression of toxins that could be excreted in the tumours. *Salmonella* Typhimurium has now become the paradigm of bacterial therapy of cancer. Future studies will be aimed to bring bacterial treatment of cancer to the clinic as the next generation of cancer therapy.

References

Arrach, N., Zhao, M., Porwollik, S., Hoffman, R.M., and McClelland, M. (2008). *Salmonella* promoters preferentially activated inside tumors. Cancer Res. *68*, 4827–4832.

Brown, J.M., and Giaccia, A.J. (1998). The unique physiology of solid tumors: opportunities (and problems). for cancer therapy. Cancer Res. *58*, 1408–1416.

Clairmont, C., Lee, K.C., Pike, J., Ittensohn, M., Low, K.B., Pawelek, J., Bermudes, D., Brecher, S.M., Margitich, D., Turnier, J., Li, Z., Luo, X., King, I., and Zheng, L.M. (2000). Biodistribution and genetic stability of the novel antitumor agent VNP20009, a genetically modified strain of *Salmonella* Typhimurium. J. Infect. Dis. *181*, 1996–2002.

Coley, W.B. (1906). Late results of the treatment of inoperable sarcoma by the mixed toxins of erysipelas and Bacillus prodigiosus. Am. J. Med. Sci. *131*, 375–430.

Dang, L.H., Bettegowda, C., Huso, D.L., Kinzler, K.W., and Vogelstein, B. (2001). Combination bacteriolytic therapy for the treatment of experimental tumors. Proc. Natl. Acad. Sci. U.S.A. *98*, 15155–15160.

Fox, M.E., Lemmon, M.J., Mauchline, M.L., Davis, T.O., Giaccia, A.J., Minton, N.P., and Brown, J.M. (1996). Anaerobic bacteria as a delivery system for cancer gene therapy: *in vitro* activation of 5-fluorocytosine by genetically engineered clostridia. Gene Ther. *3*, 173–178.

Gericke, D., and Engelbart, K. (1964). Oncolysis by Clostridia. Ii. Experiments on a tumor spectrum with a variety of Clostridia in combination with heavy metal. Cancer Res. *24*, 217–221.

Hayashi, K., Zhao, M., Yamauchi, K., Yamamoto, N., Tsuchiya, H., Tomita, K., and Hoffman, R.M. (2009a). Cancer metastasis directly eradicated by targeted therapy with a modified *Salmonella typhimurium*. J. Cell. Biochem. *106*, 992–998.

Hayashi, K., Zhao, M., Yamauchi, K., Yamamoto, N., Tsuchiya, H., Tomita, K., Kishimoto, H., Bouvet, M., and Hoffman, R.M. (2009b). Systemic targeting of primary bone tumor and lung metastasis of high-grade osteosarcoma in nude mice with a tumor-selective strain of *Salmonella typhimurium*. Cell Cycle *8*, 870–875.

Hoiseth, S.K., and Stocker, B.A. (1981). Aromatic-dependent *Salmonella* Typhimurium are non-virulent and effective as live vaccines. Nature *291*, 238–239.

Kimura, H., Zhang, L., Zhao, M., Hayashi, K., Tsuchiya, H., Tomita, H., Bouvet, M., Wessels, J., and Hoffman, R.M. (2010). Targeted therapy of spinal cord glioma with a genetically-modified *Salmonella* Typhimurium. Cell Proliferation *43*, 41–48.

Kimura, N.T., Taniguchi, S., Aoki, K., and Baba, T. (1980). Selective localization and growth of Bifidobacterium bifidum in mouse tumors following intravenous administration. Cancer Res. *40*, 2061–2068.

Kohwi, Y., Imai, K., Tamura, Z., and Hashimoto, Y. (1978). Antitumor effect of *Bifidobacterium* Infantis in mice. Gann *69*, 613–618.

Lemmon, M.J., van Zijl, P., Fox, M.E., Mauchline, M.L., Giaccia, A.J., Minton, N.P., and Brown, J.M. (1997). Anaerobic bacteria as a gene delivery system that is controlled by the tumor microenvironment. Gene Ther. *4*, 791–796.

Leschner, S., Westphal, K., Dietrich, N., Viegas, N., Jablonska, J., Lyszkiewicz, M., Lienenklaus, S., Falk, W.,

Gekara, N., Loessner, H., and Weiss, S. (2009). Tumor invasion of *Salmonella enterica* serovar Typhimurium is accompanied by strong hemorrhage promoted by TNF-alpha. PLoS One 4(8):e6692.

Low, K.B., Ittensohn, M., Le, T., Platt, J., Sodi, S., Amoss, M., Ash, O., Carmichael, E., Chakraborty, A., Fischer, J., *et al.* (1999). Lipid A mutant *Salmonella* with suppressed virulence and TNFalpha induction retain tumor-targeting *in vivo*. Nat. Biotechnol. *17*, 37–41.

Malmgren, R.A., and Flanigan, C.C. (1955). Localization of the vegetative form of Clostridium tetani in mouse tumors following intravenous spore administration. Cancer Res. *15*, 473–478.

Mengesha, A., Dubois, L., Lambin, P., Landuyt, W., Chiu, R.K., Wouters, B.G., and Theys, J. (2006). Development of a flexible and potent hypoxia-inducible promoter for tumor-targeted gene expression in attenuated *Salmonella*. Cancer Biol. Ther. *5*, 1120–1128.

Moese, J.R., and Moese, G. (1964). Oncolysis by Clostridia. Activity, I. of *Clostridium butyricum* (M-55) and other nonpathogenic Clostridia against the Ehrlich carcinoma. Cancer Res. *24*, 212–216.

Nagakura, C., Hayashi, K., Zhao, M., Yamauchi, K., Yamamoto, N., Tsuchiya, H., Tomita, K., Bouvet, M., and Hoffman, R.M. (2009). Efficacy of a genetically-modified *Salmonella* Typhimurium in an orthotopic human pancreatic cancer in nude mice. Anticancer Res. *29*, 1873–1878.

Pawelek, J.M., Low, K.B., and Bermudes, D. (1997). Tumor-targeted *Salmonella* as a novel anticancer vector. Cancer Res. *57*, 4537–4544.

Sznol, M., Lin, S.L., Bermudes, D., Zheng, L.M., and King, I. (2000). Use of preferentially replicating bacteria for the treatment of cancer. J. Clin. Invest. *105*, 1027–1030.

Thiele, E.H., Arison, R.N., and Boxer, G.E. (1964). Oncolysis by Clostridia. Iii. Effects of Clostridia and chemotherapeutic agents on rodent tumors. Cancer Res. *24*, 222–233.

Toso, J.F., Gill, V.J., Hwu, P., Marincola, F.M., Restifo, N.P., Schwartzentruber, D.J., Sherry, R.M., Topalian, S.L., Yang, J.C., Stock, F., *et al.* (2002). Phase I study of the intravenous administration of attenuated *Salmonella* Typhimurium to patients with metastatic melanoma. J. Clin. Oncol. *20*, 142–152.

Yam, C., Zhao, M., Hayashi, K., Ma, H., Kishimoto, H., McElroy, M., Bouvet, M., and Hoffman, R.M. (2009). Monotherapy with a tumor-targeting mutant of S. Typhimurium inhibits liver metastasis in a mouse model of pancreatic cancer. J. Surg. Res. Epub 31 March 2009.

Yazawa, K., Fujimori, M., Amano, J., Kano, Y., and Taniguchi, S. (2000). Bifidobacterium longum as a delivery system for cancer gene therapy: selective localization and growth in hypoxic tumors. Cancer Gene Ther. 7, 269–274.

Yazawa, K., Fujimori, M., Nakamura, T., Sasaki, T., Amano, J., Kano, Y., and Taniguchi, S. (2001). Bifidobacterium longum as a delivery system for gene therapy of chemically induced rat mammary tumors. Breast Cancer Res Treat *66*, 165–170.

Yu, Y.A., Timiryasova, T., Zhang, Q., Beltz, R., and Szalay, A.A. (2003). Optical imaging: bacteria, viruses, and mammalian cells encoding light-emitting proteins reveal the locations of primary tumors and metastases in animals. Anal. Bioanal. Chem. *377*, 964–972.

Yu, Y.A., Shabahang, S., Timiryasova, T.M., Zhang, Q., Beltz, R., Gentschev, I., Goebel, W., and Szalay, A.A. (2004). Visualization of tumors and metastases in live animals with bacteria and vaccinia virus encoding light-emitting proteins. Nat. Biotechnol. *22*, 313–320.

Zhao, M., Yang, M., Li, X.M., Jiang, P., Baranov, E., Li, S., Xu, M., Penman, S., and Hoffman, R.M. (2005). Tumor-targeting bacterial therapy with amino acid auxotrophs of GFP-expressing *Salmonella* Typhimurium. Proc. Natl. Acad. Sci. U.S.A. *102*, 755–760.

Zhao, M., Yang, M., Ma, H., Li, X., Tan, X., Li, S., Yang, Z., and Hoffman, R.M. (2006). Targeted therapy with a *Salmonella* Typhimurium leucine-arginine auxotroph cures orthotopic human breast tumors in nude mice. Cancer Res. *66*, 7647–7652.

Zhao, M., Geller, J., Ma, H., Yang, M., Penman, S., and Hoffman, R.M. (2007). Monotherapy with a tumor-targeting mutant of *Salmonella* Typhimurium cures orthotopic metastatic mouse models of human prostate cancer. Proc. Natl. Acad. Sci. U.S.A. *104*, 10170–10174.

Index

H

I

K

L

M

S

T

V

W

Z

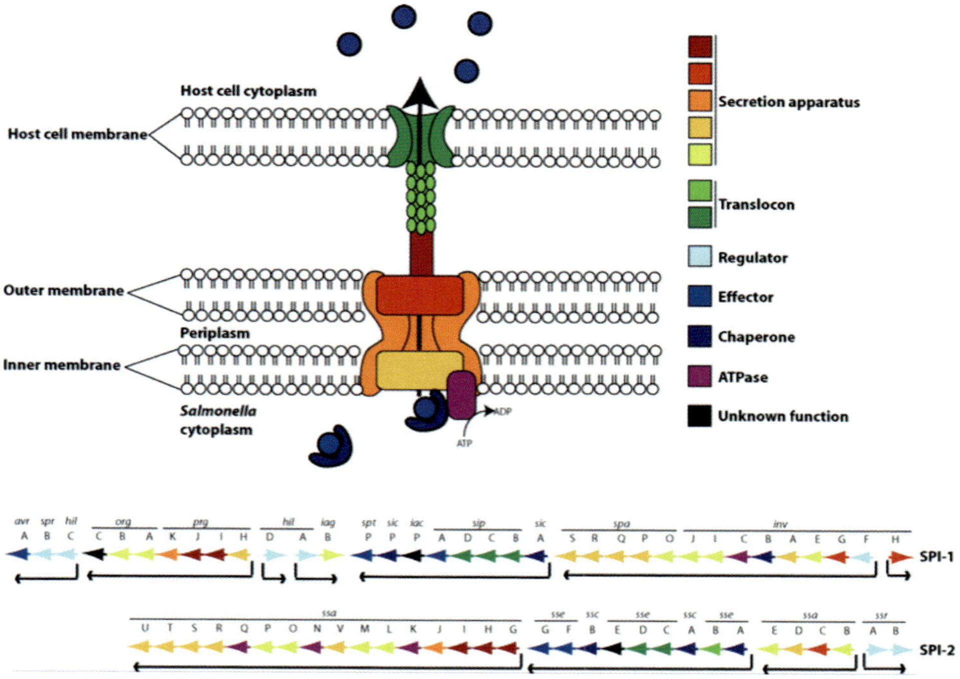

Figure 10.2 Comparison of the two type III secretion systems in *Salmonella*. Reprinted, with permission, from Kaoru Geddes (PhD thesis, Oregon Health and Science University, 2008).